Comprehensive Seismic Zonation Schemes for Regions at Different Scales

T. G. Sitharam • Naveen James
Sreevalsa Kolathayar

Comprehensive Seismic Zonation Schemes for Regions at Different Scales

Springer

T. G. Sitharam
Department of Civil Engineering
Indian Institute of Science Bangalore
Bangalore, Karnataka, India

Naveen James
Department of Civil Engineering
Indian Institute of Technology Ropar
Rupnagar, Punjab, India

Sreevalsa Kolathayar
Department of Civil Engineering
Amrita Vishwa Vidyapeetham
Coimbatore, Tamil Nadu, India

ISBN 978-3-030-07824-9 ISBN 978-3-319-89659-5 (eBook)
https://doi.org/10.1007/978-3-319-89659-5

Softcover re-print of the Hardcover 1st edition 2018

Printed on acid-free paper

This Springer imprint is published by the registered company Springer International Publishing AG part of Springer Nature.
The registered company address is: Gewerbestrasse 11, 6330 Cham, Switzerland

Preface

The practice of earthquake engineering involves identification, evaluation, and mitigation of seismic hazards. This book deals with the subdivision of regions into zones that have relatively similar exposure to various earthquake-related effects. The basis of seismic zonation is to model the rupture mechanism at the source of an earthquake, evaluate the propagation of waves through the earth to the top of the bedrock, determine the effects of local soil profile and thus develop a hazard map indicating the vulnerability of the area to the potential seismic hazards. Many earthquakes in the past have left many lessons to be learned which are essential to plan infrastructure and even mitigate such calamities in future. India has been facing threats from earthquakes since ancient times. Urbanization of cities is growing and expanding due to increasing migration of people to cities. The role of geological and geotechnical data is becoming very important in the urban planning of city infrastructure, which can recognize, control, and prevent geological hazards.

Chapter 1 presents an overview of earthquake hazards and the need for seismic zonation. Different levels of seismic zonation and issues associated with it are also discussed in this chapter. Chapter 2 presents the basic concepts of earthquake and seismicity highlighting theory of plate tectonics and global seismicity. The seismicity analysis of earthquake catalog, seismic source models, and preparation of seismotectonic atlas are discussed in this chapter. Chapter 3 presents the state-of the-art seismic hazard assessment using deterministic and probabilistic approaches, along with a few case studies of hazard assessment. Chapter 4 presents several site characterization techniques and case studies of site characterization at micro- and macro-levels. Chapter 5 discusses the effect of local site conditions on ground motion and methods for estimating local site effects. Chapter 6 presents the liquefaction mechanism and factors affecting the liquefaction susceptibility of the soil. The evaluation of liquefaction potential is elaborated with case studies. The chapter also discusses several schemes of mitigating the liquefaction. Chapter 7 presents integration of various hazards and several case studies of liquefaction studies in India at micro-, meso-, and macro-levels. The chapter also summarizes principles and practices of seismic zonation, along with brief guidelines for the same.

Study of seismic hazard and preparation of microzonation maps will provide an effective solution for urban planning. Seismic hazard and microzonation of cities enable engineers to characterize potential seismic vulnerability/risk that needs to be taken into account while designing new structures or retrofitting existing ones. This will even help in designing buried lifelines such as tunnels, water and sewage lines, gas and oil lines, and power and communication lines. This book can serve as a guideline for seismic zonation for practicing engineers and policy makers. The technicalities presented in the book will be highly beneficial to researchers and students.

Bangalore, India T. G. Sitharam
Rupnagar, India Naveen James
Coimbatore, India Sreevalsa Kolathayar
February, 2018

Contents

Acronyms

AHP	Analytical hierarchy process
CPT	Cone penetration test
CRR	Cyclic resistance ratio
CSR	Cyclic stress ratio
DEM	Digital elevation model
DSHA	Deterministic seismic hazard analysis
EHP	Earthquake hazard parameter
ENA	Eastern northern America
FS	Factor of safety
GIS	Geographic Information Systems
LPI	Liquefaction potential index
MARE	Mean annual rate of exceedance
MASW	Multichannel analysis of surface waves
MFS	Magnitude scaling factor
NEHRP	National earthquake hazard reduction program
NPP	Nuclear power plant
PHA	Peak horizontal acceleration
PI	Peninsular India
PSHA	Probabilistic seismic hazard analysis
SPT	Standard penetration test
V_s	Shear wave velocity
σ	Total stress
σ'	Effective stress
u_p	Pore pressure
γ_t	Threshold strain
γ_c	Cyclic strain
σ'	Effective stress
λ_M	Mean annual rate of exceedance of earthquakes with magnitude "M"
M	Magnitude
M_w	Moment magnitude

M_{cut}	Cutoff magnitude
a and b	Seismicity parameters
$\hat{n_i}$	Smoothed number of earthquakes in ith cell
$m_{\max}$	Maximum magnitude
$\nu(z)$	Mean annual rate of exceedance of ground motion parameter Z
V_{s30}	Average shear wave velocity for top 30 m overburden
β	Damping ratio
G	Shear modulus
$G_{\max}$	Shear modulus at very low strain
γ	Shear strain
γ_{eff}	Effective shear strain
$\gamma_{\max}$	Peak shear strain
Y_{br}	Bedrock level acceleration
FS_L	Factor of safety against liquefaction
$a_{\max}$	Surface level peak ground acceleration
σ_{v0}	Total overburden pressure
σ'_{v0}	Effective overburden pressure
r_d	Depth reduction factor
$(N_1)_{60cs}$	Corrected SPT value for energy efficiency, overburden pressure, and fine content
N_{req}	SPT value required to resist liquefaction
q_c	CPT value required to resist liquefaction
$m_{\min}$	Minimum magnitude
X_m	Rupture length
$N_i(m_o)$	Weightage factor for an individual fault
u_s	Displacement at surface
u_r	Displacement at rock
Y_s	PHA at the ground surface
F_s	Amplification factor
$W(z)$	Depth weightage factor
N_a and N_m	Number of acceleration and magnitude increment
a_i and m_i	Incremental values in acceleration and magnitude
$\Delta\lambda_{a_i,m_j}$	Incremental annual frequency of exceedance for a_i and m_j
$\text{CSR}_{eq,i}$	Cyclic stress ratio for an acceleration a_i
P_a	Atmospheric pressure
θ	Regression coefficients
R_i	Rank assigned to each category of a single theme
$R_{\min}$ and $R_{\max}$	Minimum and the maximum rating value of that theme
ϕ_{diff}	Phase difference
f	Frequency
C_f	Phase velocity
C_N	Overburden pressure correction
C_E	Hammer energy correction
C_B	Borehole diameter correction
C_S	Liner correction

C_R	Rod length correction
C_{FC}	Correction for fine contents
$S_T(\omega, \beta)$	Target response spectrum
$G(\omega)$	Power spectrum
$S_a(\omega, \beta)$	Observed response spectrum
$X(t)$	Stationary time history
$E(t)$	Non-stationary intensity envelope

Chapter 1
Introduction and Overview

1.1 Introduction

Natural calamities like earthquakes can neither be predicted nor be prevented. Although earthquakes have been reported from all parts of the world, they are mainly associated with the boundaries of the plates, which form the Earth's crust. The earth's crust is being slowly displaced at the boundaries of the plates. Differential displacements give rise to elastic strains, which eventually exceed the strength of the rocks involved, and faulting occurs. The strained rocks rebound along a fault under the elastic strains until the strain is partly or wholly dissipated.

Past earthquakes have demonstrated that Indian subcontinent is highly vulnerable to earthquake hazards. It has been estimated that about 59% of the land area of the Indian subcontinent has potential risk from moderate to severe earthquakes [237]. Major earthquakes in the last 20 years such as Khillari (30th September 1993), Jabalpur (22nd May 1997), Chamoli (29th March 1999) and Bhuj (26th January 2001) earthquakes have resulted in more than 23,000 deaths and extensive damage to infrastructure [237]. Although it is well known that the major earthquake hazard prone areas in India are the Himalayan region (inter-plate zone) and the northeast region (subduction zone), the seismicity of Peninsular India cannot be underestimated. Many studies [102, 275, 304] have proved that the seismicity of Peninsular India is significantly high and may lead to earthquakes of sizeable magnitude. This necessitates a seismic zonation for the country, as well as various regions in it. Seismic zonation is the first step towards an effective earthquake risk mitigation study.

Seismic zonation is a process in which a large region is demarcated into small zones based on the levels of earthquake hazard. Seismic Zonation is delineation of regions having diverse potentials for hazardous effects from impending earthquakes. Within a seismic zone, the seismicity rate is assumed to be consistent. Seismic

T. G. Sitharam et al., *Comprehensive Seismic Zonation Schemes for Regions at Different Scales*, https://doi.org/10.1007/978-3-319-89659-5_1

zonation is a process in which a large region is demarcated into small zones based on the levels of earthquake hazard. A seismic zoning map for engineering use is a map that defines the levels of ground motions for earthquake-resistant design, and thus it differs from a seismicity map, which provides only past earthquake data and seismic sources. The task of seismic zoning is multidisciplinary that involves the contribution from geologist, seismologist, geotechnical, earthquake, and structural engineers. Seismic zonation can act as a roadmap for designers and policy makers to identify the regions with potential seismic threat.

1.2 Earthquake Damages

Earthquakes do not pose great danger to mankind directly as public cannot be shaken to death by an earthquake. The casualties due to earthquake happen because of associated hazards and infrastructure failures.

1.2.1 Ground Shaking

The first main earthquake hazard is the effect of ground shaking. Buildings can be damaged by the ground shaking or by the ground subsidence where ground settles to a different level during the earthquake. The extent of ground shaking will depend upon the scale of an earthquake, distance from the epicenter, and nature of the material.

1.2.2 Liquefaction

Liquefaction is the mixing of soil and groundwater during the shaking because of an earthquake because of which soil loses its shear strength and behaves like a fluid. If liquefaction occurs under a building, it may sink into the ground. Liquefaction is a phenomenon seen in sandy soils with groundwater near the surface.

1.2.3 Landslides

The ground shaking may cause instability of slopes that can lead to landslides, mudslides, and avalanches, which can damage buildings located nearby.

1.2.4 *Ground Displacement*

If a structure (a building, road, etc.) is built across a fault, the ground displacement during an earthquake could seriously damage or rip apart that structure.

1.2.5 *Flooding*

An earthquake can break dams which will lead to flooding of the area downstream, damaging buildings, and sweeping away people.

1.2.6 *Tsunami*

A tsunami is a huge wave caused by an earthquake under the ocean. Tsunamis can be tens of feet high when they hit the shore and can do massive damage to the coastline. Seiches occur on lakes that are shaken by the earthquake and are usually only a few feet high, but they can still flood the surroundings.

1.2.7 *Fire*

These fires can be started by broken gas lines and power lines during an earthquake. They can be a serious problem, especially if the water lines that feed the fire hydrants are broken, too. After the Great San Francisco Earthquake in 1906, the city burned for 3 days leading to massive destruction of the city leaving 250,000 people homeless.

Most of the hazards to people come from man-made structures themselves and the shaking they receive from the earthquake. The real dangers to people are being crushed in a collapsing building, drowning in a flood caused by a broken dam or levee, getting buried under a landslide, or being burned in a fire.

1.3 Seismic Zonation

Seismic zonation is a process of dividing a large region into small zones based on the expected level of earthquake hazard. Seismic zonation helps to identify vulnerable regions and also provide necessary outputs for the earthquake resistant design. Hence it is very much required in the modern world in order to minimize the casualty and economic losses during an earthquake. The large and rapidly growing urban seismic risk, particularly in developing countries like India is a problem that needs to be quickly solved. This can be done through a comprehensive seismic microzonation of the area. The ISSMGE Technical Committee for Geotechnical Earthquake Engineering proposed three grades of zoning along with the accepted scales of mapping [144]. The scale of zonation depends mainly on the available database and on the quality of the zonation map required. Several inputs regarding seismicity, geology, geomorphology, and geotechnical characteristics are needed for doing seismic microzonation of an area.

1.4 Need for Seismic Zonation

Seismic microzonation is the first step in earthquake risk mitigation study and requires multidisciplinary approach with major contributions from the fields of geology, seismology, geophysics, geotechnical, and structural engineering. This is very important to identify the tectonic and geological formations in the study area which is essential for determining the seismic sources and also for establishing a realistic earthquake hazard models for the investigation.

The very important issue affecting the applicability and the feasibility of any microzonation study is the reliability of the parameters for doing it. It is very clear based on the earthquake damage and strong motion records that there are numerous source and site factors i.e., near field effects, directivity, duration, focusing, topographical and basin effects, etc. which are important in assessing ground motion characteristics. The national seismic zoning maps which are prepared in small scales like 1:1,000,000 or less neglecting all the above factors and they do not consider the geological and geotechnical site conditions.

Seismic microzonation involves a very detailed field investigation to evaluate the hazard. It is very effective in delineating the spatial variations in the seismic hazard. They are also useful to evaluate the risk scenarios in the study area. This has been the most widely used method to map earthquake hazard at local scales which may incorporate a wide variety of information, including seismic response of different surface geological formations, liquefaction potential, topographic amplification of seismic waves, landslides, tsunamis, etc. Seismic microzonation maps are very useful in urban planning because they help to predict the impact of future earthquakes and can also be used to locate key facilities like hospitals, fire stations, emergency operation centers, etc. Microzonation studies are also very useful to save the heritage and important structures from future major earthquakes.

The Turkish codes insist that the seismic microzonation has to be done in vulnerable cities/regions with a population of 30,000 or more. However, since the population in India is very high, the microzonation need to be carried out for the cities based on population and macro hazard maps available in the country. For example, microzonation can be taken up for the urban centers with a population of 50,000 to one lakh in an area and which falls in seismic zone five. While selecting the regions for microzonation, the population of the region, its socio-economic characteristics, and the strategic importance, etc. also need to be considered.

1.5 Level of Seismic Zonations and Methodology

Seismic zonation is generally carried out at three different levels based on the aerial extent of the region, the importance of the site and the population. They are micro-level, meso-level, and macro-level. The macro-level zonation is generally carried out for large landmass such as a state or a country. The earthquake hazard parameters used for macro-level zoning are generally evaluated with less reliability. The typical example of a macro-level zonation is the seismic zonation map of India prepared by IS-1893, where the entire India is demarcated into four seismic zones based on past seismicity and tectonic conditions. Generally the macro-level seismic zonation is carried out based on peak horizontal acceleration (PHA) estimated at bedrock level without giving emphasis on the local soil conditions. Seismic zonation at the meso-level is carried out for cities and urban centers with a population greater than 500,000. The earthquake hazard parameters, for the meso-level zonation are evaluated with greater degree of reliability, compared to the macro-level zoning. The micro-level zonation is carried out for sites which host critical installations such as nuclear power plants (NPPs). As the NPPs are considered as very sensitive structures, the earthquake parameters, for the micro-level zonation of the NPP sites are estimated with a highest degree of reliability. The local soil conditions and site effects are properly counted for carrying out the micro as well as the meso-level zonation. Several researchers have carried out meso-level zonation considering the effects of all major earthquake hazards such as PHA, site amplification, and liquefaction [223, 235, 328].

The local site effects during earthquakes are related to geotechnical characteristics such as amplification, liquefaction, land slide, mudflow and fault movements. To assess these geotechnical characteristics three grades of approaches to zonation was suggested by ISSMGE Technical Committee for Geotechnical Earthquake Engineering along with the accepted scales of mapping (Table 1.1). The scale of zonation depends mainly on the available database and on the quality of the zonation map required. Several inputs regarding seismicity, geology, and geotechnical characteristics are needed for doing seismic microzonation of an area. The national seismic zoning maps are mostly at small scales such as 1:1,000,000 or less and are mostly based on seismic source zones defined at similar scales. However, seismic microzonation for a city/town requires 1:5000–1:25,000 scale studies and needs to be based on seismic hazard studies at similar scales. The country wide macrozonation maps which are produced by national experts will go through a careful review process. However it may not be possible to adopt such a review system for the large number of seismic microzonation studies in the country. Geological Survey of India maps are available at 1:50,000 scale with regard to base map, contours, geomorphology, geology, etc. It is advisable to explore these maps for microzonation studies immediately. For some important mega cities this exercise can be repeated at large scales like 1:10,000. This is presently being done for National Capital Region Delhi. One possible solution for this scale incompatibility is to increase the scales of seismic macrozonation maps

Table 1.1 Three levels of zonation (modified from [144])

Geotechnical phenomenon	Grade-1	Grade-2	Grade-3
Ground motions	Historical earthquakes and existing information geological maps interviews with local residents	Microtremor simplified geotechnical	Geotechnical investigation ground response analysis
Slope stability	Historical earthquakes and existing geological and information geomorphological maps	Air photos and remote sensing field studies vegetation and precipitation data	Geotechnical investigation analysis
Liquefaction	Historical earthquakes and existing information geological and geomorphological maps	Air photos and remote sensing field studies interviews with local residents	Geotechnical investigation analysis
Accepted scale of mapping	1:1,000,000 to 1:50,000	1:100,000 to 1:10,000	1:125,000 to 1:5000
Grid size for testing geophysical survey and bore hole	Homogeneous sub-surface—2 km × 2 km–5 km × 5 km Heterogeneous Sub-surface—0.5 km × 0.5 km–2 km × 2 km	Homogeneous sub-surface—1 km × 1 km–3 km × 3 km Heterogeneous Sub-surface—0.5 km × 0.5–1 km × 1 km	Homogeneous sub-surface—0.5 km × 0.5 km–2 km × 2 km Heterogeneous Sub-surface—0.1 km × 0.1 km 0.5 km × 0.5 km

steadily with the accumulation of geological and seismological data as implemented in USA and Japan. Geological formations, local site classification, equivalent shear wave velocity, spectral acceleration, spectral amplification, and their variation are some of the parameters studied during a seismic microzonation studies. A consistent approach has to be implemented to assess each parameter with respect to all other parameters. The objective of seismic zonation is to establish a seismic hazard map at a scale of 1:10,000 taking into account earthquake source and local site conditions. Thus estimation of the earthquake induced forces and their variation in the investigated area are the main target in seismic microzonation.

1.6 Issues Related to Seismic Zonation

The government of India has already initiated the seismic zonation of major cities in India. However, the micro-level zoning cannot be performed for all these cities as very large resources are required for an extensive geotechnical and geophysical investigation and site effect estimation. Hence there is a need to define the level of zonation to be adopted for various study areas in the country. As there are many methodologies available for the site characterization and site effect estimation, the suitability of these methodologies for each levels of zoning also needs to be assessed in order to optimize the resources for carrying out seismic zonation.

Even though the above definitions and descriptions are available for various levels of zonation, the key issue lies in the adoption of the suitable one for a given region. There are only a few guidelines available regarding the use of a particular level of zonation for a given study area. Based on the recommendation of the disaster management authority, the government of India has initiated the seismic zonation of all major cities in India. As it is evident that large resources are required in order to carry out seismic site characterization and site effect estimation, both the micro and meso-level zonations cannot be carried out for all these cities. Hence there is a need to propose appropriate guidelines to define the suitability of each level zonation for various regions in the country. Moreover, there are many methodologies available for site characterization and estimation of site effects such as site amplification and liquefaction. The appropriateness of these methodologies for various levels of seismic zonations also needs to be assessed in order to optimize use of resources for seismic zonation. Hence, in this book, appropriate techniques for site characterization and earthquake hazard estimation for regions at different scale levels were determined. Using the appropriate techniques, the seismic zonation was carried out both at the micro and macro-level, incorporating all major earthquake hazards.

The zonation shall be graded based on the scale of the investigation and details of the studies carried out. The technical committee on earthquake geotechnical engineering (TC4) of the International society of soil mechanics and foundation engineering [144] states that the first grade (Level I) map can be prepared with scale of 1:1,000,000–1:50,000 and the ground motion to be assessed based on the historical earthquakes and existing information of geological and geomorphological maps. If the scale of the mapping is 1:100,000–1:10,000 and ground motion is assessed based on the microtremor and simplified geotechnical studies, then it is called second grade (Level II) map. In the third grade (Level III) map ground motion is to be assessed based on the complete geotechnical investigations and ground response analysis with a scale of 1:25,000–1:5000.

The existing macro-level zonation map of India, used for the design practice is developed by BIS-1893 [28]. However the major drawback of the seismic zonation map of India developed by BIS-1893 [28] is that it is based on the past seismic activity and not on scientific seismic hazard analysis. The BIS code always subjected to revision only after major earthquake. Such a zonation map is not appreciated in an engineering point of view and hence it is desirable to carry out macro-level seismic zonation based on parameters such as peak horizontal acceleration (PHA), which can be estimated scientifically. A macro-level zoning refers to the zonation of a large region such as a state or a country. Several studies have been carried out for macro-level seismic hazard assessment and zoning based on the peak horizontal acceleration (PHA) at the bedrock [25, 149, 260, 376]. However still the influence of site conditions is not considered in the macro-level zoning. Similarly, the micro-level zonations are also required especially for the NPP sites. As there are lot of constraints for the selection of a site to host NPP, seismic zonation helps to identify the least hazardous sector within the site where the critical structures of NPP can be sited. Seismic zonation of existing NPP site is also required for assessing the seismic vulnerability and risk associated with various structures of NPP. Hence in the present study an attempt has been made to review the feasibility of various methodologies for site characterization and site effect estimation and thus carry out the seismic zonation, both at the micro and macro-level using the appropriate methodology. The micro-level zonation was carried out for the Kalpakkam NPP site, while the macro-level zonation was performed for the state of Karnataka. As the Kalpakkam NPP site area does not lie within the state of Karnataka, the results of the macro-level study were validated with the existing meso-level study carried out for the Bangalore city. Based on this, recommendations have been made regarding the suitability of various methodologies for different level of zonation.

1.7 Global Trends

The first attempt of seismic microzonation of any urban area, i.e. an industrial as well as population center was carried out in the city of Yokohama, Japan in 1954 considering various zones, corresponding soil conditions and design seismic coefficients for different types of structures located in that different zones. Subsequently, in view of the immense usefulness of microzonation studies were conducted in few earthquake prone areas of the world [7, 14, 64, 89, 91, 199, 208].

Slob et al. [338] presented a technique for microzonation for the city of Armenia in Colombia. In this study they used a 3D layer model in GIS, combined with a 1D calculation of seismic response using SHAKE to get the spatial variation in seismic response which was checked with the damage assessment of Armenia. They concluded that the amplification was high in the range of frequency of 5 Hz for houses with 2 stories, it became true after the earthquake, in which low rise building experienced more damage than high rise building. Topal et al. [357] considered various parameters for microzonation such as geological, geotechnical, seismotectonic, and hydrogeological conditions and on the basis of these, four different zones were proposed for the Yeneshir, an urban area in Turkey. Ansal et al. [14] adopted a probabilistic approach in a microzonation study for the city of Siliviri, Turkey and estimated the local site effects on the basis of the available borehole data and laboratory results on collected samples from the area. For site characterization the average shear wave velocity was used, that was determined from seismic refraction tests.

Seismic microzonation in India is recognized as a tool for disaster mitigation and for providing guidelines to the construction engineers. Some states of country have made it mandatory for the new constructions to be declared seismically safe and for this prior approval of the experts is necessary. After the devastating earthquake in Gujarat in 2001, the Government of India has paid serious attention to seismic microzonation and it has accepted its importance as a guiding tool in land use planning and safe construction practices to avoid the loss from the future earthquakes. As a result of this, seismic microzonation of various urban areas, like Delhi, Jabalpur, Chennai, Bangalore, Lucknow, Ahmedabad, etc., are being carried out by different researchers in the country.

Chapter 2
Earthquake and Seismicity

2.1 Plate Tectonics

The science of earthquakes requires knowledge of plate tectonics, a theory that describes the movement of major and many smaller geological deformations that make up the earth's crust. These interlocking crustal plates cover the surface of the earth, and are moved by radioactivity-driven currents in the earth's mantle. The heavier ocean-floor plates, expanding from central ridges, meet, and sink under, the lighter continental plates. As the plates meet, they often lock up, instead of sliding by each other. This builds up huge stresses; when the plates suddenly unlock with one powerful jerk the release of this tension is what causes the earthquake. A subduction earthquake is caused when one edge of a plate is forced down under another edge; it is considered to be potentially the most dangerous earthquake.

The surface of the earth is made up of several wide, thin, and rigid plate-like blocks (Fig. 2.1). These are constantly rubbing against each other; and sometimes, their rubbing is strong enough to shake up the earth, release tremendous amounts of energy, and cause havoc. The current movement of the Indian plate is estimated to be around 50 mm/year. Of course, in our day to day lives, this kind of movement is negligible. However, a plate beneath the earth's surface, moving at this rate for a long enough time, can have a strong influence on the structure of the earth itself. The earth's crust is made up of seven major plates and different minor plates and these plates are moving relative to each other. Based on the types of relative movement of the plates, the plate boundaries are divided into three types: spreading ridge boundaries, subduction zone boundaries, and transform fault boundaries. The earthquake activity is very high along the subduction and transform plate boundaries. Based on the location of earthquakes, it can be divided into

T. G. Sitharam et al., *Comprehensive Seismic Zonation Schemes for Regions at Different Scales*, https://doi.org/10.1007/978-3-319-89659-5_2

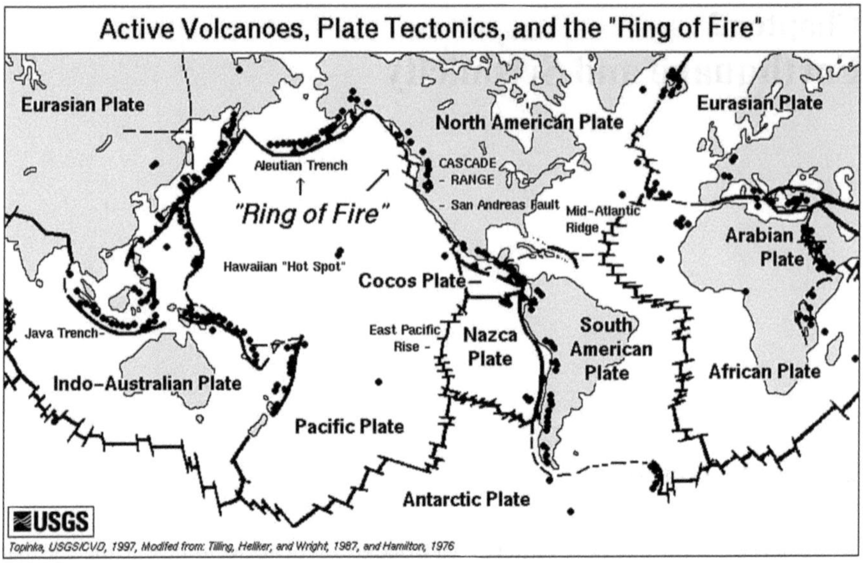

Fig. 2.1 Major tectonic plates (source: USGS)

two types—interplate earthquakes and intra plate earthquakes [282]. The main difference between the intra plate earthquakes and the inter plate earthquakes are:

- The recurrence time will be more for intra plate earthquakes
- The faults are rarely visible at the surface for intra plate earthquakes
- The energy released will be more than that of an inter plate earthquake
- The seismic energy dissipates very slowly and the seismic waves travel a longer distance for intra plate earthquakes

2.1.1 *Faulting of Rocks*

Faults are localized areas of weakness in the surface of the Earth, sometimes the plate boundary itself. A fault can be a crack in the earth's crust along which rock on one side is displaced relative to that on the other side (Fig. 2.2). Movements in the plates create stresses in the plate and if the stress reaches ultimate strength of the rock, the rock breaks and this results in sudden release of stored strain energy. This breakage of strata over large areas is known as faulting. Different variety of faults will result i.e., strike slip fault, normal fault and reverse or thrust faults depending upon stress causing the faulting. As the broken surface is planar in nature, the fault plane designated with three parameters, namely strike, dip, and dip direction. Movement along the fault of the two broken surfaces (usually called hanging wall and footwall) is called slip and there are several varieties of slips possible.

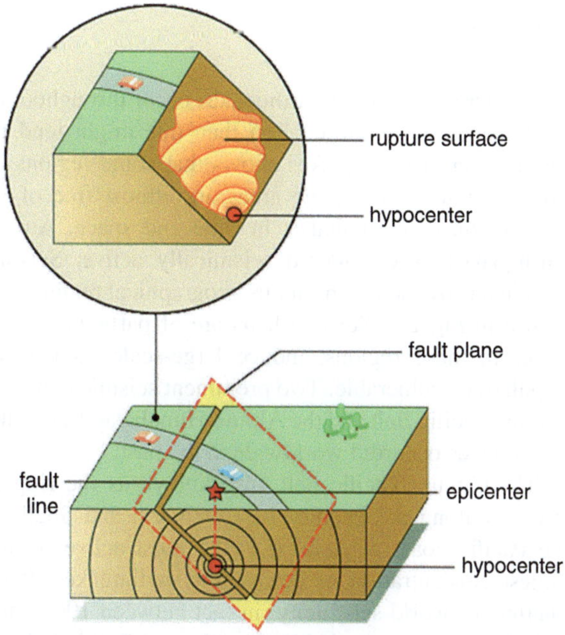

Fig. 2.2 Rupture of rock along a fault (source: USGS)

2.1.2 Earthquakes: Elastic Rebound Theory

Earthquake is associated with sudden rapid shaking of the earth causing fracture in the earth's crust, leading to release of strain energy stored in rocks which spread in all directions causing ground shaking. As stated, the plates of the earth are in constant motion and plate tectonics indicates that the majority of their relative movements occur near the boundaries. As the relative movement of the plates occur, elastic strain energy is stored in the material near the boundary. When the shear stress reaches the shear strength of the rock along the fault, the rock faults and the accumulated strain energy is released. Rupture of the rock will release the stored energy explosively, partly in the form of heat and partly in the form of the stress waves that are felt as earthquakes. The theory of elastic rebound describes the process of successive build up and release of strain energy in the rock. The elastic rebound theory explains the spreading of energy during earthquakes. When plates on opposite sides of a fault are subjected to force and shift, they accumulate energy and slowly deform. As the developed stresses exceed the internal strength of the rock, a sudden movement occurs along the fault, releasing the accumulated energy, and the rocks snap back to their original undeformed shape.

This theory was discovered by making measurements at a number of points across a fault. Prior to an earthquake it was noted that the rocks adjacent to the fault were bending. These bends disappeared after an earthquake suggesting that the energy stored in bending the rocks was suddenly released during the earthquake.

2.2 Global Seismicity

A very large number of earthquakes occur throughout the world every year; in fact, earthquakes occur more often than one might tend to believe. However, spatial distribution of earthquakes shows that some regions have more earthquakes than other regions, while large areas are almost free of seismicity. Seismicity is the distribution of earthquakes in time and space. Any region, which has frequent earthquakes, is considered seismically active. Seismicity is concentrated along certain narrow, semicontinuous geographical regions called seismic belts. These are shown in Fig. 2.3. Seismic belts are of particular interest as frequent earthquakes occur in these regions, induce large-scale damage repeatedly, and make large populations vulnerable. Two prominent seismic belts can be identified on the globe; Circum Pacific Belt and the Alpine-Himalayan Belt. Table 2.1 lists major destructive earthquakes occurred worldwide in the past.

The Circum Pacific belt, also known as the ring of fire, is long and narrow. It exists along the Pacific coast of North and South America and continues into the Pacific coast of Asia. It is the most active of all seismic belts and has the largest concentration of devastating earthquakes. It contributed more than three quarters of world seismicity; in fact between 1904 and 1952, it gave off 75.6% of global seismic energy [113]. The Circum Pacific belt is very complex and includes special topographic features such as island arcs, oceanic trenches, and mountain ranges. It has intermediate and deep focus earthquakes, together with shallow focus earthquakes.

The Alpine-Himalayan belt is the next most active belt. It contributed 22.1% of seismic energy given off on the globe between 1904 and 1954. This seismic belt is

Fig. 2.3 Map showing worldwide seismic hazard (source: USGS)

Table 2.1 Lists of 20 largest earthquakes in the world (source: [80, 154, 164–167, 198, 246, 247])

Sl. no.	Mag	Location	Alternative name	Date (UTC)	Time UTC	Lat	Long
1	9.5	Bio-Bio, Chile	Valdivia earthquake	5/22/196	19:11	38.14°S	73.41°W
2	9.2	Southern Alaska	1964 Great Alaska Earthquake, Prince William Sound Earthquake, Good Friday Earthquake	3/28/1964	3:36	60.91°N	147.34°W
3	9.1	Off the West Coast of Northern Sumatra	Sumatra-Andaman Islands Earthquake, 2004 Sumatra Earthquake and Tsunami, Indian Ocean Earthquake	12/26/2004	0:58	3.30°N	95.98°E
4	9.1	Near the East Coast of Honshu, Japan	Tohoku Earthquake	3/11/2011	5:46	38.30°N	142.37°E
5	9.0	Off the East Coast of the Kamchatka Peninsula, Russia	Kamchatka earthquake	11/4/1952	16:58	52.62°N	159.78°E
6	8.8	Offshore Bio-Bio, Chile	Maule Earthquake	2/27/2010	6:34	36.12°S	72.90°W
7	8.8	Near the Coast of Ecuador	1906 Ecuador—Colombia Earthquake	1/31/1906	15:36	0.96°N	79.37°W
8	8.7	Rat Islands, Aleutian Islands, Alaska	Rat Islands Earthquake	8/15/1950	14:09	28.36°N	96.45°E
9	8.6	Eastern Xizang-India Border Region	Assam, Tibet	8/15/1950	14:09	28.36°N	96.45°E
10	8.6	Off the West Coast of Northern Sumatra		4/11/2012	8:39	2.33°N	93.06°E
11	8.6	Northern Sumatra, Indonesia	Nias Earthquake	3/28/2005	16:10	2.09°N	97.11°E
12	8.6	Andrean of Islands, Aleutian Islands, Aleutian Islands,		3/9/1957	14:23	51.50°N	175.63°W
13	8.6	South of Alaska	Unimak Island Earthquake, Alaska	4/1/1946	12:29	53.49°N	162.83°W
14	8.5	Banda Sea		2/1/1938	19:04	5.05°S	131.61°E
15	8.5	Atacama, Chile	Chile-Argentina Border	11/11/1922	4:33	28.29°S	69.85°W
16	8.5	Kuril Islands		10/13/1963	5:18	44.87°N	149.48°E

(continued)

Table 2.1 (continued)

17	8.4	Near the East Coast of Kamchatka Peninsula, Russia	Kamchatka, Russia	2/3/1923	16:02	54.49°N	160.47°E
18	8.4	Southern Sumatra, Indonesia		9/12/2007	11:10	4.44°S	101.37°E
19	8.4	Near the Coast of Southern Peru	Arequipa, Peru Earthquake	6/23/2001	20:33	16.27°S	73.64°W
20	8.4	Off the East Coast of Honshu, Japan	Sanriku, Japan	3/2/1933	17:31	39.21°N	144.59°E

more diffuse than the Circum Pacific belt. Topographic features associated with this belt are mountain ranges on continents and island arcs and deep trenches in oceans. It includes the mountainous regions of Alps in Europe, Zagros in Iran, Sulaiman and Kirthar ranges in Pakistan, Hindu Kush and Pamir regions, the Himalayas in Asia, and extends toward the East Indies, via the Arakan Yoma mountain ranges and continues eastward into Indonesia and Philippines. It includes the mountain ranges that radiate from the Pamir knot, such as Karakoram, Kunlun, Altyn Tagh, and those that stretch into Tibet, China, and Mongolia.

Besides the Circum Pacific Belt and the Alpine-Himalayan Belt, other regions of reduced seismicity also exist on the globe. These comprise of mid-oceanic ridges, continental rifts, marginal areas, regions of old seismicity, and stable masses. Regions of old seismicity refer to pre-Cambrian shields of Africa, India, Siberia, Fenoscandia, Australia, Canada, and Brazil.

2.3 Indian Seismicity

The earthquake zoning map of India divides India into four seismic zones (Zone II to V) as shown in Fig. 2.4, unlike its previous version which consisted of five or six zones in the country. According to the present zoning map, Zone-V expects the highest level of seismicity whereas Zone-II is associated with the lowest level of seismicity.

The tectonic framework of the Indian subcontinent is diverse with respect to space and time and hence complex. The rapid drifting of the Indian plate towards the Himalayas in the northeastern direction with a high velocity along with its low plate thickness [185] might be the cause for the high seismicity of the Indian region. When continents converge, large amounts of shortening and thickening take place, like in the Himalayas and Tibet. Due to this massive collision, the Himalayas were formed and large numbers of earthquakes happen in the region. A similar process, involving the Indian Plate and the Burmese micro-plate, results in earthquakes in

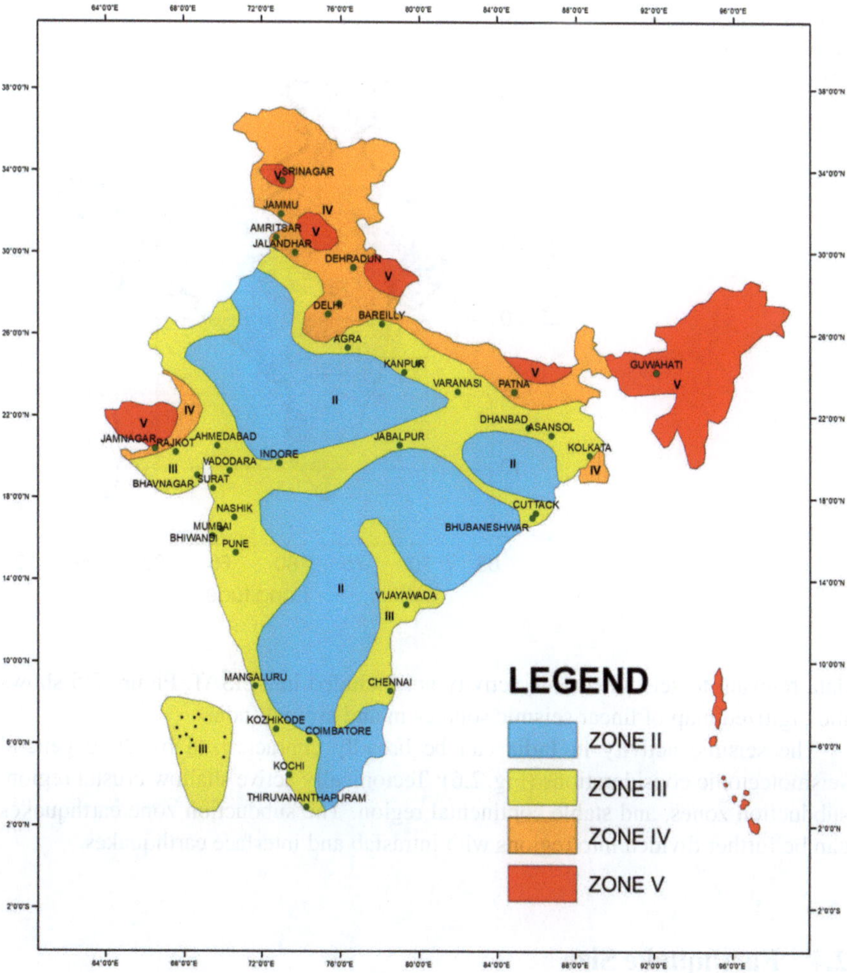

Fig. 2.4 Latest seismic zonation map of India [28]

the Andaman & Nicobar Islands. The plate boundary areas along the Himalayas and North East India are characterized by a very high level of seismicity. There are earthquakes occurring within the Indian shield region, in the Indian peninsula, and in adjoining parts of the Arabian Sea or the Bay of Bengal, which are intraplate earthquakes.

Geological Survey of India published Seismotectonic Atlas, SEISAT (2000) with details of linear seismic sources in India and adjoining areas with all available data related to earthquakes. It is a compilation of multi-thematic database comprising of 43 maps covering India and adjacent regions of neighboring countries on 1:1 million scale. Various details regarding geophysical, structural, seismicity, and geothermal

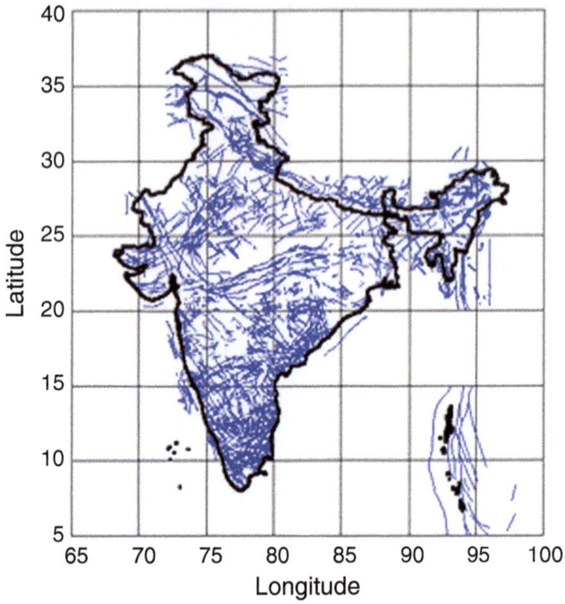

Fig. 2.5 Linear seismic sources identified in India [316]

data relevant to seismotectonic activity are included in SEISAT. Figure 2.5 shows the digitized map of linear seismic sources in and around India.

The seismic activity in India can be broadly characterized by three general seismotectonic considerations (Fig. 2.6): Tectonically active shallow crustal region, subduction zones, and stable continental region. The subduction zone earthquakes can be further divided into regions with intraslab and interface earthquakes.

2.4 Earthquake Size

The news reports of earthquakes refer to an earthquake with a certain magnitude, or that with a certain number on the Richter scale. For example, 2001 Bhuj earthquake was a magnitude 7.7 earthquake and 2015 Nepal earthquake was a magnitude 7.7 earthquake. Most of us understand that the higher the number, the larger the quake; but how are these numbers arrived at, and is this the only way to measure an earthquake? The size of an earthquake depends on the size of the fault and the amount of slip on the fault. Scientists use an instrument called the seismograph to measure the magnitude of an earthquake. Seismographs record the movement of vibrations beneath the earth, as zig-zag traces on a sheet of paper. Some seismographs are sensitive enough to detect earthquakes happening anywhere on earth.

Earthquake size can be expressed qualitatively (non-instrumental) or quantitatively (instrumental). It is commonly expressed in terms of Intensity or Magnitude. Intensity is a measure of how strong the earthquake feels to an observer. It is the

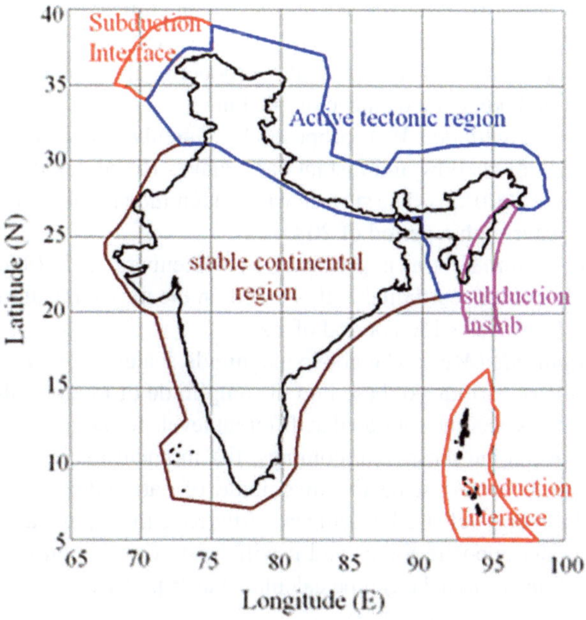

Fig. 2.6 Tectonic provinces in and around India [178]

qualitative assessment of the kind of damage done by an earthquake. Intensity depends on distance to epicenter, strength of earthquake, and local geology. It is determined from the intensity of shaking and the damage from the earthquake. Magnitude is the quantitative measurement of the amount of energy released by an earthquake. Such a quantitative measure is needed to compare the size of earthquakes worldwide, which is independent of the density of population and the type of construction. The magnitude of an earthquake is a very important parameter in seismic hazard analysis and this has been reported using different scales. Some of the important earthquake magnitude scales are described below.

1. *Intensity scale* (I): Intensity is a qualitative measurement of earthquake size as it reflects the strength of an earthquake from the observer's point of view. Intensity value also indicates the kind of damage done by an earthquake and depends neither on the seismological attributes of the earthquake nor on local geology. As seen, the magnitude of an earthquake gives an indication of the total energy released but it cannot give a clear picture of the damaging effects. The intensity scales is a measure of the damaging effect of an earthquake produced at a given location. Moreover this scale can be used to indicate the size of earthquakes for regions where there are no instrumental records are available. There are several intensity scales developed by various researchers such as Mercalli-Cancani-Selberg (MCS), Modified Mercalli (MM), Medvedev-Sponheuer-Karnik (MSK), Japanese Meteorological Agency (JMA), Rossi-ForelIntensity scale.

2. *Local magnitude* (M_L): or the Richter scale magnitude proposed by Richter [283] for measuring shallow earthquakes. A magnitude of 3 in the Ritcher scale magnitude corresponds to an earthquake, producing an amplitude of $A = 1$ mm on a seismograph kept at a distance of 100 km.
3. *Surface wave magnitude* (M_S): proposed by Gutenberg and Richter [111], for measuring size of shallow and distant earthquakes (focal depth less than 70 km and farther than 1000 km). The surface wave magnitude is based on the amplitude of Rayleigh wave with a period of 20 s.
4. *Body wave magnitude* (m_b): proposed by Gutenberg and Richter [114] for measuring deep focus earthquake. It is based on the maximum amplitude of first few cycles of P waves with a period of 1 s.
5. *Moment magnitude* (M_W): The above magnitude scales (M_L, M_S, m_b) measure the magnitude of earthquake based on the amplitude of motion. Moreover these magnitude scales become saturated at different levels for large earthquakes [395]. In order to overcome these shortcomings, the moment magnitude (MW) scale was introduced. In this scale the magnitude of earthquake is measured with respect to the energy released. During an earthquake the rupture along a fault will create equal and opposite forces and it will create a force couple. The moment magnitude of an earthquake can be calculated as Eq. (2.1).

$$M_W = \frac{2}{3} \log_{10} M_o - 10.7 \tag{2.1}$$

$$M_o = \mu A D \tag{2.2}$$

where M_o—seismic moment, μ—modulus of rigidity of the rock mass, A—area of the fault along the fault, D—average displacement, where M_o is in dyne-cm and calculated using Eq. (2.2). In the above equation the value of M_o should be used in dyne-cm. Nowadays this is the most widely used magnitude scale for most of the scientific applications.

As shown in Fig. 2.7 all other magnitude scales except M_W saturate for large earthquakes. It is apparent that m_b begins to saturate at $m_b = 5.5$ and fully saturates at 6.0, whereas, M_S does not saturate until approximately $M_S = 7.25$ and is fully saturated by 8.0. M_L begins to saturate at about 6.5. It is desirable to have a magnitude measure that does not suffer from this saturation.

2.4.1 *Magnitude Conversion*

It is a pre-requisite for complete earthquake catalogue to have a uniform magnitude scale for denoting the size of earthquakes, so that a reliable parameterization of the magnitude distribution which is homogeneous and complete with respect to time and size is used in hazard analysis. The earthquake data are generally

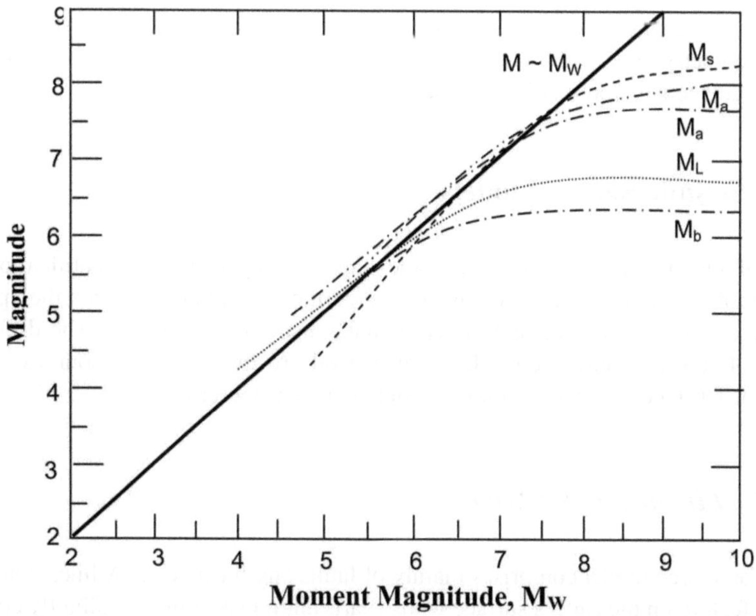

Fig. 2.7 Comparison of moment magnitude scale with other magnitude scales (after [126])

available in different magnitude scales like body wave magnitude (m_b), surface wave magnitude (M_S), local magnitude (M_L), moment magnitude (M_W), and the earthquake intensity scale (I). Unfortunately, many of the magnitude scales are all limited by saturation toward large earthquakes with $m_b > 6.0$, $M_L > 6.5$, and $M_S > 8.0$. The existence of different magnitude scales necessitates the conversion of these magnitude scales to a single magnitude scale for the analysis purposes. The moment magnitude M_W [166] can represent the true size of earthquakes because it is based on seismic moment, which in turn is proportional to the product of the rupture area and dislocation of an earthquake fault [5]. M_W is defined as

$$M_W = \frac{2}{3} \log_{10} M_o - 6.05 \tag{2.3}$$

where M_o is the scalar seismic moment in $N - m$. The homogenization of earthquake catalog involves expressing the earthquake magnitudes in one common scale. Practical problems, such as seismic hazard assessment, necessitate the use of homogenized catalogue. Since M_W does not saturate, this is the most reliable magnitude for describing the size of an earthquake [303]. As the moment magnitude (M_W) scale is the most advanced and widely used magnitude scale, it is recommended to convert the original magnitudes of earthquake events in different time periods to unified M_W magnitudes. Several relations were proposed by different researchers to convert different magnitude scales to M_W ([107, 126, 155, 173, 175, 241, 256, 262, 303, 354]; among many others). The relations between different magnitude scales will depend on the observation errors, source characters such as stress drop, fault geometry, etc. [126]. Hence it is always advisable to use the region specific magnitude conversion relations [196].

2.5 Seismic Source Models

One of the major steps in the seismic hazard analysis is the identification of vulnerable seismic sources. A seismic source model primarily defines the spatial characteristics of a source and the earthquake distribution within it. The different types of seismic sources considered in general are linear seismic sources, point sources, Gridded Seismicity source model, and areal sources.

2.5.1 Linear Source Model

A linear source model comprises mainly of faults and lineaments. A lineament is a linear feature on the earth's surface whose parts align in a straight or slightly curved and is distinctly different from adjacent features. One of the best documents listing the linear seismic sources in India and adjoining areas is the Seismotectonic Atlas [316] published by the Geological Survey of India (GSI). The seismotectonic atlas [316] was prepared after extensive studies using remote sensing technique and by geological explorations. The SEISAT maps [316] are available in A_0 size sheets with 1:1,000,000 scale and each map covers an area of $3^\circ \times 4^\circ$. SEISAT contains the details of the faults, lineaments, and shear zones in addition to the geological features in India and adjoining areas. This has been taken as an authentic reference manual for identifying the seismic sources by various researchers like [146] for Delhi, [235] for microzonation of Sikkim Himalayas, [273] for Mumbai, [33] for Chennai, and [10] for Bangalore.

Apart from Seismotectonic atlas, the major lineaments in a given study area also can be mapped using satellite data such as Indian Remote Sensing Satellite (IRS)-1D, Wide Field Sensor (WiFS), and Landsat Multi Spectral Scanner (MSS)/Thematic Mapper (TM) data on 1:1 million scale. The length and direction of each lineament was measured and the lineaments were grouped based on length and direction. The nature of each lineament was assessed using the satellite data in conjunction with the collateral data. Remote sensing data due to its synoptic nature is found to be very useful in mapping lineaments. Images taken in the near infrared (NIR) region (0.7–1.1 μm) depicts clearly more lineaments than other bands. Radar data also provides information on lineaments due to its oblique look angles. Thermal infra-red (TIR) data is found to be useful in delineating wet lineaments with moisture/water. Lineaments present in the forest areas, soil covered areas are also clearly visible on images thus enabling us to delineate better structural features.

Normally, it is difficult to decide whether the mapped lineament is a fault or not, but if there is a clear displacement/offset then the lineament can be identified as a fault. Integration of the lineament map with the available structural and geological information of the terrain plus fieldwork helps to decide the nature of the lineament. Lineaments, which are seen as linear features on satellite data were mapped. Lineaments with length more than 100 km, i.e., major lineaments were mapped first from individual scenes of Landsat data on 1:1 million scale scenes; these were transferred to base map of the study area and make a single map. This map was superimposed on physical/road network map of the study area to eliminate any cultural lineament like road, railway lines which also appear as lineaments in the satellite image. IRS-1D WiFS FCC (Fig. 2.8) on 1:1 million scale was used further to map the lineaments. Lineaments were numbered and their length and direction were measured.

Based on the correlation between major lineaments and earthquake occurrence, the seismically active lineaments were identified. Studies by Ganesha Raj and Nijagunappa [101], Sitharam et al. [333, 334] were toward the identification of lineaments using satellite data for the state of Karnataka. Earthquakes with magnitude four and above that occurred in Karnataka and regions inside 300 km from the Karnataka state boundary were considered and overlaid on major lineament map of Karnataka and adjoining regions to assess the correlation between earthquakes and major lineaments. Currently, along many lineaments rivers/streams are flowing. Parts of the river courses of Hemavathi, Tunga, Yagachi, Vedavathi, Krishna, Nethravathi, Malaprabha, etc. follow major lineaments. The major lineaments in the study area, which have some correlation with the occurrence of earthquakes, were identified and shown in Fig. 2.8. The declustered earthquake data in unified moment magnitude scale have to be superimposed on the digitized map of extracting tectonic features. To characterize the seismic sources, the maximum reported magnitude and number of earthquake events associated with each of these sources are noted. There are some earthquake events which are not falling along any of the identified faults. In those cases the linear source model alone may not be able to give the correct picture of seismic hazard levels. To overcome this limitation three alternative source models can be adopted—point sources, smoothed gridded sources, and areal sources. These are widely accepted source models for evaluation of seismic hazard for regions where the sources are not clearly delineated.

2.5.2 Smoothed Point Source Model

The point sources are considered for hazard analysis using deterministic methods. The selection of point sources is done using two steps—discretizing the observed seismic activity and then smoothing the discretized seismic events. In this method, the study area is divided into small grids and the maximum earthquake magnitude in each grid is assigned to the center of that grid. The maximum earthquake magnitudes, thus obtained are smoothed using a smoothing window. The smoothing of the data is done to account for the source dimension of the earthquake events and for location errors [73, 255]. This type of source characterization was used

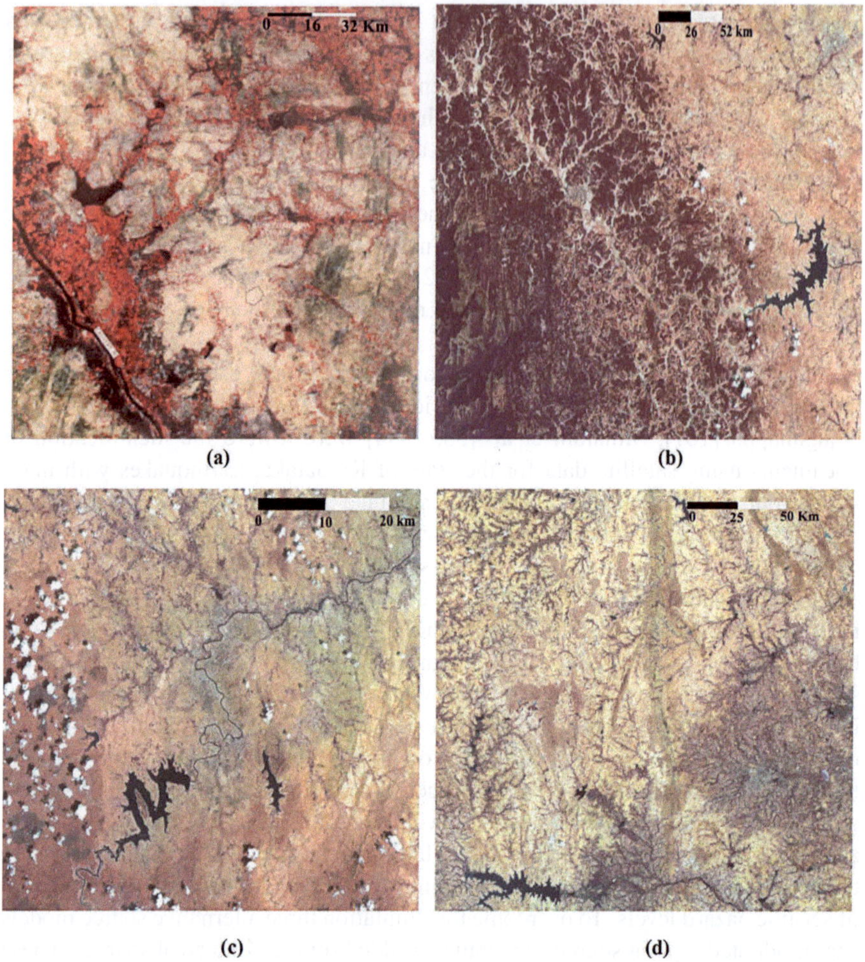

Fig. 2.8 IRS LISS-II images showing various lineaments in Karnataka (extracted from RS image procured from National Remote Sensing Agency in the year 2006 after [101]). (**a**) Cauvery lineament. (**b**) Hemavathy-Thirthahalli lineament. (**c**) Kabini lineament. (**d**) Arkavathi-Madhugiri lineament

for deterministic seismic hazard analysis (DSHA) by Costa et al. [73], Panza et al. [255], El-Sayed et al. [83]. For the deterministic seismic hazard evaluation of India, Parvez et al. [261], Kolathayar et al. [177] have modelled the seismic sources in India as point sources based on the same methodology.

2.5.3 Smoothed Gridded Seismicity Model

Gridded seismicity model [96, 386] is based on the seismic activity rate obtained from the earthquake catalogue and it is one of the most widely adopted methods to model seismic sources for the regions in the absence of clearly identified seismic sources. Some of the seismic hazard studies which considered zoneless approach for source identification are [149, 160, 189, 215, 373, 378]. In this method the study area is divided into grids and the number of earthquakes, which are having a magnitude higher than a cut-off magnitude, in each grid is counted. This will give the activity rate for that particular grid cell. Based on this value, the recurrence rates for different magnitude intervals can be calculated and these values can be smoothed using a Gaussian function (2.4) to get the final activity rate for each grid cell. The uncertainty involved in estimating the location of the earthquake event and the size of the seismic source can be accounted by this smoothing.

While considering these sources, the first step is the selection of grid size and the cut-off magnitude (M_{cut}). After selecting a suitable grid size, the number of earthquake events of magnitude greater than or equal to M_{cut} has to be calculated for each grid cell and this represents the maximum likelihood estimate of the total number of earthquakes (equal to or greater than M_{cut}) for that grid cell. In the present study the value of M_{cut} was taken as 4.0 to eliminate the effect of rock bursts, blasting. Moreover seismic events with magnitude less than 4.0 may not cause much damage also. Based on this value the recurrence rates for different magnitude intervals were calculated. These values were smoothed using a Gaussian function to get the final corrected values for each grid. This smoothing is done to account for the uncertainty associated with the location of earthquake events.

$$\hat{n}_i = \frac{\sum_j n_j e^{\frac{-\Delta_{ij}^2}{c^2}}}{\sum_j e^{\frac{-\Delta_{ij}^2}{c^2}}} \tag{2.4}$$

where n_j is the number of earthquakes in the jth grid cell and $\hat{n}_i$ is the smoothed number of earthquakes in ith cell; c is the correlation distance (to account for the location uncertainties) and Δ_{ij} is the distance between the ith and jth cells. In the present study a cut-off magnitude of $M_W = 4$ and a correlation distance of 50 km is used in smoothing the 'a' values. Since the linear source model was also considered in the present study, a single correlation distance was used.

2.5.4 Areal Source Model

For the hazard estimation using areal sources, the territory under study should be first divided into seismic sources such that, within a seismic source, earthquake-occurrence process is independent. For each seismic source, magnitude exceedance rates are estimated by means of statistical analysis of earthquake catalogs. These rates are the number of earthquakes per unit time, in which a specific magnitude is exceeded, and they represent the seismicity of the source [253].

The areal sources can be modelled as polygon-dipping areas which allow for local variations in seismicity characteristics for the other types of source models; linear sources and gridded seismicity approach (for example, changes in a, b-values, $M_{\max}$ etc.). In this method, a spatial integration process is carried out to account for all possible focal locations with an assumption that, within a seismic source, all points are equally likely to be an earthquake focus. Within a source, seismicity is evenly distributed by unit area and to correctly account for this modelling assumption, a spatial integration is performed by subdividing the original sources. Once subdivided into sub-sources, all the seismicity associated with a sub-source is assigned to a single point, and then the spatial integration adopts a summation form.

2.6 Seismicity Analysis

The seismic activity of a region is given by the Gutenberg–Richter earthquake recurrence law [112]. The recurrence rate given by this to this law is:

$$\log_{10} \lambda_M = a - bM \tag{2.5}$$

where λ_M is the total number of earthquakes with magnitude M and above which will occur in a year and a and b are the seismicity parameters of the region. The values of seismicity parameters can be evaluated using the maximum likelihood estimation technique [4, 369]. The parameter 'a' describes the productivity of volume, and b the slope of the frequency-magnitude distribution (FMD) describes the relative size distribution of events. The high-quality earthquake catalogs collected primarily in recent years and the availability of increased computing power have enabled researchers to investigate spatial variations in b with high precision. The strong difference in b is simply a reflection of the heterogeneity of the earth that emerges on all scales, once suitable datasets become available [382]. For evaluating the seismicity parameter, the completeness of the catalogue has to be analyzed and the data in the complete part of the catalogue need to be used for the analysis.

2.6.1 *Stepp's Method*

The completeness of the catalogue can be analyzed using the method suggested by Stepp [342] which will give the time interval in which the magnitude range is homogeneous. In the method proposed by Stepp [342], the sample mean is inversely proportional to the number of observations in the sample. The earthquake sequence is modelled based on a Poisson's distribution for obtaining an accurate estimate of the variance of the sample mean. Let $x_1, x_2, x_3, \ldots x_n$ be the number of events per unit time interval, then the unbiased estimate of the mean per unit interval for this sample is given by Eq. (2.6).

$$\lambda = \frac{1}{n}\sum_{i=1}^{n} X_i \tag{2.6}$$

where n is the unit time interval and its variance is $\sigma_{\lambda}^2 = \frac{\lambda}{n}$. When the time interval is taken as 1 year, the standard deviation of the above equation becomes $\sigma_{\lambda} = \frac{\sqrt{\lambda}}{\sqrt{T}}$; where T is the sample length. If this process is assumed to be stationary, it can be concluded that the standard deviation behaves as $\frac{1}{\sqrt{T}}$ in the subinterval of the complete years of the sample. Hence during this period the mean rate of occurrence in a magnitude class will be constant.

As a first step for the evaluation of the completeness period, the number of earthquakes reported during each decade for the given magnitude ranges is to be evaluated. The earthquake data is considered as complete as long as its variation is along the $\frac{1}{\sqrt{T}}$ line. The completeness periods for different ranges of magnitude can be estimated using this approach. The completeness period of the smaller magnitude ranges will be low as the globe was not well instrumented in the past. Hence lots of smaller magnitude events were not recorded properly. However the probability of events getting recorded increases with increase in earthquake magnitude. That is the reason for the increasing trend in completeness period with increase in magnitude. The seismicity parameters have to be evaluated based on the complete part of the catalogue.

2.6.2 *Magnitude of Completeness*

The magnitude of completeness is the lowest magnitude above which the earthquake recording is assumed to be complete. The magnitude of completeness (M_c) is defined as the lowest magnitude at which 100% of the events in a space-time volume are detected [294]. Below this magnitude a fraction of events is missed by the network because they are either too small to be recorded by enough stations or because they are below the magnitude of interest or because they are mixed with the coda of a larger event and therefore they passed undetected. There will be spatial and temporal variation of M_c and it will decrease with time mainly because of the increase in the number of seismographs in the region. The 'b' value can be evaluated using the maximum likelihood estimate as given below.

$$b = \frac{\log_{10}(e)}{\left[m_{\text{mean}} - \left(M_c - \frac{\Delta m_{\text{bin}}}{2}\right)\right]} \tag{2.7}$$

where m_{mean}—mean magnitude of the sample; M_c—magnitude of completeness and Δm_{bin}—magnitude bin size (incremental M_W values).

One of the most widely used methods to evaluate M_c is based on a power law fit for the frequency magnitude distribution (FMD), which was suggested by Wiemer and Wyss [381]. In this method a series of synthetic magnitude distributions are developed for each magnitude interval using maximum likely hood estimate. These '*a*' and '*b*' values obtained in the synthetic distribution are compared with the observed distribution and the goodness of fit is calculated. For calculating the goodness of fit, the absolute difference (R) between the observed and synthetic distribution has to be calculated as per Eq. (2.8).

$$R(a, b, M_i) = 100 - \left[\frac{\sum_{m_i}^{m_{\max}} (B_i - S_i)}{\sum_i B_i}\right] \tag{2.8}$$

where B_i and S_i are the observed and predicted cumulative number of events in each magnitude bin. The M_c value is then calculated at the magnitude where the goodness of fit is more than 90%. The goodness of fit can be tested for 95% also, but this level is rarely obtained for real earthquake catalogues [385]. The correct estimate of the *a* and *b* values depends critically on the completeness of the sample under investigation. The FMD deviates from a linear power law fit increasingly for smaller magnitudes which is caused by the fact that the recording network is only capable of recording a fraction of all events for magnitudes smaller than the magnitude of completeness, M_c. If M_c is raised to large values, the uncertainty in the *b* value estimate increases strongly. The situation is complicated by the fact that M_c varies as a function of space and time throughout all earthquake catalogs, hence estimating the correct M_c, while maximizing the available number of earthquakes, becomes difficult.

Figure 2.9 shows a typical FMD for Hindukush region (Source zone 30 by Kolathayar and Sitharam [175]) with *a b* value of 0.97, a value of 8.04, and Magnitude of completeness 4.7. The higher *a* value obtained implies that the average number of earthquakes per year in the region is very high. The *b* value nearing to one indicates that the size of earthquakes is evenly distributed in the region which means there is no large variation in the number of small magnitude events and large magnitude events. The magnitude of completeness of 4.7 implies that the earthquake catalog of the region is complete for all events with magnitude greater than 4.7.

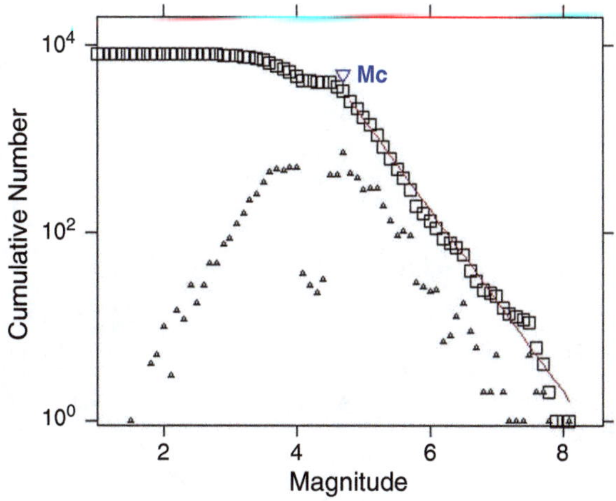

Fig. 2.9 The frequency magnitude distribution for Hindukush region (source: zone 30 of [175])

2.7 Preparation of Seismotectonic Atlas

The compilation and integration of all available data on geological, geophysical, and seismological attributes for the entire country is required for the proper evaluation of seismicity in different tectonic regions. In this context, well-defined and documented seismic sources published by authorized Geological agencies aid in identifying the potential seismic sources in a region. Maps are the representation of a geographic area on a piece of paper or canvas. These maps or Aerial photographs are never useful for computers. So in order to study the maps or aerial images it is necessary to convert them into digital form with appropriate resolutions to get high quality digital images.

The power of GIS technology makes creating and modifying maps a very simple task. Using colors and symbols, many different attributes can be highlighted. Application of GIS techniques allows inserting, extracting, handling, managing, and analyzing the data for the zoning of seismicity. The capability of Geographic Information Systems (GIS) to store and process data and images makes them very valuable in the field of seismic studies. The scanned maps can be georegistered and digitized using appropriate GIS tools. The coordinates are to be fixed based on the latitudes and longitudes of the region in the map.

2.7.1 Earthquake Catalog for India

Kolathayar and Sitharam [175] created an updated earthquake catalog for India and adjoining regions. They collected the details of earthquake events for the period from 250 BC to 1505 AD from [79]. The later portion of historic earthquakes was compiled from the work of various researchers [21, 27, 30, 62, 63, 109, 159, 170,

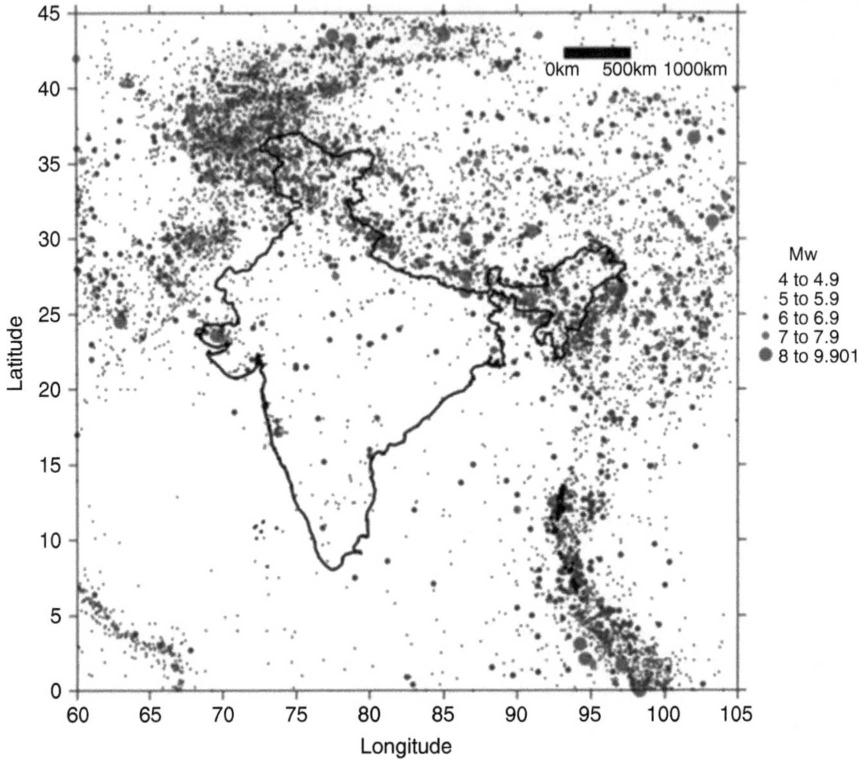

Fig. 2.10 Distribution of earthquake events in and around India (after [175])

250, 278, 280, 341, 351]. They compiled the instrumental catalog from national and international agencies like Indian Meteorological Department (IMD), National Geophysical Research Institute (NGRI) Hyderabad, International Seismological Center (ISC), data file (for the time period between 1964 and 2010), Harvard seismology and USGS/NEIC catalog (for the time period between 1973 and 2010). They have developed region specific magnitude conversion relations to homogenize the entire catalog in unified moment magnitude scale. Further, the catalog was declustered to remove aftershocks and foreshocks. The distribution of earthquake events of magnitude greater than 4, in and around India is presented in Fig. 2.10.

2.7.2 Development of Seismotectonic Map for India

Geological Survey of India has compiled all the available geological, geophysical, and seismological data for entire India. Seismotectonic [316] atlas contains 43 maps covering entire India and adjoining areas, with all available data related to

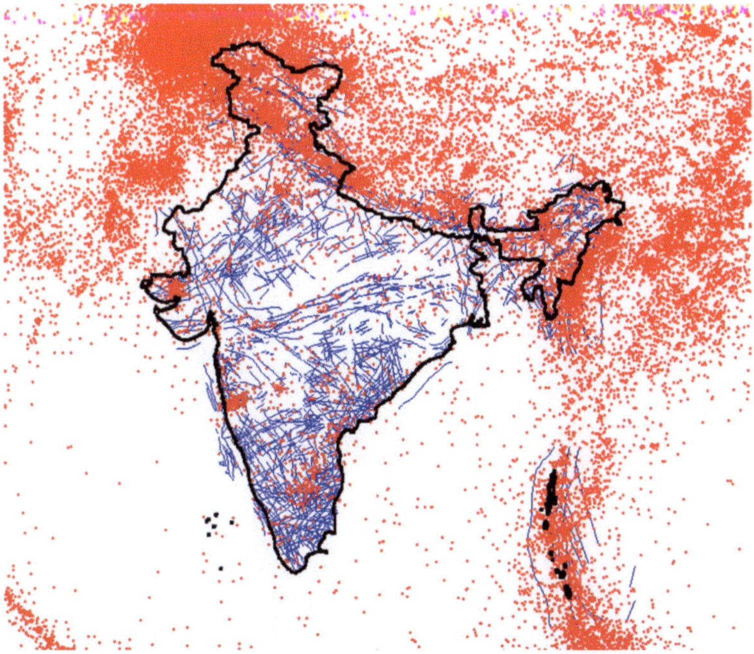

Fig. 2.11 Seismotectonic map of India showing linear sources and events of $M_W \geq 4$ (after [175])

earthquakes. It is a compilation of multi-thematic database comprising of 43 maps (presented in 42 sheets) covering India and adjacent regions of neighboring countries on 1:1 million scale. Kolathayar and Sitharam [175] developed seismotectonic map for India where all the 42 sheets in SEISAT published by GSI were scanned and the faults, shear zones, lineaments, and other seismic sources were digitized. Then the complete homogenized earthquake events [175] were superimposed on this source map to get the final digital seismotectonic map of India with all the seismic sources and earthquake events (Fig. 2.11).

Chapter 3
Seismic Hazard Analysis

3.1 Introduction

The earthquake resistant design aims to prepare the structure to withstand a certain level of ground shaking without undergoing extensive damages. Hence a reliable estimation of ground motion parameter at a given site forms a pre-requisite for any earthquake resistant design procedure. Seismic hazard analysis involves the quantitative estimation of ground shaking hazard at a particular site. This chapter describes the estimation of ground motion parameters such as peak horizontal acceleration (PHA) and spectral acceleration (S_a) for the state of Karnataka and Kalpakkam nuclear power plant (NPP) site.

3.2 Ground Motion Prediction Equations

Ground motion amplitude values will vary with variation of distance from the earthquake source. The ground motion prediction equations (GMPE or attenuation relations) are being used to predict the attenuation properties in any region. The most common parameters which are used to measure the ground motion amplitude are the peak horizontal acceleration (PHA) and peak horizontal velocity (PHV). The PHA values are more commonly used to describe the earthquake ground motion because of their direct association with the inertial forces [181]. It has been found that the PHV is a good parameter for characterization of ground motion with intermediate frequencies for structures like tall or flexible buildings, bridges, etc. The ground motion prediction equation gives the variation of ground motion parameter as a function of earthquake magnitudes and the source-to-site distance. It is typically of the form as shown in Eq. (3.1).

$$\ln y = c_1 + c_2(M - 6) + c_3(M - 6)^2 - \ln(R) - c_4(R) + \ln(\varepsilon) \tag{3.1}$$

T. G. Sitharam et al., *Comprehensive Seismic Zonation Schemes for Regions at Different Scales*, https://doi.org/10.1007/978-3-319-89659-5_3

where y, M, R, and ε refer to PGA, spectral acceleration (g), moment magnitude, hypocentral distance, and error associated with the regression, respectively.

The GMPE are broadly classified into two categories, viz. for inter plate earthquake regions and for mid-plate or stable continental shield regions. This is necessitated by the fact that the dissipation of seismic energy occurs at a faster rate at plate boundaries than at the mid plate regions. Some of the relations developed for active tectonic regions (inter plate regions) are [37, 54, 55, 156, 269, 296, 318]. Some of the attenuation relations for the stable continental shield regions are by Toro et al. [359], Atkinson and Boore [16] and Raghu Kanth and Iyengar [274]. Table 3.1 lists the popular ground motion prediction equations developed worldwide in recent times.

In India, there is a lack of strong motion data and this in turn has resulted in the development of only very few region specific GMPEs. Some of the important GMPE available in India are [318] for the Himalayan region, [146] for Delhi region; [274] for Peninsular India; [234] for Sikkim Himalaya; [236] for Guwahati and [322] for Himalayan Region. Out of these attenuation relations, the most widely used relations are [274] and [322].

Table 3.1 List of various attenuation relations proposed by various authors for different regions

Author	Region
Iyengar and Ghosh [146]	Himalayan region
Campbell and Bozorgnia [55]	ENA
Yu and Wang [401]	NE Tibetan Plateau region
Atkinson and Boore [16]	ENA
Raghu Kanth and Iyengar [274]	Southern Indian Peninsula
Malagnini et al. [205]	San Francisco
Chiou and Youngs [68]	Japan, Mexico, California
Castro et al. [60]	Mexico
Ford et al. [94]	Northern California
Bennington et al. [23]	Parkfield, California
Nath et al. [235]	Garhwal Himalaya
Boore and Atkinson [37]	Worldwide
Sharma et al. [321]	Kaahchh region, Gujarat
Chun and Henderson [70]	North Korea
Sharma et al. [322]	Himalayan region
Nath et al. [236]	Guwahati
Ghasemi et al. [106]	Iran
Koulakov et al. [180]	Turkey
Gupta [110]	Indo-Myanmar subduction zone

3.3 Deterministic Seismic Hazard Analysis

The deterministic approach considers a particular earthquake scenario, either realistic or assumed one. The DSHA approach uses known seismic sources which are near the site and available historical seismic and geological data to generate discrete, single-valued events or models of ground motion at the site. The earthquakes are assumed to occur on the source closest to the site. The deterministic seismic hazard analysis needs three input details like earthquake source, controlling earthquake at the source, and an attenuation relation to evaluate the seismic hazard. A schematic diagram of different steps involved in DSHA methodology is given in Fig. 3.1. In DSHA, the controlling earthquake is assumed to act along the source at the shortest distance from the site. The uncertainties involved in the earthquake magnitude or location are not taken into account and this method will give an upper bound values for the ground motion. Hence DSHA method is adopted in evaluation of seismic hazard for some of the critical structures like nuclear power plants, big dams, bridges, hazardous waste contaminant facilities, etc. The results obtained from deterministic analysis can be used as a cap for the probabilistic analysis.

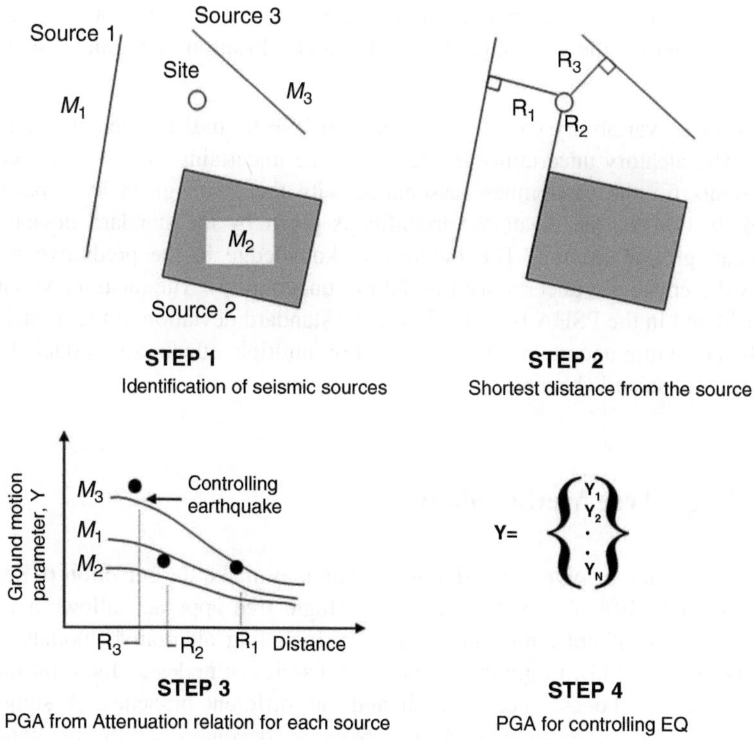

Fig. 3.1 Various steps involved in DSHA method (after [181])

3.4 Probabilistic Seismic Hazard Analysis

The evaluation of seismic hazard involves quantification of uncertainties in earthquake magnitude, location, recurrence rate, and the attenuation characteristics of seismic waves. The adoption of probabilistic approaches in seismic hazard analysis will provide a framework for identifying, quantifying, and combining these uncertainties in a rational way [181]. Probabilistic seismic hazard analysis (PSHA) was initially developed by Cornell [72]. Many researchers have adopted this methodology for evaluating hazard and recently this method has been adopted in India by Iyengar and Ghosh [146], Raghu Kanth and Iyengar [273], Anbazhagan et al. [10], Vipin et al. [376], Kolathayar and Sitharam [176] and Sitharam et al. [335] for the probabilistic seismic hazard analysis of Delhi, Mumbai, Bangalore, North West Himalaya regions, Peninsular India, Andaman & Nicobar regions, and India, respectively. The important steps involved in PSHA are shown in Fig. 3.2. The main steps involved in PSHA methods are

- Identification and characterization of earthquake sources
- Characterizing the earthquake recurrence rate
- Evaluation of ground motion using the attenuation relationships
- Determination of mean annual rate of exceedance of ground motion parameter by considering the uncertainties in earthquake location, size, and attenuation relation.

Two types of variability can be considered in PSHA studies—aleatory and epistemic. The aleatory uncertainty is because of the uncertainty in the data used and it accounts for the randomness associated with the result given by a particular model. In GMPE, the aleatory variability is given by the standard deviation of the mean ground motion. The incomplete knowledge in the predictive models causes the epistemic uncertainty (modelling uncertainty). The aleatory variability is considered in the PSHA by considering the standard deviation of the model error and the epistemic uncertainty is considered by multiple attenuation models for the evaluation of seismic hazard.

3.5 Logic Tree Methodology

The uncertainties involved in different models may make the selection of seismic hazard models difficult. In these cases the logic tree approach allows a formal characterization of epistemic uncertainty by including alternative models in the analysis [31, 52, 343]. Logic tree consists of a series of nodes and several models (hypothesis) can be assigned to each node as different branches. A subjective weightage can be given to each of these branches depending on the likelihood of being correct and the weightage for all the branches at a particular node should be equal to unity. In PSHA the different models are incorporated for magnitude

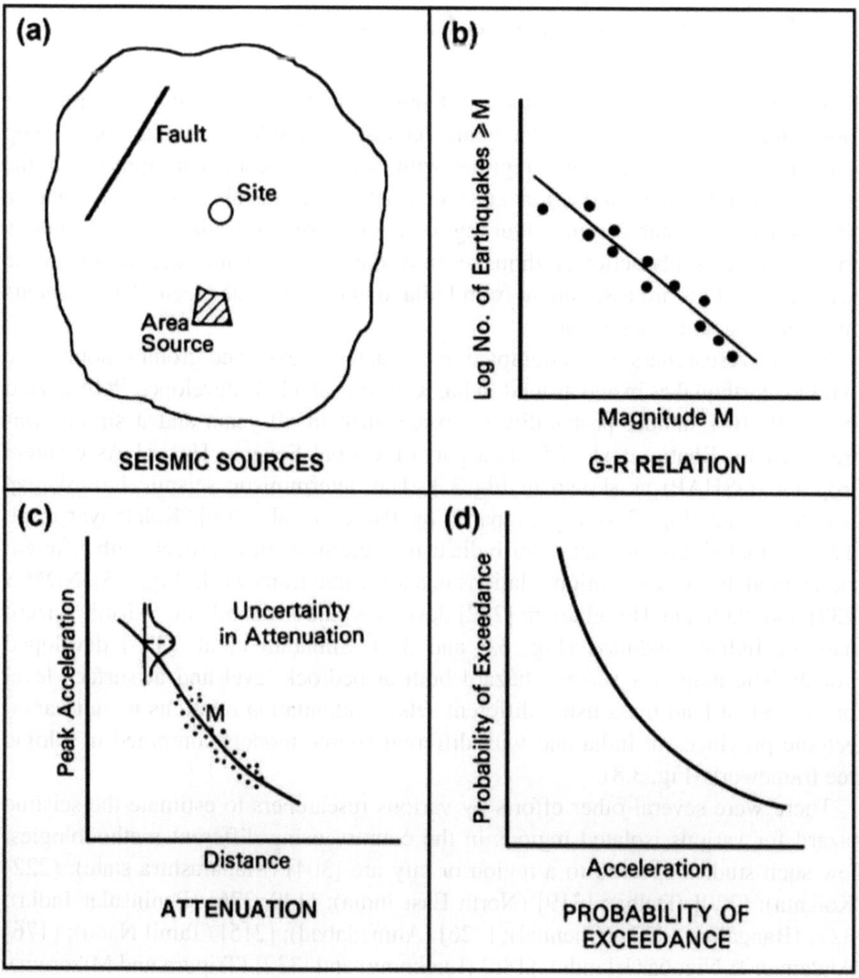

Fig. 3.2 Various steps involved in PSHA method (after [181])

recurrence rate, attenuation relation, estimation of source parameters, evaluation of maximum probable earthquake, etc. The weightage of the terminal branch can be obtained by multiplying the weightage of all the branches leading to it. Even though it is possible to include as many branches as [52] in the logic tree, this will increase the computational effort. Now almost all the PSHA studies are using a logic tree approach to capture the epistemic uncertainties.

3.6 Seismic Hazard Analysis Case Studies

This section presents the case studies of seismic hazard assessment attempted for India and adjoining areas. India being having a complex seismotectonic setup spatially with active tectonic regions with shallow crustal earthquakes in the Himalayas, Interface subduction zone in Kashmir, Intraslab subduction zone in North East India, stable continental region in southern peninsula and both shallow crustal and deep subduction earthquakes in Andaman region, it is ideal to review the case studies of hazard assessment from India to understand the recent developments on seismic hazard assessment.

Various researchers have attempted to evaluate the expected ground motion due to future earthquakes in and around India. Khattri et al. [171] developed PGA hazard map with 10% annual probability of exceedance in 50 years and a similar was presented by Bhatia et al. [25] as a part of Global Seismic Hazard Assessment Program (GSHAP) as shown in Fig. 3.3. The deterministic seismic hazard map of entire India (Fig. 3.4) was prepared by Parvez et al. [260]. Kolathayar et al. [177] estimated seismic hazard for India using deterministic approach with different source models and attenuation relations in a logic tree framework (Fig. 3.5). NDMA [237] and Nath and Thingbaijam [232] developed the Probabilistic Seismic hazard map for Indian landmass (Fig. 3.6 and 3.7). Sitharam et al. [335] developed probabilistic maps for seismic hazard both at bedrock level and at surface level for the Indian land mass using different sets of attenuation relations to suit varied tectonic provinces in India and with different source models combined in a logic tree framework (Fig. 3.8).

There were several other efforts by various researchers to estimate the seismic hazard for various isolated regions in the country using different methodologies. Few such studies specific to a region or city are [304] (Maharashtra state); [222] (Kolkata); [320] (Delhi); [319] (North East India); [149, 376] (Peninsular India); [327] (Bangalore); [33] (Chennai); [326] (Ahmedabad); [215] (Tamil Nadu); [176] (Andaman & Nicobar Islands), [186] (Lucknow) and [332] (Tripura and Mizoram).

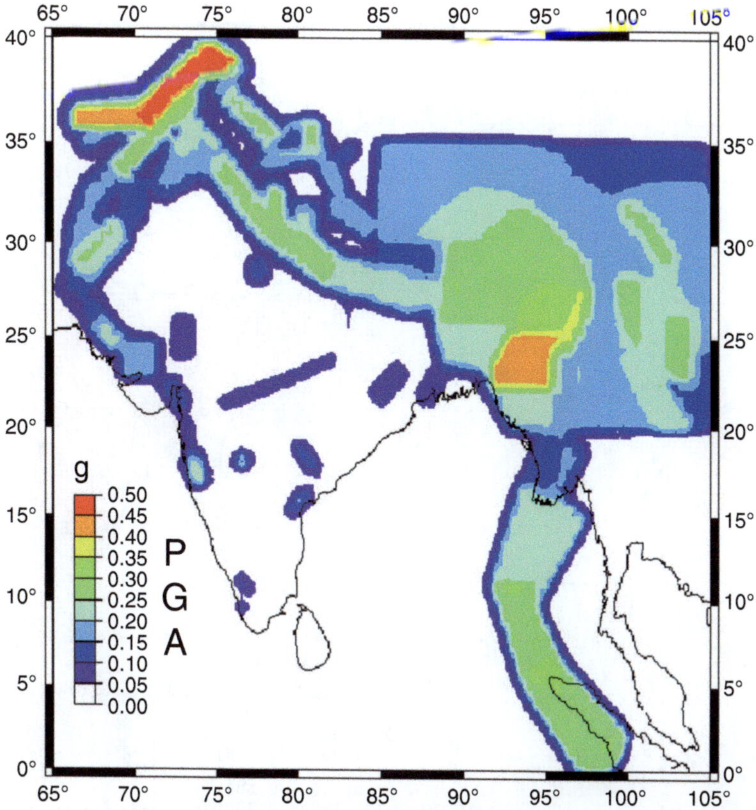

Fig. 3.3 Spatial variation of peak ground acceleration corresponding to 10% probability of occurrence after [25]

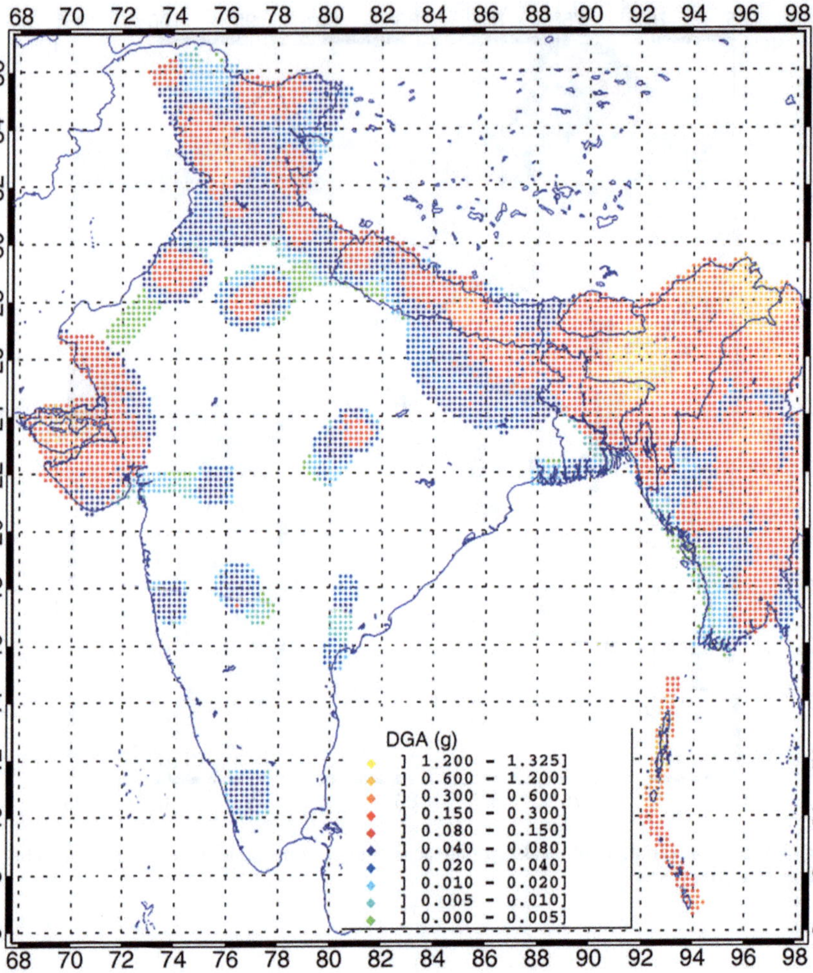

Fig. 3.4 Spatial distribution of the design ground acceleration after [260]

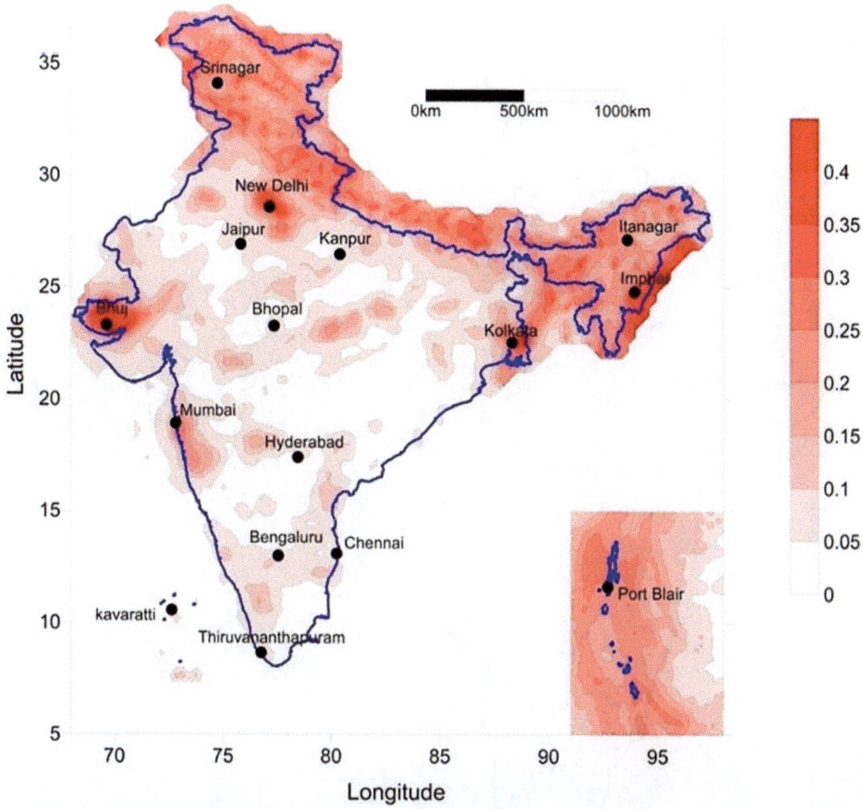

Fig. 3.5 Spatial distribution of the design ground acceleration after [177]

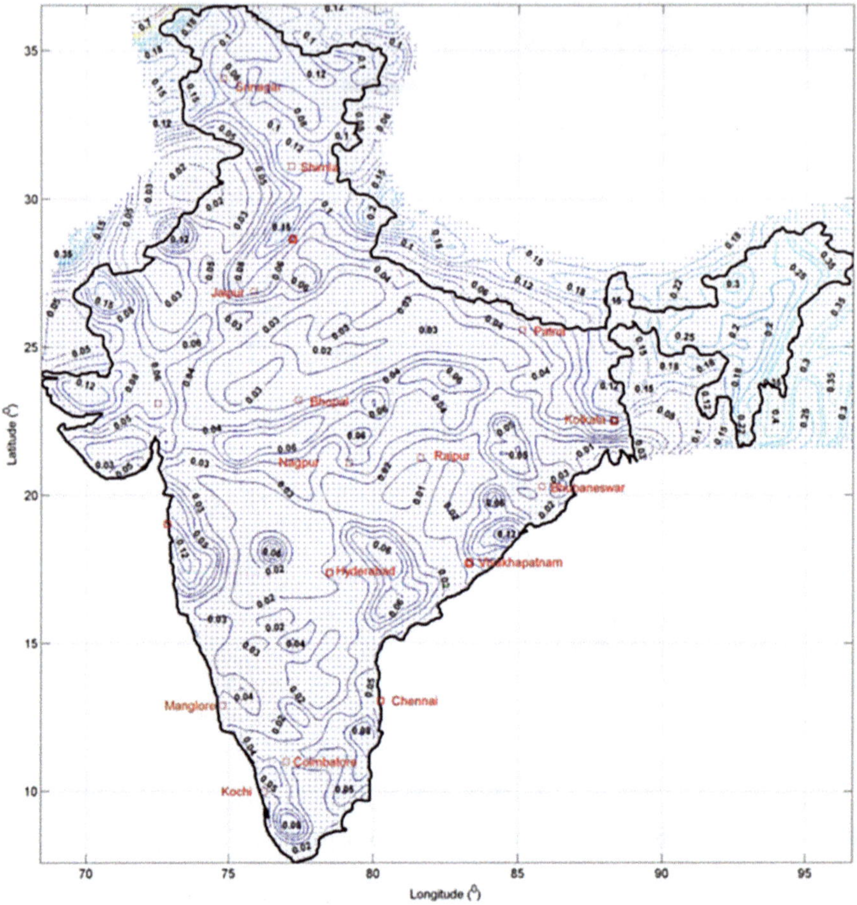

Fig. 3.6 PGA Contours with 10% probability of exceedance in 50 years (return period 500 years) on A-type sites [237]

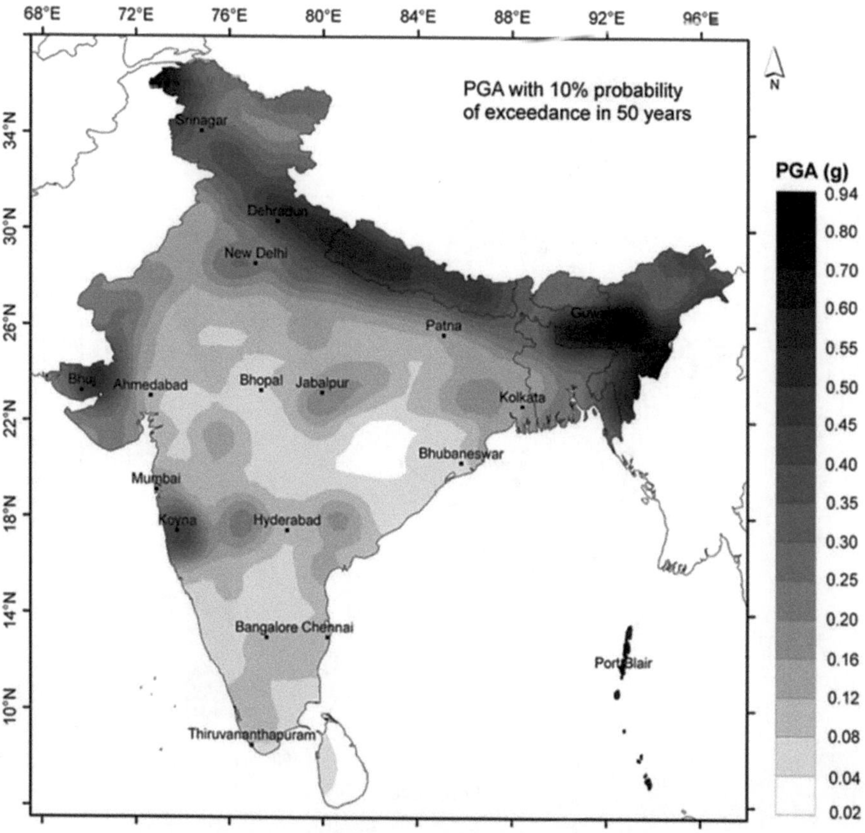

Fig. 3.7 Seismic hazard distribution in India in terms of PGA [232]

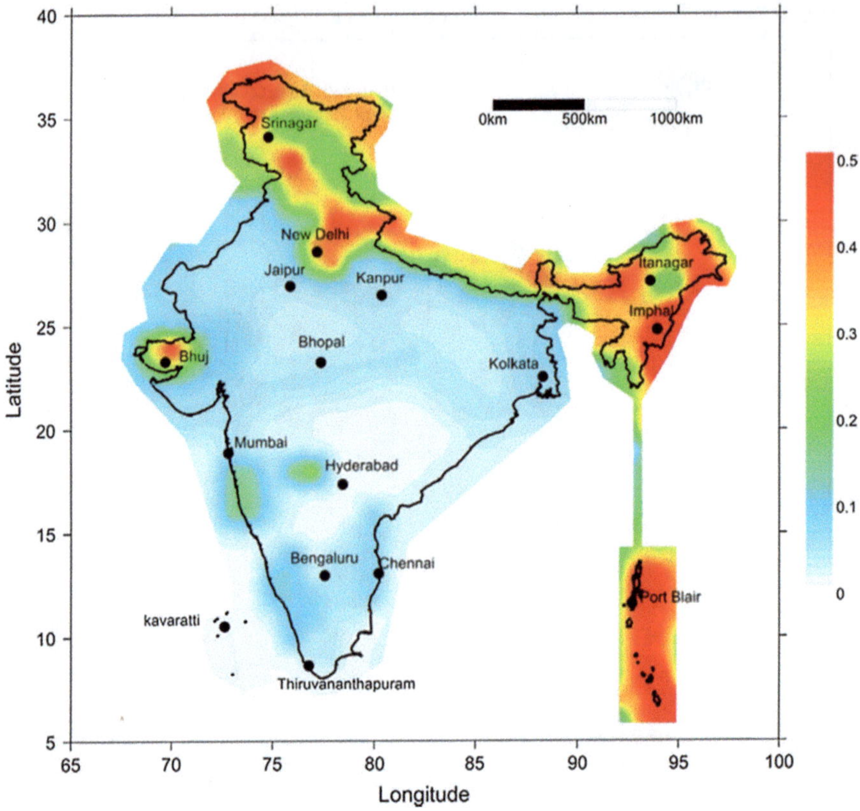

Fig. 3.8 PHA values (g) corresponding to a return period of 475 years (10% probability of exceedance in 50 years, PSHA) [335]

Chapter 4
Seismic Site Characterization

4.1 Introduction

Evidences from past earthquakes clearly show that the damages due to an earthquake and its severity are controlled mainly by three important factors i.e., earthquake source and path characteristics, local geological and geotechnical characteristics, structural design and construction features of structures. Seismic ground response at a site is strongly influenced by local geological and soil conditions. The exact details of the geological, geomorphological, and geotechnical data along with seismotectonic background and seismicity are needed to evaluate the ground response and site effects. The damage pattern during an earthquake depends on the soil characteristics at a site and it may have a major effect on the level of ground shaking. This point highlights the importance of site characterization in microzonation studies. The regional tectonic maps as well as surface geology maps and vertical geological profiles would be the essential ingredients for the seismic microzonation study. The characteristics and thickness of site's soil conditions are to be identified based on borings, in-situ geophysical and geotechnical tests. The geological, geomorphological, and geotechnical databases are needed for assessing the local site effects for site amplification as well as for liquefaction and landslide susceptibility.

The geometry of the subsoil structure, the soil types and the variation of their properties with depth, the lateral discontinuities and the surface topography influence the amplification of ground motion and hence of intensive damage during destructive earthquakes. For this the accurate knowledge of the geology, geomorphology, geophysical data, and geotechnical details are the key parameters controlling the damage severity during an earthquake. Thus the seismic site characterization has its significance in seismic microzonation studies. The present chapter describes in detail about the different methodologies for site characterization.

T. G. Sitharam et al., *Comprehensive Seismic Zonation Schemes for Regions at Different Scales*, https://doi.org/10.1007/978-3-319-89659-5_4

Further, this chapter also demonstrates the application of suitable methodology to perform seismic site characterization for two study areas both at micro and macro-level.

4.2 Methods for Seismic Site Characterization

Local geological conditions have a significant effect on the earthquake ground motion at a given site. The main aim of the site characterization study is to evaluate basic and engineering properties of soil and then categorize the soil stratum based on these soil properties. Seismic site characterization is helpful for assessing various hazards associated with the earthquake at the ground surface level. It involves site investigation, data collation, interpretation, categorizing into various soil classes and representing the spatial distribution. The site characterization methods can be broadly divided into the following categories as (1) based on the local geology (2) based on the geotechnical field test data (3) based on the geophysical test data (4) based on the topographic slope.

4.2.1 Based on Geology

The surface geology is an excellent parameter for carrying out first level site characterization. It can be correlated to the different site classes given by various codes like NEHRP [51], Eurocode [86]. Wills et al. [384] has refined the shear wave velocity classification of various geologic units and prepared a site condition map for the California. They also categorized various geological units in NEHRP site classes. The site class B consists of plutonic and metamorphic rocks, most volcanic rocks, coarse sedimentary rocks of Cretaceous age and older. The Franciscan Complex rocks other than melange and serpentine, crystalline rocks of the transverse ranges which tend to be more sheared, Cretaceous siltstones, or mudstone, etc. comes under class BC. The site class C consists of Franciscan melange and serpentine, sedimentary rocks of Oligocene to Cretaceous age, or coarse-grained and younger sedimentary rocks.

The sedimentary rocks of Miocene and younger age, unless formation is notably coarse grained, Plio-Pleistocene alluvial units, older (Pleistocene) alluvium, some areas of coarse younger alluvium are categorized in class CD. The site class D mainly consist of younger (Holocene) alluvium. The fill over bay mud in the San Francisco Bay area, fine grained alluvial, and estuarine deposits elsewhere along the coast come under DE. The site class E consists of bay mud and similar intertidal mud. Stewart et al. [344] has classified a total of 427 California recording stations based on mapped surface geology. Table 4.1 presents the classification of each site into three geologic schemes based on geologic age, depositional environment, and material texture. The site characterization based on local geology is an approximate

Table 4.1 Criteria for surface geology classifications [344]

Age	Depositional environment	Sediment texture
Holocene	Fan alluvium	Coarse
Pleistocene	Valley alluvium	Fine
	Lacustrine/marine	Mixed
	Aeolian	
	Artificial Fill	
Tertiary		
Mesozoic + Igneous		

method and very good for a first level site characterization. However there are lot of issues associated with this method such as (1) it does not address the variability of V_s within a particular geological unit and (2) it does not provide any information regarding the thickness of a particular unit which also determines the variability of V_{s30} [379].

4.2.2 Based on the Geotechnical Field Test Data

Site response and ground failure are strongly influenced by the properties of soil. Site exploration usually begins with a thorough review of the available information about the site and its surroundings. Geotechnical reports for the sites may be available from various governmental or nongovernmental agencies. Geotechnical site characterization requires a full 3D representation of stratigraphy, estimates of geotechnical parameters and hydrogeological conditions and properties. The traditional methods like drilling and undisturbed sampling can provide adequate stratigraphic details and estimates of geotechnical parameters. But they cannot provide useful estimates of hydrogeological conditions which are very important in estimating the seismic hazards like liquefaction, landslides, etc. Site response is primarily influenced by the properties that influence wave propagation, particularly stiffness and damping. Ground failure is influenced by the shear strength of the soil. Soils are highly non-linear even at very low strains. This non-linearity causes soil stiffness to decrease and damping to increase with increasing strain amplitude. Both site response and ground failure are parts of the same continuous spectrum of non-linear soil behavior.

The in-situ tests generally conducted to identify the soil stratification and engineering properties of the soil layers are penetration tests. Two methods that have been widely used are the Standard Penetration Test (SPT) and Cone Penetration Test (CPT). Standard Penetration Test is generally used to investigate cohesionless or relatively stiff soil deposits, whereas CPT is used to identify soil properties in soft soil deposits [200]. SPT is generally performed in a borehole and measures soil resistance in terms of the number of blows required for 30 cm penetration of standard sampler. The blow for the penetration is given by a 63.5 kg hammer, free falling from 75 cm height. The variability of the Standard Penetration Test equipment and procedures used has significant effects on the obtained blow counts [315, 336]. The energy delivered to the split-spoon sampler is strongly influenced by many factors such as hammer type, borehole diameter, rod length, rod diameter, tightness of the rod joints, verticality of the rod string, and type of sampler, etc. Therefore it is very important to have sufficient information to estimate the energy ratio correction for SPT blow counts before using these results for assessing the properties of soil layers. SPT is the most fundamental field test and is generally carried out as a part of subsurface investigation, prior to construction. Hence there are many correlations available relating different soil parameters to the SPT-N values. The number of blows obtained from the SPT is then subjected to corrections for overburden as well as dilatancy [29]. BIS-1893 [28] has classified the soil into three major categories: Type I, II, and III based on the SPT-N value range. Table 4.2 presents a site classification proposed by BIS-1893 [28].

Empirical relations have been proposed to correlate the penetration test results between CPT and SPT [289] as well as with the shear-wave velocities [147, 213, 245].

James et al. [153] has developed a similar correlation between shear wave velocity and SPT-N value is developed for the east coast region of South India. The N-values measured in the field were subjected to various corrections such as (1) Overburden pressure correction (C_N), (2) Hammer energy correction (C_E), (3) Borehole diameter correction (C_B), (4) liner correction (C_S), (5) Rod length correction (C_R), and (6) correction for fine contents (C_{FC}) [61, 313, 336, 400]. The SPT value corrected for 60% energy efficiency and 100 kPa overburden pressure $(N_1)_{60}$ was evaluated using Eq. (4.1) [287]. The $(N_1)_{60cs}$ provides consistent value for penetration resistance and hence become a standard for assessing liquefaction potential [291]. They have also compared V_s-SPT correlation with the correlation developed by Anbazhagan and Sitharam [9] as in Fig. 4.2.

$$(N_1)_{60} = N \times [C_N \times C_E \times C_B \times C_S \times C_R] \tag{4.1}$$

Table 4.2 Soil classification based on SPT-N values [28]

Soil type	Description	SPT-N value range
Type I: rock or hard soil	Well graded gravel and sand gravel mixtures with or without clay binder, and clayey sands poorly graded or sand clay mixtures (GB, CW, SB, SW, and SC)	>30
Type II: medium soils	All other soils like poorly graded sands or gravelly sands with little or no fines (SP)	10–30
Type III: soft soils	All soils other than SP	<10

The obtained $(N_1)_{60}$ was again corrected for fine content as suggested by Idriss and Boulanger [131] using Eqs. (4.2) and (4.3).

$$(N_1)_{60cs} = (N_1)_{60} + \Delta(N_1)_{60} \tag{4.2}$$

where

$$\Delta(N_1)_{60} = \exp\left[1.63 + \left(\frac{9.7}{FC+0.1}\right) - \left(\frac{15.7}{FC+0.1}\right)^2\right] \tag{4.3}$$

Here FC is the fine content, i.e. percentage of dry weight finer than 75 μm. The V_s values estimated from the MASW results are correlated to corrected SPT N values at respective depths. The regression equation developed between shear wave velocity V_s and corrected SPT value $(N_1)_{60}$ is having a regression coefficient of 0.63. Figure 4.1 shows the correlation between V_s and $(N_1)_{60}$ for the study area. Figure 4.2 presents a comparison between the correlation proposed by Anbazhagan & Sitharam [9] and James et al. [153].

A list of some of the relationships proposed to calculate shear wave velocity in terms of SPT 'N' value is given in Table 4.3.

Cone penetration tests (CPT) are widely used particularly for soft clays, silts, and in fine to medium sand deposits [48]. The test consists of measuring the resistance to penetration of a standard cone into the ground at a rate of 10–20 mm/s. The total resistance against penetration has two components, (1) cone side resistance and (2) cone tip resistance. A comparison of the two methods shows that the CPT method

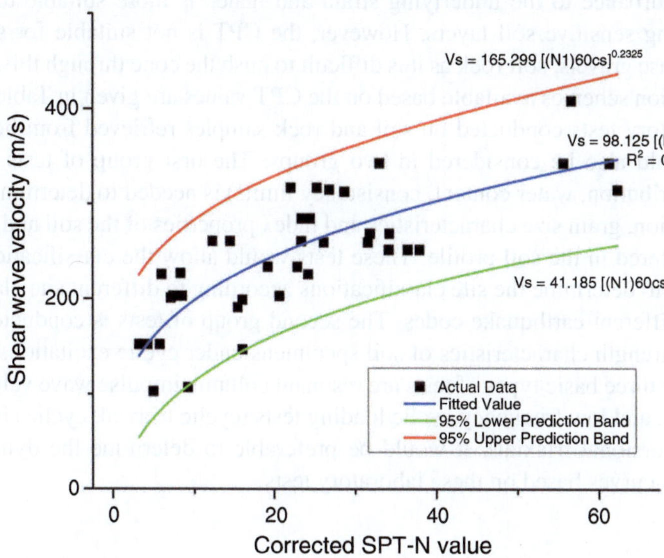

Fig. 4.1 Shear wave velocity vs corrected SPT-N value (after [153])

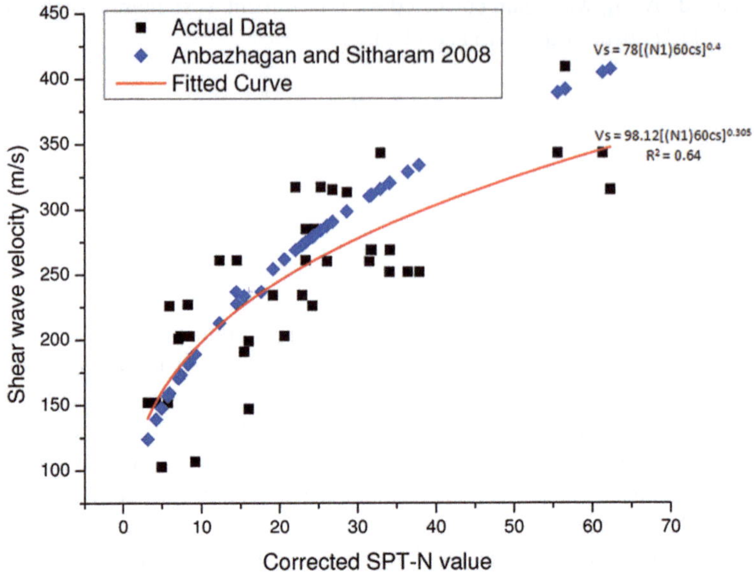

Fig. 4.2 Comparison between the correlation relations developed for corrected SPT-N values (after [153])

is advantageous over the SPT. The CPT gives a continuous profile of end resistance and side friction while the SPT only provides discrete values of blow count. The CPT also has better control over errors when compared with SPT. CPT offers a minimum disturbance to the underlying strata and hence is more suitable to site with underlying sensitive soil layers. However, the CPT is not suitable for strata containing dense gravels, soft rock as it is difficult to push the cone through this. The site classification schemes available based on the CPT values are given in Table 4.4.

The laboratory tests conducted on soil and rock samples retrieved from boring operations could also be considered in two groups. The first group of tests (i.e., grain size distribution, water content, consistency limits) is needed to determine the soil classification, grain size characteristics and index properties of the soil and rock layers encountered in the soil profile. These tests would allow the classification of the soil layers to determine the site classifications according to different site classes proposed in different earthquake codes. The second group of tests is conducted to obtain shear strength characteristics of soil specimens under cyclic excitations [11, 139, 174]. The three basic types of tests are resonant column, impulse wave velocity measurements, and low frequency cyclic loading tests (cyclic triaxial, cyclic simple shear, cyclic torsional triaxial). It would be preferable to determine the dynamic shear modulus curves based on these laboratory tests.

Table 4.3 Proposed relationships to estimate shear wave velocity from SPT value (after [238])

Author	Data	Soil type	V_s (m/s)
Kanai et al. [163]	Not known	All soils	$V_s = 19\,N^{0.6}$
Shibata [323]	Not known	Sandy soils	$V_s = 32\,N^{0.5}$
Imai and Yoshimura [135]	Not known	All soils	$V_s = 76\,N^{0.33}$
Ohba and Toriumi [242]	Not known	Alluvial soils	$V_s = 84\,N^{0.31}$
Ohta and Goto [245]	Not known	Sandy soils	$V_s = 87\,N^{0.36}$
Ohsaki and Iwasaki [244]	Not known	All soils Cohesionless soils	$V_s = 82\,N^{0.39}$ $V_s = 59\,N^{0.47}$
Imai and Yoshimura [136]	Not known	All soils	$V_s = 90\,N^{0.341}$
Imai [133]	Not known	All soils	$V_s = aN^b$ $a = 102 b = 0.29$(H.clay) $a = 81\ b = 0.33$(H.sand) $a = 114\ b = 0.29$(P.clay) $a = 97\ b = 0.32$(P.sand)
Ohta and Goto [245]	Not known	All soils	$V_s = 69\,N^{0.17}\,D^{0.2}EF$ $E = 1$(H); 1.3(P) $F = 1$ (clay); 1.09(f.sand); 1.07(m.sand); 1.14(c.sand); 1.14(c.sand); 1.15(g.sand); 1.45(gravel)
Seed and Idriss [310]	Not known	All soils	$V_s = 61\,N^{0.5}$
Imai and Tonouchi [134]	1654 sets of data (Japan)	All soils	$V_s = 96.9\,N^{0.314}$
Seed et al. [313]	Unknown	Sands	$V_s = 56.4\,N^{0.5}$
Sykora and Stokoe [349]	229 sets of crosshole data	Granular soils	$V_s = 106.7\,N^{0.27}$
Fumal and Tinsley [97]	Not known	Sands and gravelly sand Soils	$V_s = 152 + 5.1\,N^{0.27}$
Sykora and Koester [350]	186 points (7 m, 7 sites)	Holocene gravels	$V_s = 63\,N^{0.43}$
	186 points (7 m, 7 sites)	Pleistocene gravels	$V_s = 132\,N^{0.32}$
Okamoto et al. [248]	Not known	All	$V_s = 125\,N^{0.3}$ (P.sand)
Lee (1990)	Not known	Clays Sands Silts	$V_s = 114\,N^{0.31}$ $V_s = 57\,N^{0.49}$ $V_s = 106\,N^{0.32}$
Jafari et al. [148]	Not known	All soils	$V_s = 22\,N^{0.85}$
LIQUFAC	Unknown	All	$V_s = 243.8\sigma_e^{0.4}$
LIQUFAC	Unknown	All	$V_s = 152\sigma_e^{0.3}$
Anbazhagan and Sitharam [9]	SPT and MASW data	Sand	$V_s = 57(N_1)_{60,cs}^{0.44}$
		Clays	$V_s = 57\,N^{0.44}$

Where, N: SPT value; ε_e: effective vertical stress of the soils; V_s: shear wave velocity; D: depth (m); H: Holocene; P: Pleistocene; f: fine; m: medium; c: coarse; g: gravel

Table 4.4 Soil classification based on CPT N values [200]

Soil type	Cone penetration resistance q_c (MPa)
Soft clay and mud	<1
Moderately compact clay	1–5
Silt and loose sand	≤5
Compact to stiff clay and compact silt	>5
Moderately compact sand and gravel	5–12
Compact to very compact gravel	>12

4.2.3 Based on the Geophysical Test Data

Seismic tests are classified into borehole (invasive) and surface (non-invasive) methods. They are based on the propagation of body waves and surface waves, which are associated to very small strain (<0.001%). Geophysical test data give mainly the information about the shear wave velocity (V_s) profile for a soil strata. Seismic tests are also used to determine the shear modulus and the material damping ratio by measuring the spatial attenuation of body or surface waves

$$G_{\max} = \left[\frac{\gamma}{g}\right] V_s^2 \tag{4.4}$$

$$D_o = \frac{\alpha V}{2\pi f}\,[D_0 < 10\%] \tag{4.5}$$

Where, D_o—small strain damping ratio, α—attenuation coefficient; V—velocity respectively of P, S or R waves; f—frequency.

Seismic site characterization based on shear wave velocity will provide more insight into the mechanics of site amplification and liquefaction phenomena as it is an essential and fundamental parameter for characterizing the dynamic geotechnical properties of the soil. Site characterization based on geotechnical and geophysical methods are found to be most reliable. In seismic microzonation, it is required to obtain detailed subsurface profile over the region of interest. It is difficult to carry conventional geotechnical site explorations over such a large region. In addition, carrying geotechnical site explorations over a large area is very expensive. Geophysical methods are only alternative to avoid these difficulties. These methods provide lateral variability of the near-surface materials beneath a site. Shear wave velocity (V_s) is an essential parameter for evaluating the dynamic properties of soil in the shallow subsurface. A number of geophysical methods have been proposed for near-surface characterization and measurement of shear wave velocity by using a great variety of testing configurations, processing techniques, and inversion algorithms.

The most widely used borehole methods are P-S logging, down-hole logging, and cross-hole logging [207, 279]. In down-hole logging, the travel time used by vertically propagating shear-waves from a source on the surface to a subsurface receiver along a borehole is measured. Cross-hole logging is based on subsurface measurements in which the travel time is measured by horizontally propagating shear-waves from inside a borehole to neighboring boreholes [117]. Cross-hole logging also has the advantage of identifying the properties of the soil deposit between the boreholes.

Surface Wave Tests

This method is used to measure the shear wave profiles of soil. This method depends on the dispersive characteristics of Rayleigh waves traveling through layered soil medium. A dynamic source is used to create surface waves of different frequencies and these are monitored by two receivers at known distances. Using the SASW methods large area can be covered and the soil profile can be obtained. Spectral Analysis of Surface Wave (SASW) method [95, 347] and Multichannel Analysis of Surface Wave (MASW) method [258, 388] are developed to estimate shear wave velocity profile from surface wave energy. Surface wave methods is non-invasive field tests that are executed from the ground surface without drilling any boreholes. Surface wave methods are increasingly used in civil engineering applications to evaluate soil shear modulus with depth [251]. This approach has advantages over invasive subsurface measurements [389], because it can be easily implemented along linear sections to obtain a two-dimensional shear-wave velocity profile of shallow layers [124, 195]. It can be used as a tool for imaging subsurface heterogeneity [325]. The number of surface-wave profiling applications is growing, but there are questions about experimental and theoretical limitations [251]. MASW system consists of number of geophones (usually twelve or more) against two geophones in SASW. The seismic waves are created by an impulsive source (sledge hammer). These waves are captured by the geophones. The captured Rayleigh wave is further analyzed using suitable software to generate Vs data. Both SASW and MASW involve three steps, (1) data acquisition, (2) construction of a dispersion curve, and (3) inversion of the dispersion curve to get shear wave (Vs) profile.

The term "Multichannel record" indicates a seismic data set acquired by using a recording instrument with more than one channel using geode seismograph. Both SASW and MASW effectively use the highest signal-to-noise ratio (S/N) of surface waves. The surface waves having different frequencies are generated together due to impact of sledge hammer over soil half space. However, due to the variation of engineering properties of soil mass along the depth, the surface waves of different frequency will reach geophones at different time. Knowing the distance between source and geophones, as well as geophone spacing, it is possible to generate a plot which shows the variation of velocities of the waves of different frequency. This plot is typically called a dispersion curve and its pattern depends upon the engineering properties of the strata below.

Earthquake hazards like ground motion amplification and soil liquefaction, which are responsible for major catastrophes, mainly depend on the topography and dynamic geotechnical properties of soil overlying the bedrock. The physical properties of soil such as shear wave velocity (V_s), dynamic shear modulus (G), Poisson's ratio (μ), and damping ratio (β) are referred to as dynamic geotechnical properties. The site characterization based on these properties will provide more insight into the mechanics of these hazards. Nowadays, average shear wave velocity for top 30 m (V_{s30}) has been widely accepted for seismic site characterization. It provides an unambiguous site characterization and permits quantitative site amplification [39]. Most of the codes like [86], NEHRP [51], International Building Code [130], etc. specify the site characterization based on the average shear wave velocity values in the top 30 m (V_{s30}) evaluated using Eq. (4.6).

$$V_{s30} = \frac{30}{\sum_{i=1}^{n} \frac{d_i}{V_i}} \tag{4.6}$$

where d_i is the thickness of individual layers and V_i is the shear wave velocity through each layer and 'n' is the number of layers. Many literature show the use of the average shear wave velocity values in the top 30 m (V_{s30}) has become an important parameter for site classification and it is being incorporated in various codes like [86], NEHRP [51], International Building Code [130], etc. The phenomenon of site amplification is strongly influenced by the near surface shear wave velocity which can be better represented by the average shear wave velocity in the top 30 m. Table 4.5 describes the NEHRP site classification based on V_{s30}.

A site classification scheme based on V_{s30} values was proposed by Borcherdt [39] and a similar scheme was adopted by the National Earthquake Hazard Reduction Program (NEHRP) also. The NEHRP [51] site classification scheme is given in Table 4.5. Eurocode-8 [86] has also classified the site based on V_{s30}, standard penetration test (SPT), and cone penetration test (CPT) values. The classification given by Eurocode-8 is given in Table 4.6.

In many locations, the rock depth will be shallow (less than 30 m) and hence the evaluation of V_{s30} value will not be possible. In those cases, extrapolation of

Table 4.5 Site classification as per NEHRP scheme [51]

NEHRP site class	Description	V_{s30}
A	Hard rock	>1500 m/s
B	Firm and hard rock	760–1500 m/s
C	Dense soil, soft rock	360–760 m/s
D	Stiff soil	180–360 m/s
E	Soft clays, special study soils, e.g. liquefiable soil	<180 m/s

Table 4.6 Site characterization as per [86]

Ground type	Description of stratigraphic profile	V_{s30} (m/s)	SPT	Cu (kPa)
A	Rock or other rock-like geological formation, including utmost 5 m of weaker material at the surface	>800		
B	Deposits of very dense sand, gravel, or very stiff clay, at least several tens of meters in thickness, characterized by a gradual increase of mechanical properties with depth	360–800	>50	>250
C	Deep deposits of dense or medium dense sand, gravel or stiff clay with thickness from several tens to many hundreds of meters	180–360	15–50	70–250
D	Deposits of loose-to-medium cohesion less soil (with or without some soft cohesive layers), or of predominantly soft-to-firm cohesive soil	<180	<15	<70
E	A soil profile consisting of a surface alluvium layer with V_{s30} values of type C or D and thickness varying between about 5 and 20 m, underlain by stiffer material with $V_{s30} > 800$ m/s			
S1	Deposits consisting, or containing a layer at least 10 m thick, of soft C or D and thickness varying between clays/silts with a high plasticity index (PI > 40) and high water content	<10 (indicative)		10–20
S2	Deposits of liquefiable soils, of sensitive clays, or any other soil profile not included in types A-E or S1			

available V_s values has to be done to evaluate the V_{s30} values. The method proposed by Boore [36] can be used for this purpose. He has suggested different models to extrapolate the shear wave velocities, for depths less than 30 m, to get the V_{s30} value. The first method is extrapolation based on constant velocity. In this model it is assumed that the shear wave velocity remains constant from the deepest velocity measurement to the 30 m.

$$V_{s30} = \frac{30}{tt(d) + \frac{(30-d)}{V_{eff}}} \tag{4.7}$$

where $tt(d)$ is the travel time to depth d and $V_{eff} = V_{s(d)}$, $V_{s(d)}$ is the timed average velocity to a depth of d.

Even though this method is simple, it is found to underestimate the V_{s30} values, since in most of the soils, the shear wave velocity is found to increase with depth. Another relation proposed by Boore [36] was based on a power law relation, the V_{s30} value can be estimated as given in equation

$$\log V_{s30} = a + b \log \overline{V}_{s(d)} \tag{4.8}$$

where $\overline{V}_{s(d)}$ is the velocity at a depth of d m ($10 < d < 30$). The values of the regression coefficients a and b can be obtained from [36]. The extrapolation of V_s values can also be done based on the velocity statistics [36].

$$P\left(\xi > \frac{V_{eff}}{V_s(d)}\right) = a\left(\frac{V_{eff}}{V_s(d)}\right)^b \tag{4.9}$$

where $P\left(\xi > \frac{V_{eff}}{V_s(d)}\right)$ is the probability of exceedance of $\left(\frac{V_{eff}}{V_s(d)}\right)$. More details of this analysis can be had from [36].

A modified site classification system based on geotechnical data was proposed by Rodriguez-Marek et al. [290]. In this, the stiffness of soil was also taken into account for the site classification as shown in Table 4.7. The main advantage of this system is that it correlates the V_{s30} values with the geotechnical and surface geological features.

4.2.4 Based on the Topographic Slope

For small sites, V_{s30} parameter for site characterization can be evaluated using geotechnical and geophysical field tests. However for characterizing larger area,

Table 4.7 Classification based on geotechnical features [290]

Site	Description	Comments
A	Hard rock	Crystalline bedrock; $V_{s30} \geq 1500$ m/s
B	Competent bed rock	$V_{s30} > 600$ m/s or <6 m of soil Most unweathered California rock cases
C1	Weathered rock	V_{s30} 300 m/s increasing to >600 m/s, weathering zone >6 m and <30 m
C2	Shallow stiff soil	Soil depth >6 m and <30 m
C3	Intermediate depth stiff soil	Soil depth >30 m and <60 m
D1	Deep stiff Holocene soil	Soil depth >60 m and <200 m
D2	Deep stiff Pleistocene soil	Soil depth >60 m and <200 m
D3	Very deep stiff soil	Soil depth >200 m
E1	Medium thickness soft clay	Thickness of soft clay layer 3–12 m
E2	Deep soft clay	Thickness of soft clay layer >12 m
F	Potentially liquefiable sand	Holocene loose sand with high water table, $Z_w \leq 6$ m

estimation of V_{s30} based on these tests is not economically as well as physically viable. Generally in such cases, site characterization based on V_{s30} is carried out using available geological maps.

However geology-based maps are typically mapped with a goal other than the estimation of seismic site amplification, and hence the use of these maps for first order site characterization always associates with certain drawbacks [379]. In site condition map derived from geological map, a single value of V_{s30} is assigned to an individual geological unit and it does not capture the variability of V_{s30} within that unit. Studies have proved that the topographic variations are an effective index of near-surface geomorphology and lithology, with steep mountains representing rocky terrain and flat basins indicating soil, and intermediate slopes representing a transition between rock and soil [379]. The topographic-slope more accurately maps the variation of V_{s30} across a geological unit when compared with a geological map, by characterizing the presumed change in particle size with topographic gradient. Matsuoka et al. [211] have confirmed that good correlations exist between V_{s30} and slope and geomorphic indicators in Japan. A study was done by Chiou and Youngs [67] in Taiwan and they found that a good correlation exists between elevation and V_{s30}.

Wald and Allen [379] have described a technique to derive first-order site-condition maps directly from topographic data. They have used global 30 arc sec topographic data and V_{s30} measurements collected from several studies in the United States, Taiwan, Italy, and Australia. They have correlated V_{s30} values with the topographic slope to develop two sets of parameters for obtaining V_{s30}: one for active tectonic regions where the topography is steep, and one for stable shield regions where the topography is gentler. By taking the gradient of the topography and choosing ranges of slope that maximize the correlation with shallow shear-velocity observations; [379] have recovered, to first order, many of the spatially varying features of the site-condition maps developed in California (see Fig. 4.3). They found that the maps derived from the slope of the topography were correlating well with other independently derived, regional site-condition maps. Similar observations were also found in the case of other regions such as Los Angeles, Taiwan, Salt Lake City, Utah and Memphis, Tennessee [379]. Hence from all the available literature it is inferred that a reliable first order site condition map based on V_{s30} can be derived from topographic slope maps and hence has been adopted in the present study.

James and Sitharam [151] has carried out a comparison study between V_{s30} maps at the meso-level derived from the topographic slope data and V_{s30} maps prepared by various researchers using geotechnical and geophysical testing. Figure 4.4 presents the comparison between V_{s30} map of Bangalore prepared by Sitharam and Anbazahgan [329] and V_{s30} map derived from the topographic slope data. From Fig. 4.4a, b it is clear that the western part of Bangalore is in the site class C and B with V_{s30} in the range of 400–900 m/s.

Similarly the eastern part of Bangalore mainly constitutes of site class D and C with V_{s30} varying from 200 to 600 m/s. Some differences in the distribution of V_{s30} values can be observed in the south-central part of Bangalore. As per [329], this part

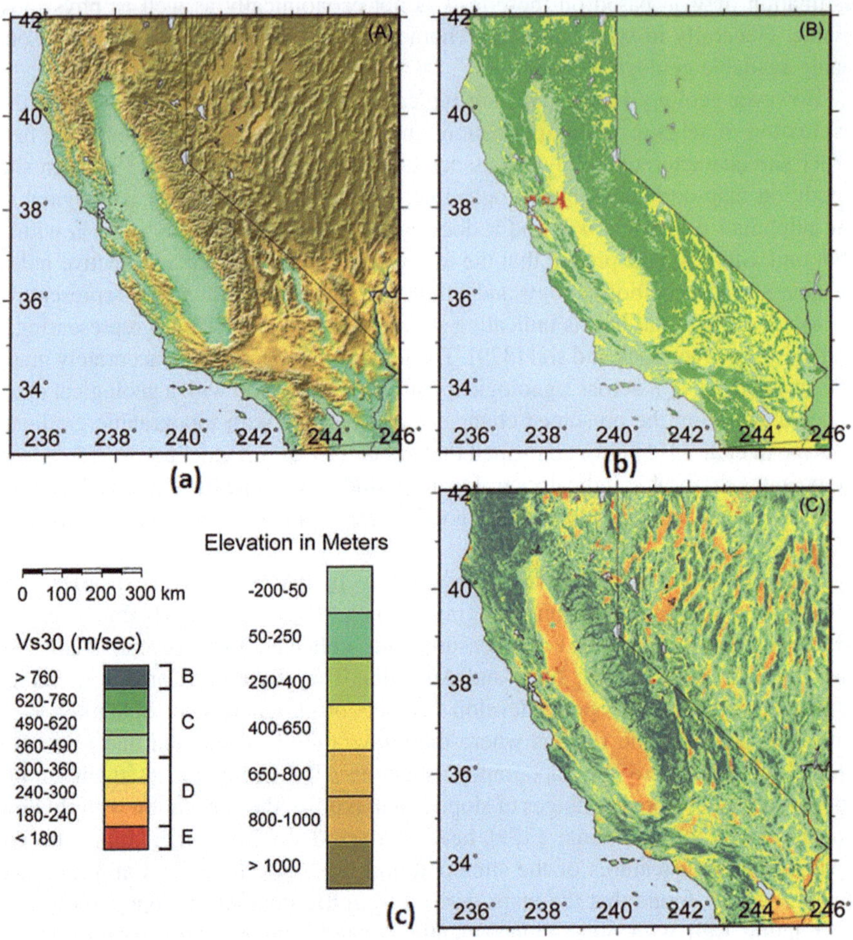

Fig. 4.3 (**a**) The topographic relief for the state of California, (**b**) Site class map for California [384], (**c**) Site class map for California derived from topographic slope map [379]

falls in the site class C and B with V_{s30} values ranging from 400 to 900 m/s, while in the present study this region comes under site class D to C (V_{s30} ranging from 200 to 600 m/s).

James and Sitharam [151] has also carried out a similar comparative study for the city of Chennai (Madras) also. The V_{s30} map derived from topographic slope map (Fig. 4.5b at a grid size of 1 km × 1 km) was compared with V_{s30} prepared by Uma Maheswari et al. [367] using MASW test data (Fig. 4.5a). Both the figures show that the major area of the city falls in site class D, with V_{s30} ranging between 180 and 360 m/s. The deviation between two maps can be observed in the southern parts of the city, which is classified as site class C and B as per [367], while the present study shows the area fall in site class D and C.

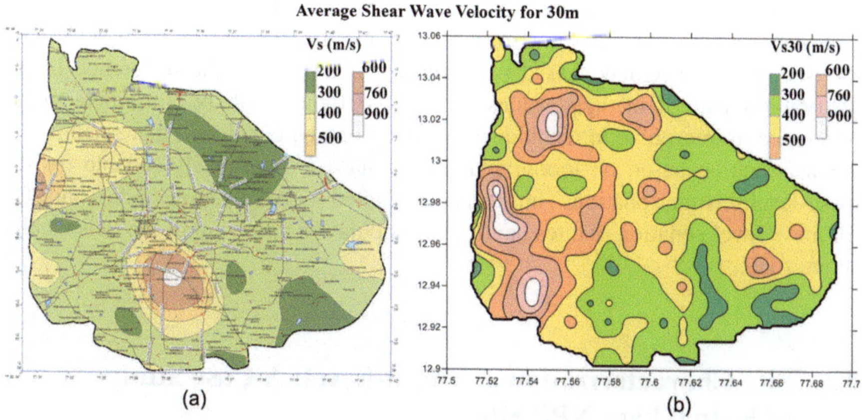

Fig. 4.4 Comparison of V_{s30} maps obtained using two different methodologies for Bangalore (**a**) developed by Sitharam and Anbazahgan [329] (**b**) derived from the topographic slope data (after [151])

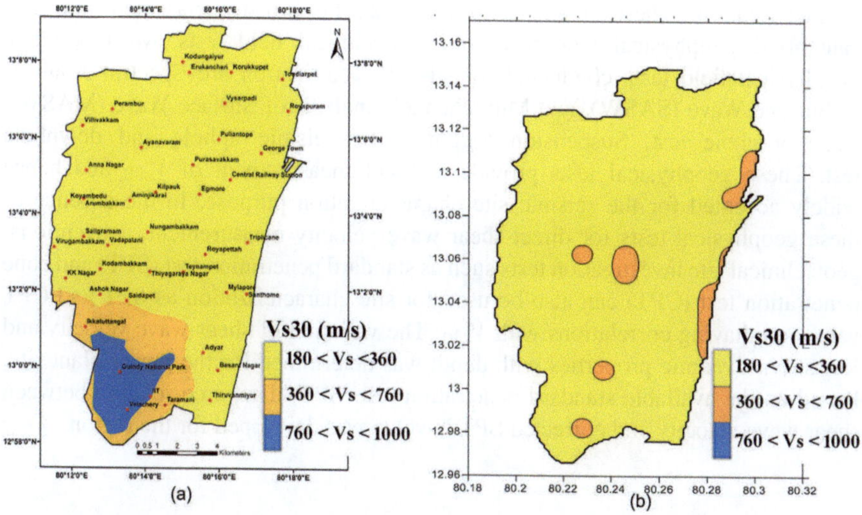

Fig. 4.5 Comparison of V_{s30} maps obtained using two different methodologies for Chennai (**a**) developed by Uma Maheswari et al. [367] (**b**) derived from the topographic slope data (after [151])

However in a broad level, it can be observed that for both Bangalore and Chennai, the V_{s30} maps derived from topographic slope data are comparable with the V_{s30} maps developed based on geophysical test data. The difference in the distribution pattern of V_{s30} values in the two figures can be attributed to the difference in interpolation techniques. Hence it can be concluded that seismic site characterization based on topographic slope maps is moderately reliable for the meso-level seismic zonation. Thus it can be adopted for macro-level studies and also for meso-level zonation of cites having low population and lying in low seismically active zone.

4.3 Site Characterization at Micro-Level: A Case Study of Kalpakkam NPP Site

James et al. [153] has carried the seismic site characterization for the Kalpakkam NPP site using geotechnical/geophysical tests. For small region such as the Kalpakkam NPP site, various in-situ field tests are available for seismic site characterization. These in-situ field tests can be broadly classified into two categories, geophysical field tests and geotechnical field tests. Most common geophysical field tests conducted for site characterization are Spectral Analysis of Surface Wave (SASW), and Multichannel Analysis of Surface Wave (MASW), Cross-borehole test, Suspension logging test, Seismic uphole and downhole test. These geophysical tests provide a direct measurement of V_{s30} and hence widely accepted for the seismic site characterization purpose. In the absence of these geophysical tests for direct shear wave velocity measurement, conventional geotechnical site investigation tests such as standard penetration test (SPT) and cone penetration test (CPT) can also be used for site characterization as SPT and CPT values are having correlations with V_{s30}. The variation of shear wave velocity and low strain dynamic properties with depth was determined for the power plant site. Based on the available standard penetration test (SPT) data, a correlation between shear wave velocity and corrected SPT-N value was developed for the region.

4.3.1 MASW Testing for Site Characterization

Shear wave and P-wave velocity (V_s, V_p) are very well related to the dynamic properties of soil at lower strains. The most widely used technique for the measurement of shear wave and P-wave velocity is by MASW (Multichannel Analysis of Surface Waves) technique. MASW is a non-destructive, indirect geophysical method, in which the variation of shear-wave velocity with depth is obtained by analyzing the surface wave propagation pattern, which is recorded using multiple channels. The MASW method was first introduced by Park et al. [258]. The surveying

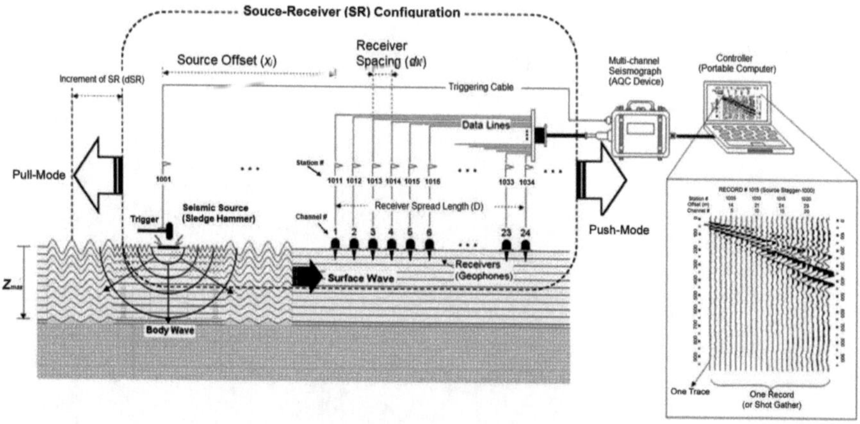

Fig. 4.6 Schematic showing typical active MASW survey (taken from http://www.masw.com/DataAcquisition.html)

procedure consists generation of surface waves using an active seismic source (e.g., a sledge hammer) and capturing generated surface wave using a linear receiver array as shown in Fig. 4.6. The MASW is used in geotechnical engineering for soil stratum identification based on shear wave velocity variations along the depth. It involves three steps: data acquisition, construction of a dispersion curve (showing the variation of phase velocity among various frequencies), and inversion of the dispersion curve to get shear wave (V_s) profile from the calculated dispersion curve. The typical MASW test setup used for the present study consists of 24-channel geophones of 4.5 Hz capacity as shown in Fig. 4.7. A propelled energy generator (PEG) (Fig. 4.8) was used as active source for generating surface waves instead of a conventional sledge hammer. The PEG was found to be more efficient than the conventional sledge hammer for generating surface wave. The PEG consists of a 40 kg hammer dropped from a height of 36–40 cm. The PEG is designed to easily mount on trucks or trolley and can be conveniently moved to various locations.

Spatial distribution of MASW test locations in the site is shown in Fig. 4.9 which were selected based on its proximity to certain borehole locations. A typical record obtained using 24 channel geophones is presented in Fig. 4.10. The acquired signal was analyzed using Surfseis software [259] and a dispersion image showing the distribution of signal intensity and phase velocity among various frequency components of surface waves was generated. The conventional spectral analysis of surface waves (SASW) method measures the phase difference (ϕ_{diff}) of surface waves for a frequency (f) between the two receivers. The phase velocity (C_f) of each frequency was then calculated from the relationship: $C_f = \frac{2\times\pi\times f}{\phi_{diff}}$. This procedure is repeated for different frequencies to construct a dispersion curve. However, in the MASW analysis, instead of calculating the individual phase velocity for each frequency, an image space (Fig. 4.11) was constructed showing

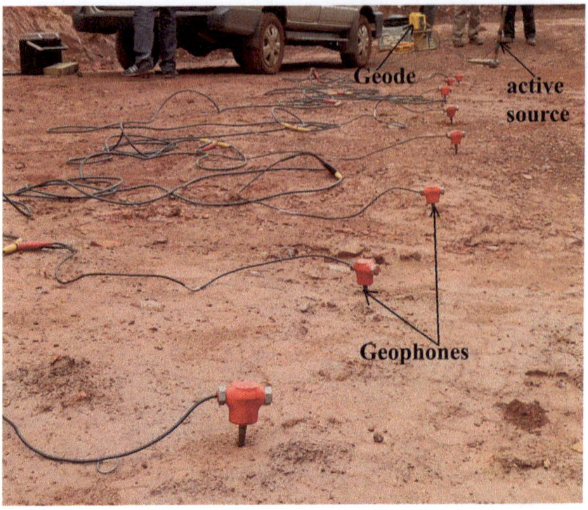

Fig. 4.7 Typical MASW setup in the field

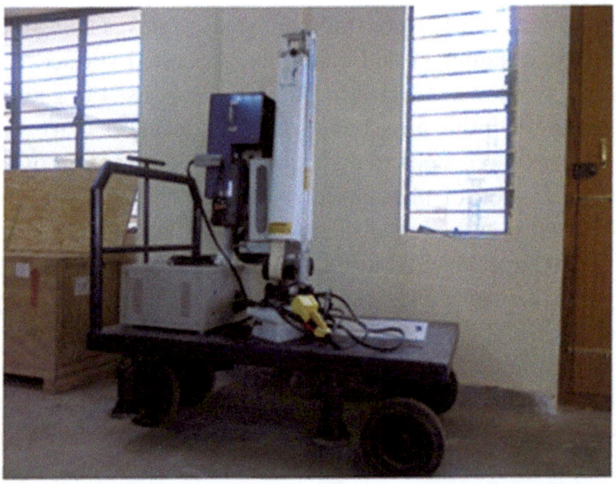

Fig. 4.8 Propelled energy generator assembly in the MASW test (after [150])

the dispersion trends which were identified from the pattern of energy accumulated in this space. The phase-shift method proposed by Park et al. [257] was used for the dispersion imaging scheme and is incorporated in the Surfseis software. Then, the necessary dispersion curves are extracted by following the image trends. A dispersion curve was fitted to this dispersion image (Fig. 4.11), through regions where the intensity of signal was found to be high (high signal to noise ratio). Finally the shear wave (V_s) variation below the surveyed area, which is most responsible for the analyzed dispersion pattern of surface waves, was deduced using an inversion technique. Inversion methodology proposed by Xia et al. [388] is incorporated in the Surfseis software. An initial earth model is assumed at the start of the inversion process which consists of information like shear and P wave velocity, density,

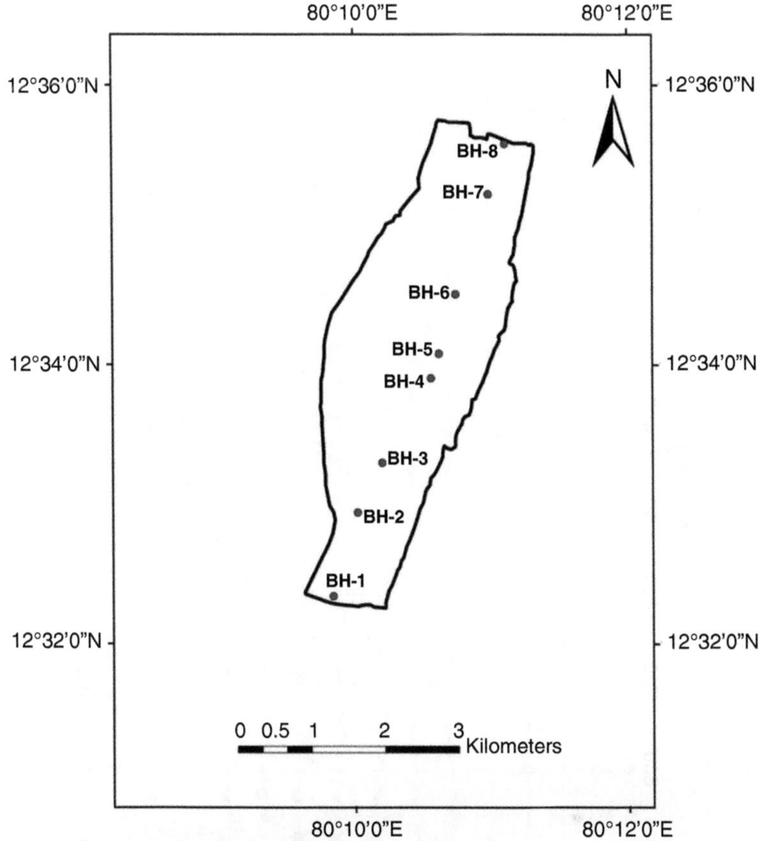

Fig. 4.9 Locations of control point where MASW testing was carried out (after [150])

Poisson's ratio, and layer thickness. Of these parameters the shear wave velocity is most significant and influential parameter which controls the changes in Rayleigh-wave phase velocities for a layered earth model [388]. Hence keeping the other parameters such as P wave velocity, density, Poisson's ratio, and layer thickness as constant, the shear wave velocity is updated in the initial earth model after each iteration till the theoretical dispersion curve matches with the actual dispersion curve obtained from the field. With known (or assumed) value of Poisson's ratio, the variation of P-wave velocity with depth (Fig. 4.12) is also being obtained.

As per the studies of [39, 274, 345], the average shear wave velocity for 30 m is decisive for site characterization, was evaluated as per Eq. (4.6). It is evident from Fig. 4.13 that the site is mainly categorized into two major site classes, site class C (dense soil or very soft rock) and site class D (stiff soil) on the basis of average shear wave velocity for top 30 m as per NEHRP [51].

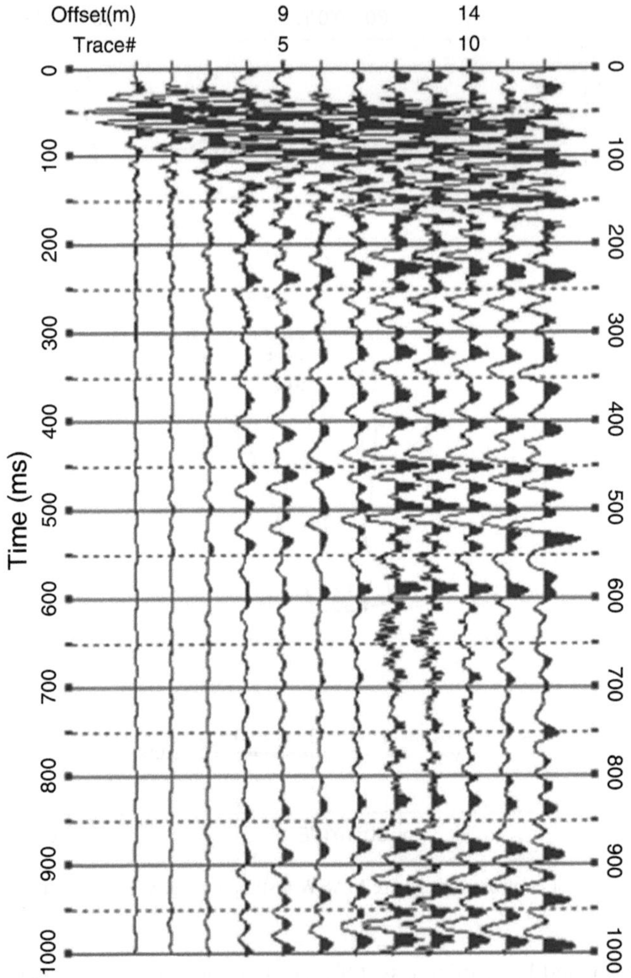

Fig. 4.10 Typical MASW recorded seismogram (after http://www.masw.com/DataAcquisition.html)

4.3.2 Evaluation of Low Strain Dynamic Properties

Low strain dynamic properties of soil includes shear modulus, damping ratio, and Poisson's ratio. Shear modulus (G) is an important soil parameter required to predict site response and liquefaction potential. Shear modulus (G) of any soil material subjected to cyclic loading (provided constant density and confining pressure) depends upon the magnitude of shear strain induced in the specimen [140, 181]. At strain level well below 10^{-5}, soil exhibits elastic behavior and the stiffness (shear modulus) of the soil at this strain range is independent of loading rate and number of load repetitions [140]. Thus the shear modulus will have its maximum

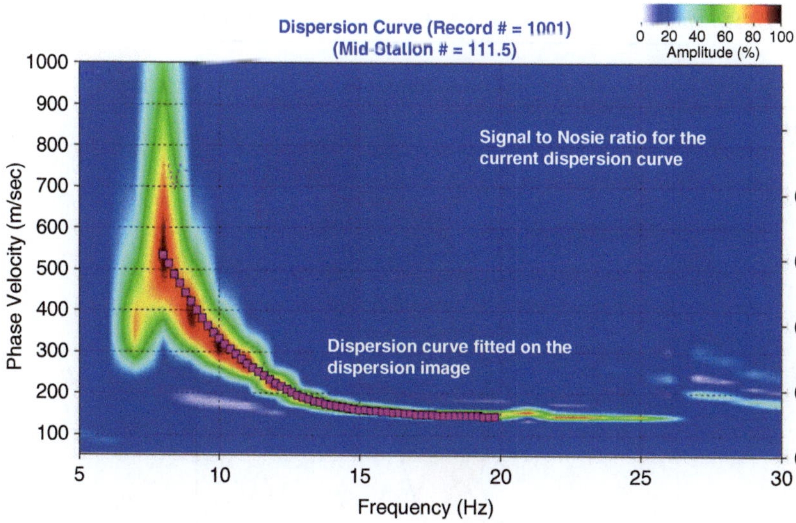

Fig. 4.11 Typical dispersion image obtained after analyzing MASW data (after [153])

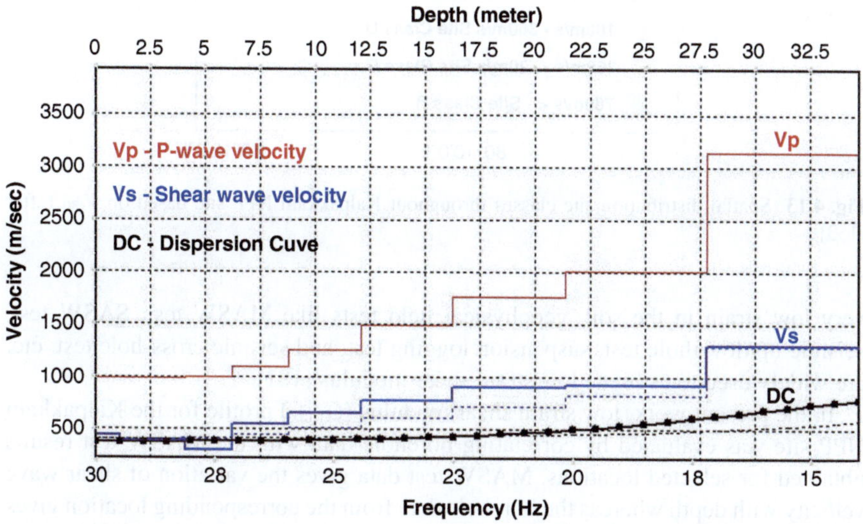

Fig. 4.12 Typical Shear wave and P-Wave velocity profile of a location in the study area (after [153])

value at a strain level less than 10^{-5} (for samples having a constant density and confining pressure) which ideally represents the in situ conditions. The degradation of shear modulus occurs with the increase in the strain level as represented by a modulus reduction curve which is an essential input for equivalent linear ground response analysis. As the phenomenon of vibrations and wave propagation induces

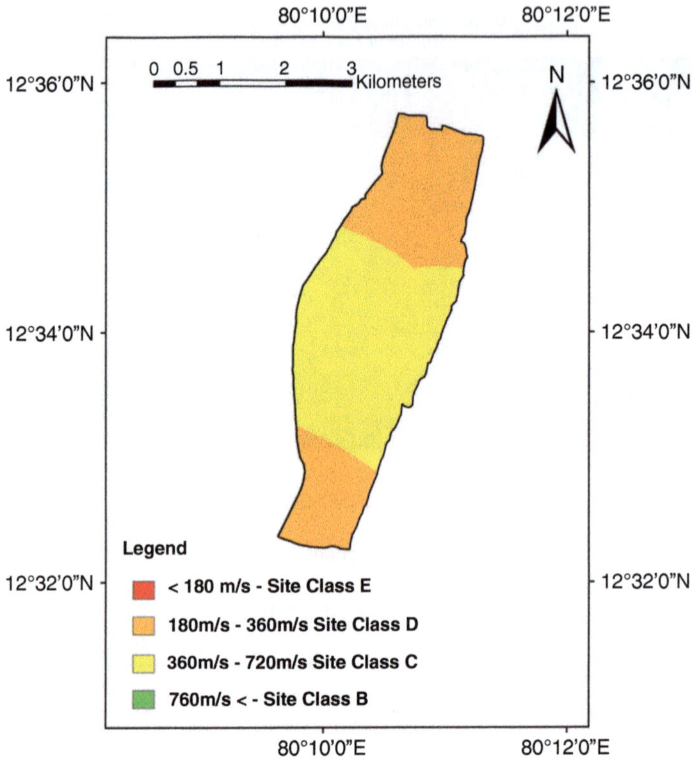

Fig. 4.13 Spatial distribution site classes throughout Kalpakkam NPP site based on V_{s30} (after [153])

very low strain in the soil, geophysical field tests like MASW test, SASW test, seismic up/downhole test, suspension logging test, and seismic cross-hole test, etc. are widely used to estimate low strain shear modulus [181].

In the present work, low strain shear modulus (G_{max}) profile for the Kalpakkam NPP site was evaluated by correlating borehole data with the MASW test results obtained for selected locations. MASW test data gives the variation of shear wave velocity with depth whereas the borehole data from the corresponding location gives soil type and in-situ density values with depth. Low strain shear modulus (G_{max}) is obtained from Eq. (4.10).

$$G_{max} = \left[\frac{\gamma}{g}\right] V_s^2 \quad (4.10)$$

where γ is the in-situ density (in kN/m^3), g—acceleration due to gravity (in m/s^2), and V_s is the shear wave velocity (in m/s). Figures 4.14 and 4.15 show the shear wave velocity and P-wave velocity variation with depth for eight major locations in the power plant site.

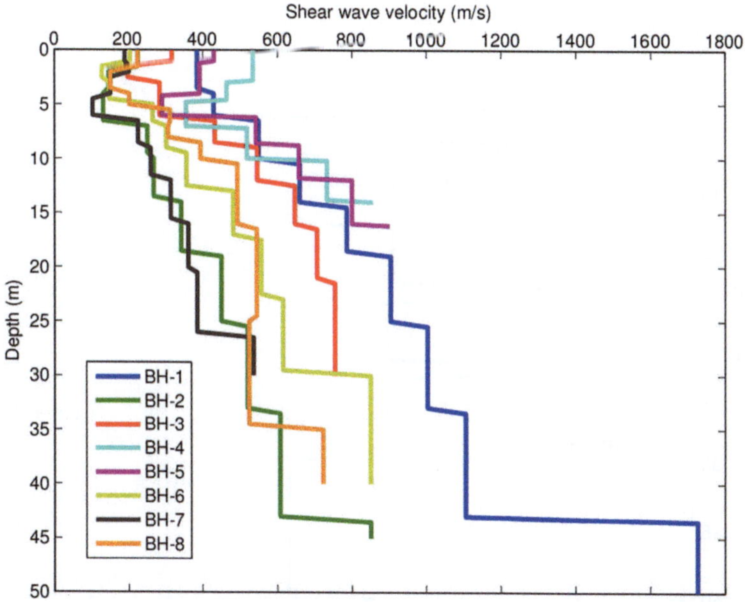

Fig. 4.14 Shear wave velocity profile for selected locations in the study area (after [150])

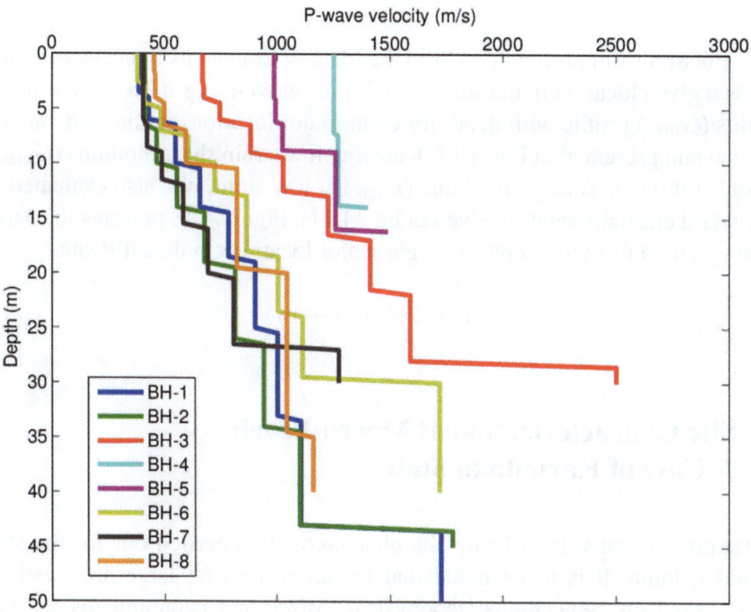

Fig. 4.15 P-wave velocity profile for selected locations in the study area (after [150])

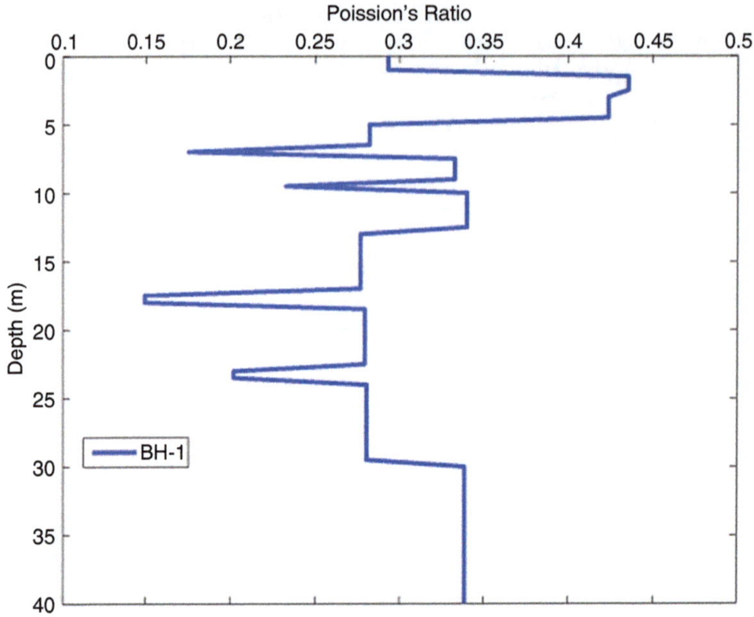

Fig. 4.16 Typical plot showing the variation in Poisson's ratio with depth (at BH-1) (after [150])

The Poisson's ratio is obtained from the MASW test and its typical variation with depth at a given location in the study area is presented in Fig. 4.16. Low strain shear modulus (G_{max}) profile with depth for eight major locations in the study area were evaluated and presented in Fig. 4.17. Knowing low strain shear modulus (G_{max}) and Poisson's ratio (ν), Young's modulus (E_{max}) at low strain was also evaluated using the standard elasticity relation given in Eq. (4.11). Figure 4.18 presents the variation of Young's modulus with depth for eight major locations in the NPP site.

$$E = 3G \times (1 + \nu) \tag{4.11}$$

4.4 Site Characterization at Macro-Level: A Case of Karnataka State

This section presents the seismic site characterization carried out for the state of Karnataka, India. It is to be noted that for characterizing large area, estimation of V_{s30} based on geotechnical/geophysical tests is not economically as well as physically viable. Hence, James et al. [151] have carried out site characterization for the state of Karnataka using the topographic slope map generated from digital elevation model (DEM) data.

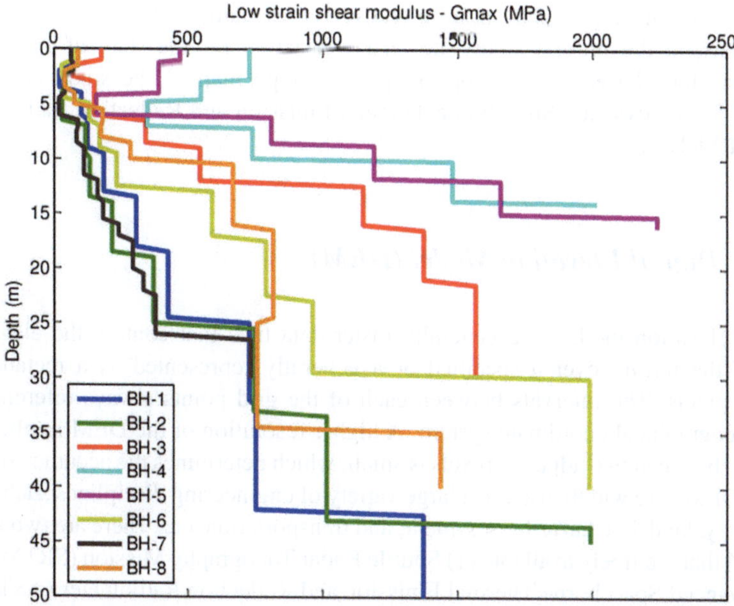

Fig. 4.17 Low strain shear modulus (G_{max}) profile for selected locations in the study area (after [150])

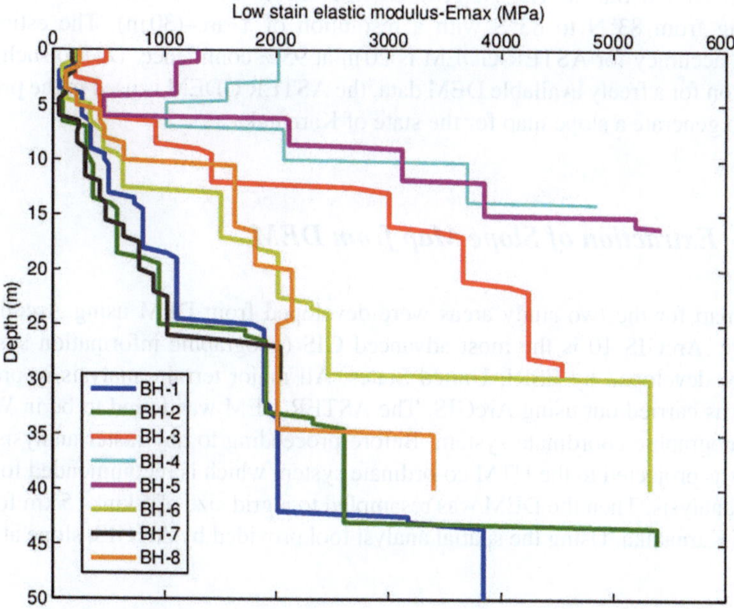

Fig. 4.18 Low strain elastic modulus (E_{max}) profile for selected locations in the study area (after [150])

As the terrain topography is an effective representation of surface geological conditions, in the present study site characterization for the state of Karnataka has been done based on the slope map. The slope map of the study area was derived from Advanced Spaceborne Thermal Emission and Reflection Radiometer (ASTER) DEM.

4.4.1 Digital Elevation Model (DEM)

Digital elevation models are generally raster data files that contain the elevation data of the terrain over a specified area generally represented as a rectangular grid of pixels. The intervals between each of the grid points always reference to some geographical coordinate system. A higher resolution of the DEM implies the distance between two adjacent pixels is small, which determines the accuracy of that DEM. DEMs are widely used in a large variety of engineering disciplines such as in hydrology, landslide hazard assessment, and transportation, etc. There are two kinds of DEM that are freely available (1) Shuttle Radar Topography Mission (SRTM) and (2) Advanced Spaceborne Thermal Emission and Reflection Radiometer (ASTER). The ASTER Global Digital Elevation Model (ASTER GDEM) (in Fig. 4.19) was developed and released by the Ministry of Economy, Trade, and Industry (METI) of Japan, in collaboration with the National Aeronautics and Space Administration (NASA) United States. The ASTER GDEM covers 99% of the total land area, spanning from 83°N to 83°S with a resolution of 1 arc–(30 m). The estimated vertical accuracy for ASTER GDEM is 20 m at 95% confidence. Due to such high resolution for a freely available DEM data, the ASTER GDEM is used in the present study to generate a slope map for the state of Karnataka.

4.4.2 Extraction of Slope Map from DEM

Slope map for the two study areas were developed from DEM using ArcGIS 10 software. ArcGIS 10 is the most advanced GIS (geographic information system) software developed by ESRI, United States. All major terrain analysis in present study was carried out using ArcGIS. The ASTER DEM was found to be in WGS-1984 geographic coordinate system. Before proceeding to any raster analysis, the DEM was projected to the UTM co-ordinate system which is recommended for any terrain analysis. Then the DEM was resampled to a grid size of 5 km × 5 km for the state of Karnataka. Using the spatial analyst tool provided by ArcGIS, slope at each

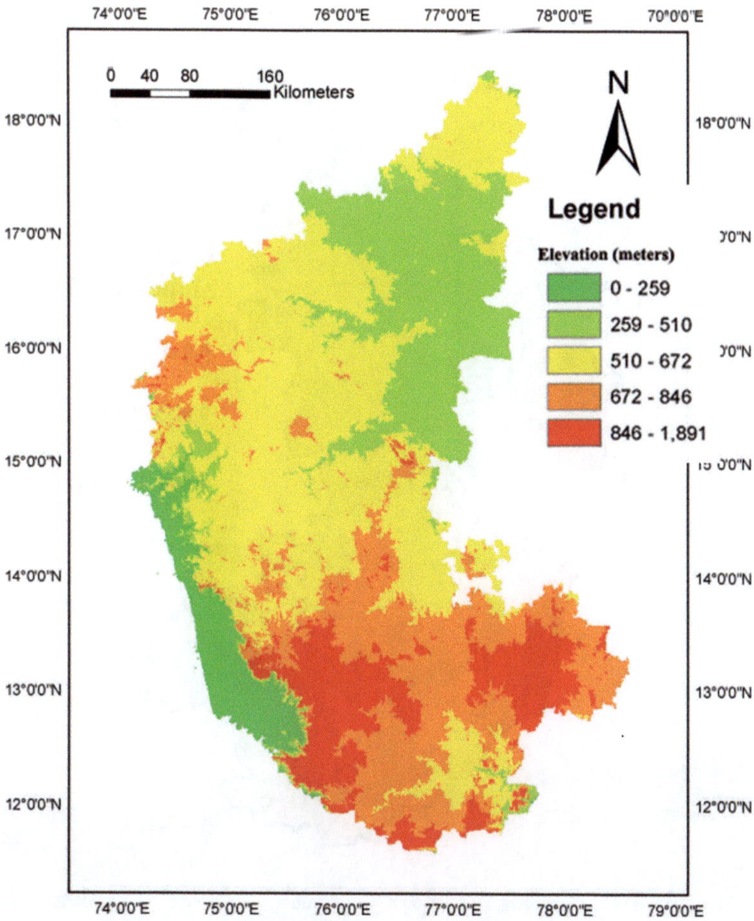

Fig. 4.19 Digital elevation model (DEM) of the Karnataka state (after [151])

grid point was evaluated. Slope map of the Karnataka state is presented in Fig. 4.20. Slope map thus obtained was converted to vector data (points) and then exported into a table form. For the state of Karnataka, there were 8691 grid points for which geographic co-ordinates and slope values were computed.

4.4.3 Site Characterization Based on Terrain Slope

Once slope maps are obtained, site characterization for the study area was carried out using methodology suggested by Wald and Allen [379] as mentioned in Sect. 4.2.4 of Chap. 4. Based on the correlation studies conducted for active tectonic and stable continental regions, Wald and Allen [379] has proposed slope ranges

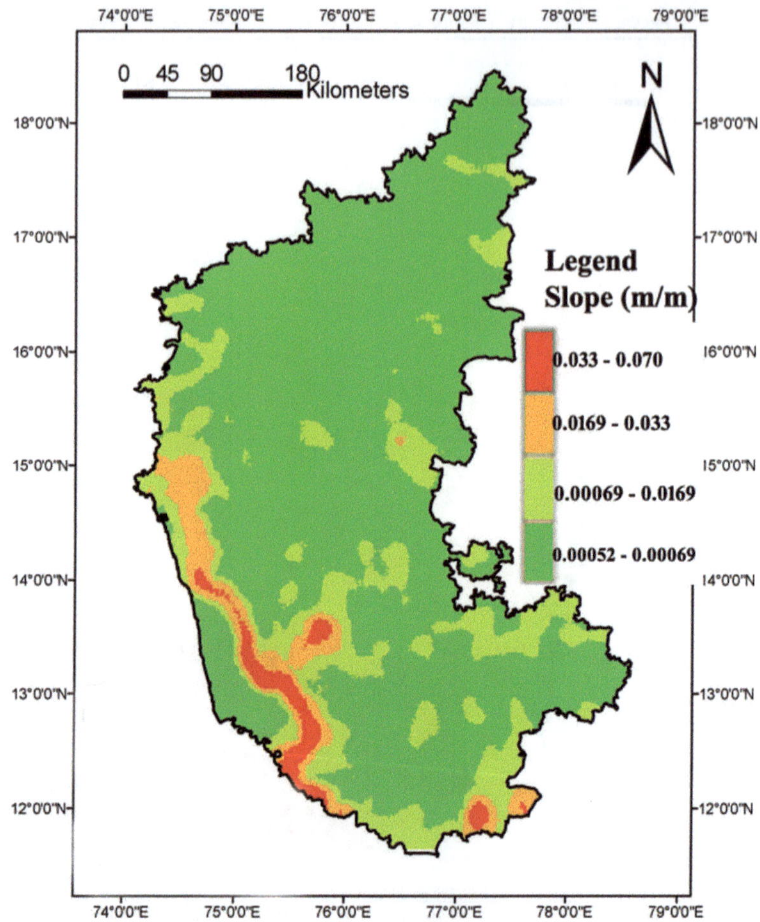

Fig. 4.20 Slope map of the Karnataka state (after [151])

corresponding to each NEHRP site class as given in Table 4.8. Based on the slope range given in Table 4.8, site class for each grid point was determined. Maps showing the spatial variation in site classes throughout Karnataka was developed and presented in Fig. 4.21. A similarly methodology has been adopted by Sitharam et al. [335] to develop the site class distribution map of entire India.

Table 4.8 Slope ranges for NEHRP site classes (after [379])

Site class (as per NEHRP)	V_{s30} range ($\mathrm{m\,s^{-1}}$)	Slope range (m/m)	
		Active tectonic	Stable continent
E	<180	<1.0E−4	<2.0E−5
D	180–240	1.0E−4–2.2E−3	2.0E−5–2.0E−3
	240–300	2.2E−3–6.3E−3	2.0E–3–4.0E−3
	300–360	6.3E−3–0.018	4.0E−3–7.2E−3
C	360–490	0.018–0.05	7.2E−3–0.013
	490–620	0.050–0.10	0.013–0.018
	620–760	0.10–0.138	0.018–0.025
B	>760	>0.138	>0.025

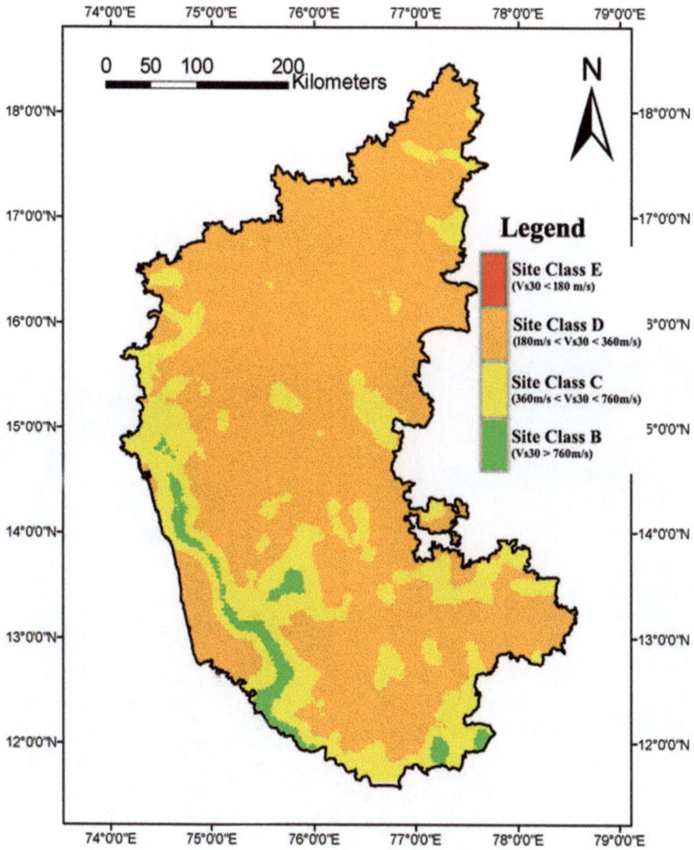

Fig. 4.21 Spatial distribution of V_{s30} values throughout the Karnataka state (after [151])

Chapter 5
Local Site Effects for Seismic Zonation

5.1 Introduction

It has been evident from the past earthquake events all over the world that the amplification of ground motion is highly dependent on the local geological, topography, and geotechnical conditions. Many researchers have worked on the estimation of local site effects [265, 266, 317, 338, 345, 357, 383]. It is observed that large concentration of damage in specific areas during an earthquake is due to site dependent factors related to surface geological conditions and local soil altering seismic motion. After the 1923 Kanto earthquake, it is very clear that major seismic damage was controlled by local geology. It is well known that each soil type responds differently, when subjected to the ground motions, imposed due to earthquake loading. Damages caused by 1985 Michoacan earthquake (M_S = 8.1), 1989 Loma Prieta earthquake (M_S = 7.1) and 2001 Bhuj earthquake (M_W = 7.7) have all reaffirmed that apart from the source characteristics, the local soil conditions also greatly influence the damage potential of an earthquake [181, 231]. The predominant influence of local soil geology/conditions on the earthquake ground motion characteristics is termed as local site effects, which have a direct impact on the response of structures during each of these earthquake events.

Hence, considering the effect of local site in the seismic design of structural and geotechnical system will improve the performance of buildings against future earthquakes. Amplification of seismic waves is one of the major local site effects, causing extensive damage to structures due to aggravated ground shaking. The amplification of seismic waves occurs when it enters into looser medium (soil) from a dense medium (rock) due to the phenomenon of impedance contrast. This chapter explains the assessment of site amplification both at the micro and macro-levels.

The response of the soil varies from type to type, when they are subjected to the earthquake excitations. When the seismic wave propagates through different medium, it tries to conserve the energy flux whenever it enters into the new medium.

T. G. Sitharam et al., *Comprehensive Seismic Zonation Schemes for Regions at Different Scales*, https://doi.org/10.1007/978-3-319-89659-5_5

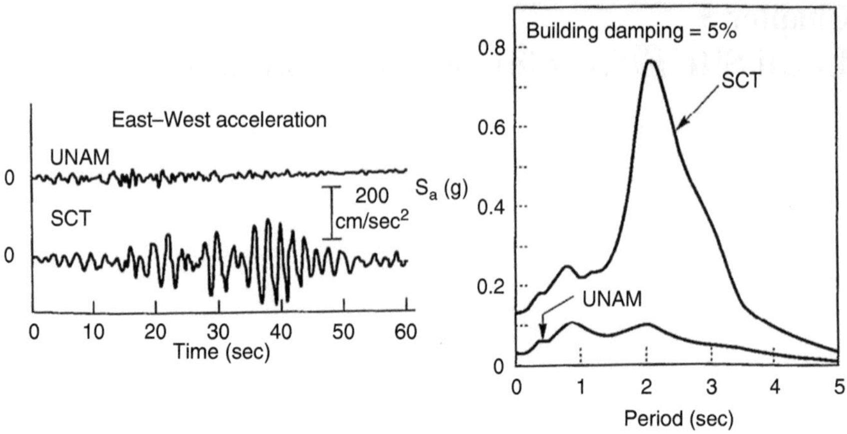

Fig. 5.1 Site amplification observed during Mexico earthquake—SCT-soft soil site and UNAM-rock site (after [348] taken from [181])

It is well established that the energy flux developed during the transmission of an earthquake wave in any medium depends upon the particle velocity and impedance (product of its density and wave velocity) of the medium. So when the wave enters a loose medium having low impedance, the particle movement gets exaggerated, so as to conserve the energy thus resulting in the amplification of seismic waves. Hence the younger loose soils often amplify the ground motion more when compared to the older compact soils or hard rock. During the 1819 Kutch earthquake, [202] observed the buildings with foundation resting on hard rock suffers less damage when compared to building founded on soil. Similar type of observations was made by Mallet [206] during the Neapolitan earthquake (of 1857), Wood [387] and Reid [281] during the San Francisco earthquake (of 1906). A well-documented case study of site effects on earthquake motion was observed during the 1985 Michoacan (Mexico) earthquake and the 1989 Loma Prieta (California) earthquake. Comparing strong motion records of 1985 Michoacan (Mexico) earthquake, at rock and soft soil site (as in Fig. 5.1) shows the amplification due to soft overlying soil [348].

5.2 Effect of Local Site Conditions on Ground Motion

Many historical evidences of the influence of local geological characteristics on the intensity of ground shaking and damage caused due to an earthquake are found. Data dating back to about 200 years supports this claim. However, it was not until the early 1970s that the local site conditions were included in building codes and provisions. Evidence for the existence of local site effects is quite overwhelming. In addition to theoretical evidence, amplification functions developed from measurements of surface and bedrock motions at the same location, and comparisons of surface motion characteristics from nearby sites with different subsurface conditions, all confirm the effects of local site conditions of earthquake ground motions.

5.2.1 *Effect of Topography*

Local site effects produce significant amplifications of the ground motion during an earthquake. One of the important local site effects is due to the effect of topography. Irregular topography can substantially affect the amplitude and frequency characteristics of seismic motion. Macro seismic observations of destructive earthquakes often show higher damage intensity at the tops of hills, ridges, and canyons than at lower elevations and on flat areas. Two general types of topography have to be distinguished

- Surface topography, mainly characterized by mountainous features, such as the presence of rock ridges or steep soil slopes,
- Subsurface (or subsoil) topography, either caused by lateral heterogeneities of the subsoil layers or by sharp basin geometry [90, 212].

The curvature of a sediment-filled basin structure in particular can capture body waves and cause some incident body waves to propagate through the alluvium as surface waves resulting in stronger shaking effects and longer duration of strong ground motion [181]. The effect of topography on the amplification has been studied by field experiments. Trifunac and Hudson [361] and Davis and West [75] recorded significant amplifications. But the amplifications recorded were much higher than those predicted by the theoretical models. The main problem involved was development of adequate soil models that can sufficiently represent the uncertainties involved in the geology and geotechnical properties of the area [19, 105]. Aki [6] pointed out that significant correlations exist between the amplification, geological and geotechnical features. Aki [6] also proposed that these complex relationships can be represented by the seismic microzonation maps. Pedersen et al. [264] found that the diffracted waves found at the reference station might explain the amplification of spectral ratios. The quantitative value of the ground motion at any particular site depends upon the source of the seismic waves.

The main effects of topography as pointed out by Brune [50] and Sánchez-Sesma [297, 298] are:

- Amplification of SH waves near the crests of canyons.
- The presence of a shear fundamental resonance in a 3–5 Hz frequency band.
- Dependence on the radiation wave field, angle of incidence, and canyon dimensions.

In contrast to subsoil topography, it is the surface topography that causes more serious effects. According to numerous reconnaissance studies after strong earthquakes, an increase of damage to buildings can be observed on steep slope situations which extend towards the plateau. Recent damaging events like Bingol earthquake of May 1, 2003 in Turkey and Northern Algerian earthquake of May 21, 2003 displayed heavy damage concentrations along the top of steep slopes. According to different scientific groups [6, 19, 105] that deal with instrumental and theoretical investigations of surface topography on ground motion characteristics, the following are stated:

- Mountain tops or ridge crests, and more generally, convex topographies (such as cliff borders), lead to an amplification of seismic ground motion, while valleys or foothills (concave topographies) tend to de-amplify the seismic signals.
- The effects of surface topography are larger on horizontal components than on
- Vertical ones, thus indicating that S motion is more affected by surface topography than P motion.
- The influence of surface topography on ground motion is directly related to the sharpness of topography. According to this theoretical model, amplification of incoming seismic waves increases as the wedge angle becomes sharper.
- Amplification and de-amplification of seismic ground motion on topographic features are both frequency-dependent and band-limited: maximum effects can be observed for wavelengths that approximately agree with the horizontal dimensions of the topographic shape.

The approach adopted by AFPS [2] was to introduce an additional empirical parameter (as in Eq. (5.1)) as the topographical amplification factor in the definition of the design spectrum to account for the topographical effects.

$$\tau = 1 + 0.8(I - i - 0.4)1.0 < \tau < 1.4 \tag{5.1}$$

where I and i are the gradients of the lower and upper slopes, respectively. During the Athens earthquake that occurred in 1999, amplification studies on sites were carried out to find the effect of topography. It was concluded that the average topographic amplification is about 50% for time periods of about 1 s at a site [45]. The effect of slope topography on the seismic ground motion is studied by Bouckovalas and Papadimitriou [46]. The following important observations are to be noted:

- Even a purely horizontal excitation, as a vertically propagating SV wave, results in considerable vertical motion at the ground surface near the slope. This component of ground motion is independent of any vertical excitation induced to the base of the slope by the earthquake itself and, consequently, it has to be superimposed to it. The results of the parametric analyses show that the vertical component of seismic motion may become comparable to the horizontal free-field motion.
- The topography aggravation of the horizontal ground motion, expressed through the peak acceleration ratio $A_h = Z_{ah}/a_{h,ff}$, fluctuates intensely with distance away from the crest of the slope, alternating between amplification ($A_h > 1.0$) and de-amplification ($A < 1.0$) within very short horizontal lengths.

- The horizontal ground motion is de-amplified at the toe of the slope and amplified near the crest. As a result, topography aggravation may be seriously overestimated, when calculated as the peak seismic ground motion at the crest over that at the toe of the slope.

The above two points can be directly associated to reflection of incoming SV waves on the inclined free surface of the slope which leads to reflected P and SV waves impinging obliquely at the free ground surface behind the crest, as well as Rayleigh waves. Topography effects become important for normalized height ratios $H/\lambda > 0.16$ and slope inclinations $i > 17$. If these conditions are met, the peak values of topography aggravation factors for the horizontal and the vertical ground acceleration behind the crest usually vary between $A_{h,\max} = 1.20$–1.50 and $A_{v,\max} = 0.10$–1.10, while free field conditions behind the crest are usually met at a distance $D_{ff} = (2$–$8)H$.

Based on the records obtained from the dense strong motion array in Taipei basin for two earthquakes ($M_L = 6.5$ and 6.57) [197] reported that there were significant differences among the peak ground accelerations, durations, and spectral accelerations in different parts of the basin as well as among the records for two earthquakes. The authors have concluded that for estimating ground motion variations for microzonation of the basin one must be very careful to draw conclusions by using only few seismic events and that it is necessary to collect more data from the array to perform a more detailed microzonation study. The last two decades have witnessed significant progress in estimation of local site effects by different researchers all over the world [43, 56, 299, 362].

The effect of Central Mountain Range in Taipei basin was evaluated by Lee et al. [191]. The Spectral Element Mesh method was adopted to simulate the seismic wave propagation. Realistic topography and complex subsurface topography can be effectively incorporated using SEM. It was concluded that in the event of a shallow earthquake in the area the central mountain range will scatter the surface waves thereby reducing the magnitude of strong ground motion. The occurrence of a deep earthquake, topography scatters the body waves which propagate as body waves resulting in increase of PGA values by over 50%.

Ma et al. [201] studied the effect of San Gabriel mountains which are bounded by the Mojave segment of the San Andreas fault on the north and by the Los Angeles Basin on the south. It was observed that effect of mountain topography reduces the value of ground motion for some areas in the basin. The topography of the mountains scatters the surface waves generated by the rupture on the San Andreas fault, leading to less-efficient excitation of basin-edge generated waves and natural resonances within the Los Angeles Basin.

It is very important to understand the local geotechnical and geological conditions when performing seismic microzonation studies. Site classification can be done based on geological units [233]. But in microzonation even further detailed study of the variations in the geological units is necessary. Wills and Silva [383] suggested and utilized the average shear wave velocity in the upper 30 m as one parameter to characterize the geologic units admitting the importance of other factors such as impedance contrast, three dimensional basin and topographical effects, and source effects such as rupture directivity on ground motion characteristics. Thus the influence of local soil stratification on earthquake characteristics is one of the major factors in evaluating the earthquake forces and thus the structural response. Local soil conditions may amplify or deamplify the earthquake forces in different regions. Sometimes an appropriate simplification may be needed to reflect the complex stratification characteristics in describing the soil amplification phenomena.

5.2.2 *Effect of Groundwater*

The relation between groundwater and earthquakes is not new and is well documented. The Chinese noticed centuries ago that water levels in wells can vary in association with earthquake activity, and used this behavior with some success to anticipate earthquakes. More recent research in the United States has attempted to use monitoring of groundwater levels in wells to help predict earthquake activity [226]. The 1983 Borah Peak earthquake in southeastern Idaho caused groundwater located near its epicenter to erupt as much as 25 ft into the air [71]. However, within the past 50 year researchers have noted that groundwater can play a direct role in earthquake occurrence and earthquake related damage. In addition, it is known that earthquakes such as the 2004 Indonesian quake can cause measurable changes on groundwater levels in places thousands of miles from the epicenter [368]. Research on earthquake mechanisms indicates that groundwater likely plays a significant role in many large earthquakes. Furthermore, groundwater can magnify the damaging effects of ground surface.

Groundwater can also play a significant role in how earthquakes affect the ground surface when an earthquake occurs. The most well-known effect is liquefaction. During the shaking cause by an earthquake, certain types of fluid saturated sediments can lose their structure and become liquefied. These phenomena led to increased structural damage. The geologic and hydrologic factors that affect liquefaction susceptibility are the age and the type of sedimentary deposits, the looseness of cohesions less sediments, and the depth to the groundwater table. The liquefaction is mostly limited to water-saturated, cohesions less, granular sediments at depths less than 15 m. Noack and Fäh [240] gave weight according to the depth of water table (Table 5.1 more weight more damage, less weight less damage).

Table 5.1 Weightage factor for assessing liquefaction susceptibility

Depth of water table	Weight
10–20 m	2
3–10 m	3
1–3 m	4

During an earthquake, base rock movements generate shear waves that propagate through overlying soils. Liquefaction results when these shear waves, passing though saturated sand layers, distort the granular structure and cause loosely packed grains to collapse. This densification causes an increase in pore pressure if drainage cannot occur. If these pore pressures exceed roughly sixty percent (60%) of the soil's effective stress, large settlements and translational deformations can occur.

Groundwater level influences the ground response significantly and cannot be neglected for site effect analyses. In the presence of confined aquifers liquefaction can take place in the subsurface. This results in attenuation of the propagation of shears waves and can reduce the ground shaking. In the presence of groundwater the shear wave cannot reach the surface as they can't pass through the water, because water offers no shear resistance.

5.2.3 *Effect of Bedrock*

In many earthquakes, the local geology and soil condition have had profound influence on site response. The term local is somewhat vague one generally meaning local compared to the total terrain transverse between the earthquake source and the site. On the assumption that the gross bedrock vibration will be similar at two adjacent sites. Local differences in geology and soil properties will cause different surface ground motions at the two sites. Factors influencing the local modification to the underlying motion are the topography and the nature of the bedrock and the nature and geometry of the depositional soils. Thus, the term local may involve a depth of a kilometer or more, and an area within a horizontal distance of several kilometers from the site. Soil conditions and local geological features affecting site response are numerous, and are now discussed as below:

- The greater the horizontal extent of the softer soils, the less the boundary effects of the bedrock on the site response.
- The depth of soil overlaying bedrock affects the dynamic response, the natural period of vibration of the ground increasing depth.

This helps to determine the frequency of the wave amplified or filtered out by the soils and is also related to the amount of soil structure interaction that will occur in an earthquake. The Mexico earthquake of 1957 and 1985 witnessed extensive damage to long-period structures in the former lake bed area of Mexico city where the flexible lacustrine deposits caused greater amplification of long period waves.

- The slope of the bedding planes of the soils overlying bedrock obviously affects the dynamic response, but it is less easy to deal rigorously with non-horizontal strata.
- Dynamic response, but it is less easy to deal rigorously with non-horizontal strata.
- Changes of soil type horizontal across a site affect the response locally within that site, and may profoundly affect the safety of a structure straddling the two soil types.

- The topography of both the bedrock and the deposited soils has various effects on the incoming seismic waves, such as reflection, refraction, focusing, and scattering.

5.3 Methods for Estimating Local Site Effects

As the local site effects play a crucial role in aggravating the damage potential of an earthquake, it must be accounted in the earthquake resistant design. There are many methods available for the estimation of site effects which can be broadly categorized as (1) Empirical methods, (2) Experimental methods, and (3) Numerical studies.

5.3.1 Empirical Methods

Many researchers have observed significant dependency of site amplification on the local surface geology based on the strong motion records of many earthquake events. Based on earthquake records for each geological formation, many empirical relations between surface geology and various ground motion parameters have been developed which can be applied in another region having same geology. These empirical relations will give only an approximate value of site amplification and hence is good for first level assessment. Intensity based empirical correlations proposed by Medvedev [214] (Table 5.2), Evernden and Thomson [87], Kagami et al. [158] and Astroza and Monge [15] are the major ones which were developed using earthquake records in Asia, California, Japan, and Chile, respectively.

Table 5.2 Intensity increment for each geological unit [87]

Lithology	MM scale
Granitic and metamorphic rock	0
Paleozoic rocks	0.4
Early mesozoic rocks	0.8
Cretaceous to eocene rocks	1.2
Undivided tertiary rocks	1.3
Oligocene to middle Pliocene rocks	1.5
Pliocene-Pleistocene rocks	2.0
Tertiary volcanic rocks	0.3
Quaternary volcanic rocks	0.3
Alluvium (water table < 30 ft)	3.0
Alluvium (30 ft < water table < 100 ft)	2.0
Alluvium (water table > 100 ft)	1.5

Table 5.3 Correlations between surface geology and relative amplification [144]

Previous works	Geology	Relative amplification
Borcherdt and Gibbs [42]	Bay mud	<11.2
	Alluvium	3.9
	Santa Clara formation	2.7
	Great valley sequence	2.3
	Franciscan formation	1.6
	Granite	1.0
Shima [324]	Peat	1.6
	Humas soil	1.4
	Clay	>1.3
	Loam	1.0
	Sand	0.9
Midorikawa [216]	Holocene	3.0
	Pleistocene	2.1
	Quaternary volcanic rock	1.6
	Miocene	1.5
	Pre-tertiary	1.0

Many researchers [42, 216, 324] have given the relative amplification factors for different geological and soil conditions (Table 5.3) based on analytical studies. Other researchers [44, 157, 216] have proposed empirical relations relating the average shear wave velocity of surficial layers to the relative amplification. Topographical effects on the earthquake ground motion characteristics have been studied by Geli et al. [105], Faccioli [88] and Chávez-García et al. [65].

Many researchers [39, 274, 345] have proposed various empirical relations for various site classes as per NEHRP scheme. These site coefficients and empirical equations for estimating PHA and S_a for each NEHRP site classes were developed using regression analysis. The assessment of site effects based on local geology is only approximate and may not incorporate deep basin effects or topographic effects. Thus this methodology provides a less reliable estimate of site amplification and hence much suitable for a first level study for a large landmass.

The most important geotechnical parameters, which can be used to estimate the amplification factors are average shear wave velocity and SPT '*N*' value. In the early days measurement of shear wave velocity using cross hole tests were expensive whereas the SPT '*N*' values were available abundantly. Shima [324] found that the analytically calculated amplification factor is linearly related with the ratio of shear wave velocity of the surface layer to that of bedrock. When the bedrock shear wave velocity is found to be relatively constant over a wide area, the relative amplification in each locality can be obtained from the shear wave velocity of the surface layer. Various researchers [44, 157, 216] proposed relations between the average shear wave velocity of surficial layers and the relative amplification as shown in Table 5.4.

Table 5.4 Correlations between average shear wave velocity and relative amplification [144]

Author	Proposed equation	
Joyner and Fumal [157]	$RA = 23(V_s)^{-0.45}$	
Midorikawa [216]	$RA = 68(V_{s30})^{-0.6}$	($V_{s30} < 1100$ m/s)
	$RA = 1.0$	($V_{s30} < 1100$ m/s)
Borchedt et al. [44]	$AHSA = 700/V_{s30}$	(for weak motion)
	$AHSA = 600/V_{s30}$	(for strong motion)

5.3.2 Experimental Methods

These methods are based on different kinds of data obtained from microtremor measurements, weak seismicity survey, and strong motion data. Detailed description for estimating site effects using the above methods is given below. Kanai[162] have presented the use of the microtremor method for site response evaluations. Microtremors are vibrations, having comparatively small periods (less than 1 s). Udwadia and Trifunac[366] compared the site amplification obtained from microtremor method and strong motion records. Nakamura [229] has proposed H/V ratio technique to predict the amplification factor for a site.

(1) Microtremor Data

The site effects are often expressed by the amplification factor and resonance or fundamental frequency. Usually there are various vibrations in the ground which are caused by natural or ambient noise like wind, sea waves, traffic, industrial machinery, etc. The range of vibration frequencies of ambient noises is from 0.1 to 10 Hz (i.e., 10–0.1 s period). The vibrations that have comparatively small periods of less than 1 s are called microtremors and those that have a larger period range is called microseisms. Due to close relation between spectral features of microtremors and site's geological conditions, these small vibrations are very useful in earthquake geotechnical engineering. Kanai [162] have explained a theoretical interpretation and practical engineering application of microtremors as a convenient tool for evaluating frequency properties of surface ground. The use of microtremor method for site response evaluations has been restricted to Japan with many controversies in USA and Europe. Applications of this method include site response analysis, natural frequency of structures, etc. However, the real generation and nature of microtremors have not yet been established.

Kanai [161] proposed a method to classify the ground into four categories, which is used by the Japan Building Code. This classification is based on the detailed comparison of microtremor results and ground conditions. A general wave pattern and period-frequency (PF) diagrams for four types are shown in Fig. 5.2. It is clear that a stiff ground/rock is inclined to have low predominant period and soft sediments have high predominant period. Kanai [162] proposed two methods based

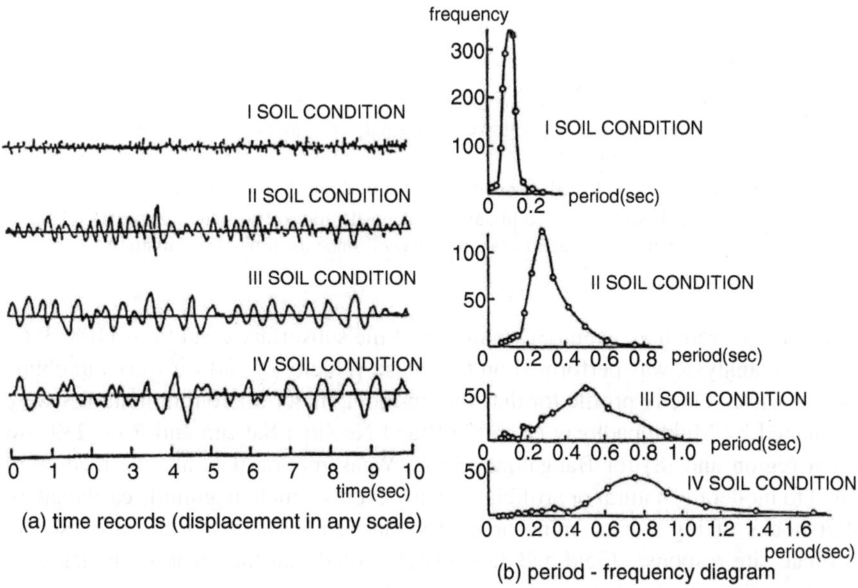

Fig. 5.2 Classification of microtremor types [163]. (**a**) Time records (displacement in any scale). (**b**) Period–frequency diagram

on microtremor records. One is based on the largest period and the mean period, the other based on the largest amplitude in microns and the predominant period. This can be used to identify site categories associated with various levels of seismic damage due to strong shaking. Table 5.5 gives the detailed description of soil type in each category. However it was found that the H/V technique is not well correlated with the S wave amplification at the site's resonant frequency [179, 243]. Though the H/V method is good to map the fundamental resonance, it cannot give a good estimation of the amplification value at a site [115]. It is highly sensitive to some parameters like Poisson's ratio near the surface. A one-to-one average correlation is developed by Konno and Ohmachi [179] on the basis of a comparison between observed H/V peaks and numerical estimates of 1D transfer functions. Bard et al. [20] compared the amplifications derived from earthquake records and H/V peak amplitudes for more than 30 sites and demonstrated that the latter is always smaller than the former. This Nakamura method provides a lower bound estimate to the actual amplification which is proved by large experimental data.

Aki [3] has measured phase velocities of surface waves by analyzing the spatial correlation of the microtremors. Then, the P and S wave velocity profiles can be derived by inversion which makes it possible to estimate the site response by 1D modelling. Several studies in Italy [69, 204], Japan [220] and Israel [108] illustrate the practical use of this technique with other methods used in geotechnical engineering in estimating velocity profiles. In particular, its coupling with the Nakamura technique (array measurements at few sites and H/V ratios at many sites)

Table 5.5 Microzones for Japan Building Code (Kanai [162])

Zones	Soil description
I	Ground consisting of rock, hard sandy soils or gravely deposits
II	Ground consisting of sandy gravel, hard sandy clay, loam or alluvial gravel with thickness of 5 m or more
III	Standard ground other than Zone I, Zone II or Zone IV
IV	Ground consisting of soft alluvial delta deposits, top soils or mud thickness of 3 m or more where less than 30 years has elapsed since the time of reclamation

may lead to a reliable 2D and 3D mapping of the subsurface conditions [108, 356]. Inversion analysis was performed on the phase velocity of surface waves to obtain a shear wave velocity profile for that site. In India, major microtremor studies were conducted by Mukhopadhyay et al. [227] and Neelima Satyam and Rao [239] for Delhi region and [8] for Bangalore region. Weak motion data are the records of small to moderate, natural or artificial seismic events (small magnitude earthquakes, aftershocks of big events, mine or quarry blasts, nuclear tests) are also used to estimate site response. Field and Jacob [92] quoted that the greatest challenge in the estimation of site response from such instrumental recordings is removing the source and path effects. Many researchers have used strong motion records of past earthquake in order to predict site amplification [24, 38, 172].

(2) Weak Motion Data

Weak motion data are the records from small to moderate, natural or artificial seismic events (small magnitude earthquakes, after shocks of big events, mine or quarry blasts, nuclear tests). Such data can be recorded by digital, high sensitive instruments identical to those used by seismologists for microseismicity and seismotectonics studies. Field and Jacob [92] quoted that the greatest challenge in the estimation of site response from such instrumental recordings is removing the source and path effects. Two techniques are developed depending on whether or not they need a reference site with respect to which the particular effects at other sites are estimated.

(a) Reference Site Technique

The three important factors which will affect the ground motion are the source, path, and the site characteristics. The identification and removal of these effects is the greatest challenge in evaluating the site response. The simplest method to evaluate the site response is to divide the response spectrum obtained at the site with that of the bed rock (reference site). If the recording in the rock is at a close distance to the soil site, then the three governing factors, which will affect the ground motion, will be the same for both the soil site and the rock. However when the reference site and the site under consideration (soil site) are not near by, then the influence of source, path, and the site characteristics will be different for these two sites. Moreover the geometric spreading of the seismic waves will also need to be accounted [40, 41, 43, 122]. To evaluate the source and site terms simultaneously, a generalized inversion scheme developed by Andrews [12] can be applied. In the generalized inversion technique, a relatively large dataset can be used [345] but the non-linear response of sedimentary deposits cannot be predicted accurately when the weak motion data are dominating the input data.

(b) Non-Reference Site Technique

In practice, adequate reference sites are not always available. For this reason, different methods without reference sites have been developed. It consists of taking the spectral ratio between the horizontal and the vertical components of the shear wave part. This technique is a combination of [188] receiver function method for determining the velocity structure of the crust from the horizontal to vertical spectral ratio (HVSR) of teleseismic P-waves and the [229] method. It was first applied to the S-wave portion of the earthquake recordings obtained at three different sites in Mexico city by Lermo and Chávez-García [192]. The same technique has been applied on various sets of weak and strong motion data [32, 65, 187, 284, 353, 402] from which it is concluded that the HVSR shape exhibits a very good experimental stability, it is well correlated with surface geology and less sensitive and comparisons with theoretical 1D computations, the absolute level of HVSR depends on the type of incident waves. Also, the determination of the absolute level of amplification from only HVSR is not straightforward. Field and Jacob [92] applied this technique and found that the method reproduces well the shape of the site response, but underestimates the amplification level. They also found very different results when applying this technique to the P-wave part of the recordings.

(3) Strong Motion Data

The development of strong motion arrays makes it possible to evaluate site effects in mega cities like Los Angeles, Tokyo, Taipei, Mexico city using the strong motion data. While using this method, even the non-linear site effects are included in the recordings. Recent studies show that there is a fairly good agreement between the old and new techniques. With the increase in strong motion data, it is possible to determine soil amplification and site response maps based on ground motion records [17, 24, 38, 121, 352]. Khoubbi-Al and Adams [172] estimated soil amplification in Ottawa, Canada using strong motion records.

5.3.3 Numerical Approach

Numerical methods of estimating site amplification mainly include ground response analysis. Ground response analysis techniques are often categorized based on the dimensionality of the problem and the complexity of the soil model employed [181]. On the basis of dimensionality there are one-, two- and three-dimensional ground response analysis. Similarly based on the soil model used, the three can be further grouped into linear, equivalent linear, and non-linear.

(1) 1D Ground Response Analysis

One-dimensional (1D) ground response analysis assumes that all soil layers are horizontal and soil response is mainly due to the vertically propagating SH-waves only [181]. Evaluation of transfer functions is the key to any 1D analysis. Transfer function relates the base motion to the surface motion in the Fourier domain. In a linear 1D analysis, dynamic soil properties such as the shear modulus (G) and damping ratio (β) are assumed to be consistent with the level of strain while evaluating the transfer function. However the equivalent linear analysis proposed by Seed and Idriss [307] considers the dependency of shear modulus (G) and damping ratio (β) with the level of strain induced. The non-linear analysis is carried out by integrating the equation of motion in small time intervals using any non-linear stress-strain models or advance constitutive models. Several computer programs such as SHAKE [132, 302] and DYNEQ [397], DEEPSOIL [123] are available in order to implement this equivalent linear approach. Similarly there are many computer codes available for carrying out non-linear 1D, ground response analysis DESRA-2 [190], DYNA 1D [270], DEEPSOIL [123].

The 1D equivalent linear approach consists of the following steps: (1) The input earthquake acceleration time history is transformed into frequency domain using a Fourier transfer algorithm. (2) The in-situ soil state is captured by assigning a low strain shear modulus ($G_{\max}$) and damping ratio ($\beta_{\min}$) each soil layer. Appropriate modulus degradation and damping curves are assigned to each layer to capture the strain dependency of shear modulus and damping ratio. (3) For the input motion (in the frequency domain), the response of each soil layer is computed considering $G_{\max}$ and $\beta_{\min}$, and shear strain time histories for each soil layer is generated. (4) Effective shear strain (γ_{eff}) for each layer is then calculated which is approximately 65% of the peak strain amplitude ($\gamma_{\max}$) occurred in the strain time history for corresponding layer. (5) For the effective shear strain (γ_{eff}) value, corresponding shear modulus and damping ratio are evaluated from the modulus reduction and damping curve which are assigned to each soil layer. (6) Step 3 is repeated with a new strain dependent value of shear modulus and damping ratio. The iteration procedure is repeated till concordant values of shear modulus and damping ratio from two successive steps are obtained. In this study, 1D equivalent linear ground response analysis for the study area is carried out using SHAKE 2000 [254] and surface motion, response spectra, predominant frequency, and spectral ratio are evaluated. In a general 1D equivalent linear analysis, the geo-materials (soil or rock) are modelled using the Kelvin–Voigt model and the wave propagation through these solids can be mathematically written as per Eq. (5.2) as given by Kramer [181].

$$\rho \frac{\partial^2 u}{\partial t^2} = G \frac{\partial^2 u}{\partial z^2} + \eta \frac{\partial^3 u}{\partial z^2 \partial t} \tag{5.2}$$

where u is displacement, t is time and z is the vertical direction along which wave propagates. The general solution for the above equation is given in Eq. (5.3).

$$u(z,t) = Ae^{i(\omega t + k^* z)} + Be^{i(\omega t - k^* z)} \tag{5.3}$$

Here, A and B are the amplitudes of the incident and the reflected waves, k^* is the complex wave number $k^* = \sqrt{\frac{\rho}{G^*}\omega^2}$ and $G^* = G(1 - 2\beta^2 + i2\beta\sqrt{1-\beta^2})$ is the frequency independent complex shear modulus. 1D ground response study analyse the vertical propagation of shear waves through soil layers lying on an elastic rock. The elastic bedrock is assumed to be extended to infinite depth. Hence Eqs. (5.4) and (5.5) (where subscripts s and r refers to soil and rock respectively) present the response of soil and elastic rock due to the upward propagation of shear waves.

$$u_s(Z_s,t) = A_s e^{i(\omega t + k^* z_s)} + B_s e^{i(\omega t - k^* z_s)} \tag{5.4}$$

$$u_r(Z_r,t) = A_r e^{i(\omega t + k^* z_r)} + B_r e^{i(\omega t - k^* z_r)} \tag{5.5}$$

Certain boundary conditions are then applied to reduce the redundancies. These conditions are (1) at free surface, the shear stress ($\tau_s = 0$) and shear strains are zero ($\gamma_s = 0$), (2) displacements and stresses at soil–rock interface are equal and continuous (i.e. $u_s(z_s = H) = u_r(z_r = 0)$, $\tau_s(z_s = H) = \tau_r(z_r = 0)$), (3) compatibility of displacements and stress along the layer boundaries, i.e. $u_m(z_m = H_m) = u_{m+1}(z_{m+1} = 0)$, $\tau_m(z_m = H_m) = \tau_{m+1}(z_{m+1} = 0)$ where m and $m+1$ are successive soil layers and as shown in Fig. 5.3.

From the compatibility conditions, amplitudes of incident waves and reflected waves in the $(m+1)$th (A_{m+1}, B_{m+1}) layer can be expressed in terms of amplitudes in mth layer (A_m, B_m) as given in Eqs. (5.6) and (5.7).

$$A_{m+1} = \frac{1}{2}A_m(1+\alpha_m)e^{ik_m^* h_m} + \frac{1}{2}A_m(1-\alpha_m)e^{-ik_m^* h_m} \tag{5.6}$$

$$B_{m+1} = \frac{1}{2}B_m(1+\alpha_m)e^{ik_m^* h_m} + \frac{1}{2}B_m(1-\alpha_m)e^{-ik_m^* h_m} \tag{5.7}$$

where $\alpha_m = \frac{k_m^* G_m^*}{k_{m+1}^* G_{m+1}^*}$ is the complex impedance ratio at the boundary between the mth layer and $(m+1)$th layer [181]. For the free surface condition, $A_1 = B_1$ can be derived. Starting with the surface layer, repeated use of recursion formulas (in Eqs. (5.6) and (5.7)) will result in Eqs. (5.8) and (5.9) relating amplitudes in mth layer with those in the surface layer.

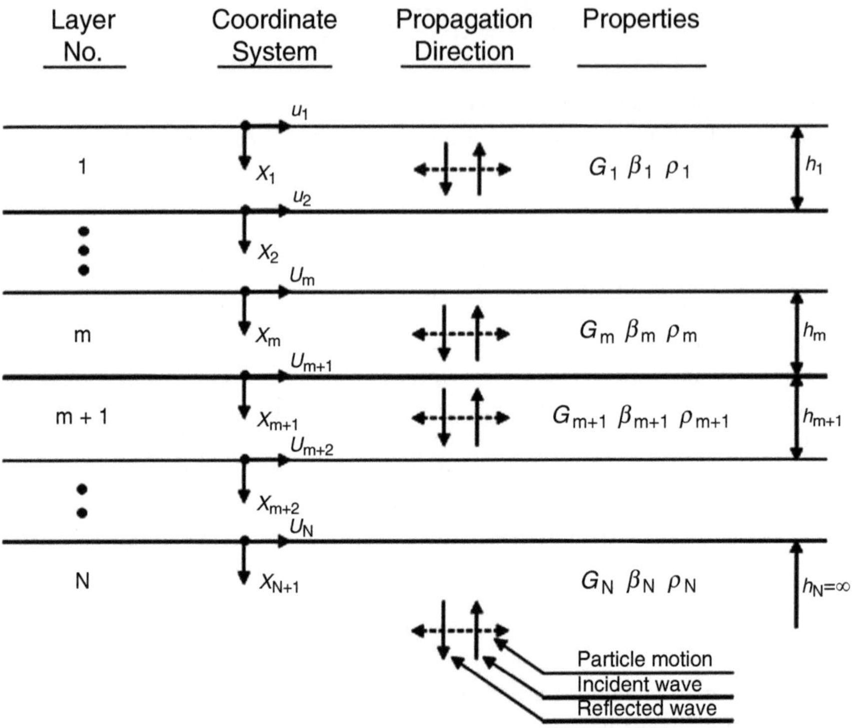

Fig. 5.3 Representations of layered soil (after [302])

$$A_m = a_m(\omega)A_1 \tag{5.8}$$

$$B_m = b_m(\omega)B_1 \tag{5.9}$$

Now the frequency dependent transfer function relating the displacement at ith and jth layer $F_{ij}(\omega)$ can be expressed as per Eq. (5.10).

$$F_{ij}(\omega) = \frac{u_i}{u_j} = \frac{a_i(\omega) + b_i(\omega)}{a_j(\omega) + b_j(\omega)} \tag{5.10}$$

Once amplitudes A and B of all layers are obtained, acceleration and strain histories of any layer can be computed using Eqs. (5.11) and (5.12), respectively.

$$\frac{\partial^2 u}{\partial t^2} = -\omega^2\left(Ae^{i(\omega t + k^* z)} + Be^{i(\omega t - k^* z)}\right) \tag{5.11}$$

$$\frac{\partial u}{\partial z} = -ik\left(Ae^{i(\omega t + k^* z)} - Be^{i(\omega t - k^* z)}\right) \tag{5.12}$$

The effective shear strain (γ_{eff}) for equivalent linear analysis is calculated from the maximum shear strain ($\gamma_{\max}$) value in the shear strain time histories of each layer. The effective shear strain is given as $\gamma_{eff} = R_\gamma \gamma_{\max}$ where the stress reduction factor R_γ is a function of earthquake magnitude. However it is found that R_γ varies from 0.5 to 0.7 and is not sensitive on response, hence in SHAKE 2000 it is taken as 0.65.

(2) 2D and 3D Ground Response Analysis

A two-dimensional (2D) analysis is normally carried out for the sites having sloping/irregular ground surface or locations where deep basin formations are present. The curvature of a soft alluvium deposited basin can trap body waves and cause them to propagate as surface waves. Hence the shaking produced will be of higher intensity and of longer duration when compared with those predicted using 1D analysis [181]. Three-dimensional (3D) analysis is the most rigorous analysis which is generally carried out to understand the response of three-dimensional structures. 2D and 3D analyses are generally carried out using dynamic finite analysis. 2D analyses can also be performed using a simplified methodology called shear beam approach [103]. Numerical modelling software like PLAXIS, FLAC, QUAKE/W, etc. can be used for modelling two-dimensional case.

The ground response analysis requires geotechnical and geophysical test data as input parameters in order to compute soil response. The 2D and 3D ground response analyses based on finite element tools require lot of computational power. Hence this method is most suitable to micro to meso-level studies.

5.4 Evaluation of Site Amplification at Micro-Level: A Case Study of Kalpakkam NPP Site

Ground response studies at the micro-level for the Kalpakkam NPP site are carried out by James et al. [153] to obtain ground surface motion, surface level response spectrum, and predominant frequency of surface motion. As the nuclear power plant structures are critical, rigorous studies are needed to be carried out for the estimation of site amplification. The amplifying effect that the site can have on earthquake motions at bedrock can reasonably be quantified by conducting ground response analyses. In the present study, 1D equivalent linear ground response analysis for the Kalpakkam NPP site was carried out using SHAKE 2000 [254].

SHAKE 2000 is a Windows based computer program for ground response analysis for the evaluation of the earthquake effects on soil deposits. SHAKE 2000 [254] is an upgraded version of SHAKE program developed by Schnabel et al. [302] at the University of California Berkeley. SHAKE 2000 is a software package that integrates ShakEdit and SHAKE. ShakEdit was originally developed as a 16-bit,

Windows 3.1 application that provides a graphical interface for SHAKE. SHAKE computes the response of the soil layers subjected to earthquake excitation at the base in the frequency domain. The non-linear behavior of the soil is approximated using an equivalent linear approach incorporating an iterative procedure to obtain the strain dependent values for modulus and damping in each of the soil layers.

5.4.1 Generation of Synthetic Earthquake Strong Motion

One-dimension ground response analysis requires an earthquake acceleration time history data of the site, which has to be given as input in the program. As there are no suitable earthquake records available for the study area available in any major database, a synthetic acceleration time history was generated by James et al. [153] for the site response analysis. There are many methodologies available for the generation of synthetic acceleration time history like empirical green function technique [120, 137] and stochastic simulation technique [35]. The above two methods require the knowledge of complex seismological parameters and hence in this work, a simple method is used, which synthesizes the artificial ground motion compatible with the target acceleration response spectrum. A MATLAB code was developed to generate by James et al. [153] artificial earthquake time history based on the methodology proposed by Zhao and Zhang [403]. In this method a target response spectrum was first selected. The response spectrum obtained from deterministic hazard analysis of the study area (Fig. 5.4) was used as target response

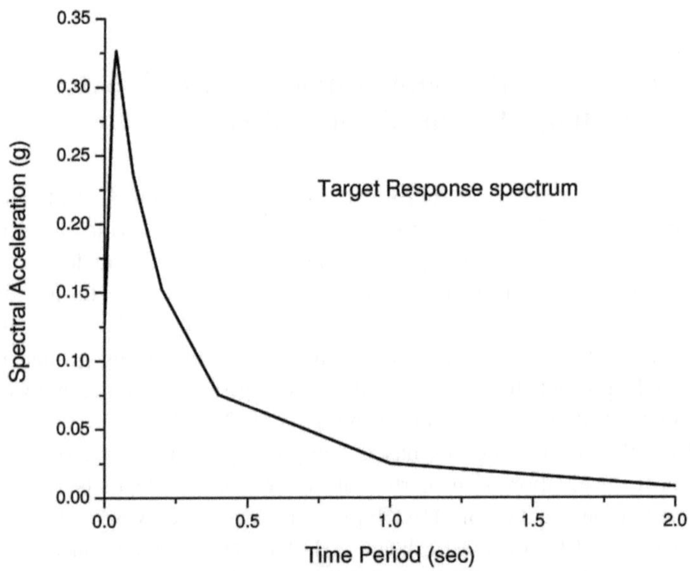

Fig. 5.4 Target response spectrum obtained from DSHA (after [153])

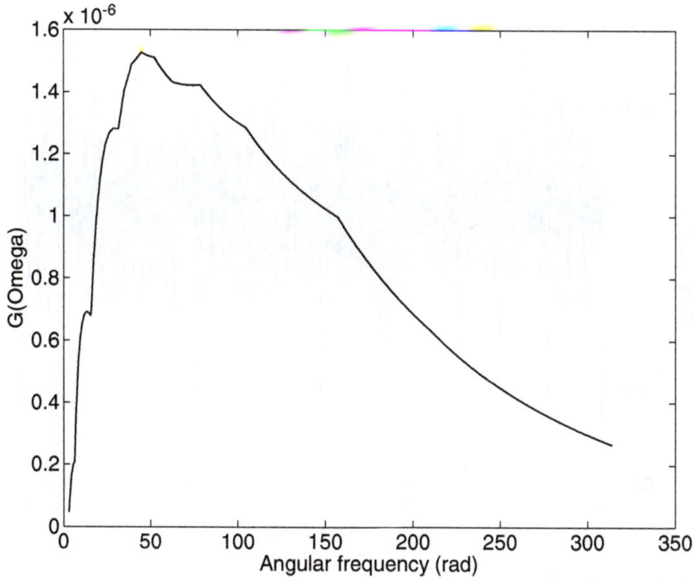

Fig. 5.5 Power density spectrum (after [153])

spectra. The target response spectrum was transformed into the corresponding power spectrum (Fig. 5.5) by the following approximate transfer formula [169] described by Eq. (5.13).

$$G(\omega) = \left[\frac{\beta}{\pi\omega}\right]\left[\frac{S_T^2(\omega, \beta)}{\ln(\omega T) - \ln[-\pi \ln(l - r)]}\right] \tag{5.13}$$

where $G(\omega)$ is the power spectrum, $S_T(\omega, \beta)$ is the target response spectrum, β is the damping ratio, T is the duration of seismic ground motion, and r is the probability of exceeding the target response spectrum. A stationary signal (Fig. 5.6) of duration ($T = 15$ s) was generated as per Eq. (5.14).

$$X(t) = \sum_{n=0}^{N-1} A(\omega_n) \cos\left[\omega_n t + \phi(\omega_n)\right] \tag{5.14}$$

where $A(\omega_n) = [4G(\omega_n)\Delta\omega]^{0.5}$, $\Delta\omega = (2\pi/T)$ is the frequency-domain sampling distance, and the initial phase $\phi(\omega_n)$ is a random variable with uniform distribution within the interval of $[0, 2\pi]$. The stationary time history $X(t)$ was then multiplied with a prescribed non-stationary intensity envelop $E(t)$ in order to transform it into non-stationary acceleration time history (Eq. (5.15)).

$$a_g(t) = E(t)X(t) \tag{5.15}$$

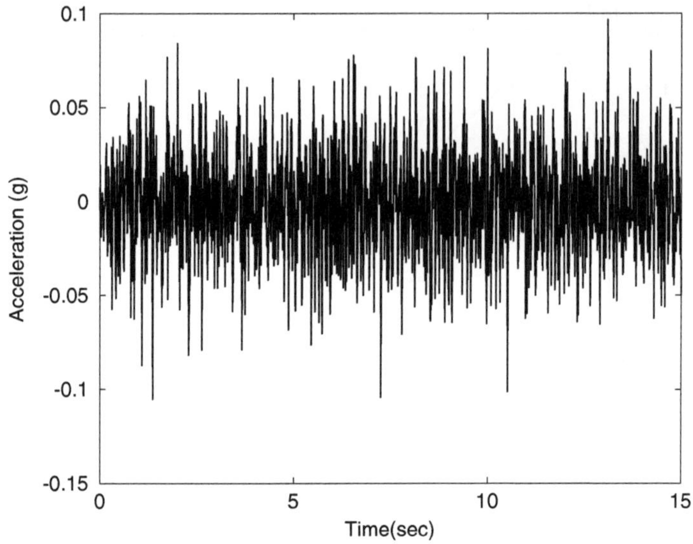

Fig. 5.6 Stationary signals

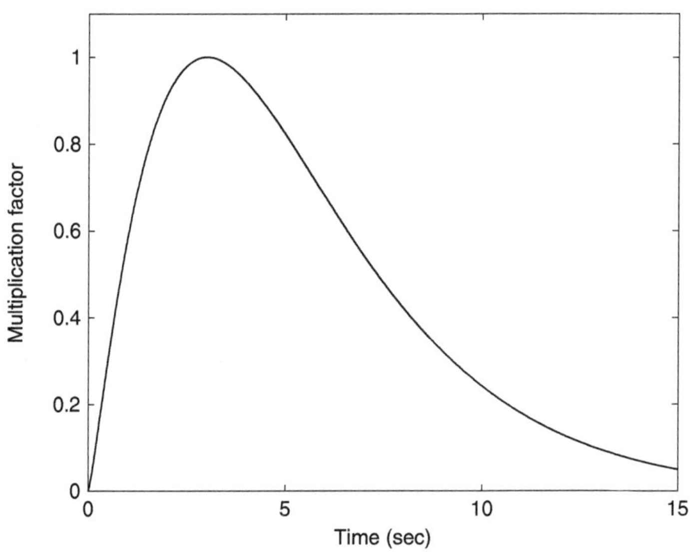

Fig. 5.7 Envelop function

where $E(t)$ is the envelope function (Fig. 5.7) given by Boore [34] and $a_g(t)$ is non-stationary acceleration time history (Fig. 5.8) as in Eq. (5.16). Same envelop function was employed by Kumar [184].

$$E(t) = at^b e^{-ct} \tag{5.16}$$

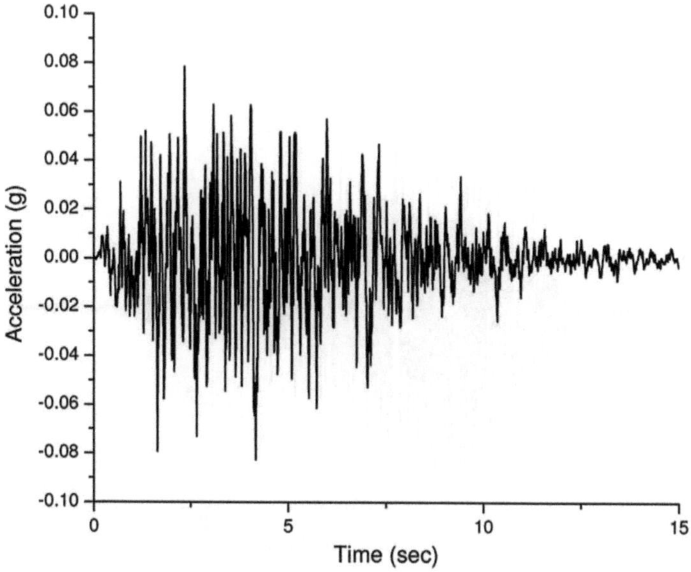

Fig. 5.8 Non-stationary time signal

where $a = \left[\frac{e}{\varepsilon T_d}\right]^b$ $b = \left[\frac{-\varepsilon \ln \beta}{1+\varepsilon(\ln \varepsilon - 1)}\right]$ and $c = \left[\frac{c}{\varepsilon T_d}\right]$ and in this work, ε was taken as 0.2 and β was taken as 0.05 and T_d is the total duration of ground motion (assumed as 15 s). The transformation of response spectrum to power spectrum using Eq. (5.13) is only approximate, so the response spectra $S_a(\omega, \beta)$ obtained from the generated time history (in step 2) rarely matches with the original or target response spectra $S_T(\omega, \beta)$. Hence an iterative procedure was employed to explicitly match the observed response spectra with target response spectra so as to obtain a compatible time history. The commonly-used iterative method modifies the Fourier amplitude spectrum in Eq. (5.17).

$$A^{i+1}(\omega_j) = \frac{S_T(\omega, \beta)}{S_a(\omega, \beta)} A^i(\omega_j) \tag{5.17}$$

where $A^i(\omega_j)$and $A^{i+1}(\omega_j)$ are the results of the ith and the $(i+1)$th iteration, respectively, and $S_T(\omega, \beta)$ and $S_a(\omega, \beta)$ are the target spectrum and the calculated spectrum at the matching frequency of ω_j. A number of iterations are performed and the observed response spectrum is matched to the target. The final acceleration time history for the site is presented in Fig. 5.9. Plot showing the target response spectrum and observed response spectrum is given in Fig. 5.10.

This methodology is very effective as it that modifies the Fourier amplitude spectrum of time history to fit the target acceleration spectrum. By superimposition of narrow-band time histories in the time domain, this method improves the fitting precision of the artificial ground motion to the target response spectrum. The

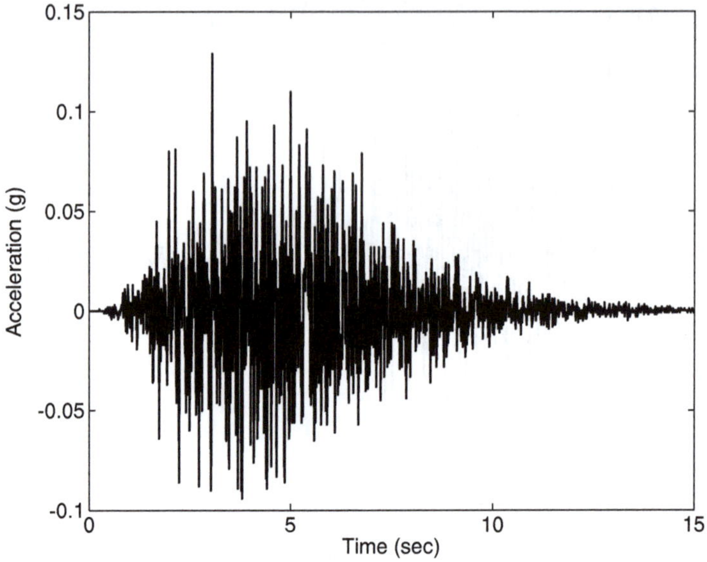

Fig. 5.9 Synthetic strong motion for the study area, compatible with TRS (after [153])

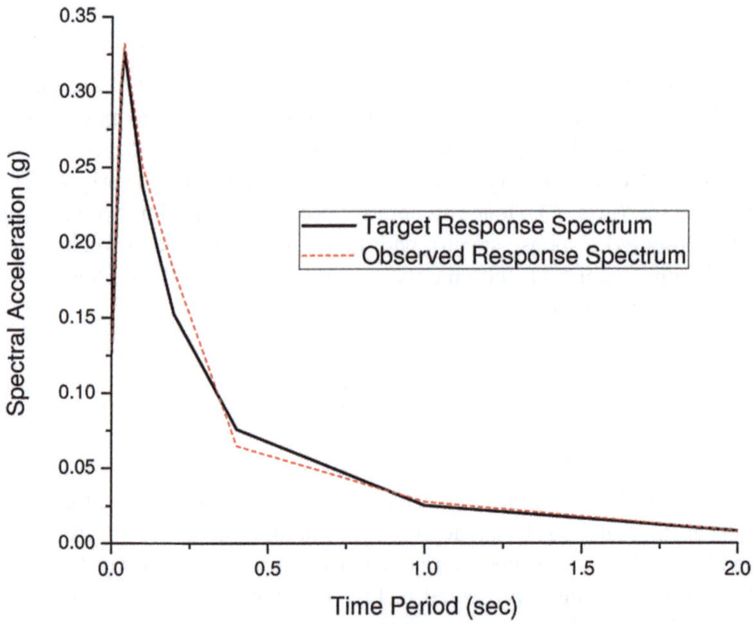

Fig. 5.10 Compatibility between observed and target response spectrum (after [153])

Table 5.6 Details of typical subsurface profile in the study area (after [153])

Strata name	In-situ density (kN/m^3)	Particle size				N-value observed	Index properties	
		C	M	S	G		W_L	IP
Sand	17–19	0	2	98	0	10–40	–	–
Soft clay	17–20	64	36	0	0	2–20	91	65
Filled up soil	20–22	5	20	75	0	>50	35	20
Residual soil	19–23	0	4	57	39	>50	–	–
Weathered rock	25–27	–	–	–	–	>100 (RQD 0–50%)	–	–
Rock	27	–	–	–	–	>100 (RQD > 100%)	–	–

synthetic ground motion generated (in Fig. 5.9) has a peak acceleration of 0.129 g at 3.07 s, peak velocity of 2.630 cm/s at 5.48 s, and peak displacement of 0.435 cm at 5.39 s.

5.4.2 *Modelling the Soil Profile*

Stratifying and assigning appropriate modulus reduction and damping model for the soil overburden is an important step in ground response analysis. Stratum identification was done by James et al. [153], based on the borelog data and shear wave velocity data. Field test data revealed the existence of five types of soil layers which were identified mainly as sand, soft silty clay, residual soil (gravelly Sand), weathered rock and bedrock (Charnockite rock type). Typical characteristics of each of the soil samples are described in Table 5.6. Borelog data also reveal the presence of dense filled up soil between BH-5 and BH-6 boreholes (Fig. 4.9), having brownish coloration which is a mixture of clay and sand.

After the characterization of soil stratum the important task was to assign appropriate modulus reduction curve and damping curve. SHAKE 2000 provides different types of modulus reduction and damping curves proposed by many researchers for different types of geo-materials. Details of modulus reduction and damping curves used for each stratum are given in Figs. 5.11 and 5.12 and Table 5.7. Modulus reduction and damping curve proposed by Vucetic and Dobry [377] for soil having an overconsolidation ratio up to 15 is used for modelling the dynamic behavior of the filled up soil.

5.4.3 *Ground Response Analysis*

One-dimensional equivalent ground response analysis was carried out using SHAKE software packages. Analysis was carried out at eight major locations in the Kalpakkam NPP site and the results are briefly presented in Table 5.8. The spatial

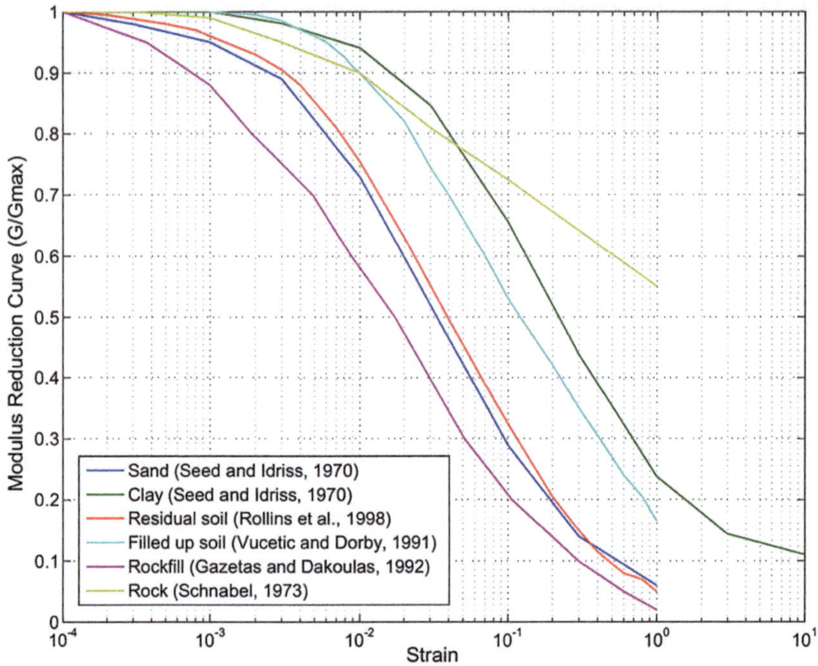

Fig. 5.11 Modulus reduction curves used for various soil layers in the study area (after [150])

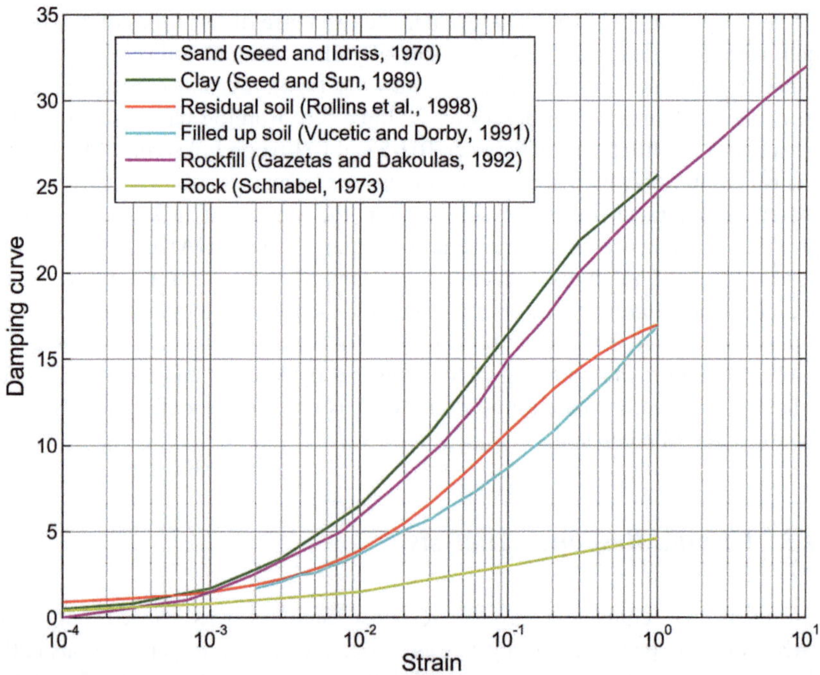

Fig. 5.12 Damping curves used for various soil layers in the study area (after [150])

Table 5.7 Dynamic properties assigned to each soil type (after [153])

Sl. no.	Soil type	Modulus reduction curve	Damping curve
1	Sand	Sand Ave [308]	Sand Ave [308]
2	Clay	Clay upper range [308]	Clay upper range [312]
3	Residual soil	Gravel mean [292]	Gravel mean [292]
4	Weathered rock	Rockfill [104]	Rockfill [104]
5	Hard rock	Rock [301]	Rock [301]

Table 5.8 Results of 1D equivalent ground response analysis (after [153])

Location description	PHA (g)	Predominant frequency (PF)	Spectral acceleration for PF (g)	Amplification factor
BH-1	0.22	2.98	0.89	1.9
BH-2	0.19	3.61	0.67	1.48
BH-3	0.45	6.59	2.39	3.52
BH-4	0.34	9.33	2.21	2.64
BH-5	0.33	9.27	2.07	2.55
BH-6	0.34	3.71	1.40	2.64
BH-7	0.15	4.05	0.54	1.19
BH-8	0.25	5.03	0.75	1.79

variation of surface level PHA values obtained from SHAKE 2000 is presented in Fig. 5.13. The maximum PHA at the ground surface of the study area is 0.45 g, close to the location BH-3. The corrected SPT-N value for this location was found to be less than 20 at 3 m depth from the bore log data. As the depth of groundwater table is very low such a high value of the PHA aggravates liquefaction hazard in the location. Figure 5.14 presents the spatial variation of the amplification factor for PHA (corresponding to zero period) over the entire the Kalpakkam NPP site, where maximum amplification occurred at the location close to BH-3. Figures 5.15 and 5.16 present the amplification spectra for location BH-1 and BH-6 in the study area.

Predominant frequency is the frequency at which the maximum amplitude is observed for a ground motion in the frequency domain [181]. In this study the predominant frequency of the surface ground motion was also evaluated after converting it into frequency domain using Fast Fourier Transform (FFT) algorithm. The Fourier spectrum obtained from FFT fluctuates significantly and to avoid the influence of individual spikes, predominant frequency was obtained from a smoothened spectrum as shown in Fig. 5.17 [181, 360]. Hanning window [360]

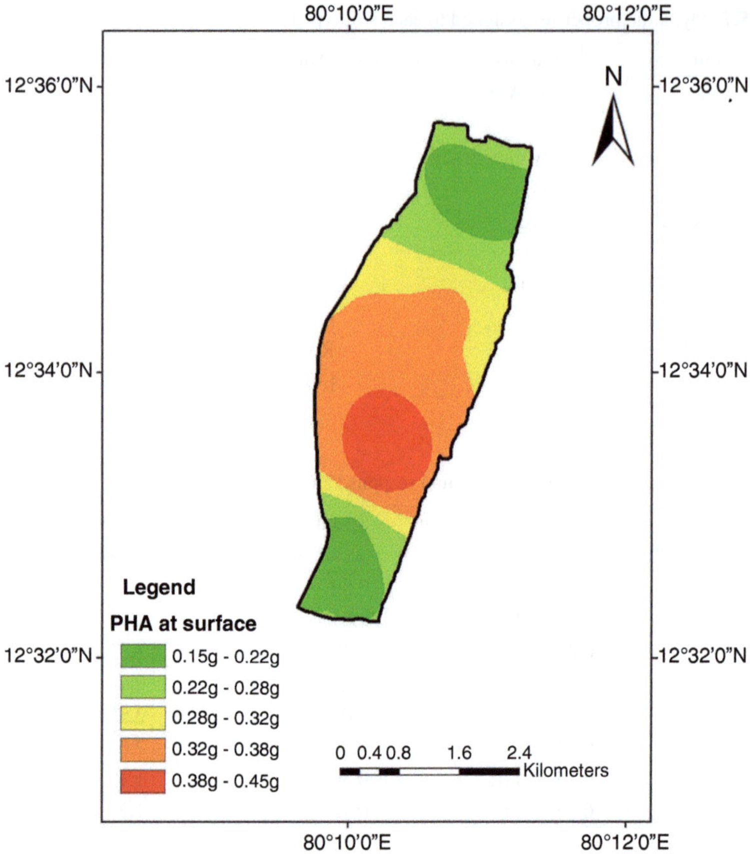

Fig. 5.13 Spatial variation of surface level PHA value throughout the Kalpakkam NPP site (after [153])

was adopted for smoothening the Fourier spectrum. Towhata [360] also suggested that significant amplification can be observed at places where the natural frequency of the soil is matched with the predominant frequency in the earthquake motion, leading to the extensive shaking and building damage.

The predominant frequency in the study area ranges from 2.9 to 9.3 Hz as given in Table 5.8 and its spatial variation is presented in Fig. 5.18. This chapter also presents response spectra of SDOF (single degree of freedom) for the amplified surface motion corresponding to 5% damping (Fig. 5.19) which is an important tool for a structural designer.

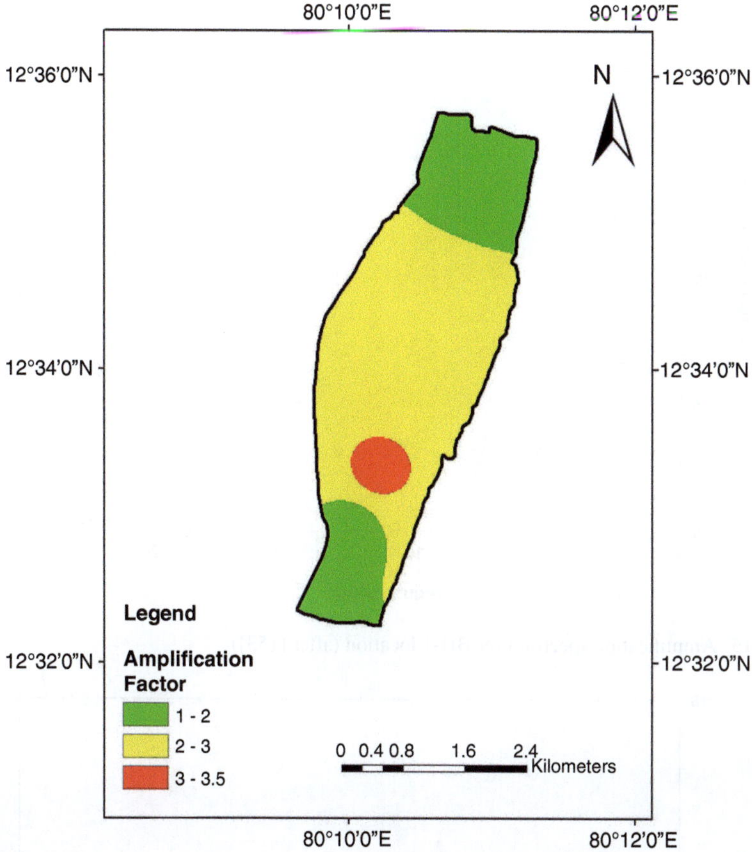

Fig. 5.14 Spatial variation of amplification factor throughout the Kalpakkam NPP site (after [153])

5.5 Evaluation of Site Amplification at Macro-Level: A Case Study of Karnataka State

For a small region, ground response analysis techniques are available for evaluating PHA at ground surface, however for a larger landmass such as a state or a country, this technique is not viable. Hence in the present study, for the state of Karnataka, the seismic hazard at ground surface was estimated using a non-linear site amplification technique proposed by Raghu Kanth and Iyengar [274]. Raghu Kanth and Iyengar [274] has developed a regression equation for estimating amplification factor for each NEHRP site class. Seismic site characterization for

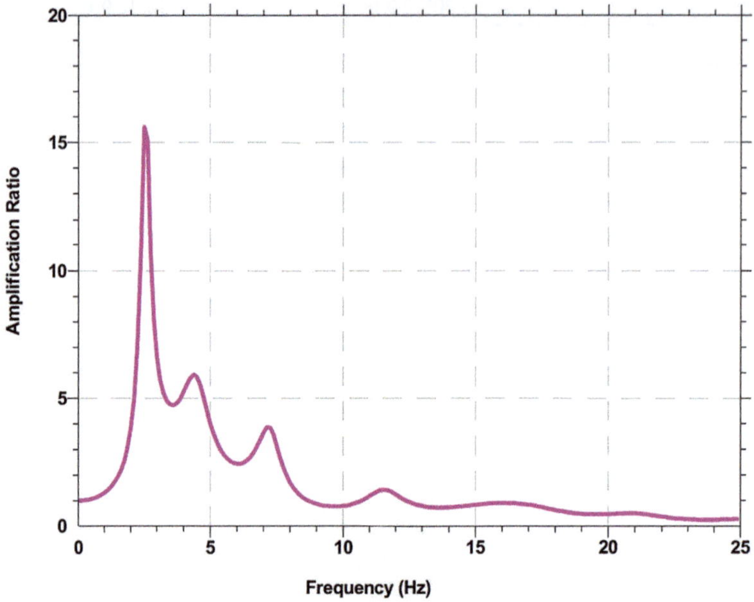

Fig. 5.15 Amplification spectrum for BH-1 location (after [153])

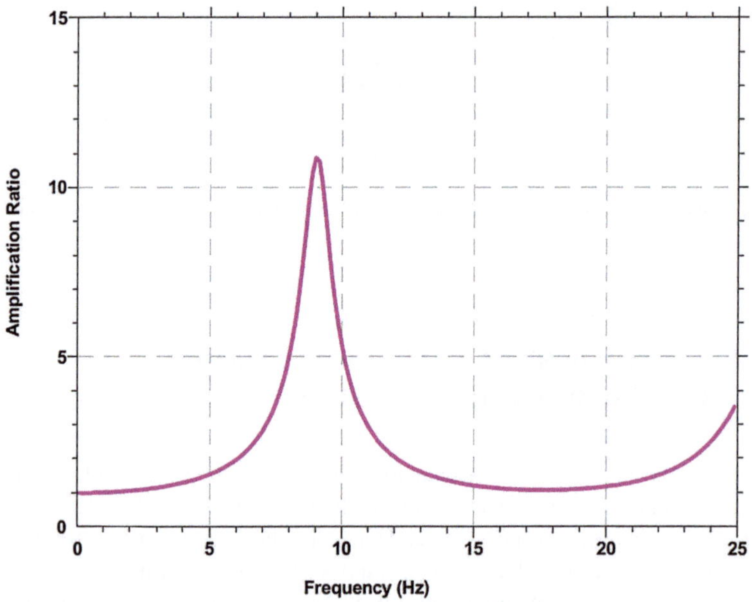

Fig. 5.16 Amplification spectrum for BH-6 location (after [153])

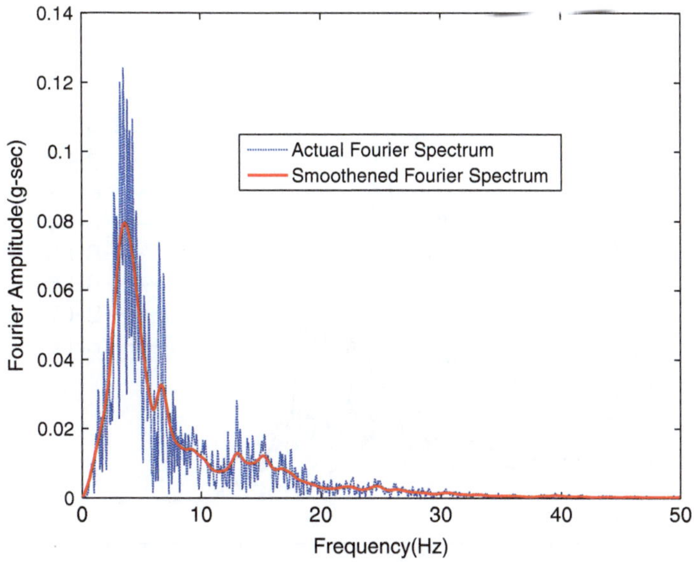

Fig. 5.17 Actual and smoothened Fourier spectrum of the surface ground motion (after [153])

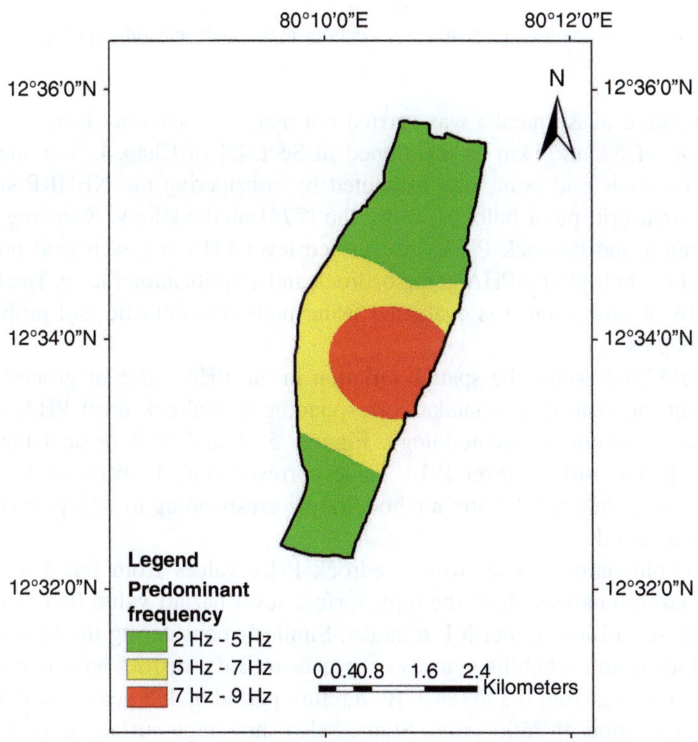

Fig. 5.18 Spatial variation of predominant frequency throughout the Kalpakkam NPP site (after [153])

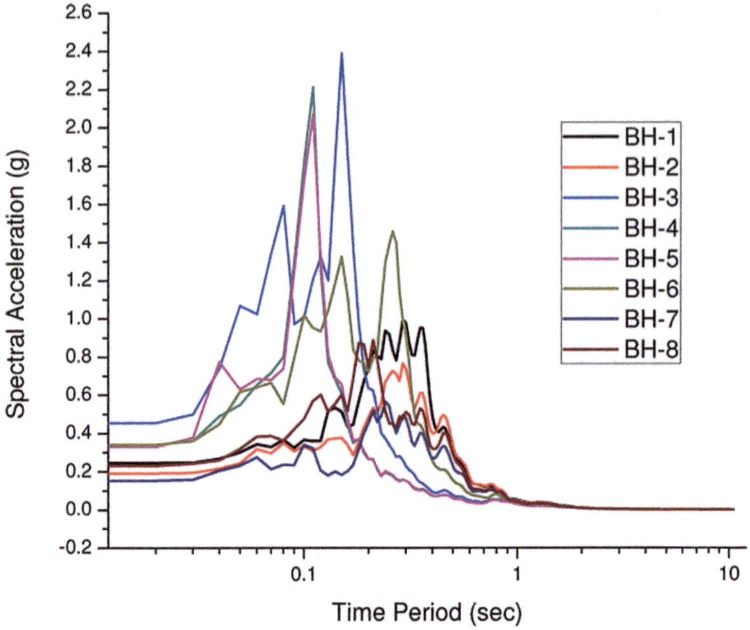

Fig. 5.19 Response spectrum plots at major locations in the study area (after [153])

the entire state of Karnataka was carried out using the satellite data, considering a grid size of 5 km × 5 km as mentioned in Sect. 4.4 of Chap. 4. The site amplification for each grid point was estimated by considering the NEHRP site class to which that grid point belongs, using the [274] methodology. Knowing the site amplification and bedrock PHA, the surface level PHA for each grid point was estimated by multiplying PHA at the bedrock and amplification factor. The bedrock level PHA at grid point was evaluated using both deterministic and probabilistic methodologies.

Figure 5.20 presents the spatial variation in the PHA value at ground surface throughout the state of Karnataka corresponding to bedrock level PHA obtained using the deterministic methodology. Figures 5.21 and 5.22 present the spatial variation in the surface level PHA value corresponding to bedrock level PHA obtained using the probabilistic methodology corresponding to 475 year and 2500 year return period.

Site amplification studies using bedrock PHA values from the deterministic seismic hazard analysis show the high surface level hazard value (0.4–0.55 g) for regions close to Bidar in north Karnataka. Similarly considering the bedrock level PHA values from probabilistic analysis, places in Bidar district have hazard value at surface ranging from 0.3 to 0.4 g, for a return period of 475 years and 0.4–0.52 g for a return period of 2500 years. Studies also show high surface level hazard in

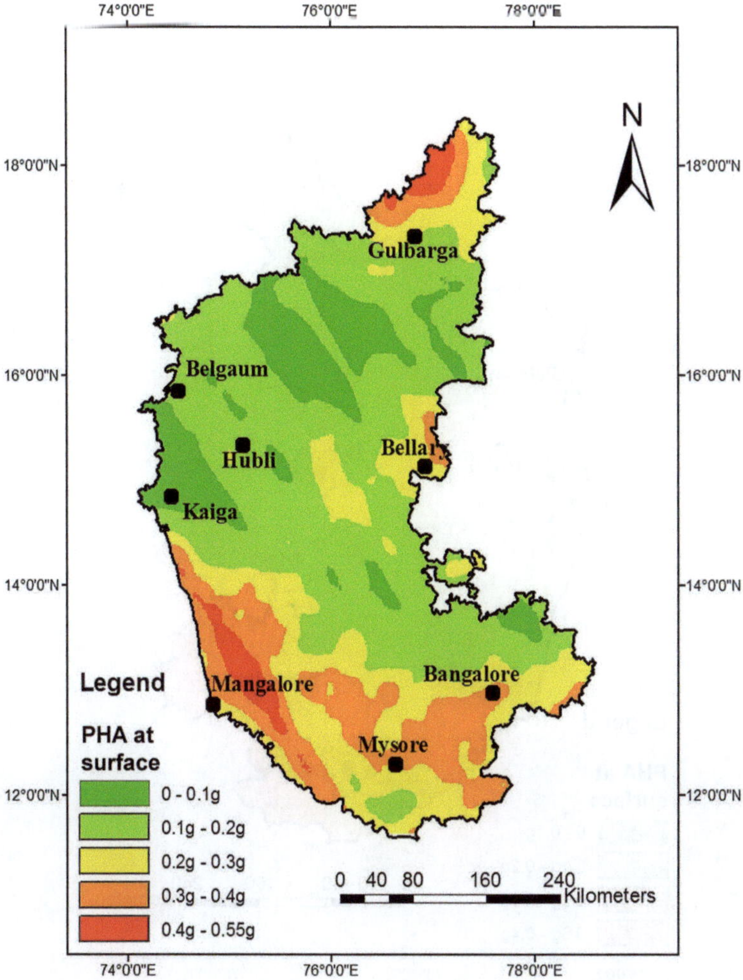

Fig. 5.20 Spatial variation of surface level PHA value throughout the Karnataka state from DSHA after [151])

the range of 0.4–0.55 g for the Mangalore–Udupi regions considering deterministic PHA values at bedrock. However considering the bedrock level PHA values from probabilistic method, the same region has moderately high surface level hazard ranging from 0.2 to 0.3 g for a return period of 475 years and 0.3–0.4 g for a return period of 2500 years.

The region between Bangalore and Mysore is also found to have moderately high value of surface PHA of about 0.3–0.4 g considering the bedrock level shaking from deterministic analysis. But, considering the bedrock level PHA values from

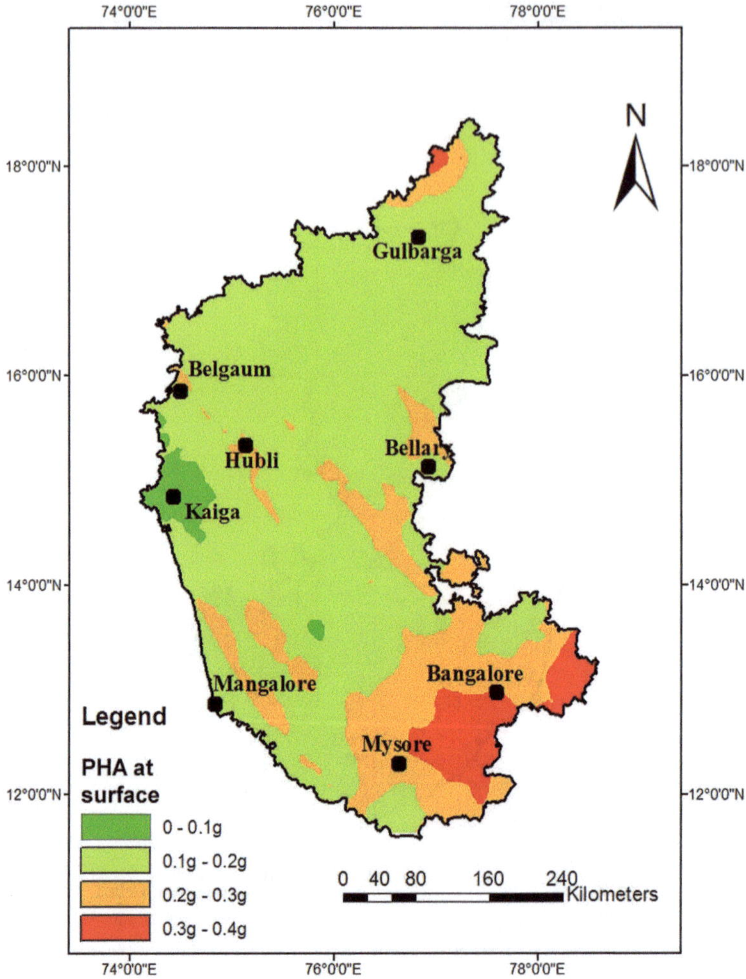

Fig. 5.21 Spatial variation of surface level PHA value throughout the Karnataka state from PSHA (for a return period of 475 year after [150])

probabilistic method, the same region is found to have high value of surface PHA of about 0.3–0.4 g for a return period of 475 year and 0.4–0.52 g for a return period of 2500 years. The interior parts of Karnataka state are found to have low surface level hazard value ranging from 0.01 to 0.2 g considering bedrock level PHA from deterministic method. The regions close to Bellary have moderate surface level hazard ranging from 0.2 to 0.3 g. Similar trends for PHA is also observed from

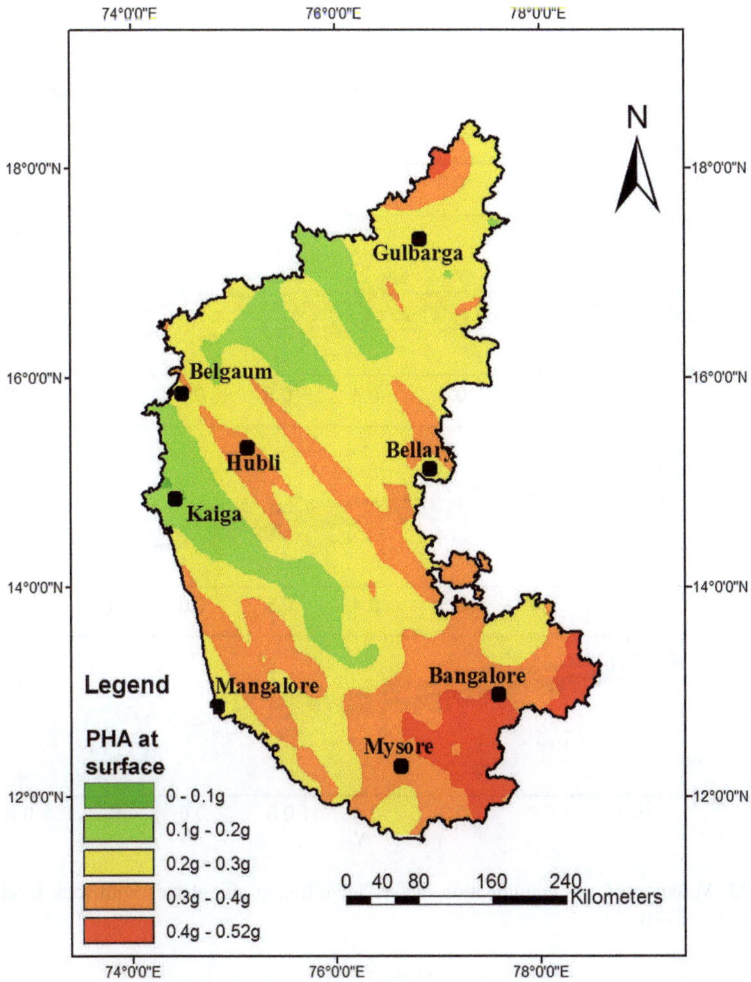

Fig. 5.22 Spatial variation of surface level PHA value throughout the Karnataka state from PSHA (for a return period of 2500 year after [151])

the site amplification studies considering the bedrock level PHA from probabilistic method having 475 year return period. However considering the bedrock level PHA having 2500 year return period, the surface level hazard value is in the category of moderately high to high.

Macro-level site amplification studies also show a low surface level hazard value for Kaiga, which is the location for nuclear power plant. The surface level PHA

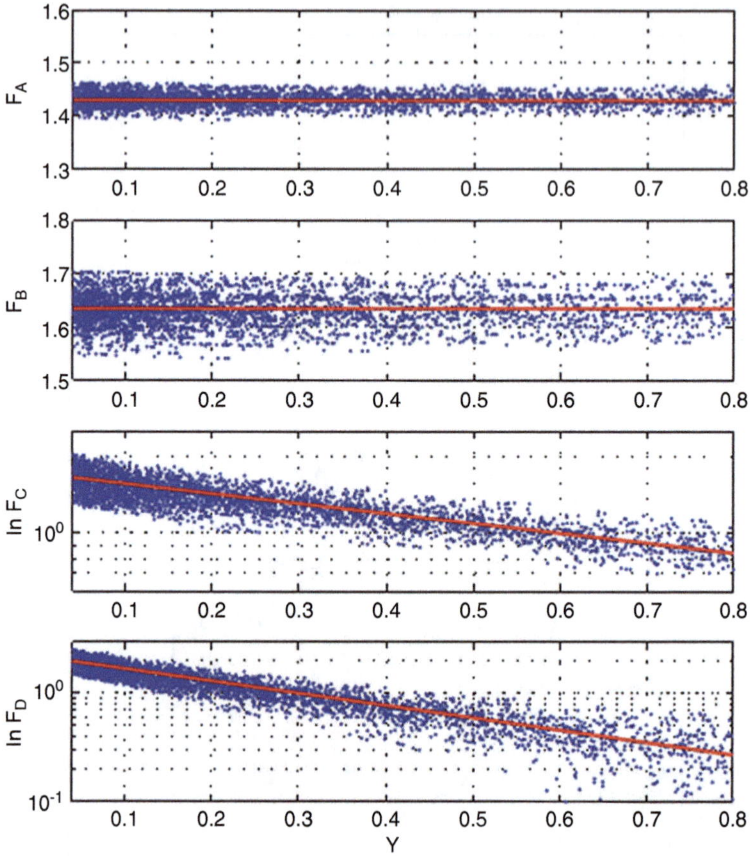

Fig. 5.23 Variation of site amplification factors for different site classes with rock level PHA values (after [274])

value for this region is 0.1 g. Considerable variations in the surface level PHA values for site class C and D is because the site coefficients (F_C and F_D) for C and D-type sites exhibit strong dependence on bedrock values when compared to (F_A and F_B) A and B type as shown in Fig. 5.23 given by Raghu Kanth and Iyengar [274].

Chapter 6
Liquefaction

6.1 Introduction

During an earthquake, soil can fail due to liquefaction with devastating effect such as land sliding, lateral spreading, or large ground settlement. The phenomenon of liquefaction of soil had been observed for many years, but was brought to the attention of researchers after Niigata (1964) and Alaska Earthquakes (1964). In India, classic scenario of earthquake induced soil liquefaction was observed during the Bhuj earthquake of 2001. Liquefaction is a phenomenon in which the strength and stiffness of a soil is reduced by earthquake shaking or by other dynamic loading. Liquefaction and related phenomena have been responsible for tremendous amounts of damage in historical earthquakes around the world [224, 391]. During the Bhuj earthquake on 26th January 2001 (M_W—7.7) lot of damages had been occurred due to liquefaction and other ground failures [276].

A large number of investigations have been carried out for understanding the phenomenon of soil liquefaction in the last four decades. From these investigations it was observed that a vast majority of liquefaction occurrences were associated with sandy soils and silty sands of low plasticity.

6.2 Mechanism of Soil Liquefaction

It is necessary to understand the mechanism of soil liquefaction, where it occurs and why it occurs so often during earthquakes. Figure 6.1 clearly depicts the mechanism of soil liquefaction. Liquefaction of soil is a process by which sediments below water table temporarily lose shear strength and behaves more as a viscous liquid than as a solid. The water in the soil voids exerts pressure upon the soil particles. If the pressure is low enough, the soil stays stable. However, once the water pressure exceeds a certain level, it decreases the effective stress and eventually the shear

T. G. Sitharam et al., *Comprehensive Seismic Zonation Schemes for Regions at Different Scales*, https://doi.org/10.1007/978-3-319-89659-5_6

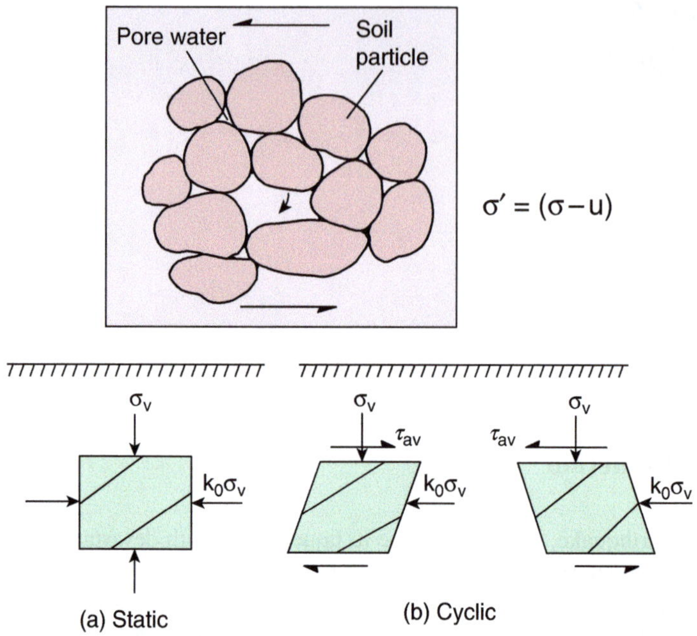

Fig. 6.1 Mechanism of soil liquefaction (after [238])

strength of the soil is reduced. During an earthquake, the sudden load is taken by the water in the form of excess pore water pressure. If increased pore water pressure cannot be released, it will continue to build up and it may lead to reduction of effective stress to zero. In this state, the soil layer loses its shear strength and it behaves like a viscous liquid and it cannot take the load coming from overlying soil or structures. Thus, the upper layers of soils move down and this will lead to the collapse of structures on the ground. Casagrande [57] proposed the concept of critical void ratio (CVR) line which is the locus of critical void ratios achieved by soils subjected to large strain shearing under drained, strain-controlled condition with different confining pressures. Saturated soils with initial states above CVR line (loose state) show tendency for contraction under drained loading hence produce positive excess pore pressure under undrained loading, and soils with initial states plotting below CVR line (dense state) show tendency for dilation under drained loading hence produce negative excess pore pressure under undrained loading. Thus CVR line was considered to mark the boundary between the soil states which are susceptible to flow liquefaction (loose), and soils which are nonsusceptible to flow liquefaction (dense). But, from investigations of partial failure of Fort Peck dam (September 1938) in Montana, it was found that sand located below the CVR line can also liquefy and it showed the inability of the type of test (strain controlled test) used for determining the critical void ratio to define correctly the boundary between sands that are safe and unsafe against flow liquefaction.

Later, Casagrande hypothesized the structure of flowing sand as that different from structure of sand in static state. In liquefied sand, each grain is constantly rotating in relation to all surrounding grains so as to offer a minimum of frictional resistance [58]. He termed this the flow structure. Castro [59] developed the flow structure of sand by conducting stress controlled undrained static and cyclic triaxial tests. He modified the critical void ratio line, termed as e_F line, as the locus of void ratios in which liquefaction with a flow structure developed which was plotted below and roughly parallel to the CVR line obtained from strain controlled tests. This difference in strength during the flow in strain controlled and stress controlled tests was attributed to the difference in strain rate. In stress controlled test, the constant load produces large strain rate during the flow and a perfect flow structure is developed, while in strain controlled test, the relatively low strain rate causes local groups of sand grains to lose temporarily their flow structure. Castro [59] observed limited liquefaction failure in medium dense sand. The specimen experienced a sudden failure with only a slight decrease in resistance. Then it exhibited a dilative behavior from this transformation point [143]. The test results obtained by Castro [59], showing liquefaction behavior (a), dilation behavior (c), and limited liquefaction (b) are illustrated in Fig. 6.2. Poulos [267, 268] termed the CVR line modified by Castro as steady state line (SSL). He defined the steady state of deformation as the state in which the soil flows continuously at constant void ratio, constant effective minor principal stress, and constant shear stress. The location of soil state with respect to SSL decides the susceptibility of soil to liquefaction. Soil whose state can be plotted above SSL is not susceptible to flow liquefaction and soil whose state is plotted below SSL is susceptible to flow liquefaction only if the static shear stress exceeds its steady state strength.

Seed and Lee [311] discovered that when a triaxial specimen of saturated sand is subjected to cyclic loading in an undrained test in such a manner that the principal stresses pass in each cycle through a hydrostatic state of stress (which means that

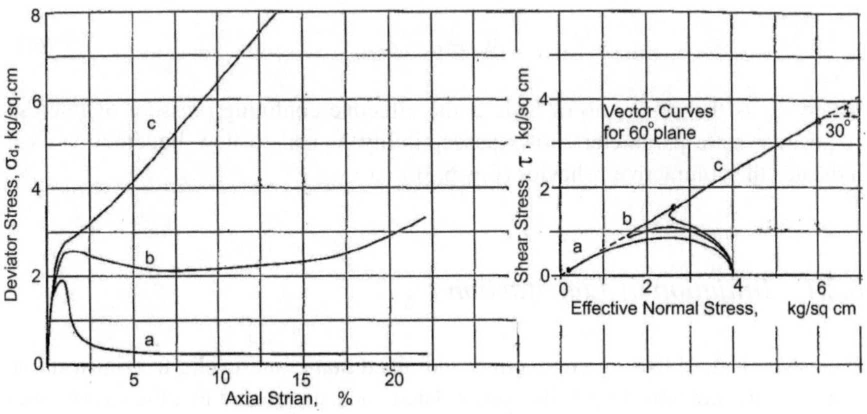

Fig. 6.2 Typical stress-strain curves from stress-controlled undrained static triaxial test (after [59])

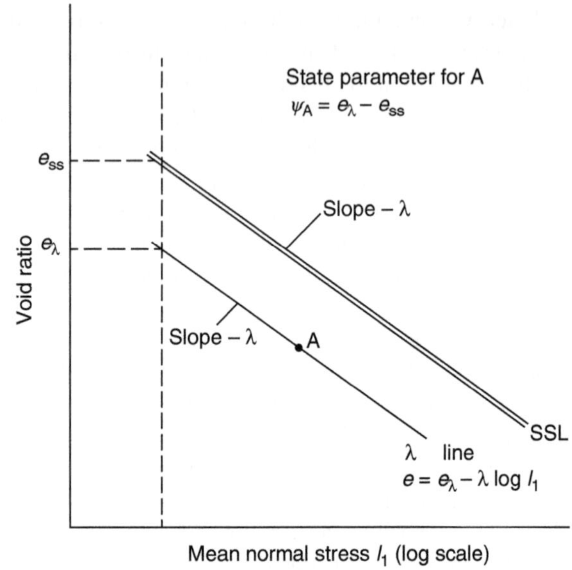

Fig. 6.3 Definition of state parameter Ψ (after [22])

all shear stresses disappear), even a dense and highly dilative sand will develop cyclically high pore pressures and deformations. The stage at which pore pressure equals the confining pressure for the first time was called the initial liquefaction. Casagrande [58] called this type of liquefaction failure under cyclic loading as cyclic liquefaction or cyclic mobility.

From the concepts of critical-state soil mechanics, [22] coined the term state parameter that combines the influence of void ratio and stress level with reference to the ultimate (steady) state to describe sand behavior. The behavior of a cohesionless soil is more closely related to the proximity of its initial state to the critical-state line (steady state line). They defined the state parameter as per Eq. (6.1)

$$\Psi = e - e_{ss} \tag{6.1}$$

where e_{ss} is the void ratio of SSL at the effective confining pressure of interest. A positive state parameter shows susceptibility of soil to flow liquefaction as it indicates the contractive behavior (Fig. 6.3).

6.2.1 *Initiation of Liquefaction*

Hanzawa et al. [118], by conducting undrained static and cyclic triaxial tests on isotropically and anisotropically consolidated soils, concluded that the stress states corresponding to the peak strengths (stress states at which the samples suddenly collapse) fall on a straight line in stress path space passing through the origin, which

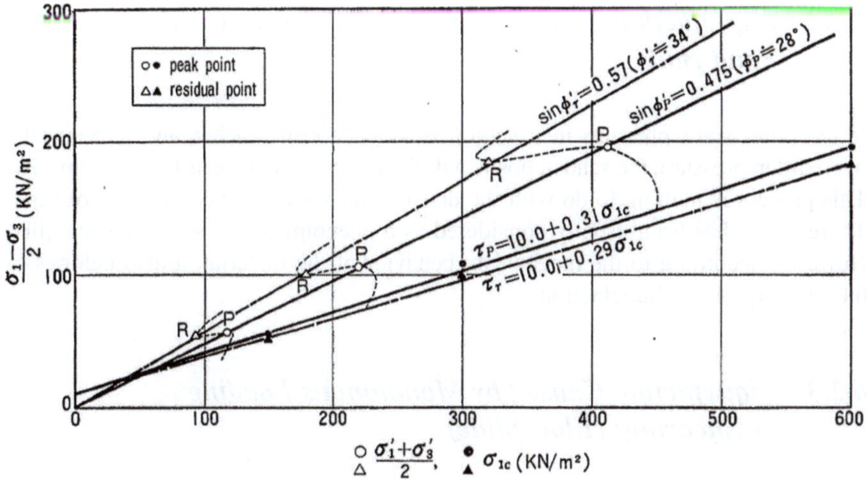

Fig. 6.4 FLS and SSL obtained from undrained static triaxial test conducted on anisotropically consolidated soil (after [118])

is termed as flow liquefaction surface (FLS) and is illustrated in Fig. 6.4. In Fig. 6.4, the peak points show the flow liquefaction line and residual points represents the steady state of deformation. The same results were produced by Vaid and Chern [370, 371]. Once the soil collapses it develops a flow structure and fall to steady state of deformation possessing its residual strength. The closer the initial stress state of soil to the FLS, the susceptible is the soil to flow liquefaction. An analogous conceptual surface was proposed by [337] and was named as collapse surface which passes through the steady-state point. Yang [392] experimentally proved that the flow liquefaction line is not a material constant but rather varies with the state of soil. In the framework of critical-state soil mechanics, he suggested an explicit relationship between the slope of FSL and the state parameter that accounts for both stress level and the density of soil.

Sitharam and Dinesh [330] and Dinesh et al. [77] numerically generated qualitatively realistic macroscopic cyclic behavior of granular materials using discrete element model (DEM) introduced by [74]. The DEM analysis exhibited an increase in resistance to liquefaction with increasing confining pressure since the separation between the individual particles is prevented. It was also observed that loose samples liquefy during cyclic loading under undrained conditions at much smaller cyclic deviator stress level than the deviator stress they carry under monotonic shearing. El Shamy and Abdelhamid [84] conducted similar numerical studies by idealizing solid phase using discrete element method and fluid phase using lattice Boltzmann method. They suggested that liquefaction is due to the reduction in void space during shaking that leads to build up in pore-fluid pressure.

6.2.2 Liquefaction Caused by Seepage Pressure Only: Sand Boils

If the pore water pressure in a saturated sand deposit reaches and exceeds the overburden pressure, the sand deposits will float or boil and lose its bearing capacity. This process is nothing to do with the density and volumetric contraction of sand. Therefore, it has been usually considered as a phenomenon of seepage instability. However, according to the mechanism behavior of the material, it also belongs to the category of soil liquefaction.

6.2.3 Liquefaction Caused by Monotonous Loading or Shearing: (Flow Slide)

The concept of critical void ratio has been suggested by Casagrande. The skeleton of loose saturated sand exhibits irreversible contraction in bulk volume under the action of monotonous loading or shearing, which will cause increase of pore water pressure and decrease of effective stress and finally brings about an unlimited flow deformation.

6.2.4 Liquefaction Caused by Cyclic Loading or Shearing: Cyclic Mobility

With various experimental techniques and testing apparatus it has been found out that cohesionless soil always show volumetric contraction at low shear strain level, but may dilate at higher shear strain level depending upon the relative density of soil. Therefore, under the action of cyclic shearing a saturated cohesionless soil could show liquefaction at time intervals when shear strain is low, but may regain shear resistance in time intervals when the shear strain level is higher. A sequence of such sort of intermittent liquefaction would bring about the phenomenon of cyclic mobility with limited flow deformation. If the saturated cohesionless soil was loose enough to keep contraction at high shear strain level, then it also could come out to be an unlimited flow deformation.

6.3 Liquefaction Susceptibility

There must have been certain conditions in order to soil to liquefy during an earthquake. Hence it is the task of geotechnical engineers to determine whether the given soil strata is susceptible to liquefaction during any earthquake. The liquefaction susceptibility of any soil strata is determined by considering the in-situ soil properties only. Following are the important factors that govern the liquefaction susceptibility of any soil.

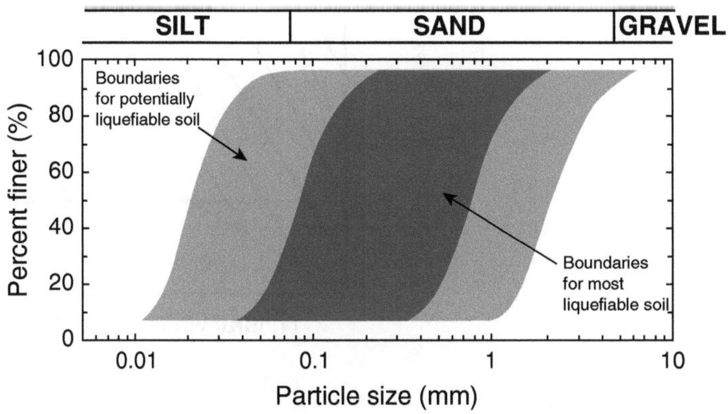

Fig. 6.5 Grain size boundary for defining most liquefiable and potentially liquefiable soils (after [363])

6.3.1 *Type of Soil (Index Properties of Soil)*

The phenomenon of liquefaction is most commonly observed in loose, cohesionless soils. Earlier it was believed that liquefaction was limited to sands only. As the grain size becomes coarser the liquefaction resistance increases due to increase in the weight of soil particles and improved drainage. However, now it has been found out that liquefaction takes place in gravelly soils as well [398]. Consequently, clarifying the gradation curve of liquefiable soil is an important approach to liquefaction susceptibility of a ground. Tsuchida [363] already showed the ranges of grain size distribution of liquefiable soil in 1970. These ranges are used in the Technical standards for Port and Harbour Facilities published by the Japan Port and Harbour Association (Fig. 6.5). Further, it was considered that fine grained soils were incapable of generating the high pore pressure. Lab experiments have shown that majority of clays remain unsusceptible to liquefaction. Sensitive clays can exhibit strain softening behavior which makes it vulnerable to liquefaction. Ishihara [138] suggested that liquefaction susceptibility of fine grained soils like silts and clays is mainly influenced by liquid limit and water content which was presented in Fig. 6.6 as per [209]. If the soil samples do not fall in the safe zone (as in Fig. 6.6) based on its natural moisture content and liquid limit, the liquefaction susceptibility of such soil sample must be verified using undrained cyclic loading tests.

6.3.2 *Shape of Soil Particles*

Rounded particles are known to density more easily than angular shaped particles. Hence, the ease of densification is directly proportional to the susceptibility of soil to liquefaction. This implies that rounded particles are more prone to liquefaction than angular shaped particles, during an earthquake.

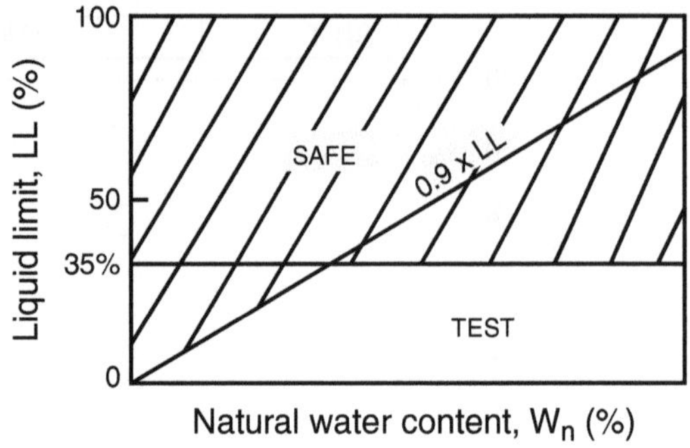

Fig. 6.6 Graphical representation of liquefaction criteria for silts and clays from studies by [380] in China (after [209])

6.3.3 *Permeability of Soil*

Permeability of soil mass leads to dissipation of pore water pressure very quickly. Hence, if this dissipation of pore water pressure occurs very quickly, then the soil may not liquefy. Based on the observations, it can be concluded that presence of highly permeable soil layers can reduce liquefaction potential of adjacent soil layers. In other words, the liquefaction susceptibility of a loose sand layer will be reduced if it is surrounded by a highly permeable gravel layer.

6.3.4 *Presence of Seismic Waves*

An earthquake is characteristically a kind of vibration energy that includes impacts or shocks. The vibration it produces varies according to the characteristics of the ground through which the vibrations transmit. Earthquake on the surface of the ground is strongly dependent on the vibration characteristics of the layers underneath.

It is known that the most common cause for liquefaction is the presence of vibration/seismic energy released during an earthquake. Based on the observations, it has been concluded that the potential for liquefaction increases with the increase in seismic energy. The energy thus produced is the cause for the cyclic shear stress that acts on the soil mass and this causes the acceleration of the soil mass in the horizontal direction. The acceleration is known as the peak acceleration exerted by earthquake at the ground surface. It has been stated that the shaking threshold required for liquefaction is a local shaking magnitude of about five and a peak acceleration of 0.1 g.

6.3.5 *Depth of Ground water Table*

Another criterion which makes soil susceptible to liquefaction is its degree of saturation. This clearly indicates the importance of depth of water table and it must be near the surface of the ground. The soil located above the water table is unsaturated and hence it will not liquefy. For those locations where the depth of water table is very deep, the liquefaction susceptibility will be less. Moreover if the level of the groundwater table keeps on changing the liquefaction susceptibility of the soil will also fluctuate.

6.3.6 Historical Environment

Observations from earlier earthquakes in the region can offer a great amount of information regarding the liquefaction susceptibility of the soil mass at the site. If the soil has already liquefied in the past due to an earthquake, then it can liquefy again in the future earthquakes. The information on the past liquefaction can be obtained in the form of maps, which provides information on the previous earthquakes (and the earthquakes expected to occur in the future at the site) which caused the liquefaction of soil. It has also been observed that liquefaction effects are historically confined to a zone that lies within a particular distance from the tremor source. It is also obvious that the distance to which the effect of liquefaction can be expected is directly related to the magnitude of the earthquake.

6.3.7 Age of Soil

The age of the soil plays a pivotal role in determining the liquefaction susceptibility. The liquefaction resistance of a soil increases, with the age of the soil. New soils liquefy more easily than old soil deposits, because the old soil deposits have more stable particle arrangement due to the compression of the soil particles with time. Moreover the particle bonding in the case of old deposits will be more. Thus, old soil deposits have a greater liquefaction resistance comparatively.

6.3.8 Confining Pressure

Soils at depths of more than 15 m generally do not liquefy due to the high confining pressure on it. The confining pressure is the overburden pressure acting at a point in the soil mass due to the surcharge by the soil layers above that point. More the confining pressure, lesser is the probability of the soil to liquefy. As already mentioned, if the groundwater table depth is more than 15 m from the ground surface, the soil lying below the water table generally does not liquefy. But this doesn't stop us from determining the liquefaction susceptibility of the soil layer that is at a depth of more than 15 m from the ground surface. In case of a sloping dam, the soil deposit has to be analyzed for its liquefaction susceptibility for a thickness which lies in water, even if this thickness lies at a depth of more than 15 m.

6.3.9 Relative Density

The soil layer which has a low relative density is more susceptible to liquefaction than dense soil. Loose soils density easily during earthquake and this will cause an increase in pore water pressure which leads to the liquefaction of soil, where as dense sands will dilate during earthquake and this will reduce the pore water pressure and hence the liquefaction susceptibility will be less.

6.3.10 Natural Soil Deposits in Water Bodies

Soils in lakes, rivers, or oceans are highly prone to liquefaction due to their loose and segregated structure. Hence soils which are susceptible to liquefaction are formed in marine depositional environments.

6.4 Liquefaction Susceptibility Analysis Based on SPT and CPT

Major in-situ tests such as the standard penetration test (SPT), cone penetration test (CPT) are most useful to determine the liquefaction susceptibility of the soil. Based on the studies of [285, 288], a soil classification based on CPT for assessing liquefaction susceptibility of fine grained soil as in Fig. 6.7. The quantities normalized CPT penetration resistance Q (dimensionless) versus normalized friction ratio F was generated first, where Q and F are defined as per Eqs. (6.2) and (6.3).

$$Q = \frac{q_c - \sigma_{vo}}{\sigma'_{vo}} \tag{6.2}$$

$$Q = \frac{f_s}{q_c - \sigma_{vo}} \tag{6.3}$$

where q_c and f_s are the tip resistance and sleeve friction from the cone penetration test. The zone A in Fig. 6.7 represents the region where cyclic liquefaction is possible depending on size and duration of cyclic loading. Soil samples whose tip resistance and sleeve friction combination lies in zone B is unlikely to liquefy, check other criteria; zone C, flow liquefaction and (or) cyclic liquefaction possible, depending on soil plasticity and sensitivity as well as size and duration of cyclic loading

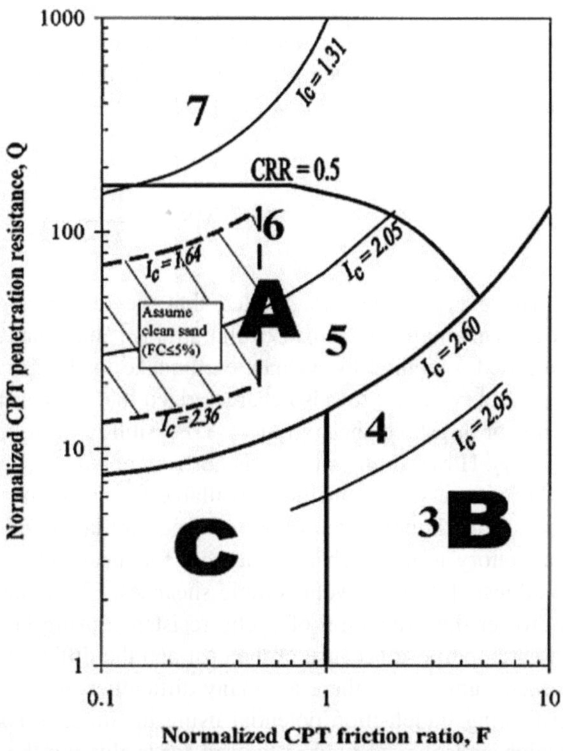

Fig. 6.7 Soil classification chart by Robertson and Wride [288]

6.5 Evaluation of Liquefaction Potential

The liquefaction potential of soil significantly depends on the initial state of the soil as well as on the intensity of earthquake loading. The earthquake loading is the driving force causing liquefaction, while the soil resistance against liquefaction mainly depends upon the initial state of the soil. Hence the evaluation procedure for liquefaction potential requires two steps,

- Characterizing earthquake loading
- Characterizing soil resistance

Seed and Idriss [309] has proposed a simplified approach for characterizing the earthquake loading which is most popular. In this approach, the earthquake loading is characterized in terms of cyclic shear stress (or cyclic stress ratio) as presented in Eq. (6.4).

$$\mathrm{CSR} = 0.65\left(\frac{a_{\max}}{g}\right)\left(\frac{\sigma_{v0}}{\sigma_{v0}^{\prime}}\right)\left(\frac{r_d}{\mathrm{MSF}}\right) \tag{6.4}$$

where a_{max} is the surface level peak ground acceleration (PHA) at the given borehole location obtained using deterministic hazard analysis; σ_{v0} and σ'_{v0} are the total and effective overburden pressure; r_d—depth reduction factor used to account for the flexibility of the soil as per [193] and MSF is the magnitude scaling factor (Eq. (6.5)),

$$\text{MSF} = \frac{10^{2.44}}{M_W^{2.56}} \tag{6.5}$$

However [78] have suggested that the tendency of pore pressure generation is fundamentally related to cyclic strains rather than cyclic stresses. Hence they proposed a method to evaluate earthquake loading in terms of cyclic shear strain (γ_c). Dobry et al. [78] also characterized the soil resistance against liquefaction in terms of threshold shear strain (γ_t). Possibility of liquefaction will be maximum if $\gamma_c > \gamma_t$. However this method is not very popular.

There are several methods available for evaluating the cyclic resistance of the soil against liquefaction. Important ones are laboratory tests and in-situ field tests. Laboratory tests which are available for measuring soil resistance are cyclic triaxial tests [59, 93], cyclic simple shear test [272], and cyclic torsional test [142]. However the evaluation of cyclic resistance using laboratory tests is restricted to research purpose because of the cost and the difficulties involved in getting undisturbed samples. As there are many difficulties and the associated laboratory tests, evaluating liquefaction potential using in-situ tests has become more convenient. Major geotechnical field tests used for evaluating the cyclic resistance of soil are (1) Standard Penetration Test [131, 313, 315], (2) Cone penetration tests [286, 306], and (3) Based on shear wave velocity [346, 355].

6.5.1 *Liquefaction Potential Based on Lab Tests*

Cyclic Triaxial Test

The cyclic triaxial test is the most commonly used test for the measurement of dynamic soil properties at high strain levels. This test simulates the liquefaction phenomenon during earthquakes by applying cyclic shear to the saturated sandy soil under undrained condition. Axial loading is applied in steps to the specimen and the shear strain and shear stress at various intervals of loading are measured. Five to ten levels of shear strain amplitudes are chosen from the range of 10^{-5} to 10^{-2} for testing. Dynamic deformation characteristics are influenced by the effective confining pressure during the test. When an undisturbed sample of normally consolidated soil is obtained, the effective vertical pressure at the depth of sampling is isotropically applied by cell pressure to avoid the influence of over consolidation. In order to obtain in-situ shear modulus of the soil from the laboratory test results, correction of these results is necessary so that a shear modulus corresponding to the average effective principal stress at the sample depth is obtained. Seed and Lee [311] were the first to reproduce liquefaction in a cyclic triaxial test on loose and dense sands and concluded that liquefaction occurs more easily in sandy soils having higher void ratios. In addition to this they have proved that if the effective confining pressure is less than the liquefaction susceptibility will be high.

Cyclic Simple Shear Test

The cyclic direct simple shear test is capable of reproducing earthquake stress conditions much more accurately than the cyclic triaxial test. It is most commonly used for liquefaction testing. In this test, a short cylindrical specimen is restrained against lateral expansion by rigid boundary platens, a wire reinforced membrane or with a series of stacked rings. By applying cyclic horizontal shear stresses to the top or bottom of the specimen, the test sample is deformed in the same way as an element of soil subjected to vertically propagating S waves. In recent years, simple shear devices that allow independent control of vertical and horizontal stresses have been developed. To simulate the actual earthquake conditions, [272] used a large-scale simple shear apparatus. It was found that cyclic strength is related to the relative density of the soil and cyclic stresses that cause liquefaction in simple shear were less than those causing liquefaction in triaxial shear.

Cyclic Torsional Shear Test

Many of the difficulties with cyclic triaxial and cyclic shear test can be overcome with cyclic torsional shear test. This is mostly used to determine stiffness and damping characteristics over a wide range of strain levels. It allows isotropic or an isotropic initial stress conditions and can impose cyclic shear stresses on horizontal planes with continuous rotation of principal axes. Dobry et al. [78] used strain controlled cyclic torsional loading along with stress controlled axial loading of solid specimens and has proven effective for measurement of liquefaction behavior. Torsional testing of solid specimens, however, produces shear stresses that range from zero along the axis of the specimen to a maximum value at the outer edge. To increase the radial uniformity of shear strains, a hollow cylindrical cyclic torsional shear apparatus was also developed. Hollow cylinder tests offer perhaps the best uniformity and control over stresses and drainage. Ishihara and Li [142] developed a torsional triaxial shear test and conducted strain controlled tests on solid cylinders of saturated sands. These tests helped in establishing relationship between cyclic triaxial tests, cyclic simple shear tests, and the torsional triaxial test.

Shake Table Test

Shake table tests of many sizes are being used for liquefaction studies on saturated soil samples. The sample is prepared in a container, fixed to a shaking platform and vibrated at the desired frequency for a prescribed time. A surcharge is placed on the sample to provide the confining pressure. The measurements of acceleration, pore water pressure, and settlements are made during the test.

6.5.2 *Liquefaction Potential Based on Field Tests*

The difficulties and the high cost involved in the laboratory test make the in-situ-tests convenient method for modelling liquefaction potential. The four major in-situ test methods which are considered for the liquefaction potential evaluation are given below.

1. The Standard Penetration Test (SPT)
2. The cone penetration test (CPT)
3. Measurement of in-situ shear wave velocity (Vs)
4. The Becker penetration test (BPT).

SPT Based Methods

Standard penetration test is widely used as an economical, quick, and convenient method for investigating the penetration resistance of non-cohesive soils. The use of SPT as a tool for evaluation of liquefaction potential began to evolve in the wake of a pair of devastating earthquakes that occurred in 1964; the 1964 Great Alaskan Earthquake ($M = 9.2$) and 1964 Niigata Earthquake ($M = 7.5$), both of which produced significant liquefaction related damage. The SPT values obtained from the site investigation must be corrected to remove the errors. The N-values measured in the field were subjected to various corrections such as (1) Overburden pressure correction (C_N), (2) Hammer energy correction (C_E), (3) Borehole diameter correction (C_B), (4) liner correction (C_S), (5) Rod length correction (C_R) and (6) correction for fine contents (C_{FC}) [61, 313, 336, 400]. The SPT value corrected for 60% energy efficiency and 100 kPa overburden pressure $(N_1)_{60}$ was evaluated using Eq. (6.6) [287]. The $(N_1)_{60cs}$ provides consistent value for penetration resistance and hence become a standard for assessing liquefaction potential [291].

$$(N_1)_{60} = N \times [C_N \times C_E \times C_B \times C_S \times C_R] \tag{6.6}$$

The obtained $(N_1)_{60}$ was again corrected for fine content as suggested by Idriss and Boulanger [131] using Eqs. (6.7) and (6.8).

$$(N_1)_{60cs} = (N_1)_{60} + \Delta(N_1)_{60} \tag{6.7}$$

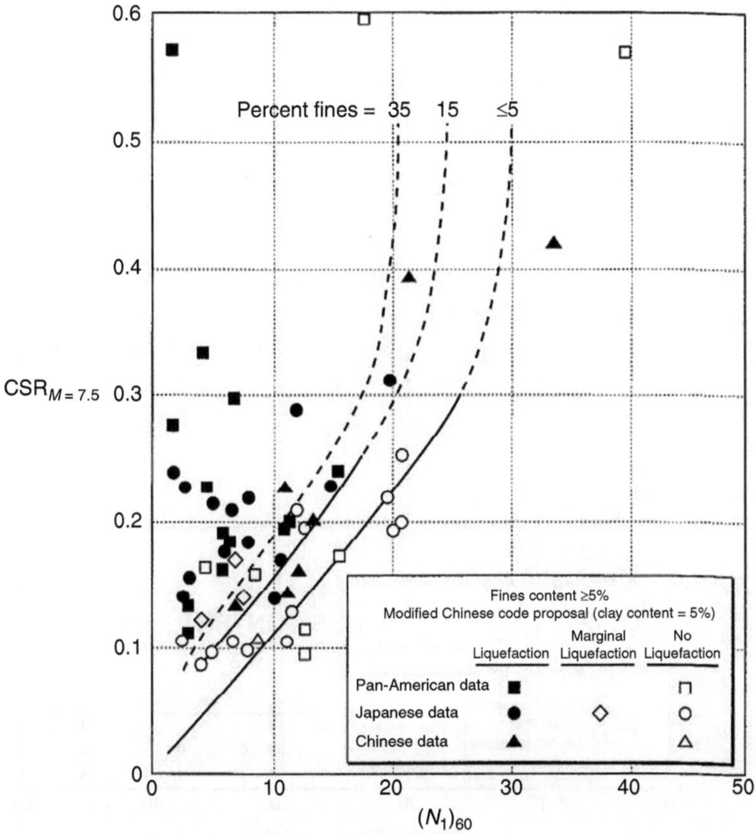

Fig. 6.8 Relationship between cyclic stress ratios causing liquefaction and $(N_1)_{60}$ values for clean sands for $M_7.5$ earthquakes (Seed et al. [305])

where

$$\Delta(N_1)_{60} = \exp\left[1.63 + \left(\frac{9.7}{FC+0.1}\right) - \left(\frac{15.7}{FC+0.1}\right)^2\right] \tag{6.8}$$

Here FC is the fine content, i.e. percentage of dry weight finer than 75 μm.

The most widely used methods to evaluate the liquefaction potential based on SPT values were proposed by Seed et al. [315]. This relationship is between corrected SPT-N values, corrected for overburden stress, instrument errors and other factors that affect the SPT testing, vs. intensity of cyclic loading that is expressed in the form of uniform cyclic stress ratio (CSR) (see Fig. 6.8). This was further modified by Youd et al. [400] and the curves between cyclic resistance ratio (CRR) and the corrected SPT values are given in Fig. 6.9. In this work it was noted that the CRR value will increase with fines content. Although this method is widely used, it is outdated and doesn't accommodate conditions by earthquakes that produce very

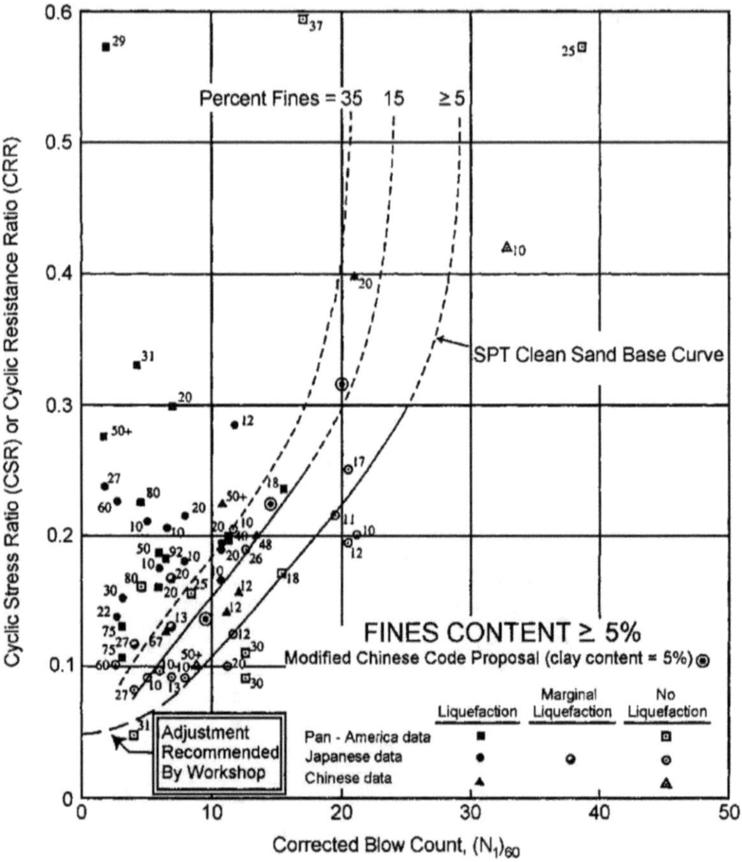

Fig. 6.9 Deterministic cyclic resistance curves proposed by Youd et al. [400]

high peak ground shaking levels (CSR > 2.5). The relationship does not have a formal probabilistic approach as well; i.e. the relationship does not provide the probability or uncertainty of occurrence of liquefaction.

A number of researchers have developed correlations based on probabilistic methods to evaluate the liquefaction potential. One of the first to develop a relation was [194] (Fig. 6.10) and more recently [399] (Fig. 6.11), and [358] (Fig. 6.12). The relationships provided by these researchers are in the form of probability contours (probability of triggering soil liquefaction). In the following figures these curves are superimposed over the [314] deterministic relationship for comparison.

The [399] correlation is based on the seismic data that were used by Seed et al. [314], which at that time were of very low quality. This relationship was developed by using the binary regression of logistic models method, for probabilistic regression. The likelihood function prepared, overstated the overall variance or uncertainty of the proposed correlation. An additional inadequacy was that [194]

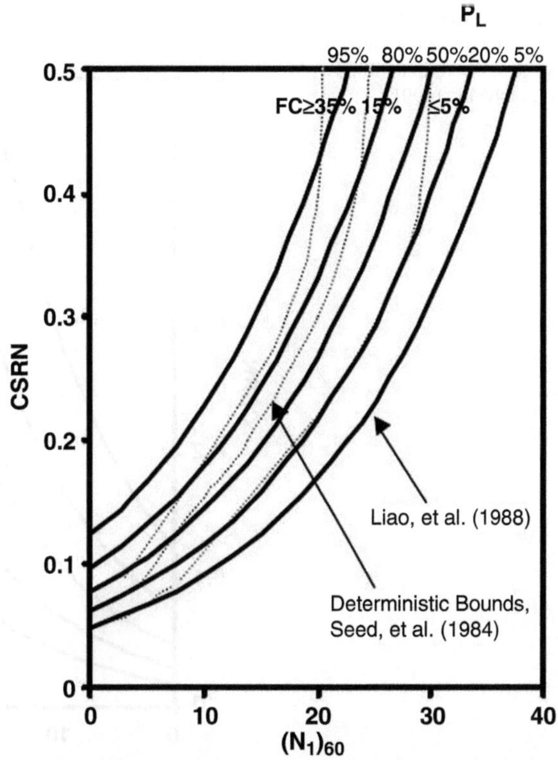

Fig. 6.10 Correlation between equivalent uniform cyclic stress ratio and SPT $(N_1)_{60}$ value [194]

couldn't find the effect of fines content in the soil on the regression relationship between SPT-N values and liquefaction resistance. So the relation developed was for sandy soils with less than 12% fines.

Youd et al. [399] correlation takes into account a number of field case history data points from the past earthquakes which have occurred since the earlier relationships have been developed. The correlation also excludes the limitations of the [194] relationship as discussed above. This correlation is applicable to soils of varied fine contents and hence can be used for silty soils as well.

The correlation as proposed [358] also makes use of an extended field case history database, and excludes the most of the questionable data used by Liao et al. [194]. The regression method used here is also that of binary regression, and as a result the overall uncertainty is again very large. However in spite of all these shortcomings, all the above relationships discussed here are being widely used.

Idriss and Boulanger [131] have revised SPT and CPT based semi-empirical relations for evaluating the liquefaction potential of saturated cohesionless soils. Further, they have also modified stress reduction factor (r_d), earthquake magnitude scaling factor for cyclic stress ratios (MSF) and overburden correction factor for cyclic stress ratios (K_σ), and the overburden normalization factor for penetration resistances (C_N).

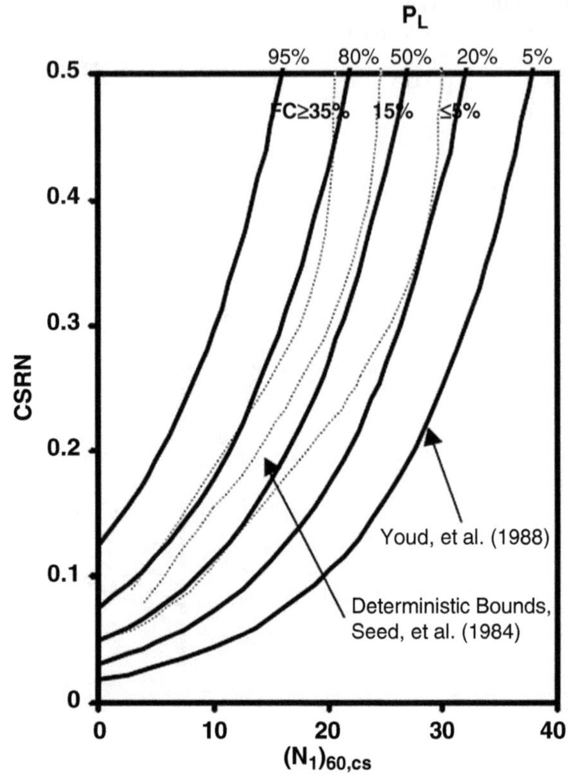

Fig. 6.11 Correlation between equivalent uniform cyclic stress ratio and SPT N1, 60-Value [399]

The cyclic resistance ratio (CRR) for each layer was evaluated using a methodology proposed by Idriss and Boulanger [131] given by Eq. (6.9)

$$\mathrm{CRR} = \exp\left[\left(\frac{(N_1)_{60cs}}{14.1}\right) + \left(\frac{(N_1)_{60cs}}{126}\right)^2 - \left(\frac{(N_1)_{60cs}}{23.6}\right)^3 + \left(\frac{(N_1)_{60cs}}{25.4}\right)^4 - 2.8\right] \tag{6.9}$$

where$(N_1)_{60cs}$ is the SPT-N value corrected for 60% energy efficiency, 100 kPa overburden pressure and fine content.

CPT Based Methods

In the previous section the discussion was centered around the correlations to evaluate the liquefaction potential based on SPT values. Due to the better accuracy and repeatability, several correlations are available for estimating the CRR values based on CPT values. The first to propose liquefaction triggering model based on CPT values was [286, 306]. One of the most widely used correlations nowadays is the one proposed by Robertson and Wride [288] (Fig. 6.13).

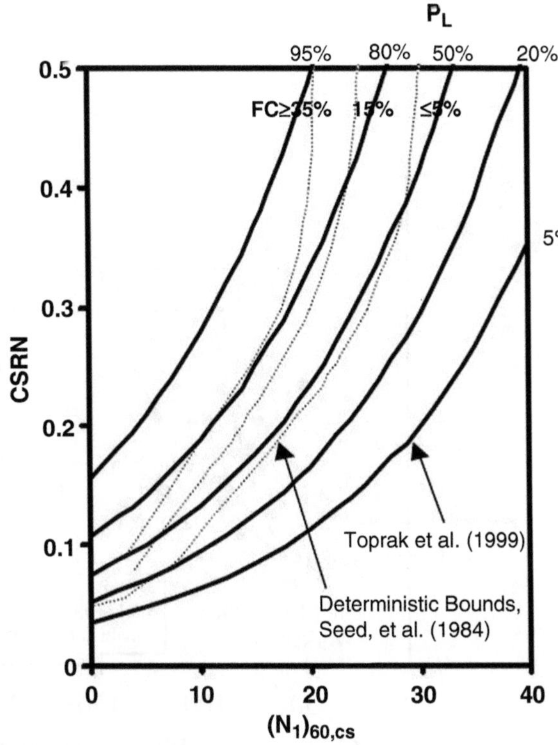

Fig. 6.12 Correlation between equivalent uniform cyclic stress ratio and SPT N1, 60 value [358]

6.5.3 Probabilistic Methodology

One of the major shortcomings in the deterministic method is that it considers only one critical magnitude-acceleration scenario for the evaluation of liquefaction potential. As it is evident that the liquefaction hazard at a site is not solely function of single ground acceleration and earthquake magnitude (a_{max} and M_W), but several magnitudes acceleration scenarios contribute towards the liquefaction hazard and their percentage of contribution varies. Hence, a performance based approach suggested by Kramer and Mayfield [182] was also adopted for the evaluation of liquefaction hazard at micro-level, where the contributions from all magnitudes and all acceleration levels were considered. Based on the available SPT data, the annual probability of exceedance for a given factor of safety value at a particular depth in a borehole (Δ_{FS^*}) is estimated using Eq. (6.10) [182].

$$\Delta_{FS^*} = \sum_{i=1}^{N_a}\sum_{j=1}^{N_m} P[FS > FS^*|a_i, m_j]\Delta\lambda_{a_i,m_j} \tag{6.10}$$

where N_a and N_m are the number of acceleration and magnitude increment; a_i and m_j are the incremental values in acceleration level and earthquake magnitude,

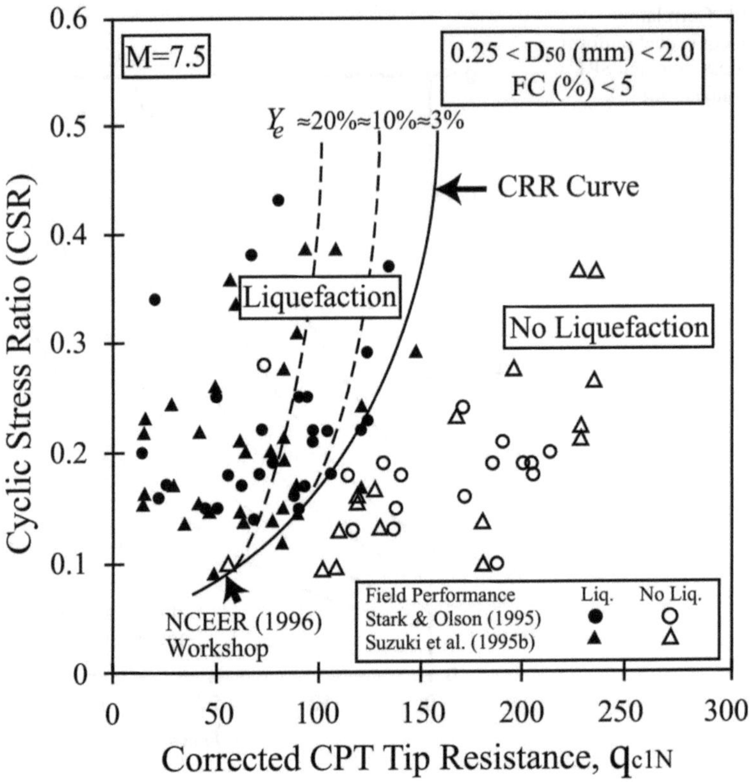

Fig. 6.13 Calculation of CRR from CPT Data along with Empirical Liquefaction Data (after 282)

$\Delta\lambda_{a_i,m_j}$ is the incremental annual frequency of exceedance for a_i and m_j obtained by deaggregating the seismic hazard curve with respect to magnitude and $P[\mathrm{FS} > \mathrm{FS}^*]$ is the conditional probability that obtained factor of safety (FS) will exceed a target value (FS*) as per Eq. (6.11) proposed by Cetin et al. [61].

$$P[\mathrm{FS} > \mathrm{FS}^*|a_i, m_j] = \Phi\left(-\frac{1}{\sigma_\varepsilon}\left[\begin{array}{c}(N_1)_{60}\,(1+\theta_1 FC) - \theta_2 \ln(\mathrm{CSR}_{eq,i}\mathrm{FS}^*) \\ -\theta_3 \ln(m_j) - \theta_4 \ln(\sigma'_{v0}/P_a) + \theta_5 FC + \theta_6\end{array}\right]\right) \quad (6.11)$$

Here FS is the factor of safety and FS* is the target value of factor of safety. $\mathrm{CSR}_{eq,i}$ is the cyclic stress ratio for an acceleration a_i corresponding to site class D obtained using probabilistic hazard analysis and magnitude m_i. $\mathrm{CSR}_{eq,i}$ will be calculated for all the acceleration and magnitude level. θ_1 to θ_6—are the regression coefficient. $(N_1)_{60}$ is the corrected N value (for energy, and overburden pressure). P_a and σ'_{v0} are atmospheric pressure and effective overburden pressure (in the same unit), FC is the fine content in percentage and Φ is the cumulative normal distribution function. From the MARE curve, FS values corresponding to 475 year and 2500 year return period (having 10% and 2% probability of exceedance in 50 years) were evaluated. This FS values were integrated over depth to obtain the LPI for that particular bore hole.

Liquefaction potential can also be characterized in terms of SPT resistance required to prevent liquefaction at a particular depth, N_{req} and $N*_{req}$ are the targeted SPT N values. This method is best suited for assessing liquefaction from meso- to macro-level. It is to be noted that both N_{req} and $N*_{req}$ are the SPT N values corrected for energy, overburden pressure, rod length, sampler, borehole diameter, and percentage of fines. $\lambda_{N*_{req}}$ is the mean annual rate of exceedance with which N_{req} exceeds $N*_{req}$ and is given in Eq. (6.12) [374].

$$\lambda_{N*_{req}} = \sum_{i=1}^{N_a} \sum_{j=1}^{N_m} P[N_{req} > N*_{req} | a_i, m_j] \Delta\lambda_{a_i m_j} \tag{6.12}$$

Here a_i and m_j are the incremental values in earthquake magnitude and acceleration level and the probability of N_{req} exceeding a predefined value of $N*_{req}$ $P\left[N_{req} > N*_{req}\right]$ is given by Eq. (6.13). $\Delta\lambda_{a_i m_j}$ is the incremental annual frequency of exceedance for acceleration a_i and magnitude m_j and $P[N_{req} > N*_{req}|a_i, m_j]$ is given by Eq. (6.14) as per [374]. A mean annual rate of exceedance curve for N_{req} is developed from which SPT N value required to resist liquefaction corresponding to desired return period can be estimated.

$$P[N_{req} > N*_{req}|a_i, m_j] = \Phi\left[-\frac{N*_{req} - \theta_2 \ln(\mathrm{CSR}_{eq,i}) - \theta_3 \ln(m_j) - \theta_4 \ln\left({\sigma'_{vo}}/{P_a}\right) + \theta_6}{\sigma_s}\right] \tag{6.13}$$

Similar to SPT-based liquefaction assessment, CPT-based analysis can also be carried out for any region. The main advantage of using CPT data is that the percentage of errors is considerably less than that of SPT method. The mean annual rate of exceedance $\left(\lambda_{q*_{c1-req}}\right)$ of CPT value (q_c) required to resist liquefaction is given as per Eq. (6.14).

$$\lambda_{q*_{c1-req}} = \sum_{i=1}^{N_a} \sum_{j=1}^{N_m} P\left[q_c > q*_{c1-req} |\, a_i, m_j\right] \Delta\lambda_{a_i m_j} \tag{6.14}$$

Here $q*_{c1-req}$ is the targeted value; N_m—number of magnitude increments; N_a—number of peak ground acceleration; $\Delta\lambda_{a_i m_j}$ is the incremental annual frequency of exceedance for acceleration a_i and magnitude m_j. Now $P\left[q_{c1}>q*_{c1-req}|a_i, m_j\right]$ is given by Eq. (6.15) as per [225, 375]. From the mean annual exceedance curve, CPT values required to prevent liquefaction for any return period can be estimated.

$$P\left[q_{c1} > q *_{c1-req} |a_i, m_j\right] = \Phi\left[-\frac{\begin{array}{l}(q*_{c1-req})^{1.045} + q *_{c1-req}(0.110R_f) + 0.001R_f \\ +c\left(1+0.850R_f\right) - 7.177\ln(\mathrm{CSR}_{eq,i}) \\ -0.848\ln m_j - 0.002\ln\left(\sigma'_{v0}\right) - 20.923\end{array}}{1.632}\right] \tag{6.15}$$

6.6 Liquefaction Potential Index

The methods described in the previous sections measure the liquefaction potential at specified depths (the depth selected in the present study was 3 and 6 m). However by using the Liquefaction Potential Index (LPI, proposed by Iwasaki et al. [145]) this shortcoming can be over come. In this method it is assumed that the severity of liquefaction should be proportional to the thickness of the liquefied layer, proximity of the liquefied layer to the surface, and the factor of safety of the liquefied layer. The LPI, which was proposed by Iwasaki et al. [145], is given by:

$$\mathrm{LPI} = \int_0^{20} F_z W(z) dz \tag{6.16}$$

where z is the depth, F_z is a function of factor of safety as per Eq. (6.17) [339], and $W(z)$ depth weightage factor ($W(z) = (10 - 0.5z)$ for $z \leq 20$ and $W(z) = 0$ for $z > 20$).

$$F_z = \begin{cases} 0, & \text{for FS} \geq 1.2 \\ 1 - \mathrm{FS} & \text{for FS} < 0.95 \\ 2 \times 10^6 e^{(-18.427\mathrm{FS})} & \text{for } 1.2 > \mathrm{FS} > 0.95 \end{cases} \tag{6.17}$$

Sonmez [339] proposed a classification of liquefaction potential of a site based on LPI (Table 6.1).

Table 6.1 Liquefaction vulnerability classification based on LPI [339]

LPI range	Liquefaction vulnerability
<2	Low
2–5	Moderate
5–15	High
>15	Very high

6.7 Liquefaction Hazard Assessment at Micro-Level: A Case Study of Kalpakkam NPP Site

In engineering design point of view, nuclear power plant structures are meant to survive all possible natural disasters, especially induced effects of earthquakes. Failure of any nuclear power plant structure can result in a cataclysm and the safety of such structures during natural disasters such as earthquakes is of great concern even to the common people. Subsurface investigations conducted in the Kalpakkam NPP site reveal the presence of loose sand layers at shallow depth along with high water table, making the site vulnerable to liquefaction hazard. The liquefaction hazard for the Kalpakkam NPP site (at micro-level) was evaluated in terms of liquefaction potential index (LPI) introduced by Iwasaki et al. [145], which is obtained by integrating factor of safety against liquefaction at all depths. The LPI was estimated for the Kalpakkam NPP site based on field test data, using both deterministic and probabilistic approaches by James et al. [152].

6.7.1 Deterministic Approach

In the deterministic methodology, PHA at ground surface level was first evaluated using deterministic seismic hazard analysis (DSHA). Using the surface level PHA value, a simplified methodology proposed by Seed and Idriss [309] was used to estimate cyclic stress ratio (CSR) at various depths of each borehole. Cyclic resistance ratio (CRR) of the soil was computed based on the corrected SPT value $[(N_1)_{60,cs}]$ as per the method suggested by Idriss and Boulanger [131]. For each borehole, the factor of safety at various depths were then evaluated and these values were integrated to obtain the LPI for that particular borehole. Figure 6.14 presents the spatial variation in LPI values from the deterministic methodology for the Kalpakkam NPP site. From Fig. 6.14 it is evident that the northern part of the site is having very high liquefaction hazard with LPI value of 15 and above. Central part of the site has low liquefaction hazard with LPI less than 2.

6.7.2 Probabilistic Approach

One of the major shortcomings in the deterministic method is that it considers only one critical magnitude-acceleration scenario for the evaluation of liquefaction potential. In order to overcome this shortcoming, the LPI values were evaluated using a probabilistic performance based methodology proposed by Kramer and Mayfield [182] as well. The probabilistic model captures various uncertainties involved in earthquake process such as the uncertainty in magnitude, location, and the recurrence rate. The mean annual rate of exceedance (MARE) curve, showing

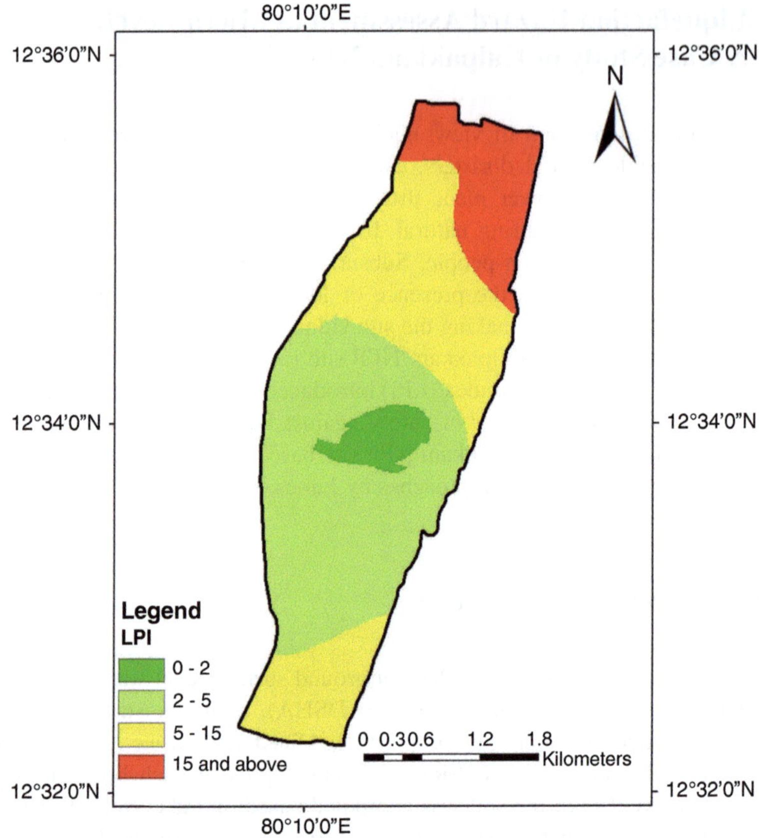

Fig. 6.14 Spatial variation in LPI for the Kalpakkam NPP site obtained from the deterministic method (after [152])

the variation factor of safety (FS) for different return periods was developed [152]. From the MARE curve, FS values corresponding to 475 year and 2500 year return period (having 10% and 2% probability of exceedance in 50 years) were evaluated. A typical MARE curve showing the variation of FS with return period at 3 m depth for a typical borehole is presented in Fig. 6.15. These FS values were integrated over depth to obtain the LPI for that particular bore hole. Figures 6.16 and 6.17 present the spatial variation of LPI value with 475 year and 2500 year return period. Similar to the results obtained from deterministic methodology, probabilistic analysis also shows the presence of highly liquefaction vulnerable region in the northern part of the Kalpakkam NPP site. It is also noted from Figs. 6.16 and 6.17 that the central area of the site has low to moderate hazard value as per the classification provided by Sonmez [339].

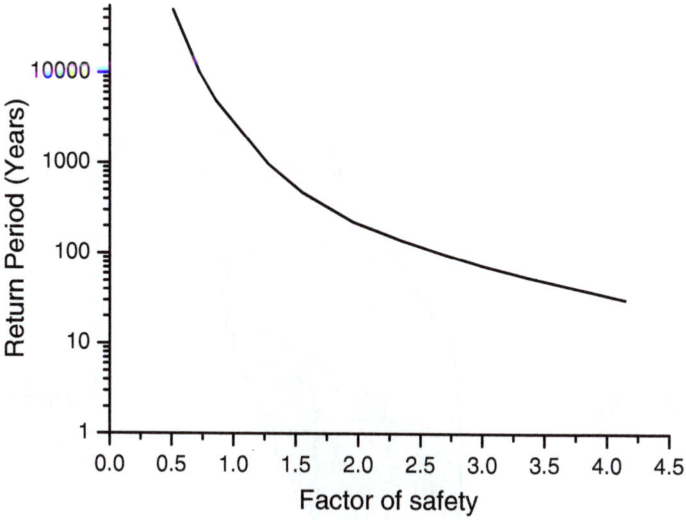

Fig. 6.15 MARE curve showing variation of FS against liquefaction with return period (after [152])

6.8 Liquefaction Hazard Assessment at Macro-Level: A Case Study of Karnataka State

Assessment of liquefaction hazard at macro-level is also essential in order to identify vulnerable regions within the state of Karnataka. However it is very difficult to generate database containing the borelog details for such large region for assessing liquefaction hazard. Hence, the liquefaction hazard must be estimated in terms of parameter which can be easily obtained for any given location using most common geotechnical tests. The liquefaction hazard for the state of Karnataka is estimated in terms of SPT and CPT values required to resist liquefaction by [150]. The standard penetration test (SPT) is the most widely used geotechnical field tests, especially in India for sub-surface exploration. In this study, a probabilistic methodology presented by Kramer and Mayfield [182] is adopted in order to predict the liquefaction hazard for the state of Karnataka in terms of SPT and CPT values at 3 m depth (considering to be the probable depth of a shallow foundation). The state of Karnataka was divided into small grids of 0.05° × 0.05° (approx. 5 km × 5 km) and seismic site characterization of each grid area was carried out using the topographic slope map as mentioned in this chapter. The liquefaction hazard was estimated for the grid points belonging to site class D. Spatial variations in the SPT and CPT values required to resist liquefaction for a definite return period (475 year and 2500 year), throughout the Karnataka state, were evaluated and are presented.

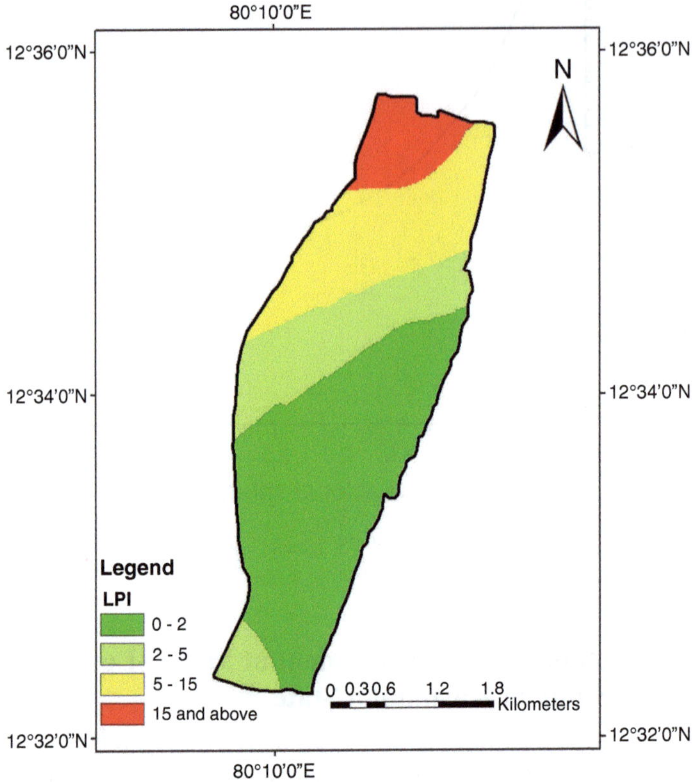

Fig. 6.16 Spatial variation in LPI values for Kalpakkam NPP site corresponding to return period of 475 years (after [152])

6.8.1 Prediction of SPT-N value Required to Resist Liquefaction

In India, engineers still depend on SPT for the site investigation, prior to any construction project and hence the estimation of liquefaction hazard in terms of SPT-N value will be most advantageous.

Figures 6.18 and 6.19 present the spatial variations in the SPT values required to resist liquefaction at 3 m depth, for 475 year and 2500 year return period, throughout the Karnataka state. From Figs. 6.18 and 6.19, it is evident that the southern (regions close to Bangalore–Mysore) and the northern parts (region close to Bidar district) are expected to have high liquefaction hazard. The SPT value required to prevent liquefaction ranges from 10 to 20, for a return period of 475 years and 20 to 30, considering a return period of 2500 year. That is, any given site in this region has less chance of liquefy, if the actual SPT value at 3 m depth is more than the predicted SPT value corresponding to the respective return period.

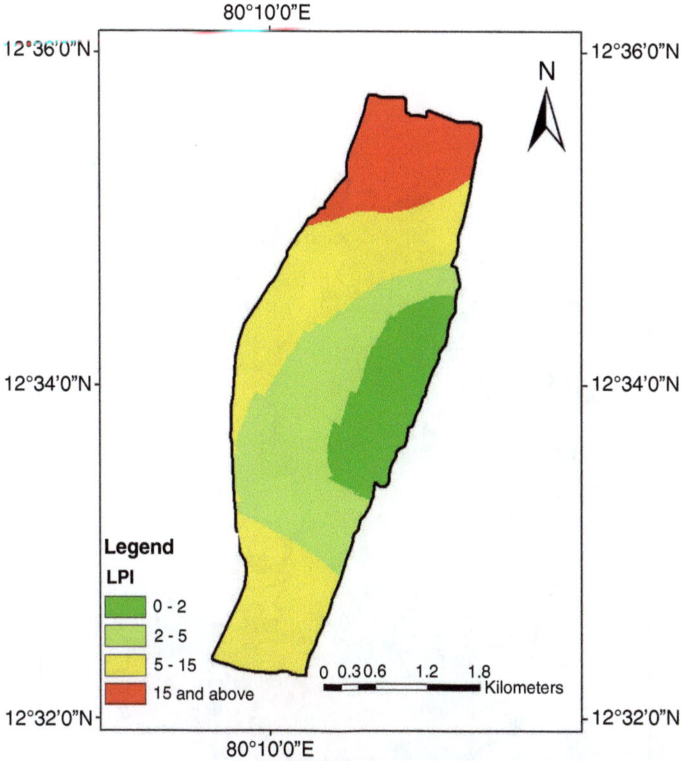

Fig. 6.17 Spatial variation in LPI values for Kalpakkam NPP site corresponding to return period of 2500 years(after [152])

Considering the return period of 475 years, the other parts of Karnataka state have low to moderate liquefaction hazard where SPT values required is less than 10. Similarly considering a return period of 2500 year, the SPT values required to prevent liquefaction is less than 20 for the interior regions of the Karnataka state. Kaiga, where a nuclear power plant site is located is found to have a very low liquefaction hazard. The SPT value required to resist liquefaction for this region is less than 5 considering both 475 year and 2500 year return period.

6.8.2 Prediction of CPT Value Required to Resist Liquefaction

In spite of being the most popular geotechnical field test for the site investigation, the SPT contains a lot of errors and thus require various corrections. Hence the liquefaction hazard for the region was also expressed in terms of CPT value as the percentage of errors is considerably less in the CPT compared to SPT

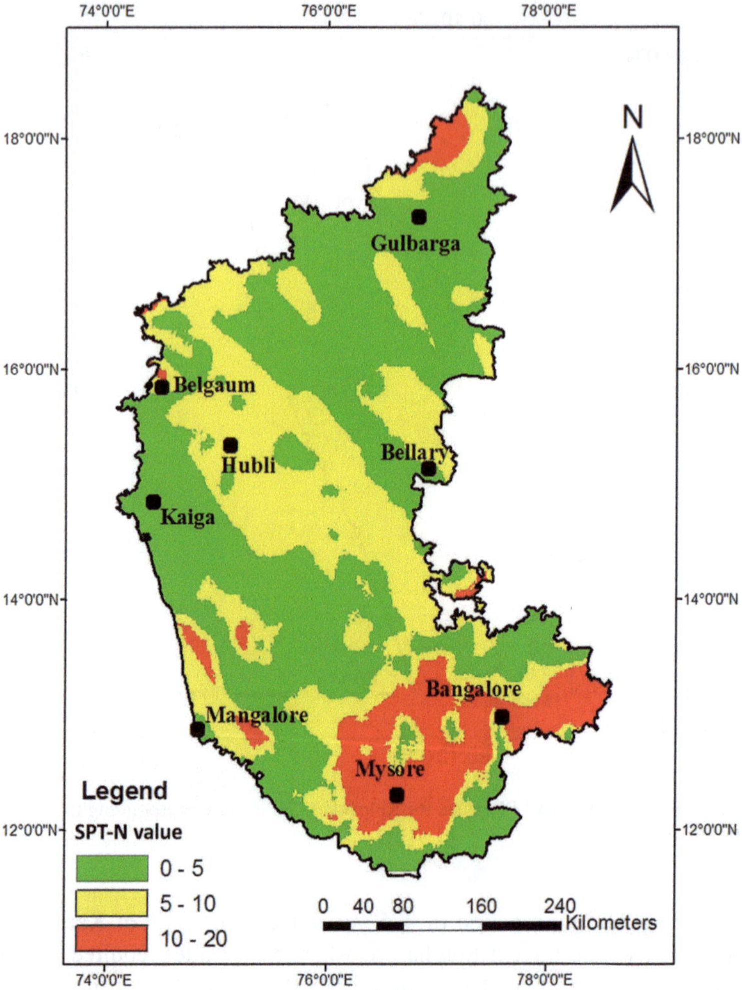

Fig. 6.18 Spatial variation in SPT values required to resist liquefaction at 3 m depth throughout the Karnataka state for 475 year return period (after [150])

method. The probabilistic methodology for the estimation of liquefaction hazard in terms of CPT values required to resist liquefaction is described in Sect. 6.5.3. Figures 6.20 and 6.21 present the spatial variations in the CPT values required to resist liquefaction at 3 m depth, for 475 year and 2500 year return period, throughout the Karnataka state. As evident from the above study (Figs. 6.20 and 6.21) that the southern (regions close to Bangalore–Mysore) and the northern parts (region close to Bidar district) of the state are expected to have a high liquefaction hazard. For both the regions, the CPT value required to resist liquefaction ranges from

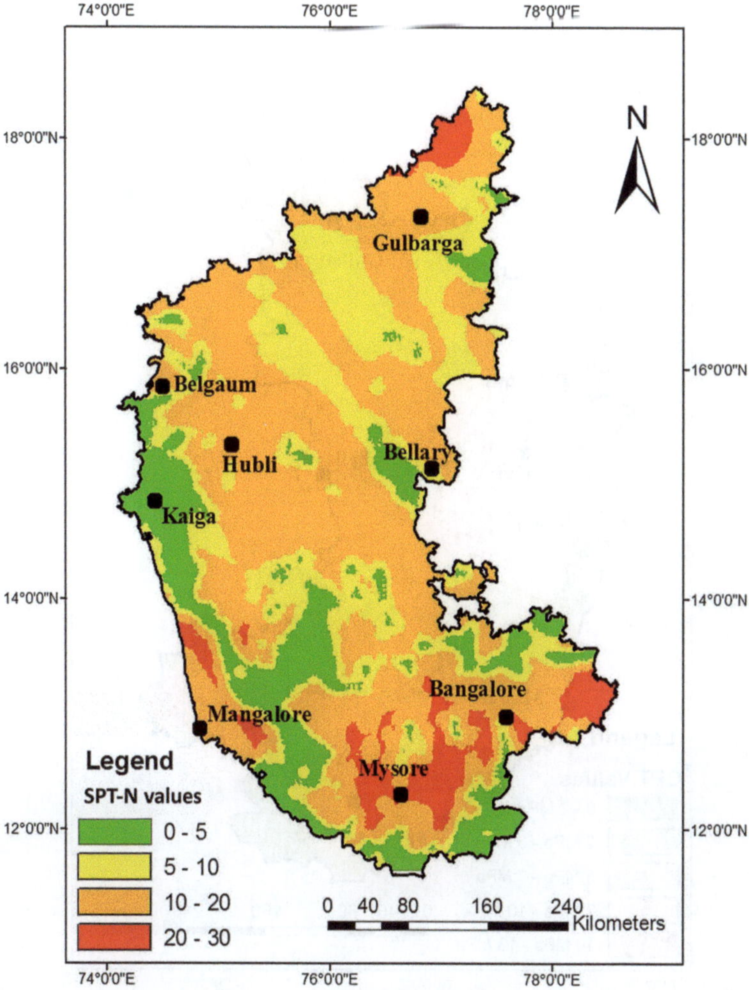

Fig. 6.19 Spatial variation in SPT values required to resist liquefaction at 3 m depth throughout the Karnataka state for 2500 year return period (after [150, 151])

7 to 13 MPa, considering both 475 year and 2500 year return period. The CPT values required to resist liquefaction throughout the interior parts of Karnataka, for a return period of 475 year is found to be in the range of 2–7 MPa. However, for a return period of 2500 year the CPT value required to resist liquefaction ranges from 5 to 13 MPa. Hence the interior region of the Karnataka state is found to have moderate to high liquefaction hazard. Such a high value of liquefaction hazard is due to high surface level PHA, as only linear source is considered.

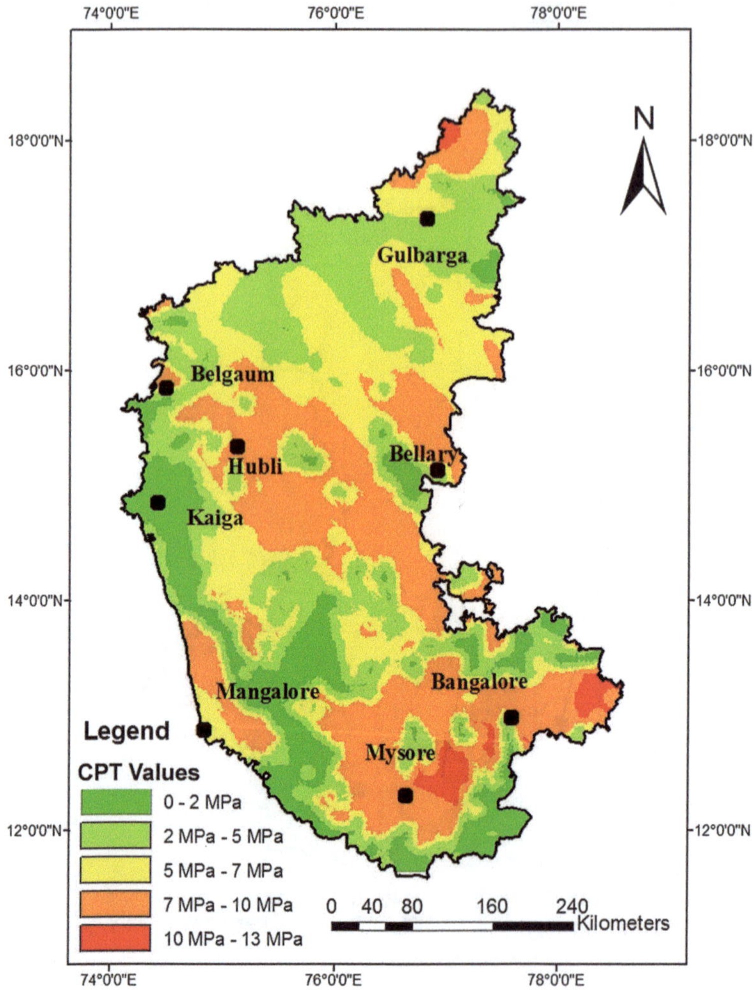

Fig. 6.20 Spatial variation in CPT values required to resist liquefaction at 3 m depth for 475 year return period (after [150])

6.9 Liquefaction Mitigation

Soil liquefaction is one of the worst seismic hazards that can occur during an earthquake. Several studies in the past have shown that the earthquake induced liquefactions of soils can cause significant damages to the man-made infrastructures. Possible causes, factors affecting the soil liquefaction along with its mechanism and evaluation procedure have been discussed in the above sections of this chapter. This section will discuss about the different methodologies developed to reduce or eliminate damage due to liquefaction. In recent years, several techniques have been

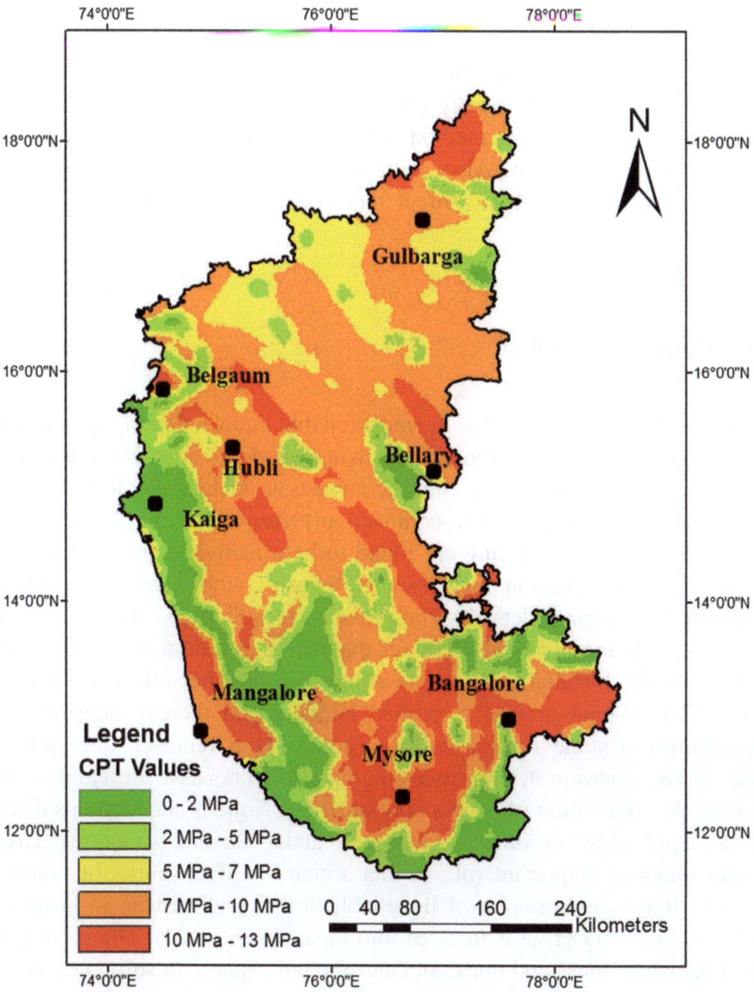

Fig. 6.21 Spatial variation in CPT values required to resist liquefaction at 3 m depth for 2500 year return period (after [150])

developed by many researchers toward liquefaction mitigation techniques, for the new as well as the existing structures. These techniques can be broadly classified into following.

6.9.1 Direct Methods: Soil Replacement and Dewatering

Soil replacement method involves replacement of liquefiable soil with a well-graded fill which has high resistance to liquefaction. Even though the method is very reliable, it becomes difficult and expensive if liquefiable soil is found for significant depth. Also, it is inapplicable if water table lies at shallow depth. Generation of excess pore water pressure leading to liquefaction can be prevented by lowering groundwater table by continuously pumping out water from the zone of interest. The biggest drawback of this method is that pumping is to be done permanently to keep the water table low. Also, both upstream and downstream seepage cut-offs are required. Hence, these direct methods are not preferable for practical applications.

6.9.2 By Densification

Soil densification is considered as the most reliable liquefaction mitigation method. Densification methods reduce the volume change tendency of soil and thus decrease the liquefaction potential. The field compaction techniques like compaction piles, dynamic compaction, vibro-compaction (vibro-rod, vibro-floatation, vibro-replacement, etc.), deep blasting are found to be effective liquefaction mitigation techniques. Sand compaction piles, installed by using vibro-hammers, displace and thus densify the original deposit. The pressurized air pushed into the ground in order to help migration of sand decreases the degree of saturation and thus provide an added benefit. Installation of stone columns is another effective remediation measure. The original ground is pushed laterally and thereby densified during the installation of stone column casing. The casing is removed after filling and compacting the stones in it. The stone column also works as a vertical drain which accelerates the dissipation of excess pore pressure. None of the methods discussed above are applicable for improving ground under existing structures. Grouting technique plays an important role in this scenario. Effectiveness of compaction grouting in improving strength of liquefiable soil is well proven in history [47]. A thick, low mobility grout is injected into the soil (Fig. 6.22). The grout remains as a homogeneous mass and does not enter the pore spaces in soil. Once the grout expands, the soil particles get displaced and thus the soil is densified. The grout mix used for densification purpose generally consists of silty sand, fly ash, cement, and water [13]. Clay content is restricted in grout mix to attain proper compaction and to avoid hydraulic fracturing. The pressure and rate of injection should be maintained such that heaving of overlying ground does not occur (Fig. 6.22).

Mitsunan et al. [219] proposed the concept of superlime piles which use burnt lime for densification of loose sand. Superlime piles are bored piles filled with a mixture of burnt lime, gypsum, blast furnace slag, and sand. The mixture absorbs groundwater and swells considerably such that the surrounding soil gets densified.

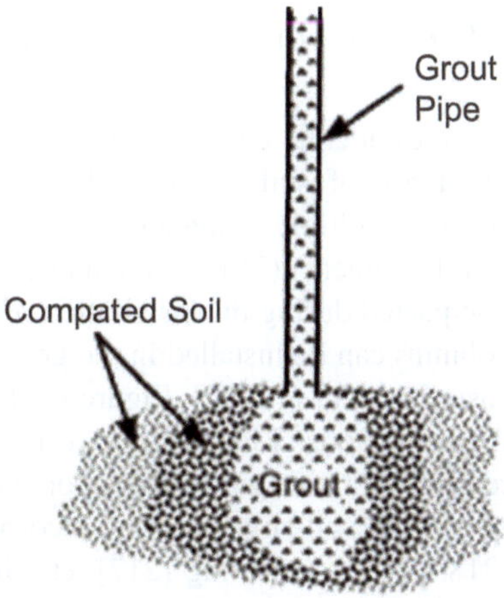

Fig. 6.22 Soil Densification by compaction grouting (after [13])

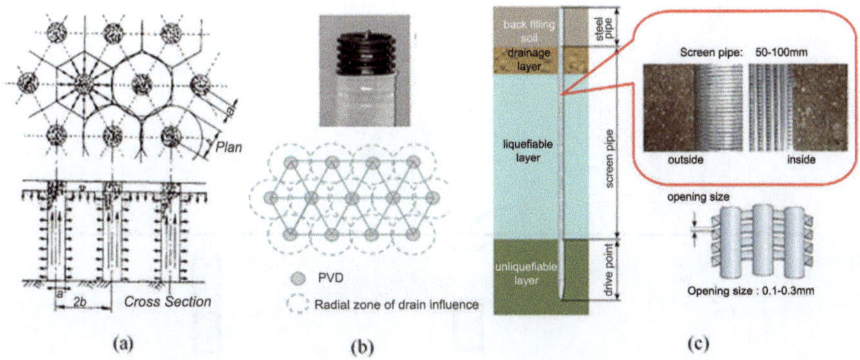

Fig. 6.23 (**a**) Arrangement of gravel drain system (after [390]) (**b**) PVD and an example of a triangular installation grid (after [127]) (**c**) Concept of screen pipe (after [119])

6.9.3 By Mechanical Intrusions for Drainage

Loss of shear strength can be reduced by accelerating dissipation of excess pore water pressure with the help of gravel drains [129, 300, 305, 390]. The vibrations produced during installation of gravel drains also help in densification of surrounding soil [1]. Prefabricated vertical drains (PVD) are also being used for providing drainage as an alternative to conventional gravel drains [127]. The shortening of drainage path inhibits rise in excess pore pressure and thus prevents liquefaction. Harada et al. [119] used screen pipes as vertical drains which can be easily installed in places with limited space. Figure shows a typical screen pipe. Different drainage schemes are shown in Fig. 6.23.

6.9.4 By Soil Reinforcements

By introducing elements which can carry more stress than soil, deformation resistance of soil can be enhanced. Installation of stone columns in the loose soil has multiple advantages such as: (1) it provides deformation restricting effect (reinforcement) (2) it accelerates pore pressure dissipation (drainage) (3) soil gets compacted during the installation of stone column casing (densification) [1]. Stone columns can be installed in the ground by vibro-replacement [1, 49] or using auger casing system [1, 252]. Figure 6.24 shows the installation of stone columns using these methods. Liquefaction resistance of natural deposits can be enhanced by reinforcing with geosynthetic or natural fibers [183, 203, 221]. Another way of improving soil strength is by incorporating structural elements such as micropiles [218, 228], soil nailing [217], etc. In the recent past, several studies were carried

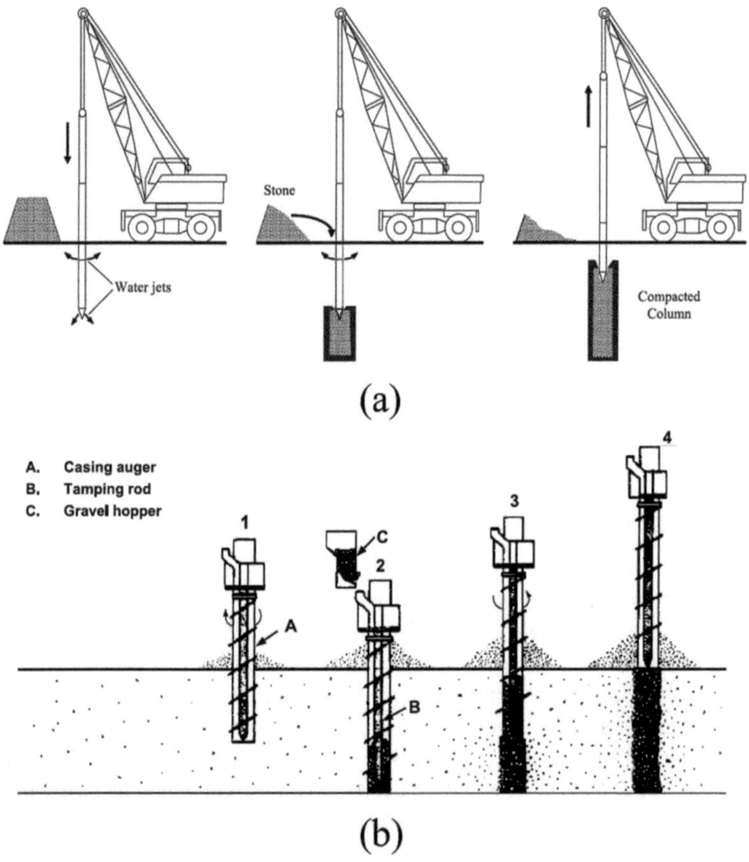

Fig. 6.24 Gravel drain installation (**a**) by vibro-replacement (after [1]) (**b**) by auger-casing system with internal gravel feeding compaction-rod system (after [340])

out toward the use of scrap tyre chips for the liquefaction mitigation in sands [210, 271, 365]. Adding tire chips to sand reduces its modulus of rigidity. Under the seismic shear force, the mixture can undergo volume contraction which tends to decrease the excess pore pressure [360]. Tire chips can also be used as a drain to dissipate the excess pore pressure rapidly [394]. Moreover, the geotechnical application of these tire chips provides a sustainable and eco-friendly solution to the problems related to the recycling of used tires.

6.9.5 By Chemical Modifications: Solidification

In this method, soils cohesive strength is improved by mixing cementitious material with it and thus prevents the soil particle movement. When injected, these grouting fluids permeates into the soil voids due its low viscosity without disturbing the soil structure. This technique is termed as permeation grouting. It strengthens the ground by cementing soil particles together. Also, it reduces the water flow by plugging soil pores. Sodium silicate grouts consisting of sodium silicates, acrylamides, lignosulfonates, and resins [168] are the most popular ones. In jet grouting, the mixing of stabilizer with the soil is accomplished by high pressure fluid jets. The grout is generally a water-cement mixture (Fig. 6.25). A small hole is drilled to the target depth and grout is jetted into the soil through the nozzles near the tip of drill rod. The rod is kept rotating during the withdrawal and the jetting is continuously done from bottom to top of the drill hole. The mechanical mixing and sometimes partial replacement of soil with cementitious materials using mixing augers is termed as in situ soil mixing. The procedure includes penetration of augers into the target depth, pumping of stabilizer through shaft and mixing of stabilizer with the soil by the rotation of

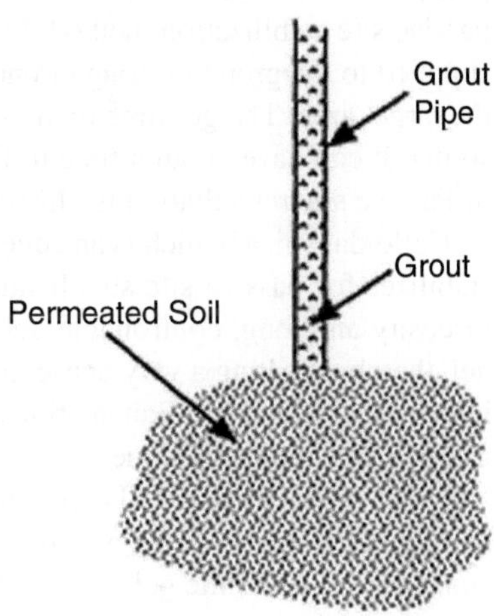

Fig. 6.25 Soil solidification by permeation grouting (after [13])

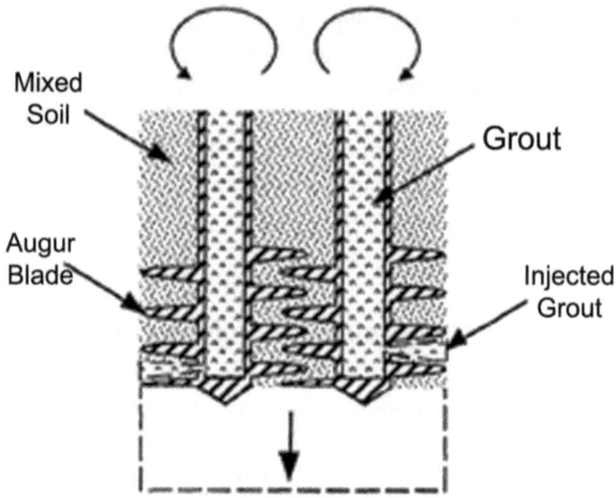

Fig. 6.26 In situ soil mixing technique (after [13])

auger (Fig. 6.26). During the withdrawal of augers a second stage of mixing is performed. As a result, strength of the mixed zone is improved and its permeability is reduced.

6.9.6 Passive Site Stabilization Techniques

In traditional methods of liquefaction mitigation, the size of treated area is limited. Vibration produced during construction creates disturbance to existing structures [128]. To overcome these limitations, Gallagher [98] proposed the concept of passive site stabilization method. In this method, the stabilizers of low viscosity are supplied to the ground through trenches and carried by natural groundwater flow to the target area. The gel time of stabilizing material should be long and controllable so that it can have enough time to flow slowly to the zone of interest. The concept of Passive site remediation is illustrated in Fig. 6.27.

Colloidal silica, which is an aqueous suspension of silica particles, is an effective stabilizer for passive site stabilization. Low concentration colloidal silica has low viscosity and long, controllable gel time [99]. On gelation it transforms into silica gel, thereby creating a very dense matrix with loose sand particles which possess an improved resistance to liquefaction [100]. Therefore this method can be categorized as a solidification technique.

Haldavnekar et al. [116] experimentally proved that addition of bentonite boosts the liquefaction resistance of sand. Addition of bentonite converts the pore fluid to concentrated bentonite gel which exhibits elastic behavior for a wide range of strain.

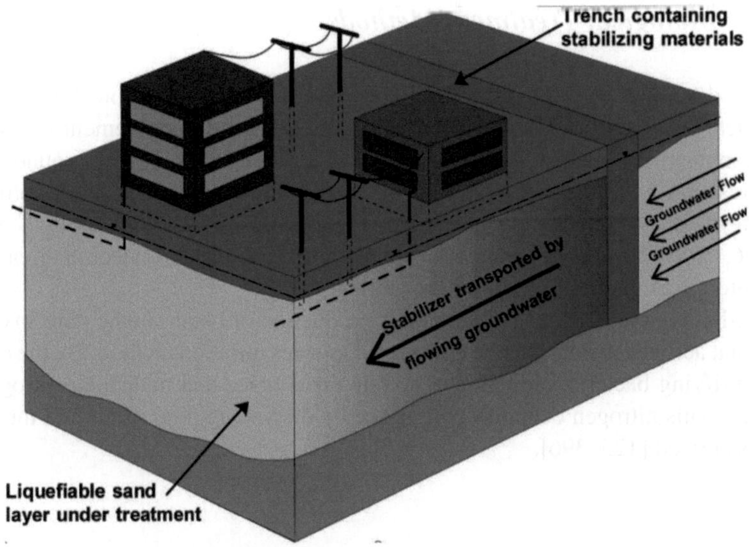

Fig. 6.27 Passive site remediation for mitigation of liquefaction risk (after [99])

The enhanced cyclic resistance is attributed to this extended elastic response of pore fluid [81, 82]. The rheological properties of bentonite suspension can be controlled by using sodium pyrophosphate (SPP). Rugg et al. [293], by examining undrained shear response of sand-bentonite mixture, observed that the mixture possesses no cohesion and its friction angle is same as that of untreated sand. It shows that the soil structure is preserved even after the permeation of bentonite.

6.9.7 By Inducing Desaturation

In recent past, desaturation has emerged as an effective technique for the liquefaction mitigation. In this method, the liquefaction resistance of the soil increased by reducing its degree of saturation by injecting air or biogas [128]. Several studies were carried out toward liquefaction mitigation by desaturation [141, 230, 249, 364, 393, 396]. The safety against liquefaction of soil can be assured by decreasing its degree of saturation [26]. A small reduction in the degree of water saturation of sand soil can result in a significant increase in shear strength against liquefaction [85]. Nakazawa et al. [230] have shown that cyclic strength in laboratory test specimens can be more than twice as high in partially saturated soil than fully saturated soil. By injecting air into the site with the help of special equipments, degree of saturation can be reduced to desired level [53, 85, 249]. Once the air injection is done for a short time period, the resulting desaturation can last for years [249].

6.9.8 Biological Treatment Methods

By the development of a new discipline called Microbial Geotechniques, the biological treatment methods evolved as effective ground improvement technique. When some microorganisms are injected along with nutrients into the ground under low pressure, as a result of some chemical reactions, products having ability to cement the sand particles are generated. This increases the shear strength of the soil [66, 76, 372]. The slow rate of chemical reactions helps to inject the treatment fluid into the ground efficiently before gelling.

Another type of biological treatment is production of biogas in the pore space by microbial activities which can be considered as a desaturation method. By the action of denitrifying bacteria, reduction of nitrate takes place and molecular nitrogen is released. This nitrogen occupies pore space by displacing pore water and thus soil get desaturated [125, 396].

Chapter 7
Principles and Practices of Seismic Zonation

7.1 Introduction

Assessment of various earthquake hazards such as the ground shaking, site amplification and liquefaction, both at micro and macro-levels, has been discussed in previous chapters. The main objective of this chapter is to describe the seismic zonation at the micro as well as macro-level, based on the spatial variation of these hazards. A detailed analysis of the spatial distribution of these hazards reveals that the distribution pattern of each hazard in a given study area differs from the other. It is worth noting that a region with a high level of shaking at the bedrock level need not exhibit high site amplification or liquefaction hazard. So it is difficult to carry out the seismic zonation if each of the earthquake hazards is treated separately. Hence there is a need to develop an index number which can effectively represent the combined effect of all these hazards at a particular location.

7.2 Integration of Various Hazards

The estimation procedures of major hazards associated with earthquakes such as ground shaking, liquefaction, landslide, etc. were discussed in the previous chapters of this book. As each of these hazards is predominantly influenced by local geological setting rather than source characteristics of the earthquake, the spatial distribution of intensity of each of these in a given region is distinct from the other. Thus it is difficult to assess the risk and the vulnerability of a region when these hazards are treated separately. Hence all major earthquake hazards are to be integrated so that their combined effect can be understood in a better way. This chapter describes the methodology for integrating various hazards associated with earthquakes. All the major seismic hazards for the two study areas are integrated to a hazard index value using the Analytical Hierarchy Process (AHP) proposed

T. G. Sitharam et al., *Comprehensive Seismic Zonation Schemes for Regions at Different Scales*, https://doi.org/10.1007/978-3-319-89659-5_7

by Saaty [295]. The hazard index effectively represents the consolidated effect of all major seismic hazards and hazard influencing parameters. The region was then demarcated based on the spatial variation of hazard index value.

The first step of hazard integration is to prioritize various themes (here seismic hazards). Hierarchical arrangement of various themes is subjected to user judgement. After prioritizing, the weightage is assigned each theme. The theme having high priority is assigned with higher weightage. After establishing hierarchy among various themes, normalized weights for each theme is then calculated using a pair-wise comparison matrix. In a pair-wise comparison matrix each theme was compared with the other as the ratio between their respective weights. The normalized weight for each theme is then calculated by averaging the values in each row of the matrix with the total weight. A second level of categorization was made within each of the themes and a rank was assigned to each category. Range of each category and number of categories within a particular theme are also based on user judgement. These ranks (for each theme category) are then normalized using Eq. (7.1).

$$X_i = \frac{R_i - R_{\min}}{R_{\max} - R_{\min}} \tag{7.1}$$

Here, R_i is the rank assigned to each category of a single theme, $R_{\min}$ and $R_{\max}$ is the minimum and the maximum rating value of that theme. Hazard index values are then estimated based on normalized weights and ranks by integrating of all themes using the following Eq. (7.2).

$$\text{HI} = \frac{\sum[(\text{EHP}_w)_i(\text{EHP}_r)_i]}{\sum w_i} \tag{7.2}$$

where HI is the hazard index, EHP_w and EHP_r are the normalized weight and rank of ith earthquake hazard parameter in a pair-wise comparison matrix, 'w' is the normalized weight of the theme. Procedures for estimating each of these earthquake hazard parameters are mentioned in the above sections of this chapter.

7.3 Seismic Zonation Studies at Micro-Level

7.3.1 Seismic Zonation of Kalpakkam NPP Site

The hazard integration at the micro-level was carried out for the Kalpakkam nuclear power plant site by James et al. [153]. Since nuclear power plants are considered as sensitive and critical installations with different components founded at different depth, seismic microzonation for these sites is of great importance. Seismic microzonation maps of the Kalpakkam NPP site were prepared based on the

Table 7.1 Parameter and weights used for generating hazard index map at the micro-level (after [153])

Earthquake hazard parameter (EHP)	Weights
Peak horizontal acceleration (PHA)	6
Amplification factor (Amp.factor)	5
Factor of safety against liquefaction (Lq.Fs)	4
Predominant frequency (PF)	3
Average shear wave velocity for top 30 m (V_{s30})	2
Overburden soil thickness/rock depth (Obr.thick)	1

spatial variation of hazard index value. These maps can be selected as a criterion for site selection for various nuclear infrastructures and also prove helpful in assessing the vulnerability for the existing facilities. Major seismic hazards considered in the evaluation of hazard index at the micro-level are (1) intensity of ground shaking at bedrock (in terms of PHA), (2) site amplification, (3) liquefaction potential, and (4) the predominant frequency of the earthquake motion at the surface. Apart from these earthquake hazards, other significant parameters representing local geology such as the average shear wave velocity for top 30 m (V_{s30}) and the thickness of overburden soil (or rock depth) were also considered in the evaluation of hazard index. A list of earthquake hazard parameters (EHP) and corresponding weights used in this study is presented in Table 7.1.

The peak horizontal acceleration (PHA) at bedrock was given highest weightage as it is the most fundamental. Subsequent priorities were allotted to hazards such as amplification factor, factor of safety against liquefaction, and predominant frequency. Other parameters such as average shear wave velocity of the top 30 m overburden (V_{s30}) and thickness of overburden soil (or rock depth) were given least weightages of all. In AHP, a matrix was constructed (Table 7.2) showing the pairwise comparisons (ratios) between the factors of earthquake hazard parameters (EHP). The pairwise comparisons will result in a normalized weight for each parameter and the sum of the normalized weights is unity. Each EHP was categorized into various ranges and for each category a rating or ranking was assigned. Table 7.3 presents the details of each EHP category and the ranks associated with each of them. These ranks (for each EHP category) were then normalized using Eq. (7.1). Once EHPs were identified and prioritized, they were integrated to a hazard index value that ranges from 0 to 1. After the estimation of normalized weight and rating for each category of earthquake hazard parameter, the hazard index for the study area was estimated using Eq. (7.3).

$$\mathrm{HI} = \left(\frac{1}{\sum w}\right)\left[\begin{array}{c} \mathrm{PHA}_w \times \mathrm{PHA}_r + \mathrm{Amp.factor}_w \times \mathrm{Amp.factor}_r \\ +\mathrm{Lq.Fs}_w \times \mathrm{Lq.Fs}_w + \mathrm{PF}_w \times \mathrm{PF}_r + V_{s30w} \times V_{s30r} \end{array}\right] \quad (7.3)$$

where HI is the hazard index, 'w' is the normalized weight of the earthquake hazard parameter (EHP), and 'r' is the normalized rank of a category in the

Table 7.2 Normalized weights for each earthquake hazard parameter (EHP) (after [153])

EHP	PHA	Amp. factor	Obr.thick	V_{s30}	Lq.Fs	PF	Weights
PHA	6/6	6/5	6/4	6/3	6/2	6/1	0.285714
Amp.factor	5/6	5/5	5/4	5/3	5/2	5/1	0.238095
Lq.Fs	4/6	4/5	4/4	4/3	4/2	4/1	0.190476
PF	3/6	3/5	3/4	3/3	3/2	3/1	0.142857
V_{s30}	2/6	2/5	2/4	2/3	2/2	2/1	0.095238
Obr.thick	1/6	1/5	1/4	1/3	1/2	1/1	0.047619

Table 7.3 Normalized rating for each EHP category for micro-level hazard integration (after [153])

EHP	Value range	Weight	Rank	Normalized rank
PHA(g)	≤0.05	0.2875	1	0
	0.05–0.09		2	0.33
	0.09–0.13		2	0.67
	≥0.13		3	1
Amp. factor	1.0–2.0	0.2381	1	0
	2.0–3.0		2	0.5
	≥3.0		3	1
Lq.Fs	≤1	0.1905	3	1
	1.0–2.0		2	0.5
	≥2.0		3	0
V_{s30} (m/s)	≤100	0.0952	6	1
	100–200		5	0.80
	200–300		4	0.60
	300–400		3	0.40
	400–500		2	0.20
	≥500		1	0
Obr.thick (m)	≤5	0.0476	1	0
	5–10		2	0.25
	10–15		3	0.50
	15–20		4	0.75
	>20		5	1

EHP. Deterministic and probabilistic seismic microzonation maps of the study area, showing the spatial variation in the hazard index values were prepared and presented in this chapter.

Figure 7.1 presents the deterministic seismic microzonation map and Figs. 7.2 and 7.3 present the probabilistic seismic microzonation map corresponding to 475 year and 2500 year return period, respectively. The hazard index for preparing the deterministic microzonation map was evaluated using EHP obtained from deterministic methodology such as PHA and factor of safety against liquefaction.

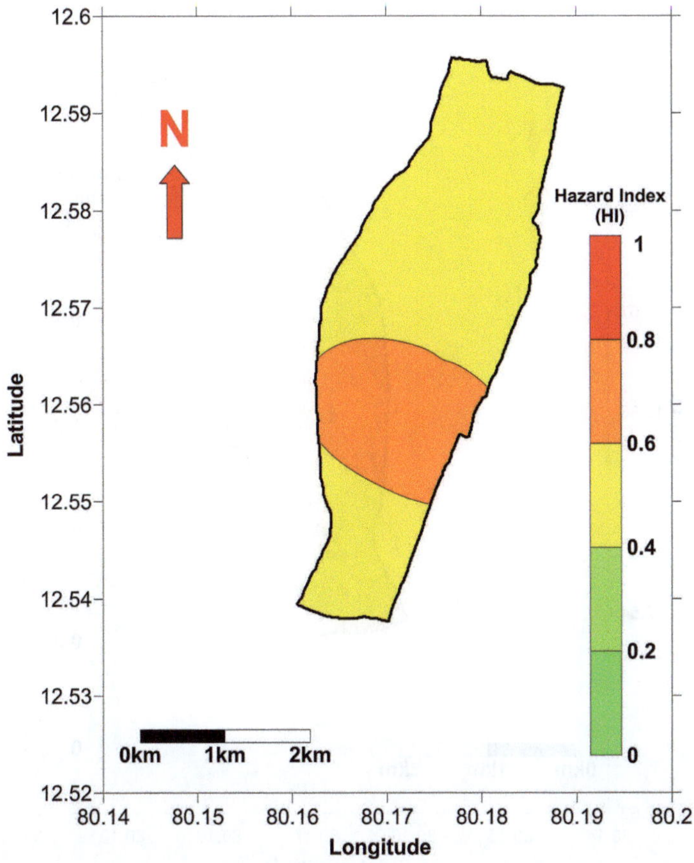

Fig. 7.1 Deterministic seismic microzonation map for the Kalpakkam NPP site (after [153])

Similarly for preparing the probabilistic hazard index map, EHPs such as PHA and factor of safety against liquefaction corresponding to a return period of 475 years and 2500 years were selected.

The deterministic microzonation map (Fig. 7.1) shows that the central sector of the Kalpakkam NPP site close to S-3 and S-4 locations has the higher hazard index (HI) value ranging from 0.6 to 0.8. Such a high value of hazard index is mainly contributed by EHPs such as PHA, amplification factor, and predominant frequency. The deterministic hazard index for the rest of the area in the Kalpakkam NPP site varies from 0.4 to 0.6. Probabilistic microzonation map for 2500 year return period also presents a similar trend in the spatial distribution of hazard index except in the northern sector of the study area, where hazard index value ranges from 0.6 to 0.8. This disparity can be attributed to the fact that the PHA for the northern sector corresponding to 2500 year return period is marginally higher than the PHA obtained from the deterministic methodology. Probabilistic microzonation map for

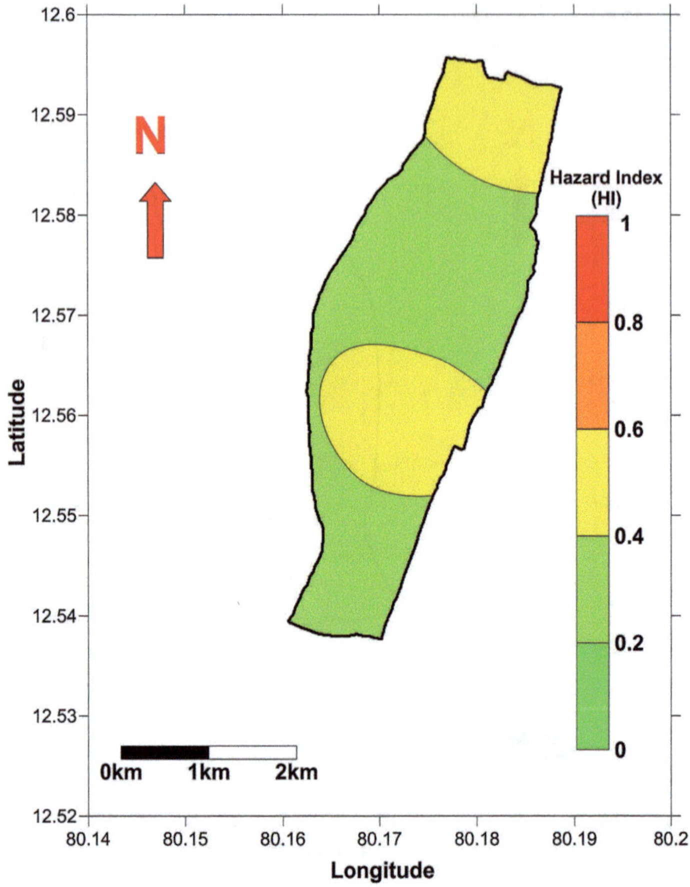

Fig. 7.2 Probabilistic seismic microzonation map of the NPP site (475 year return period after [153])

475 year return period shows that the hazard index value for a major portion of the Kalpakkam NPP site varies from 0.2 to 0.4, except for the central and northern sectors where HI varies between 0.4 and 0.6.

Hence considering the worst scenario from above three maps, the central and northern sectors can be categorized to region having high hazard as per [235], where hazard index varies from 0.6 to 0.8. The rest of the area of the Kalpakkam NPP site can be categorized as a region having moderately high hazard, where hazard index value varies from 0.4 to 0.6.

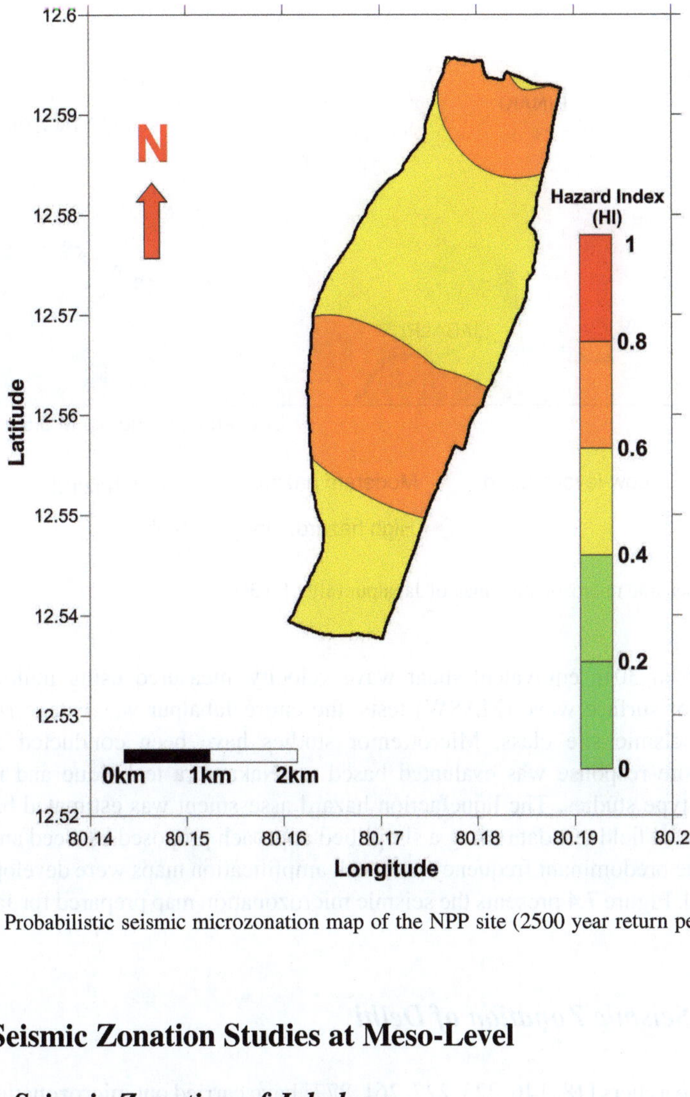

Fig. 7.3 Probabilistic seismic microzonation map of the NPP site (2500 year return period after [153])

7.4 Seismic Zonation Studies at Meso-Level

7.4.1 Seismic Zonation of Jabalpur

Jabalpur microzonation [263] was one of the earliest works in India, towards seismic microzonation with the support from the Department of Science and Technology (DST) New Delhi, Geological survey of India (GSI), Central Region Nagpur, Indian Meteorology Department (IMD) New Delhi, National Geophysical Research Institute (NGRI) Hyderabad, Central Building Research Institute (CBRI) Roorkee, and Government Engineering College, Jabalpur. Seismic hazard in terms of peak ground acceleration (PGA) was estimated at bedrock level using deterministic approach. Site characterization for the entire city was carried out based on geology, geotechnical, and geophysical investigations.

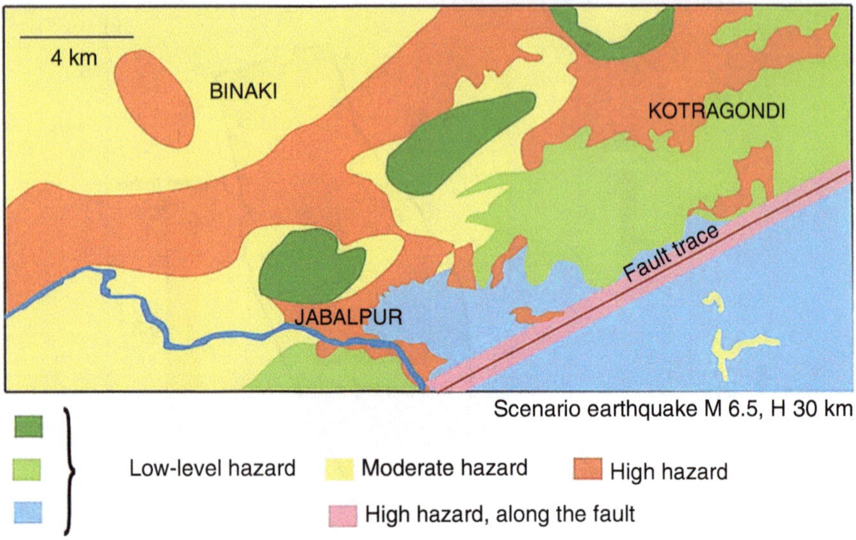

Fig. 7.4 Seismic microzonation map of Jabalpur (after [263])

Based on 30 m equivalent shear wave velocity, measured using multichannel analysis of surface wave (MASW) tests, the entire Jabalpur was categorized into various seismic site class. Microtremor studies have been conducted and the seismic site response was evaluated based on Nakamura technique and receiver function type studies. The liquefaction hazard assessment was estimated based on geotechnical field test data using a simplified approach proposed by Seed and Idriss [309]. The predominant frequency and peak amplification maps were developed and presented. Figure 7.4 presents the seismic microzonation map prepared for Jabalpur.

7.4.2 Seismic Zonation of Delhi

Many researchers [18, 146, 223, 227, 261, 277] have carried out microzonation work for Delhi region. Iyengar and Ghosh [146] evaluated PGA at bedrock level for Delhi using both deterministic and probabilistic approaches. All seismic sources and earthquake events within 300 km from the boundary of Delhi region were compiled for seismic hazard assessment. A quantitative seismic hazard map for the entire Delhi was prepared by considering a grid size of 1 km × 1 km. They have also carried out detailed 1D site response analysis based on the borelog data using SHAKE-91. Parvez et al. [261] has computed the site specific seismic ground motion in a part of Delhi City using a hybrid technique. The hybrid technique was based on the model summation and the finite-difference scheme.

Rao and Neelima Satyam [277] have generated a seismic hazard map at bedrock level for entire Delhi based on finite fault simulation technique using FINSIM computer code. Extensive geotechnical and geophysical tests were conducted and seismic site characterization was also carried out based on V_{s30} value. Microtremor studies have been carried out to compute the seismic site response. They have also carried out liquefaction assessment for Delhi based on SPT data. A first order seismic microzonation map of Delhi was prepared [304]. First they generated five thematic layers viz., Peak Ground Acceleration (PGA) contour, different soil types at 6 m depth, geology, groundwater fluctuation, and bedrock depth. The highest priority of 0.333 was given to PGA and subsequent priorities of 0.266, 0.2, 0.133, and 0.066 were given to soil type, geology, groundwater, and bedrock depth, respectively. These maps were then integrated using Analytical Hierarchy Process (AHP) in GIS platform. Figure 7.5 presents the seismic microzonation map of Delhi.

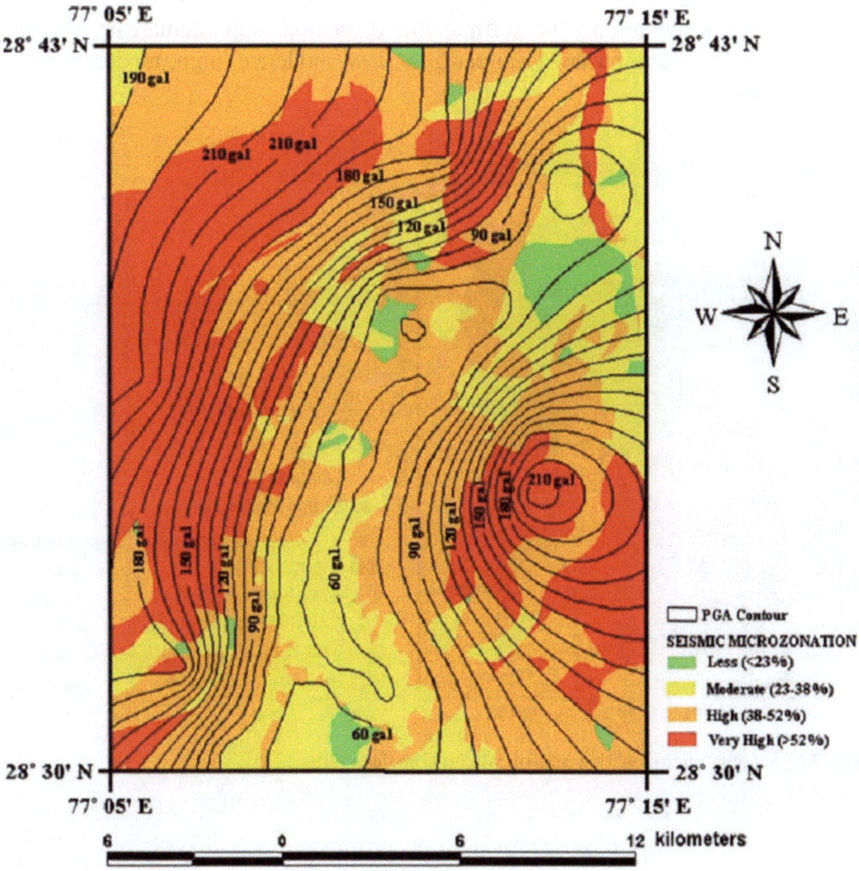

Fig. 7.5 Seismic microzonation map of Delhi (after [223])

7.4.3 Seismic Zonation of Sikkim Himalaya and Guwahati

Nath et al. [235] have presented the microzonation works carried for Sikkim Himalaya and Guwahati, which include multi-criteria seismic hazard analysis through thematic integration of influencing factors. Scenario earthquakes of M_W 8.3 and 8.7 were used for the seismic hazard analysis for two regions using dynamic finite fault simulation technique. Seismic hazard microzonation map of the Sikkim Himalaya (Fig. 7.6) was prepared after integration of geomorphological themes and seismological themes using the AHP in a GIS platform. Geomorphological theme for Sikkim Himalaya includes surface geology, soil cover, slope, rock outcrop, and landslide, while the seismological themes consist of peak ground acceleration and predominant frequency. Similarly seismic microzonation map of Guwahati (Fig. 7.7) was also prepared by integrating eight themes-geological and geomorphological, basement or bedrock, landuse, landslide, factor of safety for soil stability, shear wave velocity, predominant frequency, and surface consistent peak ground acceleration. The seismic microzonation maps demarcate the two regions into five main hazard level category, low, moderate, high, moderate high, and very high.

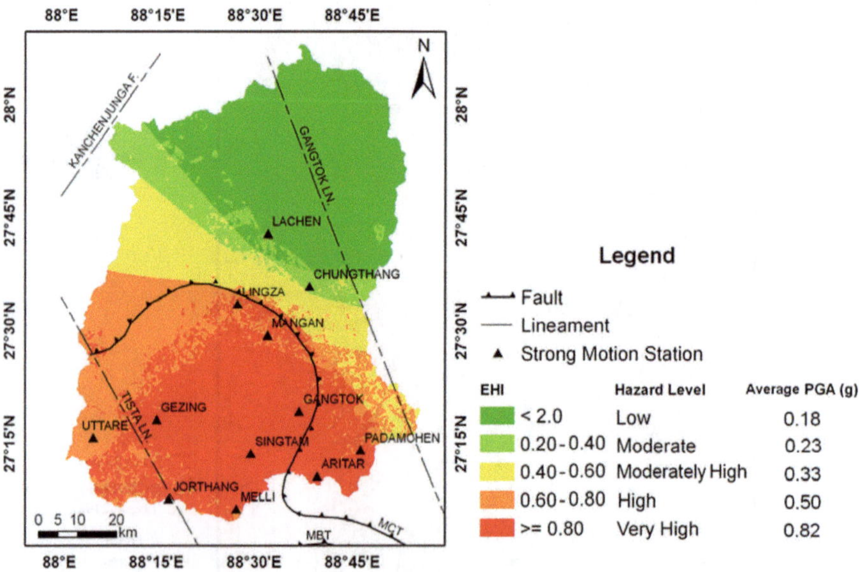

Fig. 7.6 Seismic microzonation map of Sikkim Himalaya (after [235])

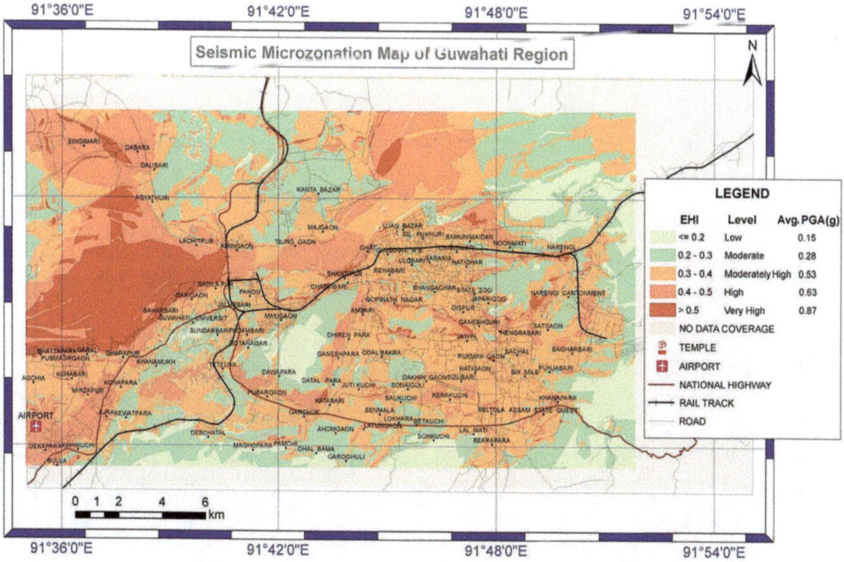

Fig. 7.7 Seismic microzonation map of Guwahati (after [235])

7.4.4 Seismic Zonation of Bangalore

Sitharam and Anbazhagan [328] have carried exhaustive studies for the seismic microzonation of Bangalore city. Seismotectonic map, comprising of earthquake events and seismic source was developed covering 350 km radius around Bangalore. Seismic hazard analysis for the study area has been carried out using deterministic as well as probabilistic approaches. Seismic site characterization for the region was carried out using both geotechnical and geophysical tests. They have also developed three dimensional sub surface borehole model for study area using geographical information system (GIS).

The study was further categorized into various NEHRP site classes based on V_{s30} measured using multichannel analysis of surface wave (MASW) tests. 1D site response analyses were carried out using synthetic earthquake time history and SHAKE, to estimate site effects. Liquefaction hazard for the entire Bangalore was estimated based on SPT N value. A final hazard index map for Bangalore was developed using Analytic Hierarchy Process (AHP) on a GIS platform by integrating various hazard parameters. Major seismological attributes considered in their study are peak Ground Acceleration (PGA) (from deterministic and probabilistic approach), amplification factor, predominant frequency, liquefaction, while the geomorphologic attributes include geology and geomorphology, rock depth/soil thickness, soil type and average shear wave velocity, drainage pattern and elevation level. Figure 7.8 presents the deterministic seismic microzonation maps for Bangalore.

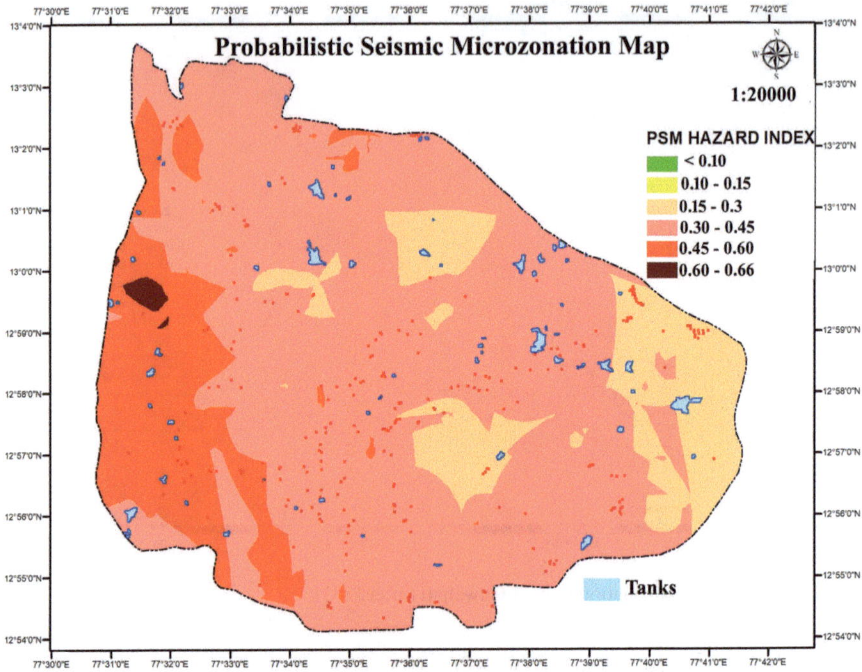

Fig. 7.8 Probabilistic seismic microzonation map of Bangalore (after [328])

7.5 Seismic Zonation Studies at Macro-Level

7.5.1 Seismic Zonation of Karnataka State

Similar to the micro-level zonation as described in the above section, a macro-level seismic zonation based on the hazard index value was also carried out for the entire state of Karnataka by James et al. [147]. Seismic macrozonation of a large area like a state or a country is an excellent tool for land-use planning. Macrozonation maps help the designers and planners to identify the sites seismically vulnerable and it can also provide inputs for site selection for critical structures nuclear/thermal power plants. These maps also provide the much needed information which can be functional in implementing appropriate mitigation works by the disaster management authority.

Similar to micro-level zonation work, the hazard index value was evaluated for the entire state of Karnataka by integrating a set of EHPs using AHP. Major EHPs considered in the evaluation of hazard index are (1) PHA at bedrock level, (2) the amplification factor, (3) SPT-N value required to prevent liquefaction at 3 m depth, and (4) the average shear wave velocity for top 30 m (V_{s30}). The highest weightage is given to PHA at bedrock level followed by amplification factor, SPT-N value

Table 7.4 Parameter and weights used for generating hazard index map at the macro-level (after [147])

Earthquake hazard parameter (EHP)	Weights
Peak horizontal acceleration (PHA)	4
Amplification factor (Amp.factor)	3
SPT-N value required to prevent liquefaction (Lq.Fn SPT)	2
Average shear wave velocity for top 30 m (V_{s30})	1

Table 7.5 Normalized weights for each (EHP) for macrozonation (after [151])

EHP	PHA	Amp. factor	Lq.Fn SPT	V_{s30}	Normalized weight
PHA	4/4	4/3	4/2	4/1	0.4
Amp.factor	3/4	3/3	3/2	3/1	0.3
Lq.Fn SPT	2/4	2/3	2/2	2/1	0.2
V_{s30}	1/4	1/3	1/2	1/1	0.1

required to prevent liquefaction and least weightage is given for (V_{s30}) as described in Table 7.4.

Once after assigning priorities to each of earthquake hazard parameters, a pairwise comparison matrix was constructed (Table 7.5). The each EHP was further sub-categorized into various ranges as in Table 7.6 and normalized ranks for each category was evaluated using Eq. (7.1). The hazard index for the entire state of Karnataka was evaluated using Eq. (7.4). Moreover the seismic landslide hazard, at a macro-level, is mainly restricted in Western Ghat sections of Karnataka, hence its effect is much concentrated in this region only.

$$\mathrm{HI} = \left(\frac{1}{\sum w}\right)\left[\begin{array}{c}\mathrm{PHA}_w \times \mathrm{PHA}_r + \mathrm{Amp.factor}_w \times \mathrm{Amp.factor}_r \\ +\mathrm{Lq.FnSPT}_w \times \mathrm{Lq.FnSPT}_w + V_{s30w} \times V_{s30r}\end{array}\right] \quad (7.4)$$

Figures 7.9 and 7.10 present the probabilistic seismic microzonation map for the state of Karnataka, for a return period of 475 years and 2500 years, respectively. Probabilistic macrozonation map for 475 year return period shows that 75–80% of Karnataka state has a hazard index value ranging from 0 to 0.4, hence falls in the category of low hazard region based on the classification proposed by Nath et al. [235]. Certain portions such as regions close to Bangalore in the south and Bidar in the north have the hazard index value ranging from 0.4 to 0.6 and hence can be categorized as regions having moderately high hazard.

Now referring to the probabilistic macrozonation map for 2500 year return period, the hazard index value for the interior regions of the Karnataka state varies from 0.2 to 0.6, falling in the category of low to moderately low hazard region. Even though the analysis shows that the Western Ghat regions are in low to moderate low hazard region, it should be kept in mind that this region has very high seismic landslide hazard. Similarly the hazard index for the regions close to Bangalore and

Table 7.6 Normalized rating for each EHP category for the macro-level hazard integration (after [150])

EHP	Value range	Weight	Rank	Normalized rank
PHA(g)	≤0.1	0.4	1	0
	0.10–0.20		2	0.25
	0.20–0.25		3	0.5
	0.25–0.30		4	0.75
	≥0.30		5	1
Amp. factor	1.0–2.0	0.3	1	0
	2.0–3.0		2	0.5
	≥3.0		3	1
Lq.Fn SPT	≤5	0.2	1	0
	5.0–15		2	0.33
	15–25		3	0.67
	≥25		4	1
V_{s30} (m/s)	≤100	0.1	6	1
	100–200		5	0.80
	200–300		4	0.60
	300–400		3	0.40
	400–500		2	0.20
	≥500		1	0

Bidar varies from 0.4 to 0.8 and can be classified as regions having moderately high to high hazard.

7.6 Guidelines for Seismic Zonation

The main objective of this book is to define the level of zonation to be adopted for various study areas in the country. As the seismic zonation program for various cities in India has been initiated by the government, the suitability of different levels of zonation for various regions within the country need to be studied. It is evident that the micro-level zonation requires very large resources in order to carry out geotechnical and geophysical testing and site effect estimation, hence it is not physically and financially viable to carry out micro-level seismic zonation for all these cities. The present study shows that the micro-level zonation is most suitable for critical facility like an NPP site situated even in a low seismic zone or a densely populated city or urban area situated in a high seismic zone. Nuclear power plants are very critical structures and any slightest of damages during an earthquake event will result in catastrophe. Similarly a densely populated city in a very high seismic zone escalates the risk to human lives and their property. Rigorous and extensive geotechnical and geophysical investigations are required for site characterization.

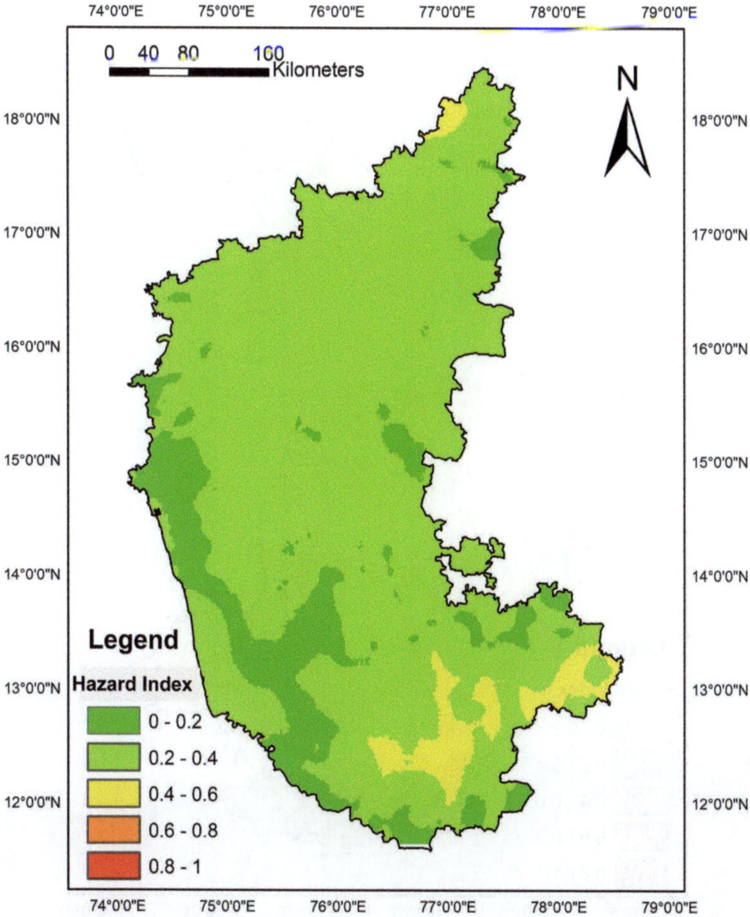

Fig. 7.9 Probabilistic seismic zonation map of Karnataka (475 year return period after [150])

Site effects should be estimated using numerical studies based on the geotechnical and geophysical test data. As the potential seismic risk is very high in the above two cases, the micro-level seismic zonation is very much required.

The meso-level zonations is preferred for the cities having population above 500,000, lying in the low to moderate seismic zones. The meso-level zonation is required as the level of seismic risk is high due to large population and infrastructure. The macro-level seismic zonation can very well be useful for analyzing the seismicity of that state or country and identifying vulnerable regions within it. This level of zonation can be adopted for cities as well, which are having very low population and infrastructures and are situated in very low seismic zones. As there are many methods available for site characterization and earthquake hazard estimation, the feasibility of various techniques for the regions at different scales

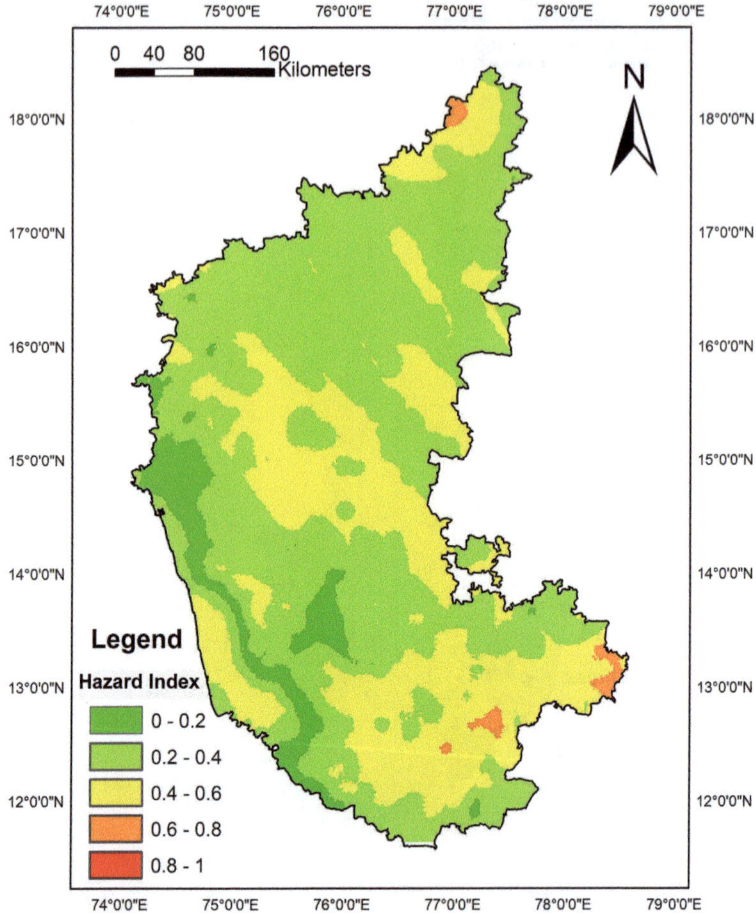

Fig. 7.10 Probabilistic seismic zonation map of Karnataka (2500 year return period after [151])

was reviewed in this study. Based on this, appropriate methods have been adopted for the micro and macro-level seismic zonation of the Kalpakkam NPP site and the Karnataka state, respectively. As the Kalpakkam NPP site area does not lie within the state of Karnataka, the results of the macro-level study were validated with the existing meso-level study carried out for the Bangalore city. The important recommendations of this book arrived after analyzing various factors are given below.

7.6.1 Delineation of Seismic Study Area

The seismic study area extending up to 300 km from the boundary of the study area should be identified and the details of the seismic events and the seismic sources need to be identified from this region.

7.6.2 Geological Consideration

Thorough regional and local geological studies including the basement configuration need to be done. Preparation of seismotectonic map considering all the active and paleo faults. The seismotectonic atlas published by GSI [316] may be taken as the base map. In addition to this the new lineaments or faults available in literature can also be considered.

7.6.3 Preparation of Earthquake Catalogue

Data need to be collected from various national and international agencies and from literature. It is recommended to decluster the data to remove the related events. The collected earthquake data may be in different magnitude scales. It has to be converted to the same scale in order to evaluate the seismicity parameters of the region. The earthquake data thus compiled need to be published in order to help compiling a national earthquake catalogue.

7.6.4 Attenuation Properties of the Region

Since the attenuation relations are not available for many regions of India, appropriate attenuation relations need to be selected. The relations developed for other parts of the world, which are having similar attenuation characteristics can be used. If the time history of regional earthquakes is available, then these attenuation relations need to be crosschecked with this data for validation.

7.6.5 Seismic Hazard Analysis

The PHA and spectral acceleration (S_a) at the bedrock level, for micro, meso, and macro-level can be evaluated using DSHA and PSHA methodologies. Since both methods are having its own advantages and disadvantages, it is better to evaluate the hazard using both the methods. Care should be taken in evaluating the maximum earthquake magnitude for each region and the recurrence rate. The initial estimates of ground motion parameters (PHA, PHV, or PHD) should be obtained at the bed rock level. The surface level values need to be calculated by considering the site response and local site conditions. Need to develop response spectra for bed rock conditions. It is advisable to get an estimate of ground motion duration also.

Seismic hazard analysis using multiple source/attenuation models along with logic tree methodology is found to be more effective for addressing epistemic uncertainties and providing realistic estimate of peak ground acceleration and spectral acceleration values.

7.6.6 Site Characterisation

Seismic site characterization based on geotechnical and geophysical tests is suitable for micro to meso-level zonation. However this methodology is not appropriate for a macro-level zonation as it is not economically and physically viable to carry out field tests for such a large area. Hence the site characterization based on topographic slope maps was found to be most suitable for the macro-level zonation. A comparison study has been made between the site characterization based on field tests and topographic slope maps at meso-level and a good correlation was found between them. Hence the study also shows that the site characterization based on topographic slope maps can be adopted for meso-level zonations also, however it is recommended for cities lying in low to moderate hazard zones.

7.6.7 Site Response Studies

Ground response analyses are very much required for micro-level zoning for estimating site amplification and surface level earthquake motion as it accurately captures the local soil effects. The empirical method of estimating site amplification based on site class coefficients and bedrock motion known as non-linear site amplification technique was found to be the most appropriate for macro-level zoning. For a meso-level zonation, the ground response analyses suitable for cities in high hazard zones having large population, and for the converse case, the non-linear site amplification techniques are preferable.

7.6.8 Liquefaction Hazard Assessment

The factor of safety against liquefaction can be evaluated based on the geotechnical data obtained from the filed. The liquefaction potential evaluation can be done

Table 7.7 Various methodologies and grid sizes suitable for the micro, meso and macro-level (after [151])

Earthquake hazards	Methodologies (grid size)		
	Micro-level	Meso-level	Macro-level
Seismic hazard at bedrock	Deterministic/ probabilistic approach (less than 1 km × 1 km)	Deterministic/ probabilistic approach (1 km × 1 km to 5 km × 5 km)	Deterministic/ probabilistic approach (5 km × 5 km to 10 km × 10 km)
Seismic site characterization	Geotechnical and geophysical testing (less than 1 km × 1 km)	Geotechnical and geophysical testing/ topographic slope map (1 km × 1 km to 5 km × 5 km)	Topographic slope map (5 km × 5 km to 10 km × 10 km)
Site amplification hazard	Ground response analyses (less than 1 km × 1 km)	Ground response analyses and Empirical method (1 km × 1 km to 5 km × 5 km)	Empirical method (5 km × 5 km to 10 km × 10 km)
Liquefaction hazard studies	Deterministic/ probabilistic method based on field test data (less than 1 km × 1 km)	Deterministic/ probabilistic method based on field test data (1 km × 1 km to 5 km × 5 km)	Performance based approach (Kramer and Mayfield [182]) (5 km × 5 km to 10 km × 10 km)

based on the deterministic or probabilistic methods. Liquefaction potential can be reported in terms of SPT and CPT values required to prevent liquefaction can also be developed (if required). Liquefaction factor of safety evaluated based on the available geotechnical data and using probabilistic techniques can then be used for the liquefaction hazard assessment. Same methodology can also be used to estimate liquefaction hazard at the meso-level for the cities lying in moderate to high hazard zone. For a macro-level study, the performance based approach for the evaluation of liquefaction hazard was found to be most appropriate and recommended.

Moreover, recommendations have also been made regarding suitability of various methodologies and the grid sizes to be adopted for site characterization and site effect studies at the micro, meso, and macro-level are presented in Table 7.7.

7.7 Summary

Seismic zonation is recognized as the most acceptable tool in seismic hazard assessment and risk evaluation. Making improvements on the conventional zonation maps and regional hazard maps, microzonation of a region generates detailed maps that predict the hazard at much smaller scales from one block of a city to another. Damage patterns of many recent earthquakes around the world, including the 1999 Chamoli and 2001 Bhuj earthquakes in India, have demonstrated that the soil conditions at a site may have a major effect on the ground shaking levels. Earthquake damage is commonly controlled by three interacting factors—source and path characteristics, local geological and geotechnical conditions, and design of the structures. Seismic zonation is the process of assessment of the first two factors to provide a basis for estimating and mapping potential damage to buildings, which in other words is quantification of the risk. Obviously, all of this would require analysis and presentation of a large amount of data consisting of geological, seismological, engineering specifications, etc. History of earthquakes, geometry and history of faults in the region, attenuation relationships, ground amplification, liquefaction susceptibility and landslide vulnerability are the other important inputs required.

Although several efforts are on for seismic zonation of selected regions in India, these studies are under development. While steps have been taken to strengthen research and development efforts on various aspects of earthquake related studies with the ultimate aim of mitigating the earthquake related disasters, efforts are also being made to prepare knowledge based products in the form of seismic hazard, vulnerability, and risk maps. The procedures are often modified or simplified and these are not adequately reflected in the final product. So also are the missing gaps in data, treatment of the data, etc. There is also an increasing tendency to do a partial or incomplete analysis of some available data, prepare some maps using such data, and refer to them as seismic zonation maps. Due to these inconsistencies in the procedures followed by various workers and the maps developed would be unacceptable to the end users.

In this book the suitability of different levels of seismic zonation for various regions with the country was assessed. The feasibility of various techniques for site characterization and estimation of site effects for regions at various scale levels were reviewed. Based on this, appropriate methods have been adopted and recommended for site characterization and estimation of local site effects for regions at different scale levels. Moreover, recommendations also have made regarding the grid size to be adopted for site characterization and site effect studies at micro, meso, and macro-level.

References

1. Adalier, K., Elgamal, A.: Mitigation of liquefaction and associated ground deformations by stone columns. Eng. Geol. **72**(3), 275–291 (2004)
2. AFPS, F.A.F.E.E.: Guidelines for Seismic Microzonation Studies. Delegation of Major Risks of the French Ministry of the Environment-Direction for Prevention, Pollution and Risks (1995)
3. Aki, K.: Space and time spectra of stationary stochastic waves, with special reference to microtremors. Bull. Earthq. Res. Inst., Univ. Tokyo **35**, 415–456 (1957)
4. Aki, K.: Maximum likelihood estimate of b in the formula logn= a-bm and its confidence limits. Bull. Earthq. Res. Inst. Tokyo **43**, 237–239 (1965)
5. Aki, K.: Generation and propagation of g waves from the Niigata earthquake of June 16, 1964: Part 1. A statistical analysis. Bull. Earthq. Res. Inst. Univ. Tokyo **44**(1), 23–72 (1966)
6. Aki, K.: Local site effects on ground motion. In: Earthquake Engineering and Soil Dynamics II-Recent Advances in Ground Motion Evaluation. Geotechnical Special Publication, vol. 20, pp. 103–155 (1988)
7. Alfaro, A., Pujades, L.G., Goula, X., Susagna, T., Navarro, M., Sanchez, J., Canas, J.: Preliminary map of soil's predominant periods in Barcelona using microtremors. In: Earthquake Microzoning, pp. 2499–2511. Springer, New York (2002)
8. Anbazhagan, P.: Site Characterization and Seismic Hazard Analysis with Local Site Effects for Microzonation of Bangalore. Indian Institute of Science, Bangalore (2007)
9. Anbazhagan, P., Sitharam, T.: Mapping of average shear wave velocity for Bangalore region: a case study. J. Environ. Eng. Geophys. **13**(2), 69–84 (2008)
10. Anbazhagan, P., Vinod, J., Sitharam, T.: Probabilistic seismic hazard analysis for Bangalore. Nat. Hazards **48**(2), 145–166 (2009)
11. Andersen, K.: Cyclic and static laboratory tests on Drammen clay. J. Geotech. Eng. Div. **106**(5), 499–529 (1980)
12. Andrews, D.: Objective determination of source parameters and similarity of earthquakes of different size. In: Earthquake Source Mechanics, pp. 259–267. American Geophysical Union, Washington (1986)
13. Andrus, R.D., Chung, R.M.: Ground Improvement Techniques for Liquefaction Remediation Near Existing Lifelines. US National Institute of Standards and Technology, Gaithersburg (1995)
14. Ansal, A., Erdik, M., Studer, J., Springman, S., Laue, J., Buchheister, J., Giardini, D., Faeh, D., Koksal, D.: Seismic microzonation for earthquake risk mitigation in turkey. In: 13th World Conference on Earthquake Engineering, Vancouver (2004)

T. G. Sitharam et al., *Comprehensive Seismic Zonation Schemes for Regions at Different Scales*, https://doi.org/10.1007/978-3-319-89659-5

15. Astroza, M., Monge, J.: Seismic microzones in the city of Santiago. Relation damage-geological unit. In: Proceedings of the Fourth International Conference on Seismic Zonation, vol. 3, pp. 25–29 (1991)
16. Atkinson, G., Boore, D.: Earthquake ground-motion prediction equations for eastern North America. Bull. Seismol. Soc. Am. **96**(6), 2181–2205 (2006)
17. Atkinson, G.M., Cassidy, J.F.: Integrated use of seismograph and strong-motion data to determine soil amplification: response of the Fraser river delta to the Duvall and Georgia Strait earthquakes. Bull. Seismol. Soc. Am. **90**(4), 1028–1040 (2000)
18. Bansal, B., Vandana, C.: Microzonation studies in India: DST initiatives. In: Proceedings of Workshop on Microzonation. Indian Institute of Science, Bangalore (2007)
19. Bard, P.Y., Tucker, B.E.: Underground and ridge site effects: a comparison of observation and theory. Bull. Seismol. Soc. Am. **75**(4), 905–922 (1985)
20. Bard, P.Y., Duval, A.M., Lebrun, B., Lachet, C., Riepl J. Hatzfeld, D.: Reliability of the H/V technique for site effects measurement: an experimental assessment.İ Seventh International Conference on Soil Dynamics and Earthquake Engineering, Istanbul (1997)
21. Basu, K.: A note on the Coimbatore earthquake of 8th February 1900. Indian J. Metamorph. Geol. **15**(2), 281–286 (1964)
22. Been, K., Jefferies, M.G.: A state parameter for sands. Géotechnique **35**(2), 99–112 (1985)
23. Bennington, N., Thurber, C., Roecker, S.: Three-dimensional seismic attenuation structure around the Safod site, Parkfield, California. Bull. Seismol. Soc. Am. **98**(6), 2934–2947 (2008)
24. Beresnev, I., Wen, K., Yeh, Y.: Nonlinear soil amplification: its corroboration in Taiwan. Bull. Seismol. Soc. Am. **85**(2), 496–515 (1995)
25. Bhatia, S., Kumar, M., Gupta, H.: A probabilistic seismic hazard map of India and adjoining regions. Ann. Geophys. **42**(6), 1153–1164 (1999)
26. Bian, H., Jia, Y., Shahrour, I.: A potential cost effective liquefaction mitigation countermeasure: induced partial saturation. In: AIP Conference Proceedings, vol. 1020, pp. 427–433. American Institute of Physics, New York (2008)
27. Bilham, R.: Earthquakes in India and the Himalaya: tectonics, geodesy and history. Ann. Geophys. **47**(2–3), 839–858 (2004)
28. BIS-1893: Indian Standard Criteria for Earthquake Resistant Design of Structures, Part 1 – Genera l Provisions and Buildings. Bureau of Indian Standards, New Delhi (2002)
29. BIS-2131: Method for Standard Penetration Test for Soils. Bureau of Indian Standards, New Delhi (2002)
30. Biswas, S., Gupta, A.D.: Some observations on the mechanism of earthquakes in the Himalaya and the Burmese arc. Tectonophysics **122**(3–4), 325–343 (1986)
31. Bommer, J., Scherbaum, F., Bungum, H., Cotton, F., Sabetta, F., Abrahamson, N.: On the use of logic trees for ground-motion prediction equations in seismic-hazard analysis. Bull. Seismol. Soc. Am. **95**(2), 377–389 (2005)
32. Bonilla, L.F., Steidl, J.H., Lindley, G.T., Tumarkin, A.G., Archuleta, R.J.: Site amplification in the San Fernando Valley, California: variability of site-effect estimation using the s-wave, coda, and h/v methods. Bull. Seismol. Soc. Am. **87**(3), 710–730 (1997)
33. Boominathan, A., Dodagoudar, G., Suganthi, A., Uma Maheswari, R.: Seismic hazard assessment of Chennai city considering local site effects. J. Earth Syst. Sci. **117**, 853–863 (2008)
34. Boore, D.: Stochastic simulation of high-frequency ground motions based on seismological models of the radiated spectra. Bull. Seismol. Soc. Am. **73**(6A), 1865–1894 (1983)
35. Boore, D.: Simulation of ground motion using the stochastic method. Pure Appl. Geophys. **160**, 635–676 (2003)
36. Boore, D.M.: Estimating vs (30)(or nehrp site classes) from shallow velocity models (depths < 30 m). Bull. Seismol. Soc. Am. **94**(2), 591–597 (2004)
37. Boore, D., Atkinson, G.: Ground-motion prediction equations for the average horizontal component of PGA, PGV, and 5%-damped PSA at spectral periods between 0.01 s and 10.0 s. Earthq. Spectra **24**(1), 99–138 (2008)

38. Boore, D., Joyner, W., Fumal, T.: Estimation of response spectra and peak accelerations from Western North American earthquakes: an Interim Report. US Geological Survey, Menlo Park (1993)
39. Borcherdt, R.: Estimates of site-dependent response spectra for design (methodology and justification). Earthq. Spectra **10**, 617–617 (1994)
40. Borcherdt, R.: Preliminary amplification estimates inferred from strong ground motion recordings of the Northridge earthquake of January 17, 1994. In: Proceeding of the International Workshop on Site Response Subjected to Strong Ground Motion (1996)
41. Borcherdt, R.D.: Empirical evidence for acceleration-dependent amplification factors. Bull. Seismol. Soc. Am. **92**(2), 761–782 (2002)
42. Borcherdt, R., Gibbs, J.: Effects of local geological conditions in the San Francisco bay region on ground motions and the intensities of the 1906 earthquake. Bull. Seismol. Soc. Am. **66**(2), 467–500 (1976)
43. Borcherdt, R.D., Glassmoyer, G.: Influences of local geology on strong and weak ground motions recorded in the San Francisco Bay region, p. 77. US Geological Survey Professional Paper (1994)
44. Borcherdt, R., Wentworth, C., Glassmoyer, G., Fumal, T., Mork, P., Gibbs, J.: On the observation, characterization and predictive GIS mapping of ground response in the San Francisco Bay region, California. In: Proceedings of 4th International Conference on Seismic Zonation, Stanford, pp. 545–552 (1991)
45. Bouckovalas, G., Kouretzis, G.: A review of soil and topography effects in Athens 09/07/199 (Greece) earthquake. In: Proceedings of Fourth International Conference on Recent Advances in Geotechnical Earthquake Engineering and Soil Dynamics, San Diego, March (in CDROM) (2001)
46. Bouckovalas, G.D., Papadimitriou, A.G.: Numerical evaluation of slope topography effects on seismic ground motion. Soil Dyn. Earthq. Eng. **25**(7), 547–558 (2005)
47. Boulanger, R.W., Hayden, R.F.: Aspects of compaction grouting of liquefiable soil. J. Geotech. Eng. **121**(12), 844–855 (1995)
48. Bowles, J.: Foundation Analysis and Design, 5th edn. McGraw-Hill, New Delhi (1988)
49. Brown, R.E.: Vibroflotation compaction of cohesionless soils. J. Geotech. Geoenviron. Eng. **103**, 1437–1451 (1977). ASCE 13415 Proceeding
50. Brune, J.N.: Preliminary results on topographic seismic amplification effect on a foam rubber model of the topography near Pacoima Dam. In: Proceeding of the 8th World Conference on Earthquake Engineering, vol. 2, pp. 663–670 (1984)
51. BSSC: NEHRP recommended provisions for seismic regulations for new buildings and other structures (FEMA 450), Part 1: Provisions. Building Seismic Safety Council for the Federal Emergency Management Agency, Washington (2003)
52. Budnitz, R.J., Apostolakis, G., Boore, D.M.: Recommendations for probabilistic seismic hazard analysis: guidance on uncertainty and use of experts. Tech. rep., Nuclear Regulatory Commission, Washington. Div. of Engineering Technology; Lawrence Livermore National Lab.; Electric Power Research Inst., Palo Alto; USDOE, Washington (1997)
53. Camp III, W.M., Camp, H.C., Andrus, R.D.: Liquefaction mitigation using air injection. In: Fifth International Conferences on Recent Advances in Geotechnical Earthquake Engineering and Soil Dynamics, San Diego (2010)
54. Campbell, K.: Empirical near-source attenuation relationships for horizontal and vertical components of peak ground acceleration, peak ground velocity, and pseudo-absolute acceleration response spectra. Seismol. Res. Lett. **68**(1), 154–179 (1997)
55. Campbell, K., Bozorgnia, Y.: Updated near-source ground-motion (attenuation) relations for the horizontal and vertical components of peak ground acceleration and acceleration response spectra. Bull. Seismol. Soc. Am. **93**(1), 314–331 (2003)
56. Campillo, M., Bard, P., Nicollin, F., Sánchez-Sesma, F.: The Mexico earthquake of September 19, 1985 – the incident wavefield in Mexico city during the great Michoacán earthquake and its interaction with the deep basin. Earthq. Spectra **4**(3), 591–608 (1988)

57. Casagrande, A.: Characteristics of cohesionless soils affecting the stability of slopes and earth fills. J. Boston Soc. Civ. Eng. 257–276 (1936). Reprinted in Contributions to soil mechanics, 1925–1940, Boston Society of Civil Engineers, Boston, Oct. 1940
58. Casagrande, A.: Liquefaction and Cyclic Deformation of Sands-A Critical Review. Harvard Soil Mechanics Series. Harvard University Press, Cambridge (1976)
59. Castro, G.: Liquefaction of sands. Harvard University Division of Engineering and Applied Mechanics (1969)
60. Castro, R.R., Condori, C., Romero, O., Jacques, C., Suter, M.: Seismic attenuation in northeastern Sonora, Mexico. Bull. Seismol. Soc. Am. **98**(2), 722–732 (2008)
61. Cetin, K., Seed, R., Der Kiureghian, A., Tokimatsu, K., Harder L. Jr., Kayen, R., Moss, R.: Standard penetration test-based probabilistic and deterministic assessment of seismic soil liquefaction potential. J. Geotech. Geoenviron. Eng. **130**(12), 1314–1340 (2004)
62. Chandra, U.: Earthquakes of peninsular India-a seismotectonic study. Bull. Seismol. Soc. Am. **67**(5), 1387–1413 (1977)
63. Chandra, U.: Seismicity, earthquake mechanisms and tectonics along the Himalayan mountain range and vicinity. Phys. Earth Planet. Inter. **16**(2), 109–131 (1978)
64. Chávez-García, F.J., Cuenca, J.: Site effects and microzonation in Acapulco. Earthq. Spectra **14**(1), 75–93 (1998)
65. Chávez-García, F.J., Sánchez, L., Hatzfeld, D.: Topographic site effects and HVSR a comparison between observations and theory. Bull. Seismol. Soc. Am. **86**(5), 1559–1573 (1996)
66. Cheng, X.H., Qiang, M., Zuan, Y., et al.: Dynamic response of liquefiable sand foundation improved by bio-grouting. Chin. J. Geotech. Eng. **35**(8), 1486–1495 (2013)
67. Chiou, B., Youngs, R.: PEER-NGA empirical ground motion model for the average horizontal component of peak acceleration and pseudo-spectral acceleration for spectral periods of 0.01 to 10 seconds. Interim Report Issued for USGS Review (2006)
68. Chiou, B.J., Youngs, R.R.: An nga model for the average horizontal component of peak ground motion and response spectra. Earthq. Spectra **24**(1), 173–215 (2008)
69. Chouet, B., De Luca, G., Milana, G., Dawson, P., Martini, M., Scarpa, R.: Shallow velocity structure of Stromboli Volcano, Italy, derived from small-aperture array measurements of Strombolian tremor. Bull. Seismol. Soc. Am. **88**(3), 653–666 (1998)
70. Chun, K.Y., Henderson, G.A.: Lg attenuation near the North Korean border with China, part II: model development from the 2006 nuclear explosion in North Korea. Bull. Seismol. Soc. Am. **99**(5), 3030–3038 (2009)
71. Commission on Geosciences, Environment and Resources (CGER): Ground Water at Yucca Mountain: How High Can It Rise? National Academies Press, Washington (1992)
72. Cornell, C.: Engineering seismic risk analysis. Bull. Seismol. Soc. Am. **58**(5), 1583–1606 (1968)
73. Costa, G., Panza, G., Suhadolc, P., Vaccari, F.: Zoning of the Italian territory in terms of expected peak ground acceleration derived from complete synthetic seismograms. J. Appl. Geochem. **30**(1), 149–160 (1993)
74. Cundall, P.A.: A computer model for simulating progressive large scale movements in blocky rock systems. In: Proceedings of ISRM Symposium, vol. 2, Nancy, pp. 129–136 (1971)
75. Davis, L.L., West, L.R.: Observed effects of topography on ground motion. Bull. Seismol. Soc. Am. **63**(1), 283–298 (1973)
76. DeJong, J.T., Fritzges, M.B., Nüsslein, K.: Microbially induced cementation to control sand response to undrained shear. J. Geotech. Geoenviron. Eng. **132**(11), 1381–1392 (2006)
77. Dinesh, S., Sitharam, T., Vinod, J.: Dynamic properties and liquefaction behaviour of granular materials using discrete element method. Curr. Sci. **87**, 1379–1387 (2004)
78. Dobry, R., Ladd, R.S., Yokel, F.Y., Chung, R.M., Powell, D.: Prediction of Pore Water Pressure Buildup and Liquefaction of Sands During Earthquakes by the Cyclic Strain Method, vol. 138. US Department of Commerce National Bureau of Standards, Washington, DC (1982)

79. Dunbar, P.K., Lockridge, P.A., Whiteside, L.S.: Catalog of Significant Earthquakes, 2150 BC-1991 AD: Including Quantitative Casualties and Damage. US Department of Commerce, National Oceanic and Atmospheric Administration, Environmental Data and Information Service, National Geophysical Data Center, Washington (1992)
80. Duputel, Z., Rivera, L., Kanamori, H., Hayes, G.: W phase source inversion for moderate to large earthquakes (1990–2010). Geophys. J. Int. **189**(2), 1125–1147 (2012)
81. El Mohtar, C.S.: Pore fluid engineering: an autoadaptive design for liquefaction mitigation. Doctoral dissertation, Purdue University (2008)
82. El Mohtar, C., Bobet, A., Santagata, M., Drnevich, V., Johnston, C.: Liquefaction mitigation using bentonite suspensions. J. Geotech. Geoenviron. Eng. **139**(8), 1369–1380 (2012)
83. El-Sayed, A., Vaccari, F., Panza, G.: Deterministic seismic hazard in Egypt. Geophys. J. Int. **144**(3), 555–567 (2001)
84. El Shamy, U., Abdelhamid, Y.: Modeling granular soils liquefaction using coupled lattice Boltzmann method and discrete element method. Soil Dyn. Earthq. Eng. **67**, 119–132 (2014)
85. Eseller-Bayat, E., Yegian, M., Alshawabkeh, A., Gokyer, S.: Prevention of liquefaction during earthquakes through induced partial saturation in sands. In: Geotechnical Engineering: New Horizons, pp. 188–194. IOS Press, Amsterdam (2012)
86. Eurocode-8: BS-EN 1998-1, Design of structures for earthquake resistance? part 1: general rules, seismic actions and rules for buildings. European Committee for Standardization, Brussels (2003)
87. Evernden, J., Thomson, J.: Predicting seismic intensities. In: Evaluating Earthquake Hazards in the Los Angeles Region – An Earth-Science Perspective, vol. 1360, pp. 151–202. US Geological Survey Professional Paper (1985)
88. Faccioli, E.: Seismic amplification in the presence of geological and topographic irregularities. In: Proceedings of the Second International Conference on Recent Advances in Geotechnical Earthquake Engineering and Soil Dynamics, pp. 11–15 (1991)
89. Faccioli, E., Pessina, V.: The Catania Project: Earthquake Damage Scenarios for High Risk Area in the Mediterranean. CNR-Gruppo Nazionale per la Difesa dai Terremoti, Roma (2000)
90. Fah, D., Iodice, C., Suhadolc, P., Panza, G.: Application of numerical simulations for a tentative seismic microzonation of the city of Rome. Ann. Geophys. **38**(5–6), 607–615 (1995)
91. Fäh, D., Kind, F., Lang, K., Giardini, D.: Earthquake scenarios for the city of Basel. Soil Dyn. Earthq. Eng. **21**(5), 405–413 (2001)
92. Field, E., Jacob, K.: The theoretical response of sedimentary layers to ambient seismic noise. Geophys. Res. Lett. **20**(24), 2925–2928 (1993)
93. Finn, W., Pickering, D., Bransby, P.: Sand liquefaction in triaxial and simple shear tests. J. Soil Mech. Found. Div. **97**(SM4), 639–659 (1971)
94. Ford, S.R., Dreger, D.S., Mayeda, K., Walter, W.R., Malagnini, L., Phillips, W.S.: Regional attenuation in northern California: a comparison of five 1d q methods. Bull. Seismol. Soc. Am. **98**(4), 2033–2046 (2008)
95. Foti, S.: Multistation methods for geotechnical characterization using surface waves. Ph.D. thesis, Politecnico di Torino (2000)
96. Frankel, A.: Mapping seismic hazard in the central and eastern United States. Seismol. Res. Lett. **66**(4), 8–21 (1995)
97. Fumal, T., Tinsley, J.: Mapping shear-wave velocities of near-surface geologic materials. In: Evaluating Earthquake Hazards in the Los Angeles Region: An Earth-Science Perspective, vol. 1360, pp. 101–126. US Geological Survey Professional Paper (1985)
98. Gallagher, P.M.: Passive site remediation for mitigation of liquefaction risk. Ph.D. thesis (2000)
99. Gallagher, P.M., Mitchell, J.K.: Influence of colloidal silica grout on liquefaction potential and cyclic undrained behavior of loose sand. Soil Dyn. Earthq. Eng. **22**(9), 1017–1026 (2002)
100. Gallagher, P.M., Pamuk, A., Abdoun, T.: Stabilization of liquefiable soils using colloidal silica grout. J. Mater. Civ. Eng. **19**(1), 33–40 (2007)
101. Ganesha Raj, K., Nijagunappa, R.: Major lineaments of Karnataka state and their relation to seismicity-remote sensing based analysis. J. Geol. Soc. India **63**, 430–439 (2004)

102. Gangrade, B., Arora, S.: Seismicity of the Indian peninsular shield from regional earthquake data. Pure Appl. Geophys. **157**(10), 1683–1705 (2000)
103. Gazetas, G.: Seismic response of earth dams: some recent developments. Soil Dyn. Earthq. Eng. **6**(1), 2–47 (1987)
104. Gazetas, G., Dakoulas, P.: Seismic analysis and design of rockfill dams: state-of-the-art. Soil Dyn. Earthq. Eng. **11**(1), 27–61 (1992)
105. Geli, L., Bard, P., Jullien, B.: The effect of topography on earthquake ground motion: a review and new results. Bull. Seismol. Soc. Am. **78**(1), 42–63 (1988)
106. Ghasemi, H., Zare, M., Fukushima, Y., Koketsu, K.: An empirical spectral ground-motion model for Iran. J. Seismol. **13**(4), 499–515 (2009)
107. Giardini, D.: Systematic analysis of deep seismicity: 200 centroid-moment tensor solutions for earthquakes between 1977 and 1980. Geophys. J. Int. **77**(3), 883–914 (1984)
108. Gitterman, Y., Zaslavsky, Y., Shapira, A., Shtivelman, V.: Empirical site response evaluations: case studies in Israel. Soil Dyn. Earthq. Eng. **15**(7), 447–463 (1996)
109. Guha, S., Basu, P.: Catalogue of earthquakes (=> m 3.0) in peninsular India. Tech. rep., Atomic Energy Regulatory Board (1993)
110. Gupta, I.: Response spectral attenuation relations for in-slab earthquakes in Indo-Burmese subduction zone. Soil Dyn. Earthq. Eng. **30**(5), 368–377 (2010)
111. Gutenberg, B., Richter, C.: Magnitude and energy of earthquakes. Ann. Geophys. **83**(2147), 183–185 (1936)
112. Gutenberg, B., Richter, C.: Frequency of earthquakes in California. Bull. Seismol. Soc. Am. **34**(4), 185–188 (1944)
113. Gutenberg, B.U., Richter, C.F.: Seismicity of the Earth and Related Phenomena. Princeton University Press, Princeton (1954)
114. Gutenberg, B., Richter, C.: Magnitude and energy of earthquakes. Ann. Geophys. **9**, 1–15 (1956)
115. Haghshenas, E., Bard, P., Theodulidis, N.: Empirical evaluation of microtremor h/v spectral ratio. Bull. Earthq. Eng. **6**(1), 75–108 (2008)
116. Haldavnekar, V., Bobet, A., Santagata, M., Drnevich, V.: Soil treatment with a thixotropic fluid: an autoadaptive design for liquefaction prevention. In: Proceedings of the 11th International Conference on Soil Dynamics and Earthquake Engineering and 3rd International Conference on Earthquake Geotechnical Engineering, vol. 2, pp. 553–560 (2003)
117. Hall, L., Bodare, A.: Analyses of the cross-hole method for determining shear wave velocities and damping ratios. Soil Dyn. Earthq. Eng. **20**(1), 167–175 (2000)
118. Hanzawa, H., Itoh, Y., Suzuki, K.: Shear characteristics of a quick sand in the Arabian gulf. Soils Found. **19**(4), 1–15 (1979)
119. Harada, N., Towhata, I., Takatsu, T., Tsunoda, S., Sesov, V.: Development of new drain method for protection of existing pile foundations from liquefaction effects. Soil Dyn. Earthq. Eng. **26**(2), 297–312 (2006)
120. Hartzell, S.: Earthquake aftershocks as Green's functions. Geophys. Res. Lett. **5**(1), 1–4 (1978)
121. Hartzell, S., Cranswick, E., Frankel, A., Carver, D., Meremonte, M.: Variability of site response in the Los Angeles urban area. Bull. Seismol. Soc. Am. **87**(6), 1377–1400 (1997)
122. Hartzell, S., Carver, D., Cranswick, E., Frankel, A.: Variability of site response in Seattle, Washington. Bull. Seismol. Soc. Am. **90**(5), 1237–1250 (2000)
123. Hashash, Y.M.A., Groholski, D., Phillips, C., Park, D., Musgrove, M.: DEEPSOIL 4.0, User Manual and Tutorial. US Geological Survey, Washington (2011)
124. Hayashi, K., Suzuki, H.: Cmp cross-correlation analysis of multi-channel surface-wave data. Explor. Geophys. **35**(1), 7–13 (2004)
125. He, J., Ivanov, V., Chu, J.: Mitigation of liquefaction of saturated sand using biogas. Géotechnique **63**, 267–275 (2013)
126. Heaton, T., Tajima, F., Mori, A.: Estimating ground motions using recorded accelerograms. Surv. Geophys. **8**(1), 25–83 (1986)

127. Howell, R., Rathje, E.M., Kamai, R., Boulanger, R.: Centrifuge modeling of prefabricated vertical drains for liquefaction remediation. J. Geotech. Geoenviron. Eng. **138**(3), 262–271 (2012)
128. Huang, Y., Wen, Z.: Recent developments of soil improvement methods for seismic liquefaction mitigation. Nat. Hazards **76**(3), 1927–1938 (2015)
129. Iai, S., Koizumi, K., Noda, S., Tsuchida, H.: Large Scale Model Tests and Analyses of Gravel Drains. Un'yushō Kōwan Gijutsu Kenkyūjo, Yokosuka (1988)
130. IBC: International Building Code. International Code Council, Washington (2009)
131. Idriss, I., Boulanger, R.: Semi-empirical procedures for evaluating liquefaction potential during earthquakes. Soil Dyn. Earthq. Eng. **26**(2), 115–130 (2006)
132. Idriss, I., Sun, J.: User's Manual for shake91. Center for Geotechnical Modeling, Department of Civil Engineering, University of California, Davis (1992)
133. Imai, T.: P and s wave velocities of the ground in Japan. In: Proceeding of the 9th ICSMFE, vol. 2, pp. 257–260 (1977)
134. Imai, T.: Correlation of n-value with s-wave velocity and shear modulus. In: Proceedings of the Second European Symposium on Penetration Testing, pp. 67–72 (1981)
135. Imai, T., Yoshimura, Y.: Elastic wave velocity and soil properties in soft soil. Tsuchito-Kiso **18**(1), 17–22 (1970)
136. Imai, T., Yoshimura, M.: The relation of mechanical properties of soils to p and s wave velocities for soil ground in Japan. Urana Research Institute, OYO Corp., Tokyo (1972)
137. Irikura, K.: Semi-empirical estimation of strong ground motions during large earthquakes. Bull. Disaster Prev. Res. Inst. Kyoto Univ. **33**(2), 63–104 (1983)
138. Ishihara, K.: Stability of natural deposits during earthquakes. In: Proceedings of the 11th International Conference on Soil Mechanics and Foundation Engineering, vol. 1, pp. 321–376 (1985)
139. Ishihara, K.: Dynamic properties of soils and gravels from laboratory tests. In: Soil Dynamics and Geotechnical Engineering, pp. 1–17. Balkema, Rotterdam (1993)
140. Ishihara, K.: Soil Behaviour in Earthquake Geotechnics. Clarendon Press/Oxford University Press, Oxford (1996)
141. Ishihara, K.: Liquefaction resistance of nearly saturated sand as correlated with longitudinal wave velocity. In: Poromechanics, pp. 583–586. Balkema, Rotterdam (1998)
142. Ishihara, K., Li, S.: Liquefaction of saturated sand in triaxial torsion shear test. Soils Found. **12**(2), 19–39 (1972)
143. Ishihara, K., Tatsuoka, F., Yasuda, S.: Undrained deformation and liquefaction of sand under cyclic stresses. Soils Found. **15**(1), 29–44 (1975)
144. ISSMGE, T.: Manual for Zonation on Seismic Geotechnical Hazard. International Society of Soil Mechanics and Geotechnical Engineering (ISSMGE). The Japanese Geotechnical Society, Tokyo (1999)
145. Iwasaki, T., Tokida, K., Tatsuoka, F., Watanabe, S., Yasuda, S., Sato, H.: Microzonation for soil liquefaction potential using simplified methods. In: Proceedings 3rd International Conference on Microzonation, pp. 1319–1330 (1982)
146. Iyengar, R., Ghosh, S.: Microzonation of earthquake hazard in greater Delhi area. Curr. Sci. **87**(9), 1193–1202 (2004)
147. Iyisan, R.: Correlations between shear wave velocity and in-situ penetration test results. Tech. J. Chamber Civil Eng. Turkey **7**, 371–374 (1996)
148. Jafari, M., Asghari, A., Rahmani, I.: Empirical correlation between shear wave velocity (vs) and spt-N value for south of Tehran soils. In: Proceedings of the 4th International Conference on Civil Engineering, Tehran (1997)
149. Jaiswal, K., Sinha, R.: Probabilistic seismic-hazard estimation for peninsular India. Bull. Seismol. Soc. Am. **97**(1B), 318–330 (2007)
150. James, N.: Site characterization and assessment of various earthquake hazards for micro and macro-level seismic zonations of regions in the peninsular India. Ph.D. thesis, Indian Institute of Science, Bangalore (2013)

151. James, N., Sitharam, T.: Seismic zonations at micro and macro-level for regions in the peninsular India. Int. J. Geotech. Earthq. Eng. **7**(2), 35–63 (2016)
152. James, N., Sitharam, T., Vipin, K.: Assessment of liquefaction potential index using deterministic and probabilistic approaches. Int. J. Geotech. Earthq. Eng. **3**(2), 60–76 (2012)
153. James, N., Sitharam, T., Padmanabhan, G., Pillai, C.: Seismic microzonation of a nuclear power plant site with detailed geotechnical, geophysical and site effect studies. Nat. Hazards **71**(1), 419–462 (2014)
154. Johnson, J.M., Tanioka, Y., Ruff, L.J., Satake, K., Kanamori, H., Sykes, L.R.: The 1957 great Aleutian earthquake. In: Shallow Subduction Zones: Seismicity, Mechanics and Seismic Potential, pp. 3–28. Springer, Berlin (1994)
155. Johnston, A.: Seismic moment assessment of earthquakes in stable continental regions, instrumental seismicity. Geophys. J. Int. **124**(2), 381–414 (1996)
156. Joyner, W., Boore, D.: Peak horizontal acceleration and velocity from strong-motion records including records from the 1979 imperial valley, California, earthquake. Bull. Seismol. Soc. Am. **71**(6), 2011–2038 (1981)
157. Joyner, W., Fumal, T.: Use of measured shear-wave velocity for predicting geologic site effects on strong ground motion. In: Proceeding of the 8th World Conference on Earthquake Engineering, vol. 2, pp. 777–783 (1984)
158. Kagami, H., Okada, S., Ohta, Y.: Versatile application of dense and precision seismic intensity data by an advanced questionnaire survey. In: Proceeding Ninth World Conference on Earthquake Engineering, vol. 8, pp. 937–942 (1988)
159. Kaila, K., Sarkar, D.: Atlas of Isoseismal Maps of Major Earthquakes in India. National Geophysical Research Institute, Hyderabad (1978)
160. Kalkan, E., Gülkan, P., Yilmaz, N., Çelebi, M.: Reassessment of probabilistic seismic hazard in the Marmara region. Bull. Seismol. Soc. Am. **99**(4), 2127–2146 (2009)
161. Kanai K., Osada, T., Tanaka, T.: An investigation into the nature of microtremors. Bull. Earthq. Res. Inst. **32**, 199–209 (1954)
162. Kanai, K.: On microtremors VIII. Bull. Earthq. Res. Inst. **39**, 97–114 (1961)
163. Kanai, K., Tanaka, T., Morishita, T., Osada, K.: Observation of microtremors. xi.: Matsushiro earthquake swarm area. Bull. Earthq. Res. Inst. Tokyo Univ. **44**(3), 1297–1333 (1967). http://hdl.handle.net/2261/12296
164. Kanamori, H.: Seismological evidence for a lithospheric normal faulting – the Sanriku earthquake of 1933. Phys. Earth Planet. Inter. **4**(4), 289–300 (1971)
165. Kanamori, H.: Re-examination of the earth's free oscillations excited by the Kamchatka earthquake of November 4, 1952. Phys. Earth Planet. Inter. **11**(3), 216–226 (1976)
166. Kanamori, H.: The energy release in great earthquakes. J. Geophys. Res. **82**(20), 2981–2987 (1977)
167. Kanamori, H., Anderson, D.L.: Theoretical basis of some empirical relations in seismology. Bull. Seismol. Soc. Am. **65**(5), 1073–1095 (1975)
168. Karol, R.H.: Chemical grouts and their properties. In: Grouting in Geotechnical Engineering, pp. 359–377. American Society of Civil Engineers, New York (1982)
169. Kaul, M.: Stochastic characterization of earthquakes through their response spectrum. Earthq. Eng. Struct. Dyn. **6**(5), 497–509 (1978)
170. Kelkar, Y.: Earthquakes experienced in Maharashtra during the last 300 years. Daily Kesari Poona (1968)
171. Khattri, K., Rogers, A., Perkins, D., Algermissen, S.: A seismic hazard map of India and adjacent areas. Tectonophysics **108**(1), 93–134 (1984)
172. Khoubbi-Al, I., Adams, J.: Local site effects in Ottawa, Canada–first results from a strong motion network. In: Proceedings of the 13th World Conference on Earthquake Engineering, Paper, 2504 (2004)
173. Kiratzi, A., Karakaisis, G., Papadimitriou, E., Papazachos, B.: Seismic source-parameter relations for earthquakes in Greece. Pure Appl. Geophys. **123**(1), 27–41 (1985)
174. Kokusho, T.: Cyclic triaxial test of dynamic soil properties for wide strain range. Soils Found. **20**(2), 45–60 (1980)

175. Kolathayar, S., Sitharam, T.: Characterization of regional seismic source zones in and around India. Seismol. Res. Lett. **83**(1), 77–85 (2012)
176. Kolathayar, S., Sitharam, T.: Comprehensive probabilistic seismic hazard analysis of the Andaman–Nicobar regions. Bull. Seismol. Soc. Am. **102**(5), 2063–2076 (2012)
177. Kolathayar, S., Sitharam, T., Vipin, K.: Deterministic seismic hazard macrozonation of India. J. Earth Syst. Sci. **121**(5), 1351–1364 (2012)
178. Kolathayar, S., Vipin, K., Sitharam, T.: Recent seismicity in India and adjoining regions. Int. J. Earth Sci. Eng. **5**(1), 51–59 (2012)
179. Konno, K., Ohmachi, T.: Ground-motion characteristics estimated from spectral ratio between horizontal and vertical components of microtremor. Bull. Seismol. Soc. Am. **88**(1), 228–241 (1998)
180. Koulakov, I., Bindi, D., Parolai, S., Grosser, H., Milkereit, C.: Distribution of seismic velocities and attenuation in the crust beneath the North Anatolian Fault (Turkey) from local earthquake tomography. Bull. Seismol. Soc. Am. **100**(1), 207–224 (2010)
181. Kramer, S.: Geotechnical Earthquake Engineering. Pearson Education, Delhi (1996). Reprinted 2003
182. Kramer, S., Mayfield, R.: Return period of soil liquefaction. J. Geotech. Geoenviron. Eng. **133**(7), 802–813 (2007)
183. Krishnaswamy, N., Isaac, N.T.: Liquefaction potential of reinforced sand. Geotext. Geomembr. **13**(1), 23–41 (1994)
184. Kumar, A.: Software for generation of spectrum compatible time history. In: 13th World Conference on Earthquake Engineering Vancouver (2004)
185. Kumar, P., Yuan, X., Kumar, M.R., Kind, R., Li, X., Chadha, R.: The rapid drift of the Indian tectonic plate. Nature **449**(7164), 894–897 (2007)
186. Kumar, A., Anbazhagan, P., Sitharam, T.: Seismic hazard analysis of Lucknow considering local and active seismic gaps. Nat. Hazards **69**(1), 327–350 (2013)
187. Lachet, C., Hatzfeld, D., Bard, P.Y., Theodulidis, N., Papaioannou, C., Savvaidis, A.: Site effects and microzonation in the city of Thessaloniki (Greece) comparison of different approaches. Bull. Seismol. Soc. Am. **86**(6), 1692–1703 (1996)
188. Langston, C.A.: Structure under mount rainier, Washington, inferred from teleseismic body waves. J. Geophys. Res. Solid Earth **84**(B9), 4749–4762 (1979)
189. Lapajne, J., Motnikar, B., Zupančič, P.: Probabilistic seismic hazard assessment methodology for distributed seismicity. Bull. Seismol. Soc. Am. **93**(6), 2502–2515 (2003)
190. Lee, M.K., Finn, W.: DESRA-2: dynamic effective stress response analysis of soil deposits with energy transmitting boundary including assessment of liquefaction potential. Department of Civil Engineering, University of British Columbia (1978)
191. Lee, S.J., Chen, H.W., Liu, Q., Komatitsch, D., Huang, B.S., Tromp, J.: Three-dimensional simulations of seismic-wave propagation in the Taipei basin with realistic topography based upon the spectral-element method. Bull. Seismol. Soc. Am. **98**(1), 253–264 (2008)
192. Lermo, J., Chávez-García, F.J.: Site effect evaluation using spectral ratios with only one station. Bull. Seismol. Soc. Am. **83**(5), 1574–1594 (1993)
193. Liao, S., Whitman, R.: A Catalog of Liquefaction and Non-liquefaction Occurrences During Earthquakes. Department of Civil Engineering, MIT, Cambridge (1986)
194. Liao, S., Veneziano, D., Whitman, R.: Regression models for evaluating liquefaction probability. J. Geotech. Eng. **114**(4), 389–411 (1988)
195. Lin, C.P., Chang, C.C., Chang, T.S.: The use of masw method in the assessment of soil liquefaction potential. Soil Dyn. Earthq. Eng. **24**(9), 689–698 (2004)
196. Liu, R.F., Chen, Y.T., Ren, X., Xu, Z.G., Sun, L., Yang, H., Liang, J.H., Ren, K.X.: Comparison between different earthquake magnitudes determined by china seismograph network. Acta Seismol. Sin. **20**, 497–506 (2007)
197. Loh, C.H., Hwang, J.Y., Shin, T.C.: Observed variation of earthquake motion across a Basin – Taipei city. Earthq. Spectra **14**(1), 115–133 (1998)
198. López, A.M., Okal, E.A.: A seismological reassessment of the source of the 1946 aleutian 'tsunami' earthquake. Geophys. J. Int. **165**(3), 835–849 (2006)

199. Lungu, D., Aldea, A., Cornea, T., Arion, C.: Seismic microzonation of the city of Bucharest. In: 6th International Conference on Seismic Zonation, Palm Springs (2000)
200. Lunne, T., Robertson, P., Powell, J.: Cone Penetration Testing. Geotechnical Practice (1997)
201. Ma, S., Archuleta, R.J., Page, M.T.: Effects of large-scale surface topography on ground motions, as demonstrated by a study of the San Gabriel mountains, Los Angeles, California. Bull. Seismol. Soc. Am. **97**(6), 2066–2079 (2007)
202. MacMurdo, J.: Papers relating to the earthquake which occurred in India in 1819. Philos. Mag. **63**, 105–177 (1824)
203. Maher, M.H., Woods, R.D.: Dynamic response of sand reinforced with randomly distributed fibers. J. Geotech. Eng. **116**(7), 1116–1131 (1990)
204. Malagnini, L., Rovelli, A., Hough, S., Seeber, L.: Site amplification estimates in the Garigliano Valley, Central Italy, based on dense array measurements of ambient noise. Bull. Seismol. Soc. Am. **83**(6), 1744–1755 (1993)
205. Malagnini, L., Mayeda, K., Uhrhammer, R., Akinci, A., Herrmann, R.B.: A regional ground-motion excitation/attenuation model for the San Francisco region. Bull. Seismol. Soc. Am. **97**(3), 843–862 (2007)
206. Mallet, R.: Great Neapolitan Earthquake of 1857. Chapman and Hall, London (1862)
207. Mancuso, C.: Damping of soil by cross hole method. In: Proceedings of the International Conference on Soil Mechanics and Foundation Engineering-International Society for Soil Mechanics and Foundation Engineering, vol. 3, pp. 1337–1337. AA Balkema, Rotterdam (1994)
208. Marcellini, A., Daminelli, R., Pagani, M., Riva, F., Tento, A., Crespellani, T., Madiai, C., Vannucchi, G., Frassineti, G., Martelli, L., et al.: Seismic microzonation of some municipalities of the Rubicone area (Emilia-Romagna region). In: Bisch, P., Labb, P., Pecker, A. (eds.) Proceedings of the Eleventh European Conference on Earthquake Engineering, Invited Lectures Volume, pp. 339–350. Balkema, Rotterdam (1999)
209. Marcuson, W., Hynes, M., Franklin, A.: Evaluation and use of residual strength in seismic safety analysis of embankments. Earthq. Spectra **6**(3), 529–572 (1990)
210. Mashiri, M., Vinod, J., Sheikh, M.N.: Liquefaction potential and dynamic properties of sand-tyre chip (STCh) mixtures. Geotech. Test. J. **39**, 69–79 (2016)
211. Matsuoka, M., Wakamatsu, K., Fujimoto, K., Midorikawa, S.: Nationwide site amplification zoning using GIS-based Japan engineering geomorphologic classification map. In: Proceedings of the 9th International Conference on Structural Safety and Reliability, pp. 239–246 (2005)
212. Mayer-Rosa, D., Jimenez, M.: Seismic zoning, recommendations for Switzerland. Landeshydrologie und-geologie, geologischer bericht (1999)
213. Mayne, P.W., Rix, G.J.: Correlations between shear wave velocity and cone tip resistance in natural clays. Soils Found. **35**(2), 107–110 (1995)
214. Medvedev, J.: Engineering Seismology, p. 260. Academia Nauk Press, Moscow (1962)
215. Menon, A., Ornthammarath, T., Corigliano, M., Lai, C.: Probabilistic seismic hazard macrozonation of Tamil Nadu in southern India. Bull. Seismol. Soc. Am. **100**(3), 1320–1341 (2010)
216. Midorikawa, S.: Prediction of isoseismal map in the Kanto plain due to hypothetical earthquake. J. Struct. Eng. B **33**, 43–48 (1987)
217. Millea, M.: Liquefaction mitigation technology. Tech. rep., Naval Civil Engineering Lab Port Hueneme (1990)
218. Mitrani, H., Madabhushi, S.: Centrifuge modelling of inclined micro-piles for liquefaction remediation of existing buildings. Geomech. Geoeng.: Int. J. **3**(4), 245–256 (2008)
219. Mitsunan, T., Isemoto, N., Yasui, M., Shimoda, M., Bessho, M., Hanada, M., et al.: A use of super lime pile method for improvement of saturated loose sandy soils. In: The Third International Offshore and Polar Engineering Conference International Society of Offshore and Polar Engineers (1993)
220. Miyakoshi, K., Okada, H.: Estimation of the site response in the Kushiro city, Hokkaido, Japan, using microtremors with seismometer arrays. In: 10th World Conference on Earthquake Engineering (1996)

221. Mochizuki, Y., Fukushima, S., Kagawa, K.: Shaking table test on reinforced sand. In: Proceedings of Geotextile Symposium, vol. 2, pp. 9–13. Japan Chapter of International Geosynthetics Society, Tokyo (1987)
222. Mohanty, W., Walling, M.: Seismic hazard in mega city Kolkata, India. Nat. Hazards **47**(1), 39–54 (2008)
223. Mohanty, W., Walling, M., Nath, S., Pal, I.: First order seismic microzonation of Delhi, India using geographic information system (GIS). Nat. Hazards **40**(2), 245–260 (2007)
224. Morales, E.M.: Liquefaction-induced damage to buildings in 1990 Luzon earthquake. J. Geotech. Eng. **121**(5), 453–454 (1995)
225. Moss, R.E., Seed, R.B., Kayen, R.E., Stewart, J.P., Der Kiureghian, A., Cetin, K.O.: CPT-based probabilistic and deterministic assessment of in situ seismic soil liquefaction potential. J. Geotech. Geoenviron. Eng. **132**(8), 1032–1051 (2006)
226. Moyle, W. Jr.: Ground-water-level monitoring for earthquake prediction; a progress report based on data collected in southern California, 1976–79. Tech. rep., Water Resources Division, US Geological Survey (1980)
227. Mukhopadhyay, S., Pandey, Y., Dharmaraju, R., Chauhan, P., Singh, P., Dev, A.: Seismic microzonation of Delhi for ground-shaking site effects. Curr. Sci. **82**(7), 877–881 (2002)
228. Naein, S., Moayed, R.Z.: Improving liquefaction potential strength by using micropiles. IAEG the Geological Society of London, pp. 333–338 (2006)
229. Nakamura, Y.: A method for dynamic characteristics estimation of subsurface using microtremor on the ground surface, vol. 30(1). Railway Technical Research Institute, Quarterly Reports (1989)
230. Nakazawa, H., Ishihara, K., Tsukamoto, Y., Kamata, T.: Case studies on evaluation of liquefaction resistance of imperfectly saturated soil deposits. In: Proceedings of International Conference on Cyclic Behaviour of Soils and Liquefaction Phenomena, Bochum, vol. 31, pp. 295–304 (2004)
231. Narayan, J., Sharma, M.: Effects of local geology on damage severity during Bhuj, India earthquake. In: 13th World Conference on Earthquake Engineering, Vancouver (2004)
232. Nath, S., Thingbaijam, K.: Probabilistic seismic hazard assessment of India. Seismol. Res. Lett. **83**(1), 135–149 (2012)
233. Nath, S., Chatterjee, D., Biswas, N., Dravinski, M., Cole, D., Papageorgiou, A., Rodriguez, J., Poran, C.: Correlation study of shear wave velocity in near surface geological formations in anchorage, Alaska. Earthq. Spectra **13**(1), 55–75 (1997)
234. Nath, S., Vyas, M., Pal, I., Singh, A., Mukherjee, S., Sengupta, P.: Spectral attenuation models in the Sikkim Himalaya from the observed and simulated strong motion events in the region. Curr. Sci. **88**(2), 295–303 (2005)
235. Nath, S.K., Thingbaijam, K.K.S., Raj, A.: Earthquake hazard in northeast India-a seismic microzonation approach with typical case studies from Sikkim Himalaya and Guwahati city. J. Earth Syst. Sci. **117**(2), 809–831 (2008)
236. Nath, S., Raj, A., Thingbaijam, K., Kumar, A.: Ground motion synthesis and seismic scenario in Guwahati city a stochastic approach. Seismol. Res. Lett. **80**(2), 233–242 (2009)
237. NDMA: Development of Probabilistic Seismic Hazard Map of India. National Disaster Management Authority, Government of India, New Delhi (2010)
238. NDMA: Geotechnical/Geophysical Investigations for Seismic Microzonation Studies of Urban Centres in India. National Disaster Management Authority, Government of India, New Delhi (2011)
239. Neelima Satyam, D., Rao, K.: Microtremor studies for seismic site characterization of Delhi region. In: The 12th International Conference of International Association for Computer Methods and Advances in Geomechanics (IACMAG), Goa, pp. 2811–2815 (2008)
240. Fäh, D., Rüttener, E., Noack, T., Kruspan, P.: Microzonation of the city of Basel. J. Seismol. **1**(1), 87–102 (1997)
241. Nuttli, O.W.: Average seismic source-parameter relations for mid-plate earthquakes. Bull. Seismol. Soc. Am. **73**(2), 519–535 (1983)

242. Ohba, S., Toriumi, I.: Dynamic response characteristics of Osaka plain. In: Proceedings of the Annual Meeting AIJ (1970)
243. Ohmachi, T., Nakamura, Y., Toshinawa, T.: Ground motion characteristics in the San Francisco bay area detected by microtremor measurements. In: Proceedings of the Second International Conference on Recent Advances in Geotechnical Earthquake Engineering and Soil Dynamics, March, pp. 11–15 (1991)
244. Ohsaki, Y., Iwasaki, R.: On dynamic shear moduli and Poisson's ratios of soil deposits. Soils Found. **13**(4), 61–73 (1973)
245. Ohta, Y., Goto, N.: Empirical shear wave velocity equations in terms of characteristic soil indexes. Earthq. Eng. Struct. Dyn. **6**(2), 167–187 (1978)
246. Okal, E.A.: Use of the mantle magnitude mm for the reassessment of the moment of historical earthquakes. Pure Appl. Geophys. **139**(1), 17–57 (1992)
247. Okal, E.A., Reymond, D.: The mechanism of great Banda Sea earthquake of 1 February 1938: applying the method of preliminary determination of focal mechanism to a historical event. Earth Planet. Sci. Lett. **216**(1–2), 1–15 (2003)
248. Okamoto, T., Kokusho, T., Yoshida, Y., Kusuonoki, K.: Comparison of surface versus subsurface wave source for p-s logging in sand layer. In: Proceedings of 44th Annual Conference of the JSCE, vol. 3, pp. 996–997 (1989)
249. Okamura, M., Teraoka, T.: Shaking table tests to investigate soil desaturation as a liquefaction countermeasure. In: Seismic Performance and Simulation of Pile Foundations in Liquefied and Laterally Spreading Ground, pp. 282–293. American Society of Civil Engineers, Reston (2006)
250. Oldham, T.: A catalogue of Indian earthquakes ed. by R.D. Oldham. Mem. Geol. Surv. India **19**, 163–215 (1883)
251. O'Neill, A.: Some pitfalls associated with dominant higher-mode inversion. In: Proceedings of the 8th International Symposium on Recent Advances in Exploration Geophysics, Kyoto University, pp. 48–55 (2004)
252. Ono, Y., Nakajima, Y., Ito, K., Oishi, H.: Efficient installation of gravel drains. In: International Conferences on Recent Advances in Geotechnical Earthquake Engineering and Soil Dynamics, 18 (1991)
253. Ordaz, M., Aguilar, A., Arboleda, J.: Crisis2007: program for computing seismic hazard. Instituto de Ingenierıa, Universidad Naconal Autónoma de México, UNAM, México (2007)
254. Ordonez, G.: SHAKE-2000, a Computer Program for the 1-D Analysis of Geotechnical Earthquake Engineering Problems, User's Manual (2000)
255. Panza, G., Vaccari, F., Cazzaro, R.: Deterministic seismic hazard assessment. In: Vrancea Earthquakes: Tectonics, Hazard and Risk Mitigation, pp. 269–286. Kluwer Academic Publishers, Dordrecht (1999)
256. Papazachos, B., Karakostas, V., Kiratzi, A., Margaris, B., Papazachos, C., Scordilis, E.: Uncertainties in the estimation of earthquake magnitudes inGreece. J. Seismol. **6**(4), 557–570 (2002)
257. Park, C., Miller, R., Xia, J.: Imaging dispersion curves of surface waves on multi-channel record. In: 1998 SEG Annual Meeting (1998)
258. Park, C., Miller, R., Xia, J.: Multichannel analysis of surface waves MASW. Geophysics **64**(3), 800–808 (1999)
259. Park, C., Heljeson, M., Ivanov, J., Brohammer, M.: Surfseis User's Manual v 2.0. Kansas Geological Survey, Kansas (2007)
260. Parvez, I., Vaccari, F., Panza, G.: A deterministic seismic hazard map of India and adjacent areas. Geophys. J. Int. **155**(2), 489–508 (2003)
261. Parvez, I., Vaccari, F., Panza, G.: Site-specific microzonation study in Delhi metropolitan city by 2-D modelling of SH and P-SV waves. Pure Appl. Geophys. **161**, 1165–1184 (2004)
262. Patton, H.J., Walter, W.R.: Regional moment: magnitude relations for earthquakes and explosions. Geophys. Res. Lett. **20**(4), 277–280 (1993)

263. PCRSMJUA: Project Completion Report of Seismic Microzonation of Jabalpur Urban Area. Department of Science and Technology, Government of India (2005)
264. Pedersen, H., Le Brun, B., Hatzfeld, D., Campillo, M., Bard, P.Y.: Ground-motion amplitude across ridges. Bull. Seismol. Soc. Am. **84**(6), 1786–1800 (1994)
265. Phillips, W.S., Aki, K.: Site amplification of coda waves from local earthquakes in central California. Bull. Seismol. Soc. Am. **76**(3), 627–648 (1986)
266. Pitilakis, K.: Site effects. In: Recent Advances in Earthquake Geotechnical Engineering and Microzonation, pp. 139–197. Springer, Dordrecht (2004)
267. Poulos, S.J.: The Stress-Strain Curves of Soils. Geotechnical Engineers Incorporated, Winchester (1971)
268. Poulos, S.J.: The steady state of deformation. J. Geotech. Geoenviron. Eng. **107**, 553–562 (1981). ASCE 16241 Proceeding
269. Power, M., Chiou, B., Abrahamson, N., Bozorgnia, Y., Shantz, T., Roblee, C.: An overview of the NGA project. Earthq. Spectra **24**(1), 3–21 (2008)
270. Prevost, J.: DYNA1D: A Computer Program for Nonlinear Seismic Site Response Analysis Technical Documentation. National Center for Earthquake Engineering Research, Buffalo (1989)
271. Promputthangkoon, P., Hyde, A.: Compressibility and liquefaction potential of rubber composite soils. In: Scrap Tire Derived Geomaterials-Opportunities and Challenges: Proceedings of the International Workshop IW-TDGM 2007 (Yokosuka, 23–24 March 2007), p. 161. CRC Press, Boca Raton (2007)
272. Pyke, R., Chan, C., Seed, H.: Settlement and Liquefaction of Sands Under Multi-Directional Shaking. Earthquake Engineering Research Center, University of California, Berkeley (1974)
273. Raghu Kanth, S., Iyengar, R.: Seismic hazard estimation for Mumbai city. Curr. Sci. **91**(11), 1486–1494 (2006)
274. Raghu Kanth, S., Iyengar, R.: Estimation of seismic spectral acceleration in peninsular India. J. Earth Syst. Sci. **116**(3), 199–214 (2007)
275. Rao, B.: Historical seismicity and deformation rates in the Indian peninsular shield. J. Seismol. **4**(3), 247–258 (2000)
276. Rao, K., Mohanty, W.: The Bhuj earthquake and lessons for the damages. IGS News **33**(2), 3–10 (2001)
277. Rao, K., Neelima Satyam, D.: Seismic microzonation studies for Delhi region. In: Symposium on Seismic Hazard Analysis and Microzonation, September 23–24, Roorkee (2005)
278. Rao, B.R., Rao, P.S.: Historical seismicity of peninsular India. Bull. Seismol. Soc. Am. **74**(6), 2519–2533 (1984)
279. Raptakis, D., Anastasiadis, A., Pitilakis, K., Lontzetidis, K.: Shear wave velocities and damping of Greek natural soils. In: Proceedings of the 10th European Conference on Earthquake Engineering, vol. 1, pp. 477–482 (1994)
280. Rastogi, B.: Earthquake mechanisms and plate tectonics in the Himalayan region. Tectonophysics **21**(1–2), 47–56 (1974)
281. Reid, H.: The California Earthquake of April 18, 1906, vol. 27. Carnegie Institution, Washington (1910)
282. Reiter, L.: Earthquake Hazard Analysis: Issues and Insights. Columbia University Press, New York (1990)
283. Richter, C.: An instrumental earthquake magnitude scale. Bull. Seismol. Soc. Am. **25**(1), 1–32 (1935)
284. Riepl, J., Bard, P.Y., Hatzfeld, D., Papaioannou, C., Nechtschein, S.: Detailed evaluation of site-response estimation methods across and along the sedimentary valley of Volvi (Euroseistest). Bull. Seismol. Soc. Am. **88**(2), 488–502 (1998)
285. Robertson, P.: Soil classification using the cone penetration test. Can. Geotech. J. **27**(1), 151–158 (1990)

286. Robertson, P., Campanella, R.: Liquefaction potential of sands using the CPT. J. Geotech. Eng. **111**(3), 384–403 (1985)
287. Robertson, P., Wride, C.: Cyclic liquefaction and its evaluation based on the SPT and CPT. In: Proceedings of the NCEER Workshop on Evaluation of Liquefaction Resistance of Soils. Technical Report NCEER-97-0022, National Center for Earthquake Engineering Research, Buffalo, pp. 41–87 (1997)
288. Robertson, P., Wride, C.: Evaluating cyclic liquefaction potential using the cone penetration test. Can. Geotech. J. **35**(3), 442–459 (1998)
289. Robertson, P., Campanella, R., Wightman, A.: SPT-CPT correlations. J. Geotech. Eng. **109**(11), 1449–1459 (1983)
290. Rodriguez-Marek, A., Bray, J.D., Abrahamson, N.A.: An empirical geotechnical seismic site response procedure. Earthq. Spectra **17**(1), 65–87 (2001)
291. Rogers, J.: Subsurface exploration using the standard penetration test and the cone penetrometer test. Environ. Eng. Geosci. **12**(2), 161–179 (2006)
292. Rollins, K., Evans, M., Diehl, N., William D. III: Shear modulus and damping relationships for gravels. J. Geotech. Geoenviron. Eng. **124**(5), 396–405 (1998)
293. Rugg, D., Yoon, J., Hwang, H., El Mohtar, C.: Undrained shearing properties of sand permeated with a bentonite suspension for static liquefaction mitigation. In: Geo-Frontiers 2011: Advances in Geotechnical Engineering, pp. 677–686. American Society of Civil Engineers, Reston (2011)
294. Rydelek, P.A., Sacks, I.S.: Testing the completeness of earthquake catalogues and the hypothesis of self-similarity. Nature **337**(6204), 251–253 (1989)
295. Saaty, T.: The Analytic Hierarchy Process. McGraw-Hill, New York (1980)
296. Sadigh, K., Chang, C., Egan, J., Makdisi, F., Youngs, R.: Attenuation relationships for shallow crustal earthquakes based on California strong motion data. Seismol. Res. Lett. **68**(1), 180–189 (1997)
297. Sánchez-Sesma, F.J.: Diffraction of elastic waves by three-dimensional surface irregularities. Bull. Seismol. Soc. Am. **73**(6A), 1621–1636 (1983)
298. Sánchez-Sesma, F.J.: Diffraction of elastic SH waves by wedges. Bull. Seismol. Soc. Am. **75**(5), 1435–1446 (1985)
299. Sánchez-Sesma, F., Chávez-Pérez, S., Suarez, M., Bravo, M., Pérez-Rocha, L.: The méxico earthquake of 19 September 1985 on the seismic response of the valley of méxico. Earthq. Spectra **4**(3), 569–589 (1988)
300. Sasaki, Y., Taniguchi, E.: Shaking table tests on gravel drains to prevent liquefaction of sand deposits. Soils Found. **22**(3), 1–14 (1982)
301. Schnabel, P.: Effects of local geology and distance from source on earthquake ground motions. University of California, Berkeley (1973)
302. Schnabel, P., Lysmer, J., Seed, H.: SHAKE: a computer program for earthquake response analysis of horizontally layered sites. Earthquake Engineering Research Center. Report No EERC, pp. 72–12 (1972)
303. Scordilis, E.: Empirical global relations converting M_s and m_b to moment magnitude. J. Seismol. **10**(2), 225–236 (2006)
304. Seeber, L., Armbruster, J., Jacob, K.: Probabilistic assessment of seismic hazard for Maharashtra. Government of Maharashtra, Unpublished Report (1999)
305. Seed, H.B., Booker, J.R.: Stabilization of potentially liquefiable sand deposits using gravel drains. J. Geotech. Geoenviron. Eng. **103**(ASCE 13050) (1977)
306. Seed, H., De Alba, P.: Use of SPT and CPT tests for evaluating the liquefaction resistance of sands. Use of in Situ Tests in Geotechnical Engineering, pp. 281–302. ASCE (1986)
307. Seed, H., Idriss, I.: Influence of soil conditions on ground motions during earthquakes. University of California, Institute of Transportation and Traffic Engineering, Soil Mechanics Laboratory (1969)
308. Seed, H., Idriss, I.: Soil moduli and damping factors for dynamic response analyses. Report No. EERC 70-10, University of California, Berkeley (1970)

309. Seed, H., Idriss, I.: Simplified procedure for evaluating soil liquefaction potential. J. Soil Mech. Found. Div. **97**(9), 1249–1273 (1971)
310. Seed, H., Idriss, I.: Evaluation of liquefaction potential sand deposits based on observation of performance in previous earthquakes. ASCE National Convention (MO), pp. 481–544 (1981)
311. Seed, B., Lee, K.L.: Liquefaction of saturated sands during cyclic loading. J. Soil Mech. Found. Div. **92**, 105–134 (1966). ASCE# 4972 Proceeding
312. Seed, H., Sun, J.: Implication of site effects in the Mexico city earthquake of 19 September 1985 for earthquake-resistance-design criteria in the San Francisco bay area of California. Report No. UCB/EERC-89/03, University of California, Berkeley (1989)
313. Seed, H., Idriss, I., Arango, I.: Evaluation of liquefaction potential using field performance data. J. Geotech. Eng. **109**(3), 458–482 (1983)
314. Seed, H., Tokimatsu, K., Harder, L., Chung, R.: The influence of SPT procedures in soil liquefaction resistance evaluations: Berkeley, University of California. Earthquake Engineering Research Center Report UBC/EERC-84/15 (1984)
315. Seed, H., Tokimatsu, K., Harder, L., Chung, R.: Influence of SPT procedures in soil liquefaction resistance evaluations. J. Geotech. Eng. **111**(12), 1425–1445 (1985)
316. SEISAT: Seismotectonic Atlas of India. Geological Survey of India, New Delhi (2000)
317. Semblat, J.F., Duval, A.M., Dangla, P.: Numerical analysis of seismic wave amplification in nice (France) and comparisons with experiments. Soil Dyn. Earthq. Eng. **19**(5), 347–362 (2000)
318. Sharma, M.L.: Attenuation relationship for estimation of peak ground horizontal acceleration using data from strong-motion arrays in India. Bull. Seismol. Soc. Am. **88**(4), 1063–1069 (1998)
319. Sharma, M., Malik, S.: Probabilistic seismic hazard analysis and estimation of spectral strong ground motion on bed rock in North East India. In: Fourth International Conference on Earthquake Engineering, Taipei, Paper, 15 (2006)
320. Sharma, M., Wason, H., Dimri, R.: Seismic zonation of the Delhi region for bedrock ground motion. Pure Appl. Geophys. **160**(12), 2381–2398 (2003)
321. Sharma, B., Gupta, A.K., Devi, D.K., Kumar, D., Teotia, S., Rastogi, B.: Attenuation of high-frequency seismic waves in Kachchh region, Gujarat, India. Bull. Seismol. Soc. Am. **98**(5), 2325–2340 (2008)
322. Sharma, M., Douglas, J., Bungum, H., Kotadia, J.: Ground-motion prediction equations based on data from the Himalayan and Zagros regions. J. Earthq. Eng. **13**(8), 1191–1210 (2009)
323. Shibata, T.: Analysis of liquefaction of saturated sand during cyclic loading. Disaster Prev. Res. Inst. Bull. **13**, 563–570 (1970)
324. Shima, E.: Seismic microzoning map of Tokyo. In: Proceedings of the Second International Conference on Microzonation, vol. 1, pp. 433–443 (1978)
325. Shtivelman, V.: Surface wave sections as a tool for imaging subsurface inhomogeneities. Eur. J. Environ. Eng. Geophys. **7**, 121–138 (2002)
326. Singh, R.: Probabilistic seismic hazard and risk analysis: a case study for Ahmedabad city. Ph.D. thesis, Indian Institute of Technology, Kanpur (2009)
327. Sitharam, T., Anbazhagan, P.: Seismic hazard analysis for the Bangalore region. Nat. Hazards **40**(2), 261–278 (2007)
328. Sitharam, T., Anbazhagan, P.: Seismic microzonation: principles, practices and experiments. EJGE Special Volume Bouquet 8, 61 (2008)
329. Sitharam, T., Anbazahgan, P.: Report on Seismic Microzonation of Bangalore Urban Centre (Vol. 1). Seismology Division, Ministry of Earth Sciences Government of India, New Delhi (2009)
330. Sitharam, T., Dinesh, S.: Numerical simulation of liquefaction behaviour of granular materials using discrete element method. J. Earth Syst. Sci. **112**(3), 479–484 (2003)
331. Sitharam, T., Kolathayar, S.: Seismic hazard analysis of India using areal sources. J. Asian Earth Sci. **62**, 647–653 (2013)
332. Sitharam, T., Sil, A.: Comprehensive seismic hazard assessment of Tripura and Mizoram states. J. Earth Syst. Sci. **123**(4), 837–857 (2014)

333. Sitharam, T., Anbazhagan, P., Ganesha Raj, K.: Use of remote sensing and seismotectonic parameters for seismic hazard analysis of Bangalore. Nat. Hazards Earth Syst. Sci. **6**(6), 927–939 (2006)
334. Sitharam, T., James, N., Vipin, K., Raj, K.: A study on seismicity and seismic hazard for Karnataka state. J. Earth Syst. Sci. **121**(2), 475–490 (2012)
335. Sitharam, T., Kolathayar, S., James, N.: Probabilistic assessment of surface level seismic hazard in India using topographic gradient as a proxy for site condition. Geosci. Front. **6**(6), 847–859 (2015)
336. Skempton, A.: Standard penetration test procedures and the effects in sands of overburden pressure, relative density, particle size, ageing and over consolidation. Geotechnique **36**(3), 425–447 (1986)
337. Sladen, J.A., D'hollander, R.D., Krahn, J.: The liquefaction of sands, a collapse surface approach. Can. Geotech. J. **22**(4), 564–578 (1985)
338. Slob, S., Hack, R., Scarpas, T., van Bemmelen, B., Duque, A.: A methodology for seismic microzonation using GIS and Shakea case study from Armenia, Colombia. In: Engineering Geology for Developing Countries – Proceedings of 9th Congress of the International Association for Engineering Geology and the Environment, Durban, pp. 16–20 (2002)
339. Sonmez, H.: Modification of the liquefaction potential index and liquefaction susceptibility mapping for a liquefaction-prone area (Inegol, Turkey). Environ. Geol. **44**(7), 862–871 (2003)
340. Sonu, C.J., Ito, K., Oishi, H.: Harry seed, liquefaction and the gravel drain. Civil Eng. **63**(12), 58 (1993)
341. Srivastava, H., Ramachandran, K.: New catalogue of earthquakes for peninsular India during 1839–1900. Mausam **36**(3), 351–358 (1985)
342. Stepp, J.: Analysis of completeness of the earthquake sample in the Puget Sound area and its effect on statistical estimates of earthquake hazard. In: Proceedings of the International Conference on Microzonation for Safer Construction Research and Application, vol. 64(4), 1189–1207 (1972)
343. Stepp, J., Wong, I., Whitney, J., Quittmeyer, R., Abrahamson, N., Toro, G., Youngs, R., Coppersmith, K., Savy, J., Sullivan, T.: Probabilistic seismic hazard analyses for ground motions and fault displacement at Yucca mountain, Nevada. Earthq. Spectra **17**(1), 113–151 (2001)
344. Stewart, J., Choi, Y., Liu, A., Baturay, M.: Amplification factors for spectral acceleration in active regions. Report No. PEER-2001/10, Pacific Earthquake Engineering Research Center, University of California, Berkeley, vol. 2001 (2001). http://www.cee.ucla.edu/Faculty/jstewart/publications.htm
345. Stewart, J., Liu, A., Choi, Y.: Amplification factors for spectral acceleration in tectonically active regions. Bull. Seismol. Soc. Am. **93**(1), 332–352 (2003)
346. Stokoe, K., Roesset, J., Bierschwale, J., Aouad, M.: Liquefaction potential of sands from shear wave velocity. In: Proceedings, 9nd World Conference on Earthquake, vol. 13, pp. 213–218 (1988)
347. Stokoe, K.H., Rix, G.J., Nazarian, S.: In situ seismic testing with surface wave: processing. In: XII International Conference on Soil Mechanics and Foundation Engineering, vol. 33, pp. l–334 (1989)
348. Stone, W., Yokel, F., Celebi, M., Hanks, T., Leyendecker, E.: Engineering aspects of the 19 September 1985 Mexico earthquake. US Department of Commerce, National Bureau of Standards, Washington (1987)
349. Sykora, D.W.: Correlations of in situ measurements in sands of shear wave velocity, soil characteristics, and site conditions. Ph.D. thesis, University of Texas at Austin (1983)
350. Sykora, D.W., Koester, J.P.: Correlations between dynamic shear resistance and standard penetration resistance in soils. In: Earthquake Engineering and Soil Dynamics II: Recent Advances in Ground-Motion Evaluation, pp. 389–404. American Society of Civil Engineers, New York (1988)
351. Tandon, A., Srivastava, H.: The stress drop and average dislocation of some earthquakes in the Indian sub-continent. Pure Appl. Geophys. **112**(6), 1051–1057 (1974)

352. Theodulidis, N., Bard, P.Y.: Horizontal to vertical spectral ratio and geological conditions: an analysis of strong motion data from Greece and Taiwan (smart-1). Soil Dyn. Earthq. Eng. **14**(3), 177–197 (1995)
353. Theodulidis, N., Bard, P.Y., Archuleta, R., Bouchon, M.: Horizontal-to-vertical spectral ratio and geological conditions: the case of Garner Valley Downhole Array in southern California. Bull. Seismol. Soc. Am. **86**(2), 306–319 (1996)
354. Thingbaijam, K.K.S., Nath, S.K., Yadav, A., Raj, A., Walling, M.Y., Mohanty, W.K.: Recent seismicity in northeast India and its adjoining region. J. Seismol. **12**(1), 107–123 (2008)
355. Tokimatsu, K., Kuwayama, S., Tamura, S.: Liquefaction potential evaluation based on Rayleigh wave investigation and its comparison with field behavior. In: Proceedings, pp. 357–364 (1991)
356. Tokimatsu, K., Nakajo, Y., Tamura, S.: Horizontal to vertical amplitude ratio of short period microtremors and its relation to site characteristics. J. Struct. Construction Eng. **457**, 11–18 (1994)
357. Topal, T., Doyuran, V., Karahanoğlu, N., Toprak, V., Süzen, M., Yeşilnacar, E.: Microzonation for earthquake hazards: Yenişehir settlement, Bursa, Turkey. Eng. Geol. **70**(1), 93–108 (2003)
358. Toprak, S., Holzer, T., Bennett, M., Tinsley, J.: CPT-and SPT-based probabilistic assessment of liquefaction potential. In: Proceedings of the 7th US–Japan Workshop on Earthquake Resistant Design of Lifeline Facilities and Countermeasures Against Liquefaction (1999)
359. Toro, G., Abrahamson, N., Schneider, J.: Model of strong ground motions from earthquakes in central and eastern North America: best estimates and uncertainties. Seismol. Res. Lett. **68**(1), 41–57 (1997)
360. Towhata, I.: Geotechnical Earthquake Engineering. Springer, Dordrecht (2008)
361. Trifunac, M.D., Hudson, D.E.: Analysis of the pacoima dam accelerogram – San Fernando, California, earthquake of 1971. Bull. Seismol. Soc. Am. **61**(5), 1393–1411 (1971)
362. Trifunac, M., Todorovska, M.: Nonlinear soil response as a natural passive isolation mechanism – the 1994 Northridge, California, earthquake. Soil Dyn. Earthq. Eng. **17**(1), 41–51 (1998)
363. Tsuchida, H.: Evaluation of liquefaction potential of sandy deposits and measures against liquefaction induced damage. In: Proceedings of the Annual Seminar of the Port and Harbour Research Institute, pp.(3-1)–(3-33) (1970; in Japanese)
364. Tsukamoto, Y., Ishihara, K., Nakazawa, H., Kamada, K., Huang, Y.: Resistance of partly saturated sand to liquefaction with reference to longitudinal and shear wave velocities. Soils Found. **42**(6), 93–104 (2002)
365. Uchimura, T., Chi, N., Nirmalan, S., Sato, T., Meidani, M., Towhata, I.: Shaking table tests on effect of tire chips and sand mixture in increasing liquefaction resistance and mitigating uplift of pipe. In: Proceedings, International Workshop on Scrap Tire Derived Geomaterials – Opportunities and Challenges, Yokosuka, pp. 179–186 (2007)
366. Udwadia, F., Trifunac, M.: Comparison of earthquake and microtremor ground motions in El centro, California. Bull. Seismol. Soc. Am. **63**(4), 1227–1253 (1973)
367. Uma Maheswari, R., Boominathan, A., Dodagoudar, G.: Seismic site classification and site period mapping of Chennai city using geophysical and geotechnical data. J. Appl. Geochem. **72**(3), 152–168 (2010)
368. USGS: Virginia Well Records Sumatra-Andaman Islands Earthquake. USGS News Release (2005)
369. Utsu, T.: A method for determining the value of b in formula log n = a-bm showing the magnitude-frequency relation for earthquakes. Geophys. Bull. Hokkaido Univ. **13**, 99–103 (1965)
370. Vaid, Y.P., Chern, J.C.: Effect of static shear on resistance to liquefaction. Soils Found. **23**(1), 47–60 (1983)
371. Vaid, Y., Chern, J.: Effect of static shear on resistance to liquefaction. Soils Found. **25**(3), 154–156 (1985)

372. Van Paassen, L., Harkes, M., Van Zwieten, G., Van der Zon, W., Van der Star, W., Van Loosdrecht, M.: Scale up of Biogrout: a biological ground reinforcement method. In: Proceedings of the 17th International Conference on Soil Mechanics and Geotechnical Engineering, pp. 2328–2333. IOS Press, Lansdale (2009)
373. Vilanova, S., Fonseca, J.: Probabilistic seismic-hazard assessment for Portugal. Bull. Seismol. Soc. Am. **97**(5), 1702–1717 (2007)
374. Vipin, K.S., Sitharam, T.G., Anbazhagan, P.: Probabilistic evaluation of seismic soil liquefaction potential based on SPT data. Nat. Hazards **53**(3), 547–560 (2010)
375. Vipin, K.S., Sitharam, T.G.: A performance-based framework for assessing liquefaction potential based on CPT data. Georisk: Assess. Manag. Risk Eng. Syst. Geohazards **6**(3), 177–187 (2012)
376. Vipin, K., Anbazhagan, P., Sitharam, T.: Estimation of peak ground acceleration and spectral acceleration for south India with local site effects: probabilistic approach. Nat. Hazards Earth Syst. Sci. **9**, 865–878 (2009)
377. Vucetic, M., Dobry, R.: Effect of soil plasticity on cyclic response. J. Geotech. Eng. **117**(1), 89–107 (1991)
378. Wahlström, R., Grünthal, G.: Probabilistic seismic hazard assessment (horizontal PGA) for Sweden, Finland and Denmark using different logic tree approaches. Soil Dyn. Earthq. Eng. **20**(1), 45–58 (2000)
379. Wald, D., Allen, T.: Topographic slope as a proxy for seismic site conditions and amplification. Bull. Seismol. Soc. Am. **97**(5), 1379–1395 (2007)
380. Wang, W.: Some Findings in Soil Liquefaction Earthquake Engineering. Department Water Conservancy and Hydroelectric Power Scientific Research Institute, Beijing (1979)
381. Wiemer, S., Wyss, M.: Minimum magnitude of completeness in earthquake catalogs: Examples from Alaska, the western united states, and Japan. Bull. Seismol. Soc. Am. **90**(4), 859–869 (2000)
382. Wiemer, S., Wyss, M.: Mapping spatial variability of the frequency-magnitude distribution of earthquakes. Adv. Geophys. **45**, 259–302 (2002)
383. Wills, C.J., Silva, W.: Shear-wave velocity characteristics of geologic units in California. Earthq. Spectra **14**(3), 533–556 (1998)
384. Wills, C., Petersen, M., Bryant, W., Reichle, M., Saucedo, G., Tan, S., Taylor, G., Treiman, J.: A site-conditions map for California based on geology and shear-wave velocity. Bull. Seismol. Soc. Am. **90**(6B), S187–S208 (2000)
385. Woessner, J., Wiemer, S.: Assessing the quality of earthquake catalogues: estimating the magnitude of completeness and its uncertainty. Bull. Seismol. Soc. Am. **95**(2), 684 (2005). http://dx.doi.org/10.1785/0120040007
386. Woo, G.: Kernel estimation methods for seismic hazard area source modeling. Bull. Seismol. Soc. Am. **86**(2), 353–362 (1996)
387. Wood, H.: Distribution of apparent intensity in San Francisco. Report of the State Earthquake Investigation Commission 1(1906), pp. 220–245 (1908)
388. Xia, J., Miller, R., Park, C.: Estimation of near-surface shear-wave velocity by inversion of Rayleigh waves. Geophysics **64**(3), 691–700 (1999)
389. Xia, J., Miller, R.D., Park, C.B., Hunter, J.A., Harris, J.B., Ivanov, J.: Comparing shear-wave velocity profiles inverted from multichannel surface wave with borehole measurements. Soil Dyn. Earthq. Eng. **22**(3), 181–190 (2002)
390. Xu, Z.Y.: Analysis of liquefiable sand deposits using gravel drains. In: International Conferences on Recent Advances in Geotechnical Earthquake Engineering and Soil Dynamics, 6 (1991)
391. Yanagisawa, E.: Damage to structures due to liquefaction in the Japan sea earthquake of 1983. Disasters **7**(4), 259–265 (1983)
392. Yang, J.: Non-uniqueness of flow liquefaction line for loose sand. Geotechnique Lond. **52**(10), 757–760 (2002)
393. Yang, J., Savidis, S., Roemer, M.: Evaluating liquefaction strength of partially saturated sand. J. Geotech. Geoenviron. Eng. **130**(9), 975–979 (2004)

394. Yasuhara, K., Komine, H., Murakami, S., Miyota, S., Hazarika, H.: Mitigation of liquefaction using tire chips as a gravel drain. In: Proceedings of the 6th International Congress on Environmental Geotechnology. New Delhi, pp. 1176–1181 (2010)
395. Yeats, R., Sieh, K., Allen, C.: The Geology of Earthquakes, vol. 568. Oxford University Press, New York (1997)
396. Yegian, M., Eseller-Bayat, E., Alshawabkeh, A., Ali, S.: Induced-partial saturation for liquefaction mitigation: experimental investigation. J. Geotech. Geoenviron. Eng. **133**(4), 372–380 (2007)
397. Yoshida, N., Suetomi, I.: DYNEQ: a computer program for dynamic analysis of level ground based on equivalent linear method. Reports of Engineering Research Institute, Sato Kogyo, pp. 61–70 (1996)
398. Yoshimi, Y., Kuwabara, F.: Effect of subsurface liquefaction on the strength of surface soil. Soils Found. **13**(2), 67–81 (1973)
399. Youd, T., Noble, S., Idriss, I.: Liquefaction criteria based on statistical and probabilistic analyses. Tech. Rep. NCEER 97, 201–15 (1997)
400. Youd, T., Idriss, I., Andrus, R., Arango, I., Castro, G., Christian, J., Dobry, R., W.D., L.F., Harder, L., Hynes, M., Ishihara, K., Koester, J., Laio, S., Marcuson, W., Martin, G., Mitchell, J., Moriwaki, Y., Power, M., Robertson, P., Seed, R., Stokoe, K.: Liquefaction resistance of soils. Summary report from the 1996 NCEER and 1998 NCEER/NSF Workshops on Evaluation of Liquefaction Resistance of Soils. J. Geotech. Geoenviron. Eng. **127**(10), 817–833 (2001)
401. Yu, Y.x., Wang, S.y.: Attenuation relations for horizontal peak ground acceleration and response spectrum in northeastern Tibetan plateau region. Acta Seismol. Sin. **17**(6), 651–661 (2004)
402. Zare, M., Bard, P.Y., Ghafory-Ashtiany, M.: Site characterizations for the Iranian strong motion network. Soil Dyn. Earthq. Eng. **18**(2), 101–123 (1999)
403. Zhao, F., Zhang, Y.: Artificial ground motion compatible with specified peak velocity and target spectrum. Acta Seismol. Sin. **19**(4), 461–471 (2006)

Index

T. G. Sitharam et al., *Comprehensive Seismic Zonation Schemes for Regions at Different Scales*, https://doi.org/10.1007/978-3-319-89659-5

T

MIX
Papier aus verantwortungsvollen Quellen
Paper from responsible sources
FSC® C105338

If you have any concerns about our products, you can contact us on
ProductSafety@springernature.com

In case Publisher is established outside the EU, the EU authorized representative is:
Springer Nature Customer Service Center GmbH
Europaplatz 3, 69115 Heidelberg, Germany

Printed by Libri Plureos GmbH
in Hamburg, Germany

H. Fan (Ed.)

Jaagsiekte Sheep Retrovirus and Lung Cancer

With 63 Figures and 14 Tables

Springer

Springer-Verlag Berlin Heidelberg GmbH

H. Fan (Ed.)

Jaagsiekte Sheep Retrovirus and Lung Cancer

With 63 Figures and 14 Tables

Springer

Prof. Dr. HUNG FAN

University of California, Irvine
Cancer Research Institute
3221 Bio Sci II
Irvine, CA 92697-3905
USA
e-mail: hyfan@uci.edu

Cover Illustration by Massimo Palmarini (this volume)

This composite shows several features of Jaagsiekte sheep retrovirus and the lung cancer that it causes (ovine pulmonary adenocarcinoma or OPA). *Upper right:* lungs from an animal with OPA are shown. The major areas of tumor are the gray regions at the top of each lung. A characteristic foamy fluid is exuding from the wind pipe, reflecting production of excess surfactant by the tumor cells. *Upper left:* A histological section from an OPA lung, showing an area of OPA disease. The tumor is positive by immunohistochemistry with an antibody directed against JSRV capsid protein (red stain). *Center:* The genetic organization of JSRV is shown. Open boxes indicate the different open reading frames, characteristic of other retroviruses (with the exception of *orf-X*). the closed boxes indicate the long terminal repeats (LTRs). *Lower left:* Scanning electron micrograph of a tumor cell in the presence of normal lung tissue. *Lower right:* Transformation of rodent fibroblasts (NIH-3T3) by JSRV *env* protein. A focus of transformed cells after DNA transfection with viral DNA is shown.

ISSN 0070-217X

ISBN 978-3-642-62897-9 ISBN 978-3-642-55638-8 (eBook)

DOI 10.1007/978-3-642-55638-8

Library of Congress Catalog Card Number 72-152360

http://www.springer.de

Library of Congress Catalog Card Number 15-12910
Softcover reprint of the hardcover 1st edition 2003

Cover Design: Design & Production GmbH, Heidelberg
Typesetting: Fotosatz-Service Köhler GmbH, Würzburg
Production Editor: Christiane Messerschmidt, Rheinau
Printed on acid-free paper SPIN: 10890431 27/3020 5 4 3 2 1 0

Preface

Retroviruses have been of great importance to biomedical science for the past half century. Initially, studies on oncogenic animal retroviruses provided important insights into molecular processes in carcinogenesis (viral or non-viral) – most notably the existence and mechanisms of action of oncogenes and proto-oncogenes. Moreover, several human diseases are caused by retroviruses, including AIDS [human immunodeficiency virus (HIV)], adult T-cell leukemia [human T-cell leukemia virus type I (HTLV-I)] and the neurological disease HAM/TSP (also HTLV-I). Several animal retroviruses cause cancers and other conditions of veterinary importance, including avian leukosis virus in chickens, feline leukemia virus in cats, maedi-visna virus in sheep and bovine leukemia virus in cattle. Finally, retroviruses have been important for biotechnology. This includes the use of reverse transcriptase in cDNA cloning, and retrovirus-based vectors in gene therapy.

The topic of this volume is a relatively unknown animal retrovirus, jaagsiekte sheep retrovirus (JSRV), the causative agent of a transmissible lung cancer in sheep – ovine pulmonary adenocarcinoma (OPA). OPA is a tumor of secretory epithelial cells of the distal airways – type II pneumocytes and Clara cells. The disease was first documented in South Africa in the 1800s, it has a wide geographical distribution, and is of economic importance in high endemic regions (e.g., Europe and Africa). However, until very recently the nature of the etiologic agent was unclear, and relatively few laboratories actively studied the disease, although it has always intrigued virologists and lung cancer researchers. In addition, it was not possible to culture the etiologic agent. In the past 10 years, great strides have been made in JSRV research. Important landmarks include deduction of the sequence of the etiologic agent JSRV in 1991, identification of highly related endogenous JSRV proviruses in uninfected sheep, demonstration that exogenous JSRV infection is always associated with OPV, and molecular cloning of an infectious and oncogenic JSRV provirus. The availability of an infectious and oncogenic JSRV clone has allowed further molecular and biological studies. Noteworthy recent results include the determination that the envelope protein of the virus appears to function as an oncogene, that JSRV expression is very specific for lung epithelial cells, and the identification of the JSRV receptor.

A colloquium on JSRV/OPA research was held in Missillac, France, in June of 2001, sponsored by the Borchard Foundation. At this workshop, the great majority of researchers in the field were brought together. A historical and biological perspective was provided, the latest results were shared, and consensus on several points was reached. As an example, all investigators agreed henceforth to refer to the JSRV-induced disease as OPA [previously referred to as sheep pulmonary adenomatosis (SPA) or ovine pulmonary carcinoma (OPC)]. This volume contains contributions from all of the workshop participants, and it should serve as the definitive reference for JSRV/OPA in the future.

The chapter by York and Querat describes early research on JSRV/OPA, including determining that the etiologic agent is a retrovirus, and culminating in deduction of the JSRV sequence from a series of overlapping cDNA clones. The chapter by De Las Heras and Gonzales reviews the pathology of OPA, including the classical and atypical forms. Sharp and DeMartini review the natural history of OPA, describing both naturally occurring and experimentally induced disease. It is noteworthy that JSRV is a potent carcinogen in the experimental setting – end-stage tumors have been observed after as little as 2 weeks post-inoculation.

The contribution by Palmarini and Fan reviews the molecular biology of JSRV, including isolation of an infectious and oncogenic JSRV clone, establishment of an in vitro infection system, and demonstration that the JSRV long terminal repeat (LTR) is transcriptionally quite specific for lung epithelial cells. The chapter by DeMartini et al. describes endogenous JSRV related viruses. These endogenous viruses are noteworthy because they are highly related to the oncogenic exogenous JSRV, they have entered the sheep genome relatively recently, and they are expressed at high level in certain normal sheep tissues (e.g., the female reproductive tract). The contribution by Fan et al. considers transformation and oncogenesis by JSRV. The most noteworthy finding is that JSRV appears to carry an oncogene, as measured by transformation of rodent or avian fibroblasts in culture. Most interestingly, the oncogene is the envelope protein of the virus – unprecedented for retroviruses. MILLER describes cloning and identification of the JSRV receptor, hyaluronidase 2 (Hyal2). It is interesting that human (but not rodent) Hyal2 can function as a JSRV receptor, which opens the way to development of JSRV-based retroviral vectors for gene therapy in the human lung. Moreover, Hyal2 as the JSRV receptor has potential implications for the mechanism of oncogenesis.

The contribution by DE LAS HERAS et al. describes another JSRV-related virus that causes nasal tumors in sheep and goats – ovine nasal adenocarcinoma virus (ONAV). This virus induces a malignancy of secretory epithelial cells in the nasal cavity. Finally, the chapter MORNEX et al. reviews human bronchioloalveolar carcinoma (BAC), a cancer with many similarities to OPA. Challenges in diagnosing BAC are discussed. There is the possibility of a JSRV-related virus in some human lung adenocarcinoma cases, which raises interest in the virus-induced sheep disease.

I would like to thank Nita Driscoll and Joana Banks for assistance in organizing and preparing the manuscript, the Borchard Foundation for financial support, and all of the contributors to this volume for their enthusiasm and patience.

September 2002 HUNG FAN

List of Contents

List of Contributors

(Their addresses can be found at the beginning of their respective chapters.)

CHAPTER 1

A History of Ovine Pulmonary Adenocarcinoma (Jaagsiekte) and Experiments Leading to the Deduction of the JSRV Nucleotide Sequence

D. F. YORK, G. QUERAT

Abstract. Jaagsiekte (JS), a contagious cancer affecting the lungs of sheep has been called many names over the years. At a recent workshop in Missilac, France it was agreed that the disease would be called ovine pulmonary adenocarcinoma (OPA). The disease is caused by an infectious retrovirus called jaagsiekte sheep retrovirus (JSRV). This chapter focuses on the early research that led up to the isolation, cloning and sequencing of the exogenous infectious form of JSRV and the demonstration that it has an endogenous counter part that is present in all sheep. As there was

D. F. YORK
Department of Virology, Nelson R. Mandela School of Medicine,
University of Natal, South Africa
e-mail: york@med.und.ac.za

G. QUERAT
INSERM U372, Campus de Luminy, BP 178, 13276 Marseille, France
e-mail: gquerat@inserm-u372.univ-mrs.fr

no in vitro production source of the virus much of the early research focused on the in vivo production and purification of the virus to obtain sufficient material to use to identify the viral proteins and purify the viral genetic material. Typically, new born lambs were inoculated intra-tracheally with concentrated lung lavage from previously infected sheep lungs. The optimal purification involved the concentration of lung lavage of freshly slaughtered sheep, an extraction with organic solvent, and final purification by both rate zonal and isopycnic centrifugation. Monoclonal and polyclonal antibodies were made against the purified fractions. The polyclonal antibodies were not very specific and the monoclonal antibodies proved to be against antigens expressed in high concentrations in response to any lung pathology. The genomic RNA of the virus was isolated from ex vivo purified materials, and cloned as a collection of cDNAs. The full length sequence was assembled by walking through the cDNA clones. The genome of the exogenous virus is 7462 bases and has the classical *gag, pol, env* genome arrangement and is flanked by a long terminal repeat (LTR) on each end. An additional open reading frame (ORF) was observed in the viral genome and has been called *orfX*. A function has not been determined for this ORF. JSRV is classified as a betaretrovirus, with *gag* and *pol* closely related to D type retrovirus, whereas *env* is related to the B type viruses such as the human endogenous retrovirus HERV-K. An interesting finding was that the exogenous infectious virus had an endogenous counter part which is present in the genomes of all sheep and goats. It is estimated that there are between 15 and 20 endogenous loci per sheep genome. No circulating antibodies have been found in OPA-affected sheep. It is suggested that the endogenous JSRV transcripts are expressed at an early age and are cause for the clonal elimination of JSRV specific T cells during T-cell ontogeny. Histopathologically the sheep disease resembles human bronchiolar alveolar carcinoma and has been identified as a natural out bred animal model that could be used to study the human disease.

1 A History of Jaagsiekte/OPA

One of the first documented reports of ovine pulmonary adenocarcinoma (OPA) is a letter written by a farmer in 1825 who complained to the magistrate in the Cape of Good Hope in South Africa that he was losing many of his sheep to a disease he called *Jaagziekte*, a Dutch word that was translated into Afrikaans as Jaagsiekte. (*Jaag* is the Afrikaans word for chase + *siekte* sickness, farmers noticed that affected sheep appeared as if they had been chased; Tustin 1969). In France the disease was referred to as *La bouhite* which derives its name from patois meaning panting (Aynaud 1926). The disease was first recognized by M'Faydean in 1888 as a rare condition in sheep lungs in England. In 1891 Hutcheon, a veterinarian in South Africa, described the pathology and symptomology in great detail and described the disease as 'something special' suggesting that it was contagious and that all sheep infected with the disease should be immediately slaughtered (Hutcheon 1891). The disease was then reported in many other countries of the world. A number of common local names have been used to describe the disease but there has also been a difference of opinion as to the correct scientific term for the disease. Names such as sheep pulmonary adenomatosis (SPA), ovine pulmonary carcinoma and ovine bronchioloalveolar carcinoma have been used. The controversy is mainly due to the variation in the incidence of metastases seen in various countries; hence the term jaagsiekte was frequently used (Verwoerd 1996). At a recent Workshop in Missilac, France, it was agreed that the term ovine pulmonary adenocarcinoma (OPA) would be used to describe the disease in future. Jaagsiekte sheep retrovirus (JSRV), would be used to name the etiologic virus. JSRV has two forms, an exogenous infectious from that alone can induce the disease (Palmarini et al. 1999) and an endogenous JSRV-related proviruses that is present in all sheep genomes. It is estimated that there are 15–20 endogenous loci in the sheep genome (Hecht et al. 1996). The prefix ex and en will be used to distinguish the exogenous infectious form from the endogenous loci. Further subclassification into types and clades will obviously evolve as more isolates are sequenced. In this chapter, if the prefix is omitted the exogenous virus is referred to. The disease and pathology will be described in detail in later chapters. At this stage OPA can simply be described as a contagious cancer affecting the lungs of sheep. The disease in sheep is similar to human bronchioloalveolar carcinoma in that both tumors originate in the lung periphery and share the same histologic and ultrastructural features (see the chapter by Mornex et al., this volume). OPA is thus regarded as an ideal natural out bred model for human bronchioloalveolar carcinoma (Bonne 1930; Perk and Hod 1982; Barsky et al. 1994; Palmarini et al. 1997).

OPA has been reported throughout the world except for Australia and New Zealand. The disease occurred in Iceland but following a successful eradication program, no cases of OPA have been recorded since 1952

(Sigurdsson 1958). The epizootic of OPA in Iceland is of particular interest because it illustrates a classic example of the nature that the disease takes when it enters a farm for the first time. In December 1933, a few Karakul rams were imported into Iceland. The rams originated from Bukara in Russia but had been kept in Halle (Germany) for many years (Palsson 1985). A year after their arrival, the first cases, of what was later identified as OPA, appeared in 1–3-year-old sheep. During the next 2–3 years, about 60% of the breeding stock in affected flocks died of the disease. Thereafter, the mortality dropped to under 10% in 3 years. As the disease declined, another respiratory disease named maedi (dispnea) presented itself. It was later determined that the disease was caused by a slow virus which became the prototypic ovine maedi-visna lentivirus (MVV). The transmissibility of OPA to in-contact sheep was first demonstrated by de Kock in 1929 and then by Dungal (1946). The incubation of Jaagsiekte was on average 6–8 months while that of Maedi was much longer, in the range of 4–6 years. It was later established that two of the imported Karakul rams were infected with MVV; one of them was also carrying the Jaagsiekte agent.

2
A Retroviral Etiology for Jaagsiekte/OPA

First hints of the involvement of a retrovirus in the etiology of OPA came from observations of retroviral particles in adenomatous lungs (Perk et al. 1974). The first report on the transmission of OPA with particles containing reverse transcriptase activity was by Martin and colleagues in 1976 (Martin et al. 1976). Verwoerd et al. soon thereafter succeeded in transmitting the disease with cytoplasmic fractions of tumor cells that had buoyant density in sucrose of around 1.175 g/ml which contained reverse transcriptase activity and viral particles that resembled those of a retrovirus (Verwoerd et al. 1980). Moreover the transmission experiments demonstrated an inverse dose relationship between the reverse transcriptase activity in the inoculum and the time between inoculation and the appearance of symptoms of the disease (Verwoerd et al. 1981) (Fig. 1). These experiments were confirmed by the Scottish group who were able to transiently produce a retrovirus from experimental OPA lung fluids and explanted tumor cells (Sharp et al. 1983). A few years later, both groups demonstrated the experimental transmission of the sheep disease to goats (Sharp et al. 1986; Tustin et al. 1988).

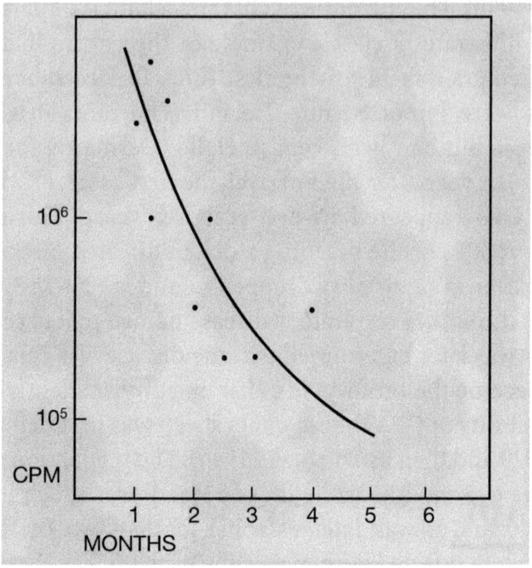

Fig. 1. Dose-response curve illustrating the decrease in incubation time (months) with increase in concentration of virus as measured by reverse transcriptase activity. (VERWOERD et al. 1981)

3
Purification of JSRV

Because the virus could not be grown in cell cultures, a major focus during the early 1980s was the development of methods to purify the virus from the lung lavage of affected sheep so as to identify the viral proteins and its genetic material (VERWOERD et al. 1983 and YORK et al. 1987). The research tools were limited to the reverse transcriptase assay, SDS–PAGE followed by Coomassie blue staining and Western immunoblotting and electron microscopy. Some excellent work on the morphology and morphogenesis of JSRV was performed on this material (PAYNE et al. 1983) (Fig. 2).

The optimal purification procedure was batch ultra centrifugation of about 1.5 l of OPA lung lavage which concentrated the virus 100 times (Table 1). Typically about one-third of this concentrate was extracted with the organic solvent Freon 113, which removed surfactant and other particulate lung material. This was followed by rate zonal centrifugation through a percoll gradient. Two visible bands were routinely observed. The lower more dense band, containing most of the virus, was then passed

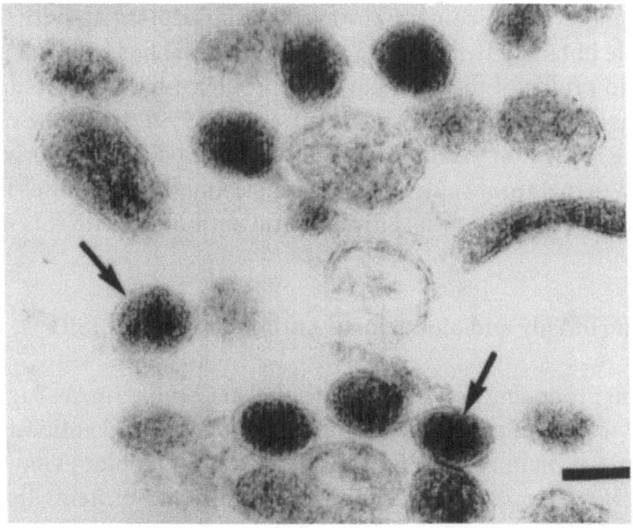

Fig. 2. Transmission electron microscopy of 'purified' JSRV particles. Bar, 100 nm

though a Sephacryl S1000 gel column. All processes were performed on ice or at 4 °C in a cold room. The peak reverse transcriptase fractions contained the 'purified virus'. Purification factors of 200 were frequently obtained. Alternative methods included the isopycnic centrifugation of the Freon or Percoll prepared virus through an acid, urea or guanidinium phase which was layered on top of a sucrose gradient. The virus, as monitored by reverse transcriptase activity, routinely banded at a density of 1.186 g/ml and comprised 7–9 proteins that were consistently present in

Table 1. Optimal purification of JSRV from lung lavage concentrate monitored at each stage for reverse transcriptase activity and protein content

Treatment	Total RT activity (c.p.m.)	Total protein content (mg)	Purification factor
Lung lavage pellet	4.7×10^6	21.2	1
Freon	6.34×10^6	12.6	2.27
Percoll gradient	4.7×10^6	7.8	2.7
Viral fractions Sephacryl S-1000 gel	7.61×10^6	0.12	286

Purification factor = total RT activity/total protein normalized to 1 at start.

purified preparations containing reverse transcriptase activity. The protein profile obtained from in vivo derived material has been compared to the protein profile of the viral particles produced by the infectious clone which was constructed by Palmarini et al. (1999). It is now clear that there were many non-viral proteins in the 'purified' in vivo derived virus compared to that from culture, thus emphasizing the difficulties that were experienced in producing specific antiviral antibodies with this material.

4 Production of Poly and Monoclonal Antibodies Against JSRV

Attempts at producing poly and monoclonal antibodies (mAbs) against the 'purified virus' were partly successful. A rabbit polyclonal antibody detected the JSp26 protein consistently on Western immunoblot (Verwoerd et al. 1983). However, there were numerous additional proteins that cross-reacted with this serum even after extensive absorption against normal lung proteins. This was evident as a high background staining on Western immunoblot. A most useful contribution to Jaagsiekte research was the finding by Sharp and colleagues that sera against the capsid proteins of Mason-Pfizer monkey virus (MPMV) and murine mammary tumor virus (MMTV) cross-reacted with JSRV in a Western immunoblot technique (Sharp and Herring 1983). This clean and specific reaction became an extremely useful research tool. It proved all the more useful when it became evident that many of our experimentally produced preparations of JSRV were also contaminated with MVV. The finding that even the 'purest preparations' of in vivo produced JSRV were complexed with immunoglobulins also contributed to the difficulties associated with early JSRV research. At the time it was felt that the only hope of overcoming the 'non-specificity' of the polyclonal antibodies was to produce mAbs. The production of mAbs against JSRV proteins involved a duplicate screening strategy. All clones were screened against 'purified JSRV' preparations and equivalently prepared normal lung antigens. This approach eliminated many clones. At the time we were not aware that JSRV had an endogenous counterpart that produced mRNAs in vivo (Palmarini et al. 1996). As the structural genes of some endogenous loci are ORFs (Palmarini et al. 2000) it is possible that proteins of these transcripts could be produced as was demonstrated for the capsid gene (Hecht et al. 1994). Some mAbs might have been discarded because of their reaction with these endogenously expressed proteins.

A number of mAbs were obtained that reacted more specifically with JS lung fluid than with normal lung preparations. Three of these mAbs, i.e., mAb 2E3, 4A10 and 11G11 were studied extensively (York et al. 1987; DeMartini and York 1997; Vinsen 2000). All reacted strongly with JS lung wash pellets and weakly or not at all with normal lung wash pellets (Table 2). All mAbs reacted with frozen and paraffin-embedded JS lung sections and weakly or not at all with equivalently prepared normal lung sections (Table 3). The mAbs also reacted with a cytoplasmic antigen in a continuous cell line 15.4 that was derived from trypsinized cultures of JS tumor cells (York 1987). The percentage of cells that reacted with the mAb peaked at 10% of cells 14 days after trypsinization. Attempts to identify the antigen against which the mAbs were directed included the preparation of an affinity column. The mAb 11G11 affinity column captured a protein of 50,000 Da that was present in soluble extracts of a 15.4 continuous cell line that was labeled with ^{35}S-methionine. This protein was also recovered from a normal cell line and was regarded as non-specific (York 1987). At the time that these mAbs were being studied, any reaction with normal lung material was difficult to explain other than that the mAb was

Table 2. EIA results showing the specificity of mAb 2E3, 4A10 and 11G11 against sheep lung lavage of OPAC and normal sheep (York 1987)

Sheep lungs	Number tested	Positive (%)		
		mAb 2E3	mAb 4A10	mAb 11G11
JS positive	17	88	88	100
Normal	76	23	4	20

Table 3. Immunohistochemical staining using mAbs 2E3, 4A10 and 11G11 against OPA and normal lung sections and an indirect immunoperoxidase technique (York 1987)

Sheep lungs	mAb 2E3		mAb 4A10		mAb 11G11	
	Number tested	Positive (%)	Number tested	Positive (%)	Number tested	Positive (%)
JS positive	13	100	7	100	16	100
Normal	5	40	3	0	5	60

against a non-specific normal lung epitope that was expressed at higher concentrations in diseased lungs.

None of the mAbs reacted using the Western immunoblot technique. Heating of the antigens tended to affect the antigenicity of the epitopes against which the mAbs were directed.

A detailed investigation into the reactivity of mAb 4A10 was made against a selection of human and sheep lung sections. The study revealed that mAb 4A10 reacted in a similar manner with the human sections as it did with the sheep sections confirming that the same epitope is present in human sections. The antigen/epitope against which mAb 4A10 is directed has not been determined. Staining of the sections with various immunohistochemical stains confirms that the epitope is not against glycogen but possibly a normal protein that is expressed in high concentrations in response to lung pathology (Vinsen 2000). Other groups have shown that antisera against the JSRV p26 capsid protein reacts with some human bronchioloalveolar lung sections, supporting the similarity between the sheep and human diseases (De las Heras et al. 2000).

5
A Dual MVV/JSRV Infection

At the start of the twentieth century, the founder and then director of the Onderstepoort Veterinary Institute, South Africa, Sir Arnold Theiler, initiated research to identify the etiological agent of the disease (Mitchell 1915). At the time it was felt that they had successfully transmitted the disease; however, a few years later it was questioned as to whether the disease described by Mitchell was indeed OPA. In 1929 de Kock published that histologically the sections looked more like Graaf Reinet disease, an ovine interstitial pneumonia, so termed because of a number of cases reported at an experimental station at Graaf Reinet, in South Africa. He described OPA as a neoplasm and called it 'papilliform cyst adenomatosis' (de Kock 1929). It was possible that the experimental sheep used in the transmission studies might have been infected with OPA but that it was the interstitial pneumonia that was being referred to in the transmission studies. The presence of both diseases, one a neoplasm and the other a pneumonia, in the same flocks and in the same sheep has been described and has posed a challenging research problem.

The morphology of the South African and Scottish particles in tumor cells and lung exudates and their biochemical characteristics were consistent with those of A and B type retroviruses. However, there were also reports that cultivable retroviruses, which could be isolated and propagated from OPA tumors, exhibited some similarities to type C retroviral particles (MALMQUEST et al. 1972). In 1974, PERK et al. presented biochemical evidence for the presence of particles with reverse transcriptase activity in lung extracts (PERK et al. 1974). However, these reports were not associated with transmission experiments and mention was made that the particles were C type. It is possible that the virus referred to was MVV. In South Africa, research on OPA was also side-tracked because of a MVV co-infection. In the early 1980s it was generally believed that MVV was not present in South Africa, even though there had been a report a few decades earlier of an outbreak of Graaf Reinet disease – which is now believed to have been MVV. Therefore the isolation of a lentivirus from both lung rinse pellets of experimentally induced cases of OPA (PAYNE et al. 1986) and from the 15.4 original tumor cell line (QUERAT et al. 1987) was most unexpected. Retrospective serological investigations of the experimental sheep revealed that sheep 15.4 was positive for MVV. Sheep 15.4 was one of the first experimentally induced cases of OPA. Lung fluid of 15.4 as well as 15.4 cell culture homogenates was used as inoculum in subsequent experimental transmissions of OPA in South Africa (COETZEE et al. 1976). In 1984, during attempts to culture JSRV using purified virus from lung fluid, one co-culture experiment excluded the use of Freon113 (organic solvent) as an early purification step. This material was seeded onto a sheep choroid plexus cell culture. When high reverse transcriptase activity counts were obtained a few weeks later there was great excitement, as we believed that we had managed to culture JSRV. However, transmission electron microscopy and SDS–PAGE analysis of the virus soon revealed that we had in fact isolated a visna-related lentivirus. We entertained studies to compare this novel isolate to the two prototypes, namely the Icelandic (1514 visna strain) and caprine arthritis encephalitis virus. It soon became evident that this lentivirus, although distinct from both prototypes, was more closely related to the visna virus (QUERAT et al. 1987). In view of the extensive genetic variability of lentiviruses, the new lentivirus was classified as a new type of ovine lentivirus (type III), and named South African ovine maedi visna virus (SA-OMVV) (QUERAT et al. 1987). Its genome was subsequently cloned and sequenced which confirmed the earlier serology results and the hybridization properties of its genomic RNA (QUERAT et al. 1990). Experimental infection of sheep with this lentivirus caused mild immunosuppression (MYER et al. 1988), but no tumors were ever produced.

6
Separation of the Two Retroviruses

Possibly one of the most useful contributions to JSRV research in the mid-1980s was the demonstration by the Scottish group that a 25,000-Da protein in OPA tumors and lung fluid was antigenically related to the capsid protein of MPMV and MMTV (Sharp and Herring 1983). Using a Western immunoblot technique and sera against MPMV and MVV it was now possible to clearly detect and distinguish between the two viruses (Fig. 3).

Western immunoblot analysis of the gradient fractions indicated that the 1.186 g/ml density fractions contained most of the JSRV capsid protein which cross reacted with an anti-MPMV p27 serum. The fractions around the lower 1.155 g/ml density peak contained a 25,000–30,000 Da protein that was antigenically related to the visna virus p30 capsid anti-sera (Fig. 3).

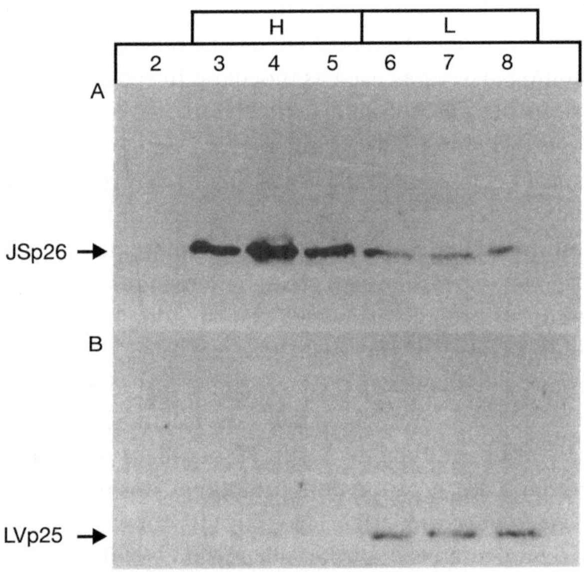

Fig. 3. Western immunoblotting of the gradient fractions when Freon-extracted sheep lung fluid (containing both JSRV and lentivirus LV) was separated to isopycnic equilibrium. A, Anti-MPMV serum was used to detect JSRV p26 capsid. B, Anti MVV (p30) serum was used to detect LV capsid

7 Cloning of the JSRV Genome

Initial attempts at screening genomic expression libraries of either genomic DNA from OPA tumor or cDNA from semi-purified JSRV RNAs, using polyclonal and monoclonal antibodies raised against semi-purified JSRV, were not successful. Success was finally obtained making cDNA from purified 1.186 g/ml density fractions, cloning it into a Lambda Zap vector, and screening the library with cDNA probes from 'purified' virus. This was achieved using previously optimized purification procedures. Briefly, lung fluid concentrate was extracted with Freon 113 to reduce the lentiviral content, the aqueous phase of the Freon 113 extract was then layered onto an isopycnic (20%–50% sucrose) gradient. Following equilibrium centrifugation the gradient was divided into 1-ml fractions, which were analyzed for reverse transcriptase activity and Western immunoblot using anti-MPMV p27 and MVV anti-p30 specific sera. Two reverse transcriptase activity peaks were seen, one at 1.186 g/ml and the other at a lower density of 1.155 g/ml expected for lentiviruses. Western immunoblot analysis further confirmed that the high density fractions contained a protein antigenically related to type B/D retroviruses (Fig. 3). Northern blot analysis of the RNA from the pools of the high and low density fractions using ^{32}P-radiolabeled cDNA probes from poly(A)+ RNAs of either the pellet or the two pooled fractions revealed two retroviral genomic-size RNAs, of 9.4 and 8.7 kb (Fig. 4). The 9.4-kb RNA, corresponding to the SA-OMVV genome, was present in the low density fractions, while the 8.7-kb RNA was mostly present in the high density fractions (Fig. 4). The 8.7-kb product was within the size range expected of replication-competent type B or D retroviruses. We therefore assumed that the 8.7-kb RNA was the genome of the retrovirus exhibiting the type B/D related capsid protein p26. We were surprised to notice that both genomic RNAs were in very good shape with few degradation products. When compared to genomic RNAs from visna virions grown in vitro, the virion-associated genomes from ex vivo material were unusually well preserved. The integrity of the genomic RNA was of prime importance to clone and sequence the complete genome.

The poly-adenylated RNA from pooled high-density (either side of the 1.186 g/ml) peak fractions, was extracted. An oligo(dT)-primed cDNA library was constructed in a lambda Zap II vector (Stratagene). The library was probed with oligo(dT) primed cDNA probes of RNA from the same

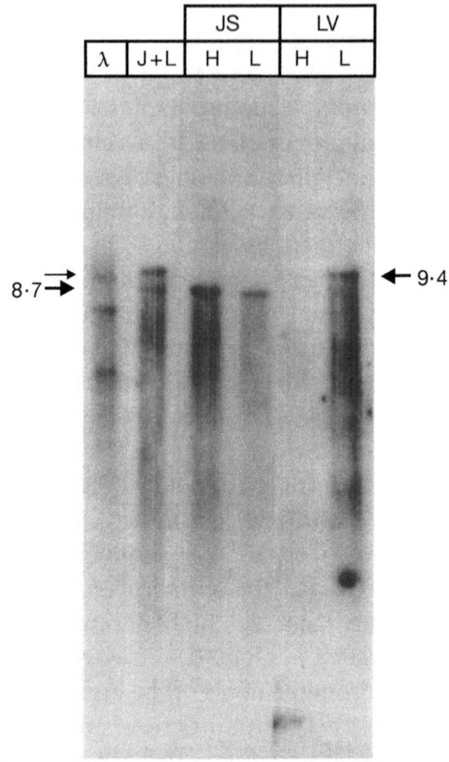

Fig. 4. Northern blot of RNA purified from the high-density (*H*) and low-density (*L*) fractions containing JSRV and LV particles. Lane *J+L*, poly (A)+ RNA of lung wash pellet not isopycnically purified and containing both JSRV and LV as shown in lanes *H* and *L*, poly (A)+ RNA from the high- and low-density fractions, respectively, of the gradient. Lanes *LV*, probe representative of SA 320, a molecular clone of a South African strain of MVV was used. Note the 9.4- and 8-kb genomes in lane *J+L* which are clearly separated in the lanes probed with *JS* and *LV*

high-density fractions. A number of nylon replicas were made of the master plates, which were probed at various times. One of the first clones isolated was a clone named JS46.1, which was sequenced. As we had no information about what the putative *env* and LTR of the JSRV might look like and because these sequences are usually the most variable in retroviruses, we chose not to use similarity searches against other retroviruses as the prime identification criteria. Instead, we looked for sequences bearing characteristic features of retroviruses, i.e., U3 and R regions, characterized by a polypurine tract upstream of the conserved TG, direct repeats with enhancer consensus sequences, and a TATA box located a few nucleotides upstream of a poly(A) signal.

Clone 46.1 was the first successful candidate; however, its structure was bizarre in that it was not colinear with a retroviral genome. This clone was a composite generated during the cloning process. *Eco*R1 adaptors were ligated to the cDNA ends to facilitate cloning into the *Eco*R1 site of the lambda Zap II multiple cloning site. To create the *Eco*R1 ends the cDNA was digested with *Eco*R1. The cDNA was not *Eco*R1 methylated prior to digestion therefore natural *Eco*R1 sites in the RNA genome were also digested. Clone JS 46.1 was the resultant ligation of a *pol* fragment with the 3′ end of the genome through the poly(A) tail which had an *Eco*R1 on its extreme 3′ end. This clone was used as a probe in Northern blots against RNA extracted from the high density fractions and a French field case of OPA.

Hybridization of this probe to the 8.7-kb RNA in both experimental and natural OPA added supportive evidence that both the 8.7-kb RNA and this clone were indeed related to the jaagsiekte retrovirus (Fig. 5). Homology searches against GenBank showed that the JSRV *pol* region was related to both type D and B retroviruses but was closer to type D, while the intracytoplasmic tail of *env* was exclusively related to B type retroviral sequences (York et al. 1991).

Using the 46.1 clone as a probe against the master plate, it was relatively easy to pick other positive clones which were then sequenced at each end to position them on the genome. In this way six clones were isolated that represented nearly the entire genome of the virus (York et al. 1992). Even though a significant fraction of the viral genome was covered, we were unable to obtain a clone covering the R-U5 and leader regions. A reverse transcription (RT)–PCR strategy was used to isolate and clone the extreme 5′ end of the genome; however it had its problems. The R region, which is common to both 5′ and 3′ ends of the genome, was only 13 nucleotides in length, not long enough for a PCR specific primer. To overcome this problem we took advantage of observations in other retroviruses which revealed that there were occasional 'read throughs' that overshot the poly(A) signal, resulting in either polyadelylation at a cryptic poly(A) signal in U5 or further transcription of downstream adjacent cellular genes. We therefore sequenced other clones representative of the 3′ end of the genome and eventually obtained some clones in which the poly(A) tail was added, not 13 nucleotides downstream of the canonical poly(A) signal, but instead 40 nucleotides downstream, in the U5 region. In this way

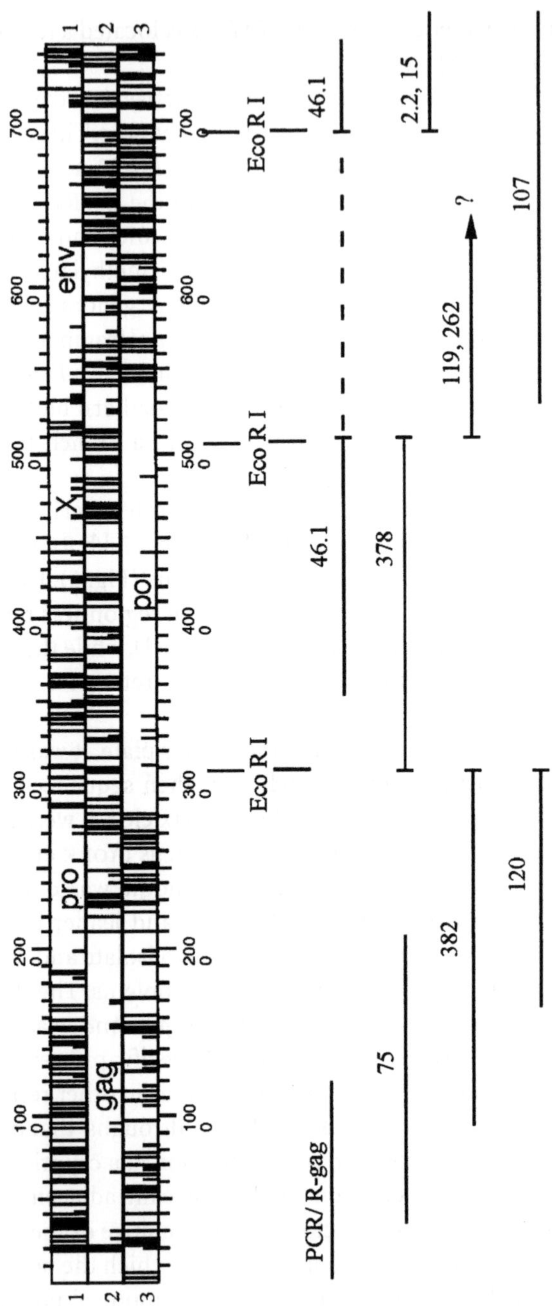

Fig. 5. Schematic representation of the clones that were isolated and sequenced that cover the entire 7,462 nucleotides of the *ex*JSRV genome. A simplified restriction map and genetic organization is also illustrated. Clone PCR-*gag* was amplified by RT–PCR and the rest of the clones were from a library of oligo dT primed cDNA

we could design a PCR primer overlapping R and U5 which was used together with a *gag* specific primer to isolate and clone the 5′ end of the JSRV genome.

8
Genetic Organization

The genetic organization of JSRV is typical of a replication-competent type D/B retrovirus with *gag, pro* and *pol* encoded by individual overlapping reading frames and the *env* reading frame slightly overlapping the end of *pol* (Fig. 5). There is another reading frame, *orfX*, which is located in a very unusual setting since it completely overlaps the integrase coding region. The position of the first AUG, 151 nucleotides downstream of the opening of the reading frame, is also unusual. Its codon usage was not in keeping with the other ORFs. Moreover, *orfX* is interrupted by stop codons in the JSRV-related enzootic nasal tumor virus (ENTV) sequence (Cousens et al. 1999). However, it remains intriguing that such a long ORF is present in the genome of all exogenous and endogenous JSRV sequences (Rosati et al. 2000).

The 5′ part of the *pro* reading frame encodes a dUTPase, and the 3′ part is a bona fide protease. In JSRV21, the start of the pro reading frame is located 53 nucleotides downstream of the *pro* reading frame of the South African JSRV clone. The true N terminus of the dUTPase domain will nevertheless depend on the position of the frame-shift between *gag* and *pro*, which, to our knowledge, is still unknown. In MMTV, dUTPase is encoded as a p30 NC-dUTPase fusion protein (Bergman 1994) and this is probably true also for JSRV. The dUTPase activity was investigated in purified JSRV virions by using an RT assay to quantify the incorporation of dUTP in poly rA/oligo dT primers/templates using either wild type viruses or dUTPase deleted mutants (IAQLL motif). The ratio of dUTP versus TTP incorporation was 4.6% for wild-type and 26% for the dUTPase deleted mutant showing that, as expected, the JSRV dUTPase was able to prevent dUTP misincorporation by the reverse transcriptase. In non-primate lentiviruses, the presence of a functional dUTPase (the gene is located in between RNaseH and integrase) is associated with their ability to replicate efficiently in macrophages. This is in keeping with the contention that in non-dividing cells such as terminally differentiated macrophages, the metabolism of the deoxynucleotides and the cellular dUTPase are repressed, resulting in low levels of TTP and dCTP and high levels of the dUTP precursor. This dNTP imbalance increases the chances of the reverse transcriptase misincorporating a uracil residue. Uracil incorporation in lieu of a cytosine in the stable rG:dU base pair is mutagenic and leads to G to A transition. The observation that JSRV heavily infects, apart from lung epithelial cells, the adherent monocytes/macrophages from lymphoid organs may explain the positive selection pressure to preserve a functional dUTPase. An alternative hypothesis could be that the very low mitotic activities of type II pneumocytes and Clara cells in adult sheep is associated with down-modulation of deoxynucleotide metabolism and therefore results in an unbalanced ratio of dUTP versus other deoxynucleotides. Then, dUTPase could be an advantage for the natural infection of these epithelial cells in the lung of adult sheep. It is interesting to note that the dUTPase active sites are 100% conserved in the three full length endogenous loci that have been sequenced (Palmarini et al. 2000).

The *gag*, *pro* and *pol* genes of JSRV are more closely related to the type D MPMV, while there is no homology at all between *env* genes of D type viruses and JSRV. Instead JSRV *env* is related to type B MMTV and HERV K (human endogenous retroviruses with Lys tRNA primer binding sites). Although the identity score is not high, the conservation of structural determinants such as cysteines and glycosylation sites is suggestive of overall structural similarities. The homology to B type *env* is clearly highlighted for the C-terminal half of *env* when looking at the skeletons of various D and B type *env* proteins.

During preparation of this manuscript, an updated FASTA search was performed, using JSRV Env protein as the query. Among the results was a new significant homology (20%–30% identity) to part of a full length cDNA from Macaque brain (*Macaca fascicularis*; Osada et al. 2001). This unannotated cDNA appears to be that of an *env* spliced mRNA from a new B type endogenous retrovirus. Its coding potential is interrupted by stop codons and frame shifts and it is distributed in at least three contiguous ORFs. Brute joining of the three major ORFs yields a pseudo-*env* gene which is 20%–30% identical to its counterpart from either JSRV, HERV K or MMTV, but exhibits no significant homology to type D *env*.

The phylogeny of reverse transcriptase and of the Env transmembrane proteins reflected the above findings. JSRV and the progenitor of the D type viruses have both evolved from a group of A and B type viruses, all of which have the ability to form intracytoplasmic Gag particles. The progenitor of D type viruses then acquired a new *env* gene, probably from primates as they have not been isolated from other species. From that

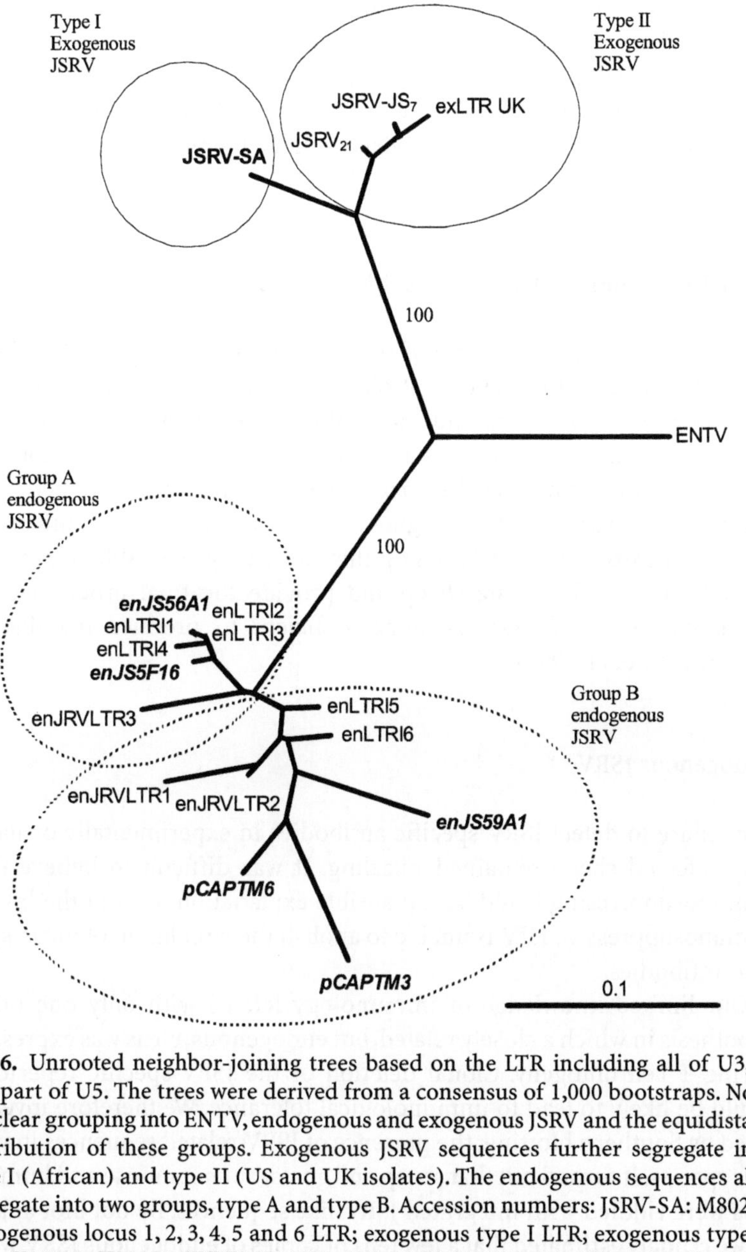

Fig. 6. Unrooted neighbor-joining trees based on the LTR including all of U3, R and part of U5. The trees were derived from a consensus of 1,000 bootstraps. Note the clear grouping into ENTV, endogenous and exogenous JSRV and the equidistant distribution of these groups. Exogenous JSRV sequences further segregate into type I (African) and type II (US and UK isolates). The endogenous sequences also segregate into two groups, type A and type B. Accession numbers: JSRV-SA: M80216 endogenous locus 1, 2, 3, 4, 5 and 6 LTR; exogenous type I LTR; exogenous type II LTR: X95445–X95452, ENTV: Y16627; JS7, 809T, 83RS28, 84RS28, 92K3: Y18301–Y18305; enJSRV-1, 2 and 3: Z66531–Z66533; JSRV-UK: Z71304; JSRV21: AF105220; enJS5.F16, 5.9A1 and 5.6A1: AF136224, AF136225 and AF153615

point of view JSRV could be classified as a B type retrovirus, because it diverged from the main branch before the capture of the new *env* gene which is the hallmark of D type retroviruses. With regards to classification, a new genus 'betaretroviruses' has been established which incorporates both the B and D type oncoviruses. JSRV therefore clearly falls into this category (Fig. 6).

9
Is the Exogenous JSRV Alone Capable of Causing the Disease?

By 1992, the complete genome of the exogenous form of JSRV had been isolated cloned and sequenced. Probes of the virus were shown to react with lung fluid from all natural cases of the disease but not against lung wash of normal sheep (York et al. 1992). However, this still did not conclusively demonstrate that the virus alone was capable of producing the disease. Palmarini and colleagues successfully isolated an infectious clone of the exogenous JSRV. Using this clone they were able to produce enough virus to inoculate sheep and provide the final proof that the exogenous form of JSRV is alone sufficient to produce the disease (Palmarini et al. 1999).

10
Endogenous JSRV

The failure to detect JSRV specific antibodies in experimentally or naturally infected sheep remained puzzling. It was difficult to believe that immunosuppression could be a plausible explanation as even the highly immunosuppressive HIV is unable to abolish the circulation of virus-specific antibodies.

Our limited knowledge of immunology left us with only one other hypothesis in which a closely related, but endogenous, virus was expressed during T cell ontogeny. Clonal deletion of the JSRV-specific repertoire would be likely to lead to immunological tolerance. We therefore investigated by Southern blotting, the presence of JSRV related sequences in various species. It turned out that sheep and goats, but not mouse and human, cells were riddled with integrated JSRV-like sequences. By dot blot analysis we crudely estimated that a few tens of copies of endogenous JSRV were present in sheep and goats (York et al. 1992). A more comprehensive and extensive study (Hecht et al. 1996) has since shown that infection by JSRV related viruses and the establishment of their endogenous counterparts were not limited to domestic sheep and goat species, since endogenous sequences were also found in bovine and antelope species. Three full length endogenous genomes have been isolated, cloned and sequenced and will be discussed in detail in the chapter by DeMartini et al., this volume.

Having the complete genome sequence of the exogenous infectious form of the etiologic agent of OPA opened many doors to study this intriguing disease in more detail. The subtle sequence differences between the endogenous and exogenous viruses enabled the development of exogenous specific PCR assays which facilitated the development of molecular diagnostic tools. Using these assays it was possible to investigate where and which cells were being infected. More recently important studies focusing on the differences between the endogenous and exogenous viral sequences have facilitated the identification of regions in the genome that are responsible for transformation. These will be discussed in greater detail in the later chapters of this volume.

References

Aynaud M (1926) Origine vermineuse du cancer pulmonaire de la brebis. Compt Rend Soc Boil 95:1540–1542

Barsky S, Cameron R, Osann KE, Tomita D, Holmes EC (1994) Rising incidence of bronchioloalveolar lung carcinoma and its unique clinicopathologic features. Cancer 73:1163–70

Bergman AC, Bjornberg O, Nord J, Nyman PO, Rosengren AM (1994) The protein p30, encoded at the gag-pro junction of mouse mammary tumor virus, is a dUTPase fused with a nucleocapsid protein. Virology 204:420–424

Bonne C (1939) Morphological resemblance of pulmonary adenomatosis (jaagsiekte) in sheep and certain cases of cancer of the lung in man. The Amer J of Cancer XXXV(4):491-501

Coetzee S, Els HJ, Verwoerd DW (1976) Transmission of Jaagsiekte (Ovine pulmonary adenomatosis) by means of a permanent epithelial cell line established from affected lungs. Onderstepoort J Vet Res 3:133–142

Cousens C, Minguijon E, Dalziel RG, Ortin A, Garcia M, Park J, Gonzalez L, Sharp JM, De las Heras M (1999) Complete sequence of Enzootic Nasal Tumour Virus, a retrovirus associated with transmissable intranasal tumours of sheep. J Virol 73:3986–3993

de Kock G (1929) Are the lesions of Jaagsiekte in sheep of the nature of a neoplasm? Onderstepoort, S Afr. Director of Vet Serv. 15th Annual Report of the Director of Veterinary Services Union of South Africa October pp 611–641

De Las Heras M, Barsky SH, Hasleton P, Wagner M, Larson E, Egan J, Ortin A, Gimenez-Mas J, Ortin A, Palmarini M, Sharp JM (2000) Evidence for a protein related immunologically to the jaagsiekte sheep retrovirus in some human lung tumours. Eur Respir J 15:330–332

DeMartini JC, York DF (1997) Retrovirus-associated neoplasms of the respiratory system of sheep and goats. Veterinary clinics of North America: food animal practice. 13(1):55–70

Dungal N (1946) Experiments with jaagsiekte. Amer J of Path 22:737–759

Hecht SJ, Carlson JO, DeMartini JC (1994) Analysis of a type D retroviral capsid gene expressed in ovine pulmonary carcinoma and present in both affected and unaffected sheep genomes. Virol 202:480–484

Hecht SJ, Sharp JM, DeMartini JC (1996) Retroviral aetiopathogenesis of ovine pulmonary carcinoma: a critical appraisal. Br Vet J 152:395–407

Hecht SJ, Stedman KE, Carlson JO, DeMartini JC (1996) Distribution of endogenous type B and Type D sheep retrovirus sequences in ungulates and other mammals. Proc Natl Acad Sci USA 93:3297–3302

Hutcheon D (1891) Reply to query no. 191 about jaagsiekte or chronic catarrhal pneumonia. Agric J Cape of Good Hope 4:87–89

M'Fadyean J (1888) Lung disease in sheep, caused by the Stronggylus rufescens. J Comp Pathol Therap 1:139–146

Martin WB, Scott FMM, Sharp JM, Angus KW (1976) Experimental production of sheep pulmonary adenomatosis (Jaagsiekte). Nature 264:183–4

Malmquist WA, Krauss HH, Moulton YE, Wandera JG (1972) Morphologic study of virus infected lung cell cultures from sheep pulmonary adenomatosis (jaagsiekte). Laboratory Investigation 26:528–533

Mitchell DT (1915) Resp Vet Rec S Afr 3:585

Myer MS, Huchzermeyer HFAK, York DF, Hunter P, Verwoerd DW, Garnet HM (1988) The possible involvement of immunosuppression caused by a lentivirus in the aetiology of Jaagsiekte and pasteurellosis in sheep. Onderstepoort. J of Vet Res 55:127–133

Ortin A, Minguijon E, Dewar P, Garcia M, Ferrer LM, Palmarini M, Gonzalez L, Sharp JM, De las Heras M (1998) Lack of a specific immune response against a recombinant capsid protein of Jaagsiekte sheep retrovirus in sheep and goats naturally affected by enzootic nasal tumour of sheep pulmonary adenomatosis Vet. Immun. and Immunopath 61:229–237

Osada N, Hida M, Kususda J, Tanuma R, Iseki K, Hirata M, Suto Y, Hirai M, Terao K, Suzuki Y, Sugano S, Hashimoto K (2001) Assignment of 118 novel cDNAs of cynomolgus monkey brain to human chromosomes. Gene 275(4): 31–37

Palmarini M, Cousens C, Dalziel RG, Bai J, Stedman K, DeMartini JC, Sharp JM (1996) The exogenous form of Jaagsiekte retrovirus is specifically associated with a contagious lung cancer of sheep. J Virology 70(3):1618–1623

Palmarini M, Dewar P, De las Heras M, Inglis NF, Dalziel RG, Sharp JM (1995) Epithelial tumour cells in the lungs of sheep with pulmonary adenomatosis are major sites of replication for Jaagsiekte retrovirus. J Gen Virol 76: 2731–2737

Palmarini M, Fan H, Sharp JM (1997) Sheep pulmonary adenomatosis: a unique model of retrovirus-associated lung cancer. Trends in Microbiology 5(12):478–482

Palmarini M, Hallwirth C, York D, Murgai C, D'Oliviera T, Spencer T, Fan H (2000) Molecular cloning and functional analysis of three type D endogenous retroviruses of sheep reveal a different tropism from that of the highly related exogenous jaagsiekte sheep retrovirus. J Virology 74(17):8065- 8076

Palmarini M, Sharp JM, Lee C, Fan H (1999) In vitro infection of ovine cell lines by jaagsiekte sheep retrovirus (JSRV) J Virol 73:10070–10078

Palmarini M, Sharp JM, De Las Heras M, Fan H (1999) Jaagsiekte sheep retrovirus is necessary and sufficient to induce a contagious lung cancer in sheep. J Virology 73:6964–6972

Palsson PA (1985) Maedi/visna of sheep in Iceland: Introduction of the disease to Iceland, clinical features, control measures and eradication. In: Sharp JM, Hoff-Jorgensen R (eds): Slow Viruses in Sheep, Goats and cattle. Brussels, commission of the European Communities, pp 3–19

Payne A-L, Verwoerd DW, Garnett HM (1983) The morphology and morphogenesis of jaagsiekte retrovirus (JSRV). Onderstepoort J Vet Res 50:317–322

Payne A-L, York DF, deVilliers EM, Verwoerd DW, Querat G, Barban V, Sauze N, Vigne R (1986) Isolation and identification of a South African lentivirus from jaagsiekte tumour tissue and cell lines. Onderstepoort J Vet Res 53: 55–62

Perk K, Hod I (1982) Sheep lung carcinoma: An epidemic analogue of a human neoplasm. J Natl Cancer Inst 69:747–750

Perk K, Michalides R, Spiegelman S, Schlom J (1974) Biochemical and morphologic evidence for the presence of an RNA tumor virus in pulmonary carcinoma of sheep (jaagsiekte). J Nat Cancer Inst 53:131–135

Querat G, Audoly G, Sonigo P (1990) Nucleotide sequence analysis of SA-OMVV, a visna-related ovine lentivirus: Phylogenetic history of lentiviruses. Virology 175:434–447

Querat G, Barban V, Sauze N, Vigne R, Payne A, York D, DeVilliers EM, Verwoerd DW (1987) Characteristics of a novel lentivirus derived from South African sheep with pulmonary adenocarcinoma (jaagsiekte). Virology 158:158–167

Rosati S, Pittau M, Alberti A, Pozzi S, York DF, Sharp JM, Palmarini M (2000) In accessory open reading frame (orf-x) of jaagsiekte sheep retrovirus is conserved between different virus isolates. Virus Research 66:109–116

Sharp JM, Herring AJ (1983) Sheep pulmonary adenomatosis: demonstration of a protein which cross-reacts with the major core proteins of Mason-Pfizer monkey virus and mouse mammary tumour virus. J Gen Virol 64:2323- 2327

Sharp JM, Angus KW, Gray EW, Scott FM (1983) Rapid transmission of sheep pulmonary adenomatosis (jaagsiekte) in young lambs. Archives of Virology 78: 89–95

Sharp JM, Angus KW, Jassim FA, Scott FMM (1986) Experimental transmission of sheep pulmonary adenomatosis to a goat. The Veterinary Record 119:245

Sigurdsson B (1958) Adenomatosis of sheep's lungs Archiv fur die gesamte virusforschung Sonderabdruck aus Band VIII, Heft 1:51–58

Tustin RC (1969) Ovine Jaagsiekte. Journal of the South African Veterinary Medical Association 40:3–23

Tustin RC, Williamson A-L, York DF, Verwoerd DW (1988) Experimental transmission of Jaagsiekte (ovine pulmonary adenomatosis) to goats. Onderstepoort J Vet Res 50:309–316

Verwoerd DW, Payne A-L, York DF, Myer MS (1983) Isolation and preliminary characterization of the jaagsiekte retrovirus. Onderstepoort J Vet Res 50:309–316

Verwoerd DW, Williamson A-L, deVilliers EM (1980) Aetiology of jaagsiekte. Transmission by means of subcellular fractions and evidence for the involvement of a retrovirus. Onderstepoort J Vet Res 47:275–280

Verwoerd DW, Williamson A-L (1981) In Yohn D, Blakeslee J (eds). Advances in comparative leukemia research Inc. Preliminary characterization of a newly isolated ovine retrovirus causing jaagsiekte, a pulmonary adenomatosis. Elsevier North Holland, p 453–454

Verwoerd DW (1996) Ovine pulmonary adenomatosis (jaagsiekte). Br vet J 152: 369–371

Vinsen C (2000) An immunohistochemical and molecular investigation into some similarities between Jaagsiekte and human bronchiolar alveolar carcinoma. MSc thesis. University of Natal, Pietermaritzburg, South Africa

York DF (1987) A biochemical and immunological comparison of the Jaagsiekte and two related retroviruses. PhD thesis. University of Natal, Pietermaritzburg South Africa

York DF, Vigne R, Verwoerd DW, Querat G (1991) Isolation, identification and partial cDNA cloning of genomic RNA of Jaagsiekte retrovirus, the etiological agent of sheep pulmonary adenomatosis. J Virol 65(9):5061–5067

York DF, Vigne R, Verwoerd DW, Querat G (1992) Nucleotide sequence of the Jaagsiekte retrovirus, an exogenous and endogenous type D and B retrovirus of sheep and goats. J Virol 66:4930–9

CHAPTER 2

Pathology of Ovine Pulmonary Adenocarcinoma

M. DE LAS HERAS, L. GONZÁLEZ, J.M. SHARP

M. DE LAS HERAS
Departamento de Patologia Animal, Facultad de Veterinaria,
Universidad de Zaragoza, Miguel Servet 177, 50013 Zaragoza, Spain
e-mail: lasheras@posta.unizar.es

L. GONZÁLEZ
Veterinary Laboratoires Agency (VLA-Lasswade), Pentlands Science Park,
Bush Loan, Penicuik EH26 OPZ, UK
e-mail: l.gonzales@vla.maff.gsi.gov.uk

J.M. SHARP
Moredon Research Institute, Pentlands Science Park Bush Loan, Penicuik EH26 OPZ, UK
e-mail: sharm@mri.sari.ac.uk

Abstract. Clinical, gross pathology, histopathology and electron microscopy of the ovine pulmonary adenocarcinoma (OPA, jaagsiekte) either natural or experimentally induced in sheep, goat and moufflon are described. OPA is caused by an oncogenic betaretrovirus, jaagsiekte sheep retrovirus (JSRV). Most natural cases of OPA appear in animals 1–4 years old. There is no evidence of sex or breed susceptibility. Sheep affected by OPA show an afebrile respiratory illness associated with loss of weight. A very characteristic clinical sign is moist rales caused by the accumulation of fluid in the respiratory airways which is discharged from the nostrils when the head is lowered. Gross lesions are confined to the lungs but occasionally thoracic or extrathoracic structures are also affected. Two pathologic forms of OPA are currently recognized, classical and atypical. In classical forms the neoplastic lesions occurs particularly in the cranioventral parts of all lung lobes. They are diffuse or nodular, light grey or light purple in colour. On the cut surface the tumour is moist, and frothy fluid may pour from the airways on slight pressure. Atypical forms tend to be more nodular in both early and advanced tumours. They are pearly white in colour, very hard in consistency, very well demarcated from the surrounding parenchyma and their surface is dry. Histology of the lung sections reveals the presence of several foci of epithelial cell neoplastic proliferation in both alveolar or bronchiolar regions. The tumours, derived from type II pneumocytes and Clara cells, proliferate into mostly papillary but also acinar or occasionally solid growths. The tumour generally shows a benign histological pattern but intra- and extrathoracic metastases have been detected in some cases. Several considerations suggest that the tumour should be classified as an adenocarcinoma of the lung. The histology of atypical OPA is similar to that of the classical disease, with an increase in the stromal reaction accompanying the epithelial proliferations. Pathological features of OPA induced experimentally in sheep, or of OPA in goats and moufflon are similar to those described in sheep. Detailed electron microscopy of tumour material confirms that type II pneumocytes and Clara bronchiolar epithelial cells are the origin of the neoplasia. Also included in this chapter is a description of the morphology of the viral particles associated with OPA.

1 Introduction

Ovine pulmonary adenocarcinoma (OPA, sheep pulmonary adenomatosis, jaagsiekte) is a contagious neoplasm of the sheep lungs. The tumour has also been recognized, although more rarely, in goats and wild moufflon, but it does not affect cattle or other animals. The disease has been described in a wide variety of breeds in more than 20 countries of Europe, Africa, America and Asia. Studies in the UK and South Africa showed that OPA accounts for almost 70% of all sheep tumours. The disease is on the B list of the Office International des Epizooties in recognition of its socioeconomic importance in the affected countries and regions, and of the threat it poses to international trade of animals and animal products. Its aetiological relationship with a betaretrovirus, jaagsiekte sheep retrovirus (JSRV) is well established (Palmarini et al. 1999) but no immune reaction to this virus can be detected in affected animals (Ortin et al. 1998). The tumour cells in OPA are derived from type II pneumocytes or Clara cells, secretory epithelial cells in the lung. A characteristic of the disease is production of excess secretions by the tumour cells, leading to fluid buildup in the lungs. The lung fluid, which contains infectious JSRV, can be harvested from animals with OPA and filtrates will induce the disease in unaffected animals. Proviral DNA of JSRV can be detected by PCR in peripheral blood leukocytes not only in clinically affected sheep, but also in apparently healthy animals in infected flocks (Garcia-Goti et al. 2000). Before these new molecular diagnostic tools became available, the only means of identifying OPA-affected sheep was the recognition of pathognomonic clinical signs and gross and histological lung lesions. Also, pathological studies have been until recently the only source of data to gain understanding of the epidemiological features of this infection. We will describe in this chapter the clinical, pathological and ultrastructural features of OPA in sheep, goats and moufflon.

2
Features of the Clinical Disease

2.1
OPA in Sheep

According to experimental transmission studies, sheep of all ages seem to be susceptible to OPA, but clinical disease is rarely seen in sheep younger than 7–9 months in natural conditions. Most natural OPA cases appear in animals aged 1–4 years (Dungal et al. 1938; Tustin 1969; Hunter and Munro 1983; González et al. 1993). The aerogenous route seems to be the most important natural mode of transmission (Tustin 1969; Wandera 1971), but there is some indication that intra-uterine infection can also occur (De las Heras et al. 2001) and other routes, such as alimentary, cannot be ruled out. Under natural conditions, the estimated incubation period in flocks where the disease is not endemic is about 6–8 months (Dungal 1938). Experimentally, clinical illness occurs within 5–12 months of intratracheal inoculation of tumour tissue extracts or lung fluid into several-months-old lambs or adult sheep (Tustin 1969; Wandera 1970; Martin et al. 1976). In very young lambs, clinical signs tend to appear in 3–6 weeks, or even more rapidly (4–6 days), after intratracheal inoculation of lung fluid from naturally affected animals (Sharp et al. 1983).

Many sheep breeds are susceptible to OPA, but it does appear that some breeds and even families may be more resistant than others (Sharp and Angus 1990a). OPA is not restricted to one sex, and small animal numbers in studies elsewhere make it difficult to form an opinion on sex-related susceptibility (Sharp and Angus 1990a). Winter seems to be the time of the year with more registered cases (Tustin 1969; Hunter and Munro 1983). The mortality rate in affected flocks varies according to the length of time that the infection has been present in them. During the OPA epidemics in Iceland and Kenya the annual mortality associated with the disease was 30%–50% in the first years after introduction the infection (Dungal et al. 1938; Shirlaw 1959), but dropped to average figures of 1%–5% (0.3%–24%) when the disease became endemic (Dungal et al. 1938; Wandera 1967; Tustin 1969).

Sheep naturally affected by OPA show signs of progressive afebrile respiratory illness associated with loss of weight. There are no clinical manifestations when the lung tumour is very small, and it is only when the tumour volume becomes sufficiently large to interfere with lung function that clinical signs can be expected. Initially, the affected sheep appear less active than normal and have a tendency to lag behind the flock when moved and appear to have been chased (hence the name 'jaagsiekte', see chapter by York and Querat, this volume). Later, the respiration becomes deeper and frequent (tachypnoea), and is often associated with noticeable movements of the abdominal wall (abdominal lift). A very characteristic clinical sign is the moist rales caused by the accumulation of fluid in the respiratory airways. These can be detected by auscultation in the early phase of the clinical disease, but as disease progresses they may be audible without the aid of the stethoscope and even at some distance from the animal. The pulmonary fluid is discharged from the nostrils and at first is only slightly increased, being sufficient to keep the nostrils glistening and moist. Later it will be noticed as a frothy sero-mucous fluid running from the nostrils when the sheep's hindquarters are raised or the head lowered (Fig. 1). The amount of fluid that can be collected from an affected animal varies from 50–60 ml to as much as 300 ml (Dungal et al. 1938; Sharp and Angus 1990a). On some occasions, the accumulation of this fluid in the airways may give rise to spasmodic coughing which can be quite prominent in advanced stages of the disease. Affected sheep remain alert,

afebrile and have a good appetite, though progressive loss of weight is obvious. Death inevitably occurs within a few months (2–3) of the start of the clinical disease (Dungal et al. 1938), but the clinical course can be shortened to a few weeks and fever appears if bacterial infections become superimposed.

Clinical presentation of the disease may vary depending on the concurrence of other diseases and on the productive stage of the affected animal. Coughing can be recorded in as much as 45 % of sheep with OPA in flocks where parasitic pneumonia is also prevalent (González 1989). Dyspnoea seems to worsen more quickly and the amount of lung fluid is greater if the disease affects lambing ewes, which also show a decrease in milk production (González 1989). Concurrence of OPA and maedi-visna has been frequently reported (Markson et al. 1983; Snyder et al. 1983; Rosadio et al. 1988; González 1989; González et al. 1993) and it results in a worsening of the clinical signs and a precipitation of the disease course.

Variations in the blood composition have been mentioned during preclinical stages (Nobel et al. 1971; Hod et al. 1974). These authors indicated a persistent elevation of the IgG concentration in the serum before the start of the disease, but the levels were not different from those in non-tumour affected animals of the same flocks. No evident elevation in circulating IgG levels has been observed by other authors in OPA-affected sheep (Sharp and Angus 1990a, De las Heras et al. 1995). The proportion of different leukocytes in the blood is altered in animals with advanced clinical signs of OPA. Various degrees of neutrophilia and lymphocytopenia have been detected in some animals (González 1989; Rosadio and Sharp 1992), and evidence of immune dysfunction with a reduction of the CD4/CD8 ratio in the peripheral blood has been reported (Rosadio and Sharp 1992).

There are some reports of a different pathological form of OPA (atypical OPA, see below), in which tumour nodules are generally small, non-confluent and dry (non-fluid producing). This form is therefore subclinical and appears as an incidental finding at necropsy or in abattoir studies (Cuba Caparo 1961; Dualde 1966, De las Heras et al. 1992, Garcia-Goti et al. 2000).

2.2 OPA in Goats and Moufflon

The clinical signs of OPA in goats have not been described, as most reports of the disease in this species come from abattoir studies or from incidental findings at post-mortem examination (Banerjee and Gupta 1979; Cuba Caparo et al. 1961; Sriraman et al. 1982; Stephanou et al. 1975). In Sardinia, several animals showing dyspnoea and copious seromucous nasal secretions were found within a group of captured wild moufflon. After necropsy, lesions compatible with OPA were observed in their lungs, and JSRV was detected in the tumours (Nieddu et al. 1987; Sanna et al. 2001).

OPA has been induced experimentally in goats, but evidence of success was not obtained until post-mortem examinations were performed, as none of the inoculated animals showed clinical signs during the 6–36-month post-inoculation period of examination (Sharp et al. 1986; Tustin et al. 1988). These and other experiments also demonstrated that goats are less susceptible to the development of clinical disease than sheep, at least by using a sheep derived inoculum (Sharma et al. 1975)

3 Gross Pathology

3.1 OPA in Sheep

Gross lesions are confined to the lungs but occasionally thoracic lymph nodes also show characteristic pathological changes. The neoplastic areas vary from small discrete nodules, measuring 0.5–2 cm to extensive tumours involving the entire ventral half of the diaphragmatic and other lobes. In clinical disease cases, the tumour usually involves both lungs, although normally not to the same extent. Two pathological forms of OPA are currently recognized, classical and atypical.

3.1.1 *Classical OPA*

After necropsy, the lungs appear considerably enlarged and they do not collapse when the chest is opened (Dungal et al. 1938; Tustin et al. 1969; Wandera 1971; Sharp and Angus 1990a). The normal weight may be

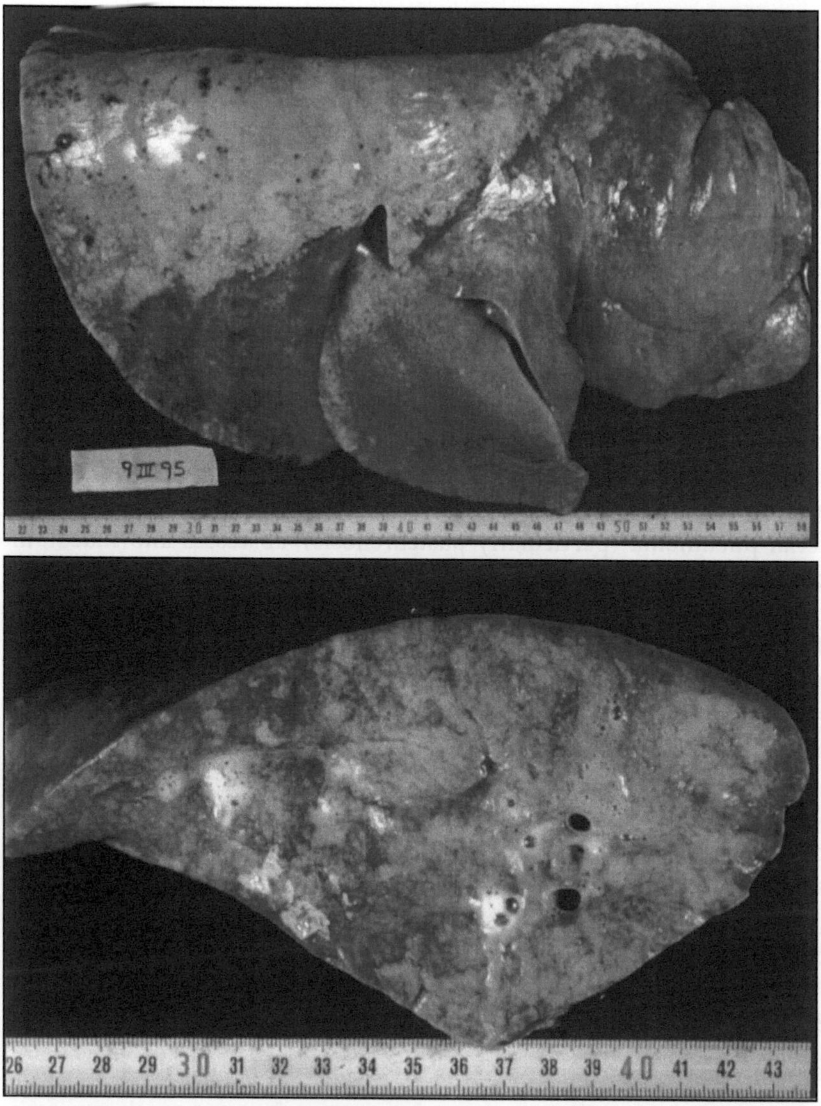

Fig. 2a, b. Classical OPA. **a** Tumour appears as purple coloured area occupying the cranioventral part of the right lung. A narrow zone of emphysematous lung separates the neoplasm from the normal parenchyma. **b** Transverse section of the same lung through the diaphragmatic lobe showing tumour-affected areas with meaty aspect and frothy fluid pouring from the bronchioles

increased three or more times, depending on the extent of the tumour. The neoplastic lesions occur particularly in the cranioventral parts of all lung lobes but any part of the lungs may be involved (Fig. 2a). They are diffuse or nodular, light grey or light purple in colour, do not protrude significantly on the surface and have an increased consistency (Dungal et al. 1938; Tustin et al. 1969; Sharp and Angus 1990a; Wandera 1971; Cutlip and Young 1982). A narrow or more extensive area of emphysema surrounds the neoplastic lung. Pleurisy, often with a chronic fibrous appearance, is a common finding over the neoplastic lung. On the cut surface, the tumour usually appears as numerous, small, slightly elevated nodules of an irregular outline; these often coalesce to give a diffuse glandular appearance. The surface is moist and frothy fluid may pour from the bronchioles and bronchi on a slight pressure (Fig. 2b). In advanced cases, the affected areas of the lung are lardaceous-white, very solid and hard as a result of fibrosis. Concurrent lesions of other lung diseases, such as bacterial pneumonia (particularly pasteurellosis), abscesses and maedi, in countries where this occurs, are relatively frequent, and they may hide or overshadow the tumour lesion pattern described (Dungal et al. 1938; Tustin et al. 1969; Wandera 1971; Cutlip and Young 1982; Rosadio et al. 1988; Sharp and Angus 1990a). In early cases, when clinical signs are not yet discernible, solitary tumour nodules may be the only evidence of the disease (Tustin et al. 1969; Sharp and Angus 1990a).

Mediastinal and tracheobronchial lymph nodes may not show any visible changes (Dungal et al. 1938; Tustin et al. 1969), or may be slightly or clearly enlarged (Sharp and Angus 1990a; Wandera 1971). Occasionally, small metastases can be grossly discernible at the surface of regional lymph nodes; more rarely, extrathoracic spread of the tumour with metastases in distant organs has been observed (Sharp and Angus 1990a).

3.1.2
Atypical OPA

A different patho-morphological form of OPA has been described (Cuba Caparo et al. 1961; Dualde 1966) and termed 'atypical' (De las Heras et al. 1992; Garcia-Goti et al. 2000) in contrast with the 'classical' form described above. This form tends to be more nodular in both early and advanced tumours, instead of diffuse. The nodules may be solitary (Fig. 3a) or multiple, and tend to appear mainly in the diaphragmatic lobes. They are pearly white in colour, very hard in consistency and some-

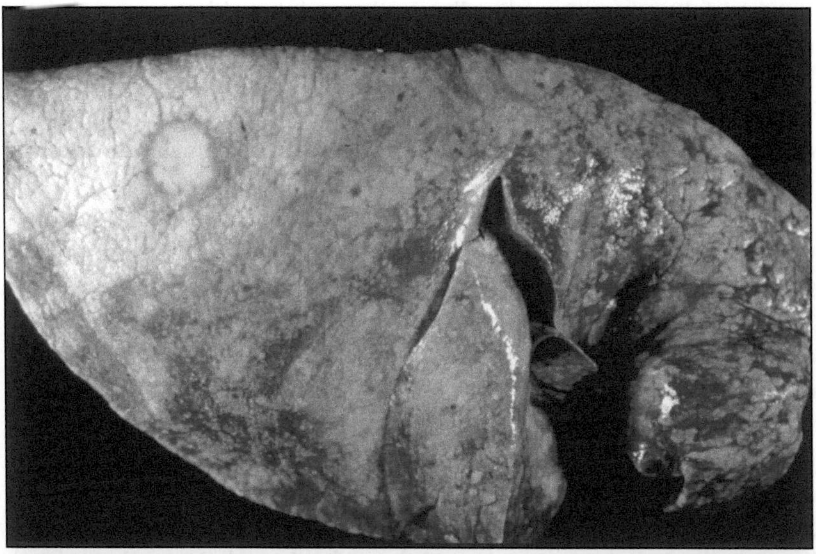

a

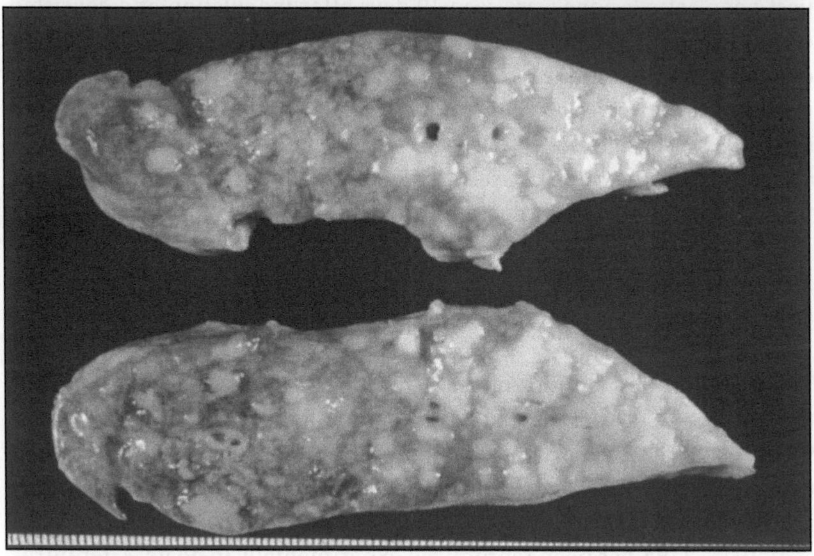

b

Fig. 3a, b. Atypical OPA. **a** Well-demarcated, white, solitary tumour nodule in the dorsocaudal area of the right lung. **b** Transverse section of a severe case of atypical OPA showing numerous, dry, confluent tumour nodules in the diaphragmatic lobe

times look like scars. Occasionally, smaller nodules with the same characteristics can be found in the periphery of the main nodule. When sectioned, the tumour nodules are white in colour, very well demarcated from the surrounding parenchyma, and their surface is dry (Fig. 3b). Some cases appear as multiple neoplastic nodules of 0.5–1 cm diameter throughout the lungs.

Atypical and classical OPA may coexist in a particular flock (Garcia-Goti et al. 2000) and in individual sheep, and intermediate and mixed forms have been described (De las Heras et al. 1992; Garcia-Goti et al. 2000). No molecular differences have been found between JSRV associated with these two forms, which probably represent two extremes of a disease spectrum rather than two separate forms (Garcia-Goti et al. 2000).

Experimental induction of OPA in sheep using lung fluid, tumour preparations, JRSV clones or tumour cell lines results in pathological changes similar to those observed in natural conditions (Wandera 1970; Sharp et al. 1983; Palmarini et al. 1999). In some experiments, the earliest neoplastic lesions were seen 10 days after inoculation and consisted of reddish-pink consolidated foci in the apical lobes, which on incision exuded a small amount of frothy fluid. By 3–4 weeks post-inoculation, most lungs showed scattered, greyish-pink, moist tumour lesions (Sharp et al. 1983). These descriptions of experimentally induced OPA in lambs are compatible with the classical form described above, whereas atypical forms have not been recorded in those experimental conditions.

3.2 OPA in Goats and Moufflon

There are few descriptions of the pathology of natural OPA in goats, but the features of the tumours in two reports in India are very similar to those for sheep. In these reports, several patterns of lesions have been described using different terms, suggesting that classical, atypical and mixed patterns are also seen in goat OPA in natural conditions (Rajya et al. 1964; Sharma et al. 1975). OPA has been reproduced experimentally in goats. After intratracheal inoculation of very young goat kids with concentrated lung fluid of sheep origin, very small, grey foci of consolidation were seen in the diaphragmatic and middle lobes of a proportion of the inoculated animals (Sharma et al. 1975; Sharp et al. 1986; Tustin et al. 1988).

The gross pathology of OPA in moufflon is also similar to that in sheep (Nieddu et al. 1987; Sanna et al. 2001). The cranioventral parts of the lungs of the affected moufflon show grey or whitish areas of consolidation. Nodules of different sizes are also observed in other superficial areas of the lungs. Foamy fluid is frequently observed in the airways and cut surface.

4 Histopathology

4.1 OPA in Sheep

Histological observation of lung sections from natural cases of OPA reveals the presence of several foci of epithelial cell neoplastic proliferation in alveolar (Fig. 4a) and bronchiolar (Fig. 4b) regions. These proliferations have mostly a papillary appearance, although sometimes they are sufficiently pronounced to give the impression of solid growths. The alveolar neoplastic nodules compress the neighbouring alveoli causing atelectasis and/or coalesce as they increase in size. Cuboidal or

a

Fig. 4a–c. Classical OPA. **a** Papillary adenocarcinoma in which neoplastic alveoli are lined by cuboidal or prismatic cells. The stroma is thin without signs of fibrosis or cellular infiltration. Adjacent para-adenomatous area showing swollen macrophages filling non-neoplastic alveoli. H&E, ×25

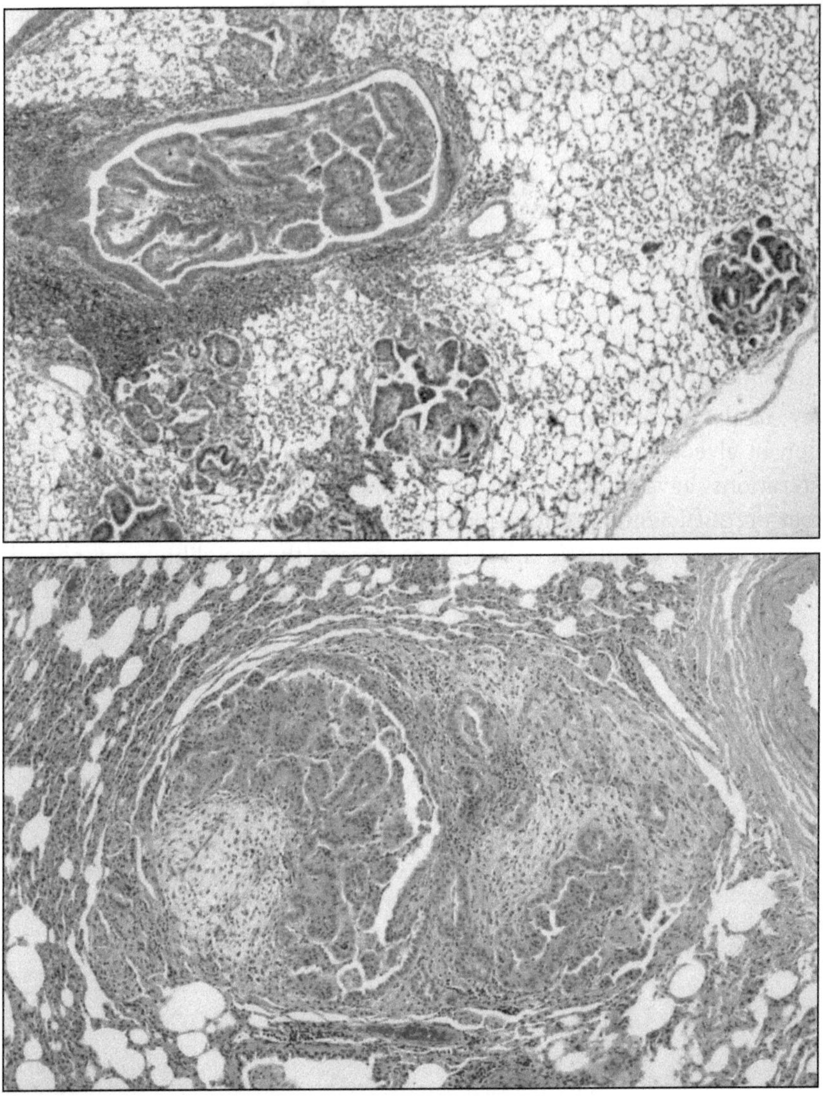

b

c

Fig. 4b,c. **b** Bronchiole showing polypoid ingrowths lined by columnar cells sustained by connective tissue core. Hyperplasia of bronchiole-associated lymphoid tissue. Alveolar tumour nodules and para-adenomatous lesions are also discernible. H&E, ×10. **c** Spiral-shaped proliferations of myxomatous tissue in association with epithelial neoplasia. H&E, ×25

columnar cells replace the normal flat type I pneumocytes, but the structure of the alveolar wall is generally maintained. The nuclei of the neoplastic cells are generally uniform and located in the basal region of the cell; mitotic figures are not numerous. The cytoplasm of the transformed cells can be either homogeneously eosinophilic in cuboidal cells or clear and vacuolated in columnar ones, and frequently it stains positively for glycogen deposits (Wandera 1971; González 1989; Sharp and Angus 1990a). The stroma of the tumour is generally thin but may be infiltrated by variable amounts of lymphocytes, plasma cells and connective tissue fibres (De las Heras et al. 1995). However, in more advanced tumours and particularly in their central areas, fibrosis of the stroma can be prominent. (González 1989; Sharp and Angus 1990a). Surrounding neoplastic alveoli and filling the lumena of normal ones, macrophages with vacuolated cytoplasm and are consistently found in variable numbers (Fig. 4a). They are very characteristic and look like a delimiting barrier of the tumour nodules and are called para-adenomatous areas (González 1989; Sharp and Angus 1990a; De las Heras et al. 1995). In some cases, neutrophils can also be found in the alveolar lumena, but they are interpreted as indicative of secondary bacterial infections.

Another histopathological feature of OPA, usually seen concurrently with the alveolar lesions, occurs in the terminal bronchioles, in which multiple polypoid ingrowths arise from the bronchiolar epithelium (Fig. 4b). These polyps are formed by a line of epithelial prismatic cells proliferating into the lumen, sometimes to the point of occluding it completely, and sustained by a connective tissue core. Their cytoplasms are clear and their nuclei round. A moderate to severe inflammatory response composed of lymphocytes, plasma cells and macrophages often surrounds the affected bronchioles, and the bronchiole-associated lymphoid tissue also may be hyperplastic. However, these changes can also appear around non-affected bronchioles, and might therefore not be related to the neoplasia, but to concurrent infections.

In some cases, nodular, spiral-shaped, more or less obvious masses of myxomatous tissue, presumably of mesodermal origin, are found in association with the epithelial neoplastic growths (Fig. 4c) (Cutlip and Young 1982; Sharp and Angus 1990a; Garcia-Goti et al. 2000). These myxomatous structures contain fusiform cells embedded in a basophilic homogenous matrix and are generally associated with neoplastic alveoli, but whether they are neoplastic or not is still subject to debate.

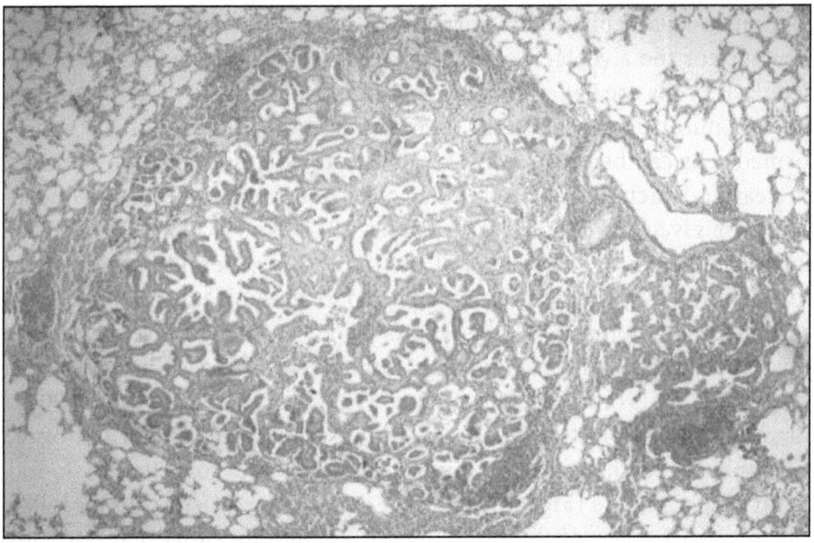

Fig. 5. Atypical OPA. Well-circumscribed papillary to acinar adenocarcinoma in which the stroma appears infiltrated by mononuclear cells and connective tissue. H&E, ×25

The histological appearance of atypical OPA (Fig. 5) is essentially the same as that of classical OPA, but the pattern of epithelial neoplasia is more often acinar than papillary and the stroma is more heavily infiltrated by inflammatory cells and connective fibres (De las Heras et al. 1992; Garcia-Goti et al. 2000). Para-adenomatous lesions and myxomatous tissue proliferation are also associated with atypical tumour nodules, while peribronchiolar infiltrates tend to be more prominent than in classical cases (Garcia-Goti et al. 2000).

In cases of concurrent OPA and maedi lesions in the same lung, the characteristic OPA lesions coexist with those of a chronic interstitial lymphoproliferative pneumonia that is more evident in the areas of the lung outside the tumour (Snyder et al. 1983; Rosadio et al. 1988; González 1989; González et al. 1993).

The histopathology of experimentally induced OPA closely resembles that of field cases (Wandera 1970; Sharp et al. 1983). The earliest lesions described in experimental infections consist of numerous neoplastic polyps in terminal bronchioles and alveoli. Later, more elaborated intrabronchiolar polyps are widespread in the tumour foci with local extension into alveoli. The cytoplasms of many epithelial cells contain large amounts

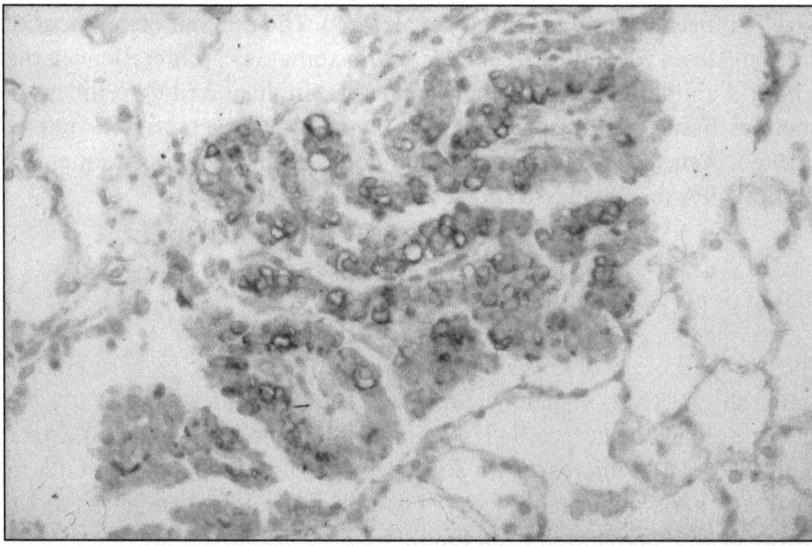

Fig. 6. Immunohistochemistry for JSRV-CA protein: immunolabelling is associated with the cytoplasm of neoplastic alveolar cells. Immuno-histochemistry and haematoxylin counterstaining, ×25

of glycogen. The fibres supporting the stroma of the polyps are embedded in an acid mucopolysaccharide ground substance (Sharp et al. 1983).

The major capsid protein of JSRV can be detected in OPA tumours by immunohistochemical procedures (Palmarini et al. 1995). This viral protein is located in the cytoplasm of transformed alveolar (Fig. 6) and bronchiolar cells, and can be shown both in classical and atypical OPA cases. Positive labelling can also be found in a few mononuclear cells with lymphocyte morphology in the paracortical areas of the regional lymph nodes.

Regional lymph nodes of OPA-affected lungs may show either depletion of cortical follicles and moderate plasmacytosis in the medullary areas (De las Heras et al. 1995), hyperplastic reactive cortices, or no specific changes (Sharp and Angus 1990a). Tumour metastases may be seen histologically in regional lymph nodes and even in extrathoracic organs and tissues (Cuba Caparo et al. 1961; Dualde 1966; Nobel et al. 1969; De las Heras et al. 1995). The lymph node metastases generally consist of epithelial neoplastic cells arranged in a similar acinar or papillary pattern as in the lung tumour, but scirrous and mixed epithelial/scirrous proliferations have also been described (Nobel et al. 1969). The scirrous component of the metastases very much resembles the myxomatous proliferations in the lung and might indicate its neoplastic nature. In the mixed type, the myxomatous tissue appears between the adenomatous structures (Nobel et al.1969). Metastases in regional lymph nodes seem to be less frequent in atypical OPA than in classical cases (Garcia-Goti et al. 2000).

4.2 OPA in Goats and Moufflon

The histopathology of OPA in goats and wild moufflon closely resembles that in sheep (Rajya et al. 1964; Sharma et al. 1975; Sanna et al. 2001). No metastases have been recorded in regional lymph nodes in goats (Sharma et al. 1975). Light microscopy features of experimentally induced OPA in goats are indistinguishable from those of sheep (Sharma et al. 1975; Sharp et al. 1986; Tustin et al. 1988).

4.3 Considerations About the Histological Classification and Disease Nomenclature

In early descriptions of the disease there was some confusion about the consideration of the disease as a true malignant neoplasm, which supported the use of the term adenomatosis for the pulmonary lesions and for the disease itself. Many of the histological features described above, such as non-invasion, low mitotic index, low degree of cellular atypia and absence of necrosis, correspond to a benign neoplasm. On the other hand, the appearance of metastases, either intra- or extrathoracic, supports the consideration of a malignant neoplasia (Mackay and Nisbet 1966) and it should be borne in mind that the incidence of metastases is probably higher than reported, as grossly indiscernible, microscopic metastases are the norm. Taking this into account, the terms carcinoma and carcinomatosis have been preferred (Nobel et al. 1969; DeMartini et al. 1987).

Once it is clear that jaagsiekte is a malignant neoplasm of the lung the next item of discussion deals with the terms carcinoma and adenocarcinoma. The neoplastic cells (type II pneumocytes and Clara cells) have secretory properties and they proliferate giving rise to gland-like structures, whereas at the same time the tumour involves alveolar and bronchiolar structures. The first features would be more consistent with the use of the term adenocarcinoma, whereas the latter would apply best to the term bronchioloalveolar carcinoma, as classified by Stunzi et al. (1974) and more recently by Dungworth et al. (1999). The latest histological classification of lung and pleural tumours in man (Travis et al. 1999) defines 'adenocarcinoma' as a malignant epithelial tumour with glandular differentiation or mucin production, showing either acinar, papillary, bronchioloalveolar or solid growth or a mixture of these patterns. The histological appearance of jaagsiekte would fit these criteria, but the problem arises as to which subtype of adenocarcinoma does jaagsiekte belong. In humans, the bronchioloalveolar subtype is currently defined as a pure bronchioloalveolar growth pattern with no evidence of stromal, vascular or pleural invasion, and never as an endobronchiolar growth or peripheral mass. This would not fit with the features of jaagsiekte, which would be better classified in the subtype of papillary adenocarcinoma or less frequently acinar adenocarcinoma, or more generally as adenocarcinoma with mixed subtypes.

Apart of the discussion about subtypes, the term adenocarcinoma seems to be more appropriate than the general term carcinoma for the histological classification of the OPA lesions. Thus it is appropriate to call the disease ovine pulmonary adenocarcinoma. We also propose to continue to use the term jaagsiekte, because it is a very unique and identifies the disease historically and its causal virus.

OPA has been proposed as a unique model of retrovirus-associated lung cancer (Palmarini et al. 1997).

5 Electron Microscopy

5.1 OPA in Sheep

Most electron microscopy (EM) studies in OPA, both on natural disease or experimentally induced cases, were conducted in the 1970s and 1980s. The main purposes of those studies were to determine the identity, morphological features and topographical arrangement of the transformed cells, and to provide evidence for the, at that time only presumed, viral aetiology of this contagious neoplasm.

Some degrees of variation in morphological features of the tumour have been described. These variations are negligible between reports on experimentally induced neoplasias in lambs, but are more marked between those dealing with natural disease. This is probably best explained by differences in the age of the lesions investigated, not only between sheep of a same or different investigation series, but also in individual lungs, which is a feature of the natural disease. Also, the relatively frequent concurrence of other pulmonary conditions might contribute to at least some of the discrepancies. As such parameters are better controlled in experimental infections, these studies tend to give a more homogeneous picture. A general overview of the ultrastructural features of the natural and experimental tumour will be described, but specific reference will be made when discrepancies arise. Other literature reviews on the subject can be found in Verwoerd et al. (1985) and in Sharp and Angus (1990a).

Observations of viral particles in lung tumour sections, lung fluid, tumour extracts and cell cultures derived from or inoculated with tumour material will also be summarized. Although this aspect rather belongs to the past history of this disease, the diversity of viruses described in association with OPA was greater in studies of the natural disease than in experimental cases. The aetiological role of JSRV in OPA is now clear, but the diversity found in earlier morphologic studies contributed to the emergence of different hypotheses on the aetiology of the tumour. Another literature review on this subject can be found in Sharp and Angus (1990b).

5.1.1
Alveolar Tumours

In incipient tumour nodules, neoplastic cells in the alveolar walls appear as small groups or even pairs, but as the tumour progresses they form flat monolayers first and papillary projections later (Payne and Verwoerd 1984). At this advanced stage, a few areas of stratified epithelial cells can be observed (Cutlip and Young 1982) as the tumour spreads to other alveoli and to respiratory bronchioles. The transformed alveolar cells are predominantly cuboidal, but also low and tall columnar, and are disposed on a thin basement membrane (Severini et al. 1980; Cutlip and Young 1982; Cutlip 1985; Angus et al. 1985). It appears that the shape of the tumour cells depends on the number packed in a neoplastic nodule, which in turn would be related to the age of the lesion, so that the younger the tumour the more cuboidal the cells.

The apical or free surface of the alveolar tumour cells presents numerous prominent microvilli. These can be seen by transmission EM (Nisbet

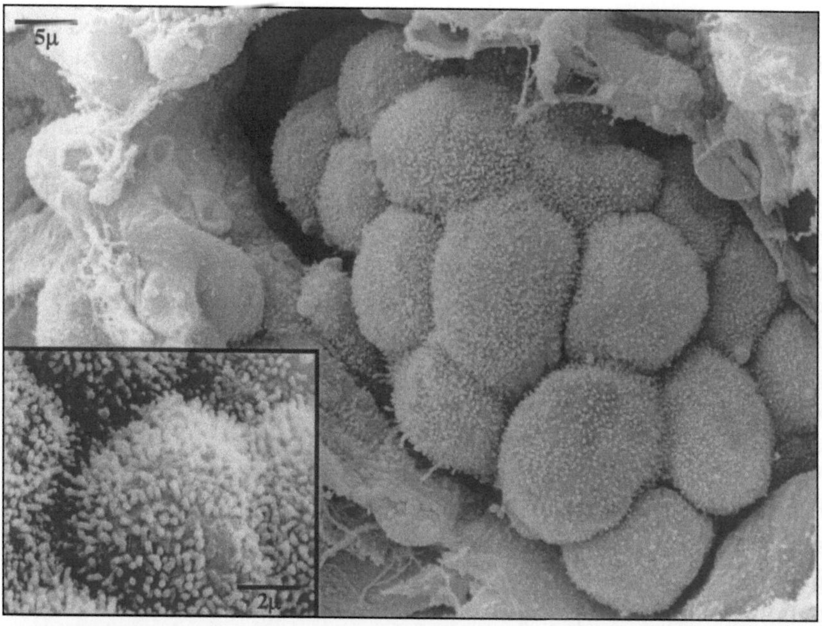

Fig. 7. Ultrastructure of alveolar tumour: scanning electron micrograph showing densely packed neoplastic type II pneumocytes forming a papillary ingrowth. *Inset* shows prominent microvilli in the apical surface of tumour cells. (From Sharp and Angus 1990a, reproduced by courtesy of Kluwer Academic Publishers)

et al. 1971; DeMartini et al. 1985, 1987), but are most evident by scanning EM (Payne and Verwoerd 1984; Angus et al. 1985). The microvilli are particularly numerous in young tumour cells (Fig. 7), in which they sometimes appear interlocked, whereas older neoplastic cells have flatter and fewer microvilli (Severini et al. 1980; Payne and Verwoerd 1984; Angus et al. 1985). Ciliated cells within alveolar tumours have been described rarely (Severini et al. 1980; Cutlip and Young 1982), and have been attributed to remnants of the original epithelium rather than to tumour cells (Cutlip and Young 1982). An electron-dense layer, believed to represent the pulmonary surfactant often covers the surface of the tumour cells (Payne and Verwoerd 1984) and multi-laminated, tubular myelin figures of the same secreted product can often be seen in the alveolar spaces (Cutlip and Young 1982; Payne and Verwoerd 1984).

The intercellular junctions between the lateral surfaces of adjacent tumour cells are tight and well developed, consisting mainly of desmosomes (Nisbet et al. 1971; Payne and Verwoerd 1984; DeMartini et al. 1985, 1987). The junctional complex at the apical border (terminal bars) appears as distinct grooves by scanning EM (Payne and Verwoerd 1984). Interdigitation of lateral membranes is more the exception than the norm (Nisbet et al. 1971; Payne and Verwoerd 1984).

In both natural and experimentally induced OPA, intracytoplasmic organelles of alveolar neoplastic cells consist of numerous mitochondria (Nisbet et al. 1971), sometimes undergoing degeneration (hyperchromatic, with sparse matrix and fragmented cristae (Severini et al. 1980; Payne and Verwoerd 1984), moderate to well developed endoplasmic reticulum (Nisbet et al. 1971; Payne and Verwoerd 1984), with numerous ribosomes attached to it (Severini et al. 1980) and more or less prominent Golgi apparatus (Severini et al. 1980; Payne and Verwoerd 1984; DeMartini et al. 1987). Lysosome-like structures have also been described in tumour cells that do not open to the alveolar space (Nisbet et al. 1971) and reference has also been made to the presence of clusters of microtubules in early lesions (Hod et al. 1977), to bundles of intracytoplasmic filaments and to centrioles (Payne and Verwoerd 1984).

The presence of cytosomes in the apical region of the tumour cells is characteristic although their numbers are variable, being in general more prominent in advanced lesions than in immature tumour cells (Hod et al. 1977). Their contents are also variable (Fig. 8): some of them appear as vacuoles, either empty or with amorphous electron-lucent contents, others contain electron-dense amorphous material and others are filled with myelinoid, lamellar structures (Nisbet et al. 1971; Severini et al. 1980; Perk et al. 1971; Cutlip and Young 1982; Payne and Verwoerd 1984; Angus et al. 1985; Cutlip 1985; DeMartini et al. 1985, 1987). These lamellar structures and the microvilli present on the apical surface (see above) are the main features that allow general recognition of the transformed cells as type II pneumocytes. Moreover, the abundance of electron-dense cytosomes is interpreted by some as reminiscent of foetal type II pneumocytes rather than mature ones, which is one criterion for malignancy.

Non-membrane bound, electron-dense granules of glycogen are variable in numbers and size and in the proportion of cells in which they appear. They are frequent and prominent in older tumour cells (Severini et al. 1980; Cutlip and Young 1982; Angus et al. 1985; Cutlip 1985), in which they are located in the basal region (in cuboidal cells) or lateral to

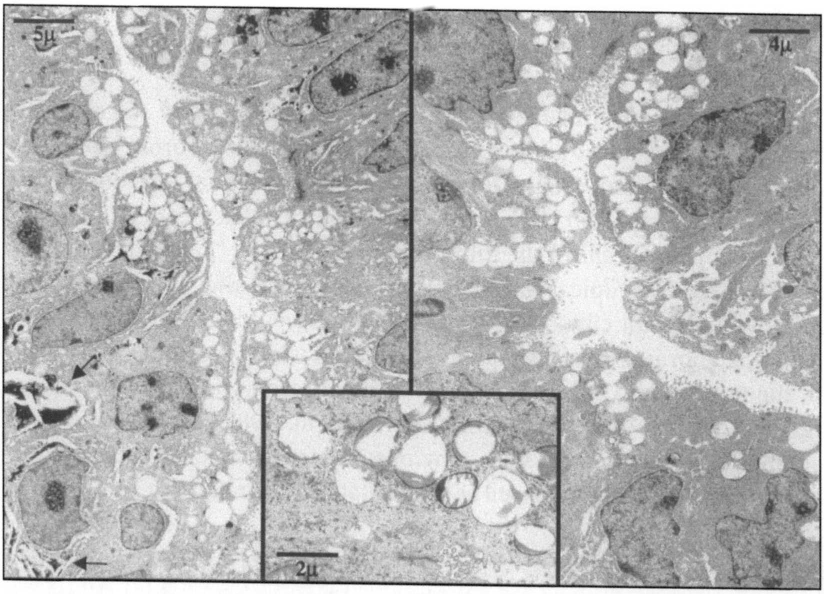

Fig. 8. Ultrastructure of alveolar tumour: transmission electron micrograph showing neoplastic cuboidal cells to columnar cells with prominent microvilli (*right panel*), abundant apical cytosomes and glycogen deposits (*arrows* in *left panel*), and basal nuclei. *Inset* shows lamellar structures filling some cytosomes. (Left micrograph from Angus et al. 1995, reproduced by courtesy of the editors of *Slow Viruses in Sheep, Goats and Cattle*. EEC Report EUR 8076 EN)

the nucleus (in more columnar cells (Fig. 8). In contrast, they are almost absent in immature neoplastic cells (Nisbet et al. 1971; Hod et al. 1977).

The nuclei of the neoplastic cells are spherical, oval, or less frequently convoluted and have a vesicular appearance, with sparse peripheral euchromatin and well-defined nuclear membrane (Fig. 8). They are situated either in a central or a basal position (in cuboidal or columnar cells, respectively) and have prominent nucleoli (Nisbet et al. 1971; Severini et al. 1980; Payne and Verwoerd 1984; Angus et al. 1985; DeMartini et al. 1987).

The appearance of the neoplastic cells in experimentally-infected goats is very similar to that described for natural and experimental OPA in sheep, the main difference according to Tustin et al. (1988) being the absence of lamellar structures in the apical cytosomes.

5.1.2
Intrabronchiolar Proliferations

These are composed of tall columnar cells, which in early stages of the tumour mainly appear at the junctions between terminal bronchioles and alveoli (Angus et al. 1985). Desmosomes and occasional interdigitations of the plasma membranes link them to each other (Nisbet et al. 1971).

The neoplastic cells of the intrabronchiolar polyps (Fig. 9) have a characteristic convex apical surface, and very short and sparse, vestigial microvilli, but not cilia (Angus et al. 1985). No cytosomes, but electron-dense, membrane-bound granular structures, which resemble those seen in the terminal bronchioles of normal sheep, are apparent in the apical region (Nisbet et al. 1971; DeMartini et al. 1987). These cells have abundant mitochondria with obvious cristae and prominent rough endoplasmic reticulum, as well as basal dilated cisternae, with amorphous,

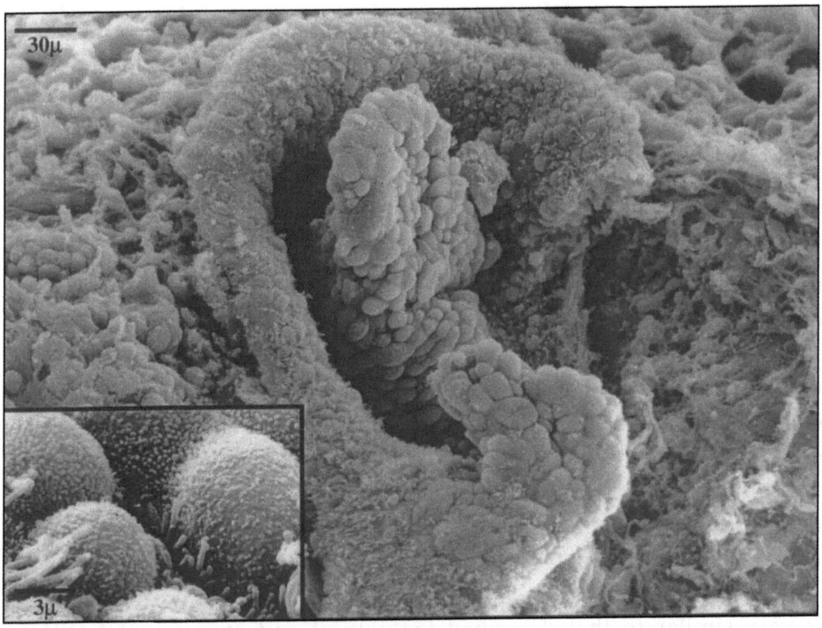

Fig. 9. Ultrastructure of bronchiolar tumour: scanning electron micrograph showing densely packed neoplastic columnar cells forming a papillary ingrowth. *Inset* shows sparse, short microvilli in the apical surface of tumour cells and intercellular cilia (presumably arising from non-transformed ciliated cells). (Inset micrograph from Sharp and Angus 1990a, reproduced by courtesy of Kluwer Academic Publishers)

electron-lucent material (Nisbet et al. 1971). The nuclei are round or oval and are situated towards the apical region of the cell (Nisbet et al. 1971).

All of these features indicate that these neoplastic cells are very similar to non-ciliated Clara cells, from which they are believed to arise.

5.1.3
Tissue Structures Related to the Tumour

The stroma of the neoplastic nodules found in the parenchyma of the lung is to a greater or lesser extent infiltrated by cells among which plasma cells are the most prominent, particularly in advanced tumours (Hod et al. 1977; Payne and Verwoerd 1984). Other lymphocytes, some fibroblasts and occasional monocytes have also been identified in EM studies (Payne and Verwoerd 1984).

The alveolar spaces adjacent to the tumour nodules are characteristically filled with large macrophages, some forming clusters in the lumena and some attached to tumour or normal cells. By EM, these cells present a highly ruffled surface, due to the presence of long microvilli, numerous lysosomes and phagosomes and myelin bodies (Hod et al. 1977; Cutlip and Young 1982; Payne and Verwoerd 1984).

It appears that only one report deals with the ultrastructure of the metastases found in the regional lymph nodes (Geisel and Bomhard 1977). The neoplastic cells also showed microvilli-like projections, similar to those seen in the lung tumour, but protruding towards intracytoplasmic vacuoles.

5.2
Morphology of Viral Particles Associated with OPA

The presence of viral particles and their morphological features has been reported not only in tissue sections of natural and experimental tumours, but also in tumour extracts and lung fluid fractions and in cell cultures inoculated with these. However, some EM studies, in which details of the ultrastructure of the tumour were given, have failed to reveal any viral particles, both in natural cases (Nisbet et al. 1971; Cutlip 1985) and in experimentally induced OPA in sheep (Severini et al. 1980) or goats (Tustin et al. 1988).

Viral particles with morphological features of herpesviruses have been observed in alveolar macrophages cultivated from lungs showing lesions of OPA after natural (Smith and Mackay 1969) or experimental (Malmquist et al. 1972) infection. Further experiments revealed that herpesviruses did not have an aetiological role in OPA (Scott et al. 1984).

Viral particles resembling lentiviruses have also been described in tissue sections of neoplastic lung (Perk et al. 1974; Hod et al. 1977; DeMartini et al. 1985), in tumour pellets (Perk et al. 1974) and in cell cultures derived from or inoculated with OPA tumour material (Malmquist et al. 1972; Cutlip and Young 1982; Cutlip 1985; Perk et al. 1985 a, b; Payne et al. 1986). Although some authors are of the opinion that these particles are more like murine type-C retroviruses (murine leukaemia and sarcoma viruses; Perk et al. 1974), there is now a general agreement that they represent ovine lentiviruses (maedi/visna virus, ovine progressive pneumonia virus). The simultaneous occurrence of the lentiviral infection and OPA has been widely documented (González et al. 1993).

Most descriptions of the morphology and morphogenesis of JSRV come from experimental infections, in which viral particles seem to be more abundant (Verwoerd et al. 1980; Payne et al. 1983; 1984; Sharp et al. 1983, 1985; Angus et al. 1985) than in natural infection (Hod et al. 1977). Several types of JSRV particles have been identified in alveolar neoplastic cells (Fig. 10) and rarely in alveolar macrophages.

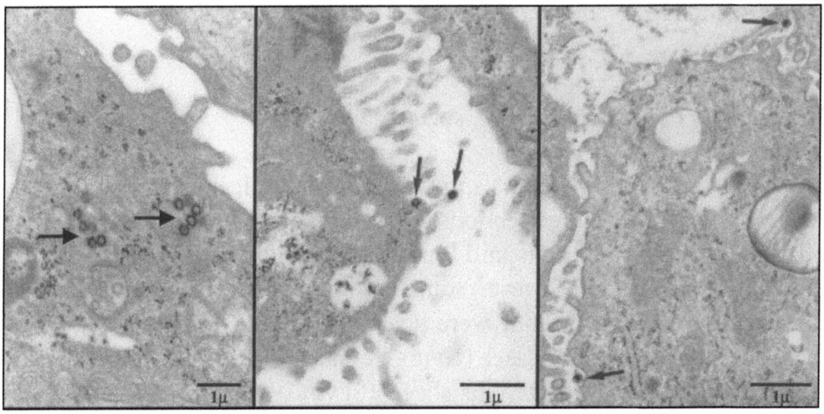

1. Intracytoplasmic type A particles: doughnut or ring-shaped, isolated or in clusters, beneath the apical membrane, of 60 nm diameter, or 74 nm for some authors (Payne et al. 1983), or 80–100 nm for some others (Hod et al. 1977), with a small electron-lucent core and a double external membrane, the inner being more electron-dense.
2. Budding particles: complete round particles (not crescent-shaped), with a more electron-dense core than intracytoplasmic particles, budding from the tips of the apical microvilli or into intracytoplasmic vacuoles.
3. Extracellular type B/D particles: round or pleomorphic, 100–120 nm diameter (90 nm for some authors; Sharp et al. 1983), with a slightly eccentric electron-dense core, sometimes cylindrical; the nucleocapsid appears surrounded by a close-fitting membrane and an outer envelope from which short, knob-like projections are visible. In negatively stained preparations of tumour fractions, these projections are 10–12 nm long and are separated 11–13 nm from each other (Verwoerd et al. 1980).

The morphological features of JSRV have been compared to those of other RNA tumour viruses (bovine leukaemia virus, mouse mammary tumour virus, murine sarcoma virus and squirrel monkey retrovirus). The main distinctive morphological features of JSRV are the lower quantity and smaller electron-lucent core of the intracytoplasmic particles, the maturation of the budding particles, and the slightly eccentric nucleoid and longer and sparser projections of the extracellular particles (Payne et al. 1983).

Acknowledgements. Funding for this work was provided by Commission of the European Communities contracts AIR 3CT94-0884, QLK2-1999-00983, QLRT-2000-02380, by Spanish Comisión Interministerial de Ciencia y Tecnología AGF96-0535-CO2-02 and by Scottish Executive, Environment and Rural Affairs Departmen ROAME MRL/043/98. We thank all practitioners of the Gabinete Tecnico Veterinario of Zaragoza (Spain), the technical staff and collaborators of the University of Zaragoza for their help.

References

Angus KW, Sharp JM, Gray EW (1985) Pathology and ultrastructure of sheep pulmonary adenomatosis tumour lesions in neonatal lambs. In: Slow Viruses in Sheep, Goats and Cattle, pp 329–332. Ed. Sharp JM and Hoff-Jørgensen R. EEC Report EUR 8076 EN

Banerjee M, Gupta PP (1979) Note on Maedi and Jaagziekte in sheep and goats in Ludbiana area of Punjab. Indian Journal of Animal Science 49:1102–1105

Cuba-Caparo A, De la Vega E, Copaira M (1961) Pulmonary adenomatosis of sheep. Metastasizing bronchiolar tumours. American Journal of Veterinary Research 22:673–682

Cutlip RC, Young S (1982) Sheep pulmonary adenomatosis (jaagsiekte) in the United States. American Journal of Veterinary Research 43:2108–2113

Cutlip RC (1985) Sheep pulmonary adenomatosis in the United States: incidence and pathology. In: Slow Viruses in Sheep, Goats and Cattle, pp 159–162. Ed. Sharp JM and Hoff-Jorgensen R. EEC Report EUR 8076 EN

De las Heras M, Calafat JJ, Jaime JM, Garcia de Jalon JA, Ferrer LM, Garcia Goti M, Minguijón E (1992) Sheep pulmonary adenomatosis (jaagsiekte) in slaugthered sheep. Variation in pathological characteristics. Medicina Veterinaria 9:52–52

De las Heras M, Minguijon E, Ferrer LM, Perez V, Bolea R (1995) Adenomatosis pulmonar ovina (Jaagsiekte): Celulas que infiltran el tumor y modificacioes en ganglios linfaticos regionales. Medicina Veterinaria 12:32–36

De las Heras M, García Goti M, Ortín A, Pascual Z, Minguijón E, Ferrer LM, García de Jalón JA, Dewar P, Sharp JM, González L (2001) Presencia del retrovirus asociado a la adenomatosis pulmonar ovina en fetos y en una neoplasia pulmonar en un cordero de tres días. Pequeños rumiantes 1:36–37

DeMartini JC, Snyder SP, Ameghino E (1985) Sheep pulmonary adenomatosis in Peru: epidemiologic, pathologic and ultrastructural studies. In: Slow Viruses in Sheep, Goats and Cattle, pp 333–343. Ed. Sharp JM and Hoff-Jorgensen R. EEC Report EUR 8076 EN

DeMartini JC, Rosadio RH, Sharp JM, Russel HI, Lairmore MD (1987) Experimental coinduction of type D retrovirus-associated pulmonary carcinoma and lentivirus associated lymphoid interstitial pneumonia in lambs. Journal of the National Cancer Institute 79: 167–177

Dualde Perez D (1966) Estudios sobre la adenomatosis pulmonar ovina en España. Instituto de Estudios Turolenses. Consejo Superior de Investigaciones Cientificas. Teruel 1966

Dungal N, Gislason G, Taylor EL (1938) Epizootic adenomatosis in the lungs of sheep. Comparisons with jaagsiekte, verminous pneumonia and progressive pneumonia. Journal of Comparative Pathology and Therapeutics 51:46–68

Dungworth DL, Hauser B, Hahn FF, Wilson DW, Haenichen T, Harkema JR (1999) Histological classification of the tumours of the respiratory system of domestic animals. World Health Organization. Armed Forces Institute of Pathology. Washinton DC. Second Series Volume VI

Garcia-Goti M, González L, Cousens C, Cortabarria N, Extramiana AB, Minguijon E, Ortin A, De las Heras M, Sharp JM (2000) Sheep pulmonary adenomatosis: Characterization of two Pathological forms assosiated with jaagsiekte retrovirus. Journal of Comparative Pathology 122:55–65

Geisel O, Bomhard, DV (1977) Zur Morphologie der spontanen Lungenadenomatose des Schafes. II. Intratorakale Lymphknotten-Metastasen. Zetralblat fur Veterinar Medicine. B 24: 114–122

González L (1989) El maedi o neumonia progresiva en el conjunto de las enfermedades respiratorias cronicas del ganado ovino en la Comunidad Autonoma Vasca. Tesis doctoral. Universidad de Zaragoza. Facultad de Veterinaria

González L, Juste RA, Cuervo LA, Idigoras I, Saez de Ocariz C (1993) Pathological and epidemiological aspects of the coexistence of maedi-visna and sheep pulmonary adenomatosis. Research in Veterinary Science 54:140–146

Hod I, Zimber A, Klopfer U, Helder AW, Nobel TA, Perk K (1974) Pulmonary carcinoma (jaagsiekte) of sheep: pathologic findings and comparison in multiple-case and case-free herds. Journal of the National Cancer Institute 53:103–110

Hod I, Herz A, Zimber A (1977) Pulmonary carcinoma (jaagsiekte) of sheep. Ultrastructural study of early and advanced tumor lesions. American Journal of Pathology 86:545–558

Hunter AR, Munro R (1983) The diagnosis, occurrence and distribution of sheep pulmonary adenomatosis in Scotland 1875 to 1981. British Veterinary Journal 139: 153–164

Mackay JMK, Nisbet DI (1966) Jaagsiekte – A hazard of intensified sheep husbandry. Veterinary Record 78:18–20

Markson LM, Spence M, Dawson M (1983) Investigations of a flock heavily infected with maedi-visna virus. Veterinary Record 112:267–271

Malmquist WA, Krauss HH, Moulton JE, Wandera JG (1972) Morphologic study of virus-infected lung cell cultures from sheep pulmonary adenomatosis (jaagsiekte). Laboratory Investigation 26:528–533

Martin BM, Scott FMM, Sharp JM, Angus KW, Norval M (1976) Experimental production of sheep pulmonary adenomatosis (Jaagsiekte). Nature 264:183–184

Nieddu A, Coda S, Loi P (1987) Adenomatosi polmonare nel muflone (Ovis Musimon) Atti della Societa Italiana delle Scienze Veterinarie 41:892–894

Nisbet DI, Mackay JMK, Smith W, Gray EW (1971) Ultrastructure of sheep pulmonary adenomatosis (jaagsiekte). Journal of Pathology 103:157–169

Nobel TA, Neumann F, Klopfer U (1969) Histological patterns of the metastases in pumonary adenomatosis of sheep (Jaagsiekte). Journal of Comparative Pathology 79:537–540

Nobel TA, Klopfer U, Neuman F, Trainen Z (1971) Clinicopathological investigations in sheep pulmonary adenomatosis (Jaagsiekte) Zentralblat fur Veterinar Medizine 18B:9–14

Ortín A, Minguijón E, Dewar P, García M, Ferrer LM, Palmarini M, Gonzalez L, Sharp JM, De las Heras M (1998) Lack of specific immune response against a recombinant capsid protein of Jaagsiekte sheep retrovirus in sheep and goats naturally affected by enzootic nasal tumour or sheep pulmonary adenomatosis. Veterinary Immmunology and Immunopathology 61:229–237

Palmarini M, Dewar P, De las Heras M, Inglis NF, Dalziel RG, Sharp JM (1995) Epithelial tumour cells in the lungs of sheep with pulmonary adenomatosis are major sites of replication for Jaagsiekte retrovirus. Journal of General Virology 76: 2731–2737

Palmarini M, Fan H, Sharp JM (1997) Sheep pulmonary adenomatosis: a unique model of retrovirus-associated lung cancer. Trends in Microbiology 5:478–482

Palmarini M, Sharp JM, De las Heras M, Fan H (1999) Jaagsiekte sheep retrovirus is necessary and sufficient to induce a contagious lung cancer in sheep. Journal of Virology 73:6964–6972

Payne A-L, York DF, De Villiers E-M, Verwoerd DW, Querat G, Barban V, Sauze N, Vigne R (1986) Isolation and identification of a South African lentivirus from jaagsiekte lungs. Onderstepoort Journal of Veterinary Research 53: 55–62

Payne A-L, Verwoerd DW, Garnett HM (1983) The morpholgy and morphogenesis of jaagsiekte retrovirus (JSRV). Onderstepoort Journal of Veterinary Research 50: 317–322

Payne A-L, and Verwoerd DW (1984) A scanning and transmission electron microscope study of jaagsiekte lesions. Onderstepoort Journal of Veterinary Research 51:1–13

Perk K, Hod I, Presentey B, Nobel TA (1971) Lung carcinoma of sheep (jaagsiekte). II. Histogenesis of the tumor. Journal of the National Cancer Institute 47:197–205

Perk K, Michalides R, Spiegelman S, Schlom J (1974) Biochemical and morphological evidence for the presence of an RNA tumor virus in pulmonary carcinoma of sheep (jaagsiekte). Journal of the National Cancer Institute 53:131–135

Perk K, Irving S, Hod I, Yaniv A, Klopfer U, Zimber A (1985a) Recent studies on the retrovirus etiology of sheep pulmonary adenomatosis. In: Slow Viruses in Sheep, Goats and Cattle, pp 163–165. Ed. Sharp JM and Hoff-Jorgensen R. EEC Report EUR 8076 EN

Perk K, Irving SG, Hod I, Gazit A, Yaniv A, Klopfer U, Zimber A (1985b) Recent studies on sheep pulmonary carcinoma. In: Slow Viruses in Sheep, Goats and Cattle, pp. 311–315. Ed. Sharp JM and Hoff-Jorgensen R. EEC Report EUR 8076 EN

Rajya BS, Singh CM (1964) The pathology of pneumonia and associated respiratory disease of sheep and goats. I. Ocurrence of jaagsiekte and maedi in sheep and goats in India. American Journal of Veterinary Research 25:61–67

Rosadio RH, Sharp JM, Lairmore M, Dahlberg J, DeMartini JC (1988) Lesions and Retroviruses associated with naturally occurring ovine pulmonary carcinoma (Sheep pulmonary adenomatosis). Veterinary Pathology 25:58–66

Rosadio R, Sharp JM (1992) Leukocyte alterations in sheep with naturally and experimentally-induced lung cancer. Medicina Veterinaria, 9(suple):49–51

Sanna MP, Sanna E, De las Heras M, Leoni A, Niedu AM, Pirino S, Sharp JM, Palmarini M (2001) Association of jaagsiekte sheep retrovirus with pulmonary carcinoma in Sardinian moufflon (Ovis musimon) Journal of Comparative Pathology 125:145–152

Scott FMM, Sharp JM, Angus KW, Gray EW (1984) Infection of specific-pathogen free lambs with a herpesvirus isolated from pulmonary adenomatosis. Archives of Virology 80:147–162

Severini M, Rampichini L, Rutili D, Fioroni A, Zampetti U (1980) Riproduzione sperimentale dell'adenomatosi polmonare della pecora (jaagsiekte). Archivo Veterinario Italiano 31: 129–135

Sharma DN, Rajya BS, Dwivedi JN (1975) Metastasizing pulmonary adenomatosis (Jaagziekte) in sheep and goats. Patho-anatomical studies. Indian Journal of Animal Science 45:363–370

Sharma DN, Rajya BS, Dwivedi JN (1975) Experimental transmission of Jaagsiekte and Maedi in sheep and goats. Indian Journal of Animal Science 45:275–281

Sharp JM, Angus KW, Gray EW, Scott FMM (1983) Rapid transmission of sheep pumonary adenomatosis (jaagsiekte) in young lambs. Archives of Virology 78:89–95

Sharp JM, Scott FMM, Herring AJ, Angus KW, Gray EW, Cuthbertson JC (1985) Studies on the epidemiology, aetiology and transmission of jaagsiekte in Scotland.

In: Slow Viruses in Sheep, Goats and Cattle, pp 167–170. Ed. Sharp JM and Hoff-Jorgensen R. EEC Report EUR 8076 EN

Sharp JM, Angus KW, Jassim FA, Scott FMM (1986) Experimental transmission of sheep pulmonary adenomatosis to a goat. Veterinary Record 119:245

Sharp JM, Angus KW (1990 a) Sheep pulmonary adenomatosis: clinical, pathological and epidemiological aspects. In: Maedi-Visna and related diseases, pp 157–175. Ed. Petursson G and Hoff-Jorgensen R. Kluwer Academic Publishers. Boston, Dordrecht, London

Sharp JM, Angus KW (1990 b) Sheep pulmonary adenomatosis: studies on its aetiology. In: Maedi-Visna and related diseases, pp 177–185. Ed. Petursson G and Hoff-Jorgensen R. Kluwer Academic Publishers. Boston, Dordrecht, London

Shirlaw JF (1959) Stuides on jaagsiekte in Kenya Bulletin of Epizootic Diseases in Africa 7:287–302

Smith W, Mackay JMK (1969) Morphological observations on a virus associated with sheep pulmonary adenomatosis (jaagsiekte). Journal of Comparative Pathology, 79:421–425

Snyder SP, DeMartini JC, Ameghino E, Caletti E (1983) Coexistence of pulmonary adenomatosis and progressive pneumonia in sheep in the central sierra of Peru. American Journal of Veterinary Research 44:1334–1338

Sriraman PK, Rama Rao P, Gopal Naidu NR (1982) Goat mortality in Andhra pradesh. Indian Veterinary Journal 59:96–99

Stefanou D, Tsangaris TH, Lekkas ST (1975) Pulmonary adenomatosis in the goats in the district of Pieria (Greece). 20th World Veterinary Congress. Thessaloniki (Greece), 495–496

Stunzi H, Head KW, Nielsen SW (1974) International Histological Classification of Tumours of Domestic Animals. Tumours of the lung. Bulletin of the World Health Organization 50:9–19

Travis WD, Colby TV, Corrin B, Shimosato Y, Brambilla E (1999) Histological typing of lung and pleural tumours. WHO International Classification of tumours. Springer Verlag. Berlin, Heidelberg, New York

Tustin RC (1969) Ovine Jaagsiekte. Journal of the South African Veterinary Medical Association 40:3–23

Tustin RC, Williamson AL, York DF, Verwoerd DW (1988) Experimental transmission of Jaagsiekte (ovine pulmonary adenomatosis) to goats. Onderstepoort Journal of Veterinary Research 55:27–32

Verwoerd DW, Tustin RC, Payne A (1985) Jaagsiekte: an infectious pulmonary adenomatosis of sheep. In: Comparative Pathobiology of Viral Diseases, pp 53–76. Ed. Olsen RG, Krakowka S and Bloukasla JR. Boca Raton CRC Press, New York

Verwoerd DW, Williamson A-L, De Villiers E-M (1980) Aetiology of Jaagsiekte: transmission by means of subcellular fractions and evidence for the involvement of a retrovirus. Ondersteeport Journal of Veterinary Research 47: 275–280

Wandera JG (1967) Pneumonia in sheep in Kenia. Part I Bacterial and parasitic pneumonia. Bulletin of epizootic disease of Africa 15:245–248

Wandera JG (1970) Clinical pulmonary adenomatosis of sheep produced experimentally. British Veterinary Journal 126:185–193

Wandera JG (1971) Sheep pulmonary adenomatosis (Jaagsiekte). Advances in Veterinary Science and Comparative Medicine 15:251–283

Chapter 3

Natural History of JSRV in Sheep

J. M. Sharp, J. C. DeMartini

Abstract. Ovine pulmonary adenocarcinoma (OPA) is a contagious lung tumour of sheep and, rarely, goats that arises from two types of secretory epithelial cell that retain their luxury function of surfactant synthesis and secretion. It is classified as a low-grade adenocarcinoma and is viewed as a good model for epithelial neoplasia because of its morphological resemblance to the human lung tumour, bronchioloalveolar adenocarcinoma.

J. M. Sharp
Moredun Research Institute, Pentlands Science Park, Bush Loan,
Penicuik EH26 OPZ, UK
e-mail: sharm@mri.sari.ac.uk

J. C. DeMartini
Department of Microbiology, Immunology and Pathology,
College of Veterinary Medicine and Biomedical Sciences, Colorado State University,
Fort Collins, CO 80523-1682, USA
e-mail: james.demartini@colostate.edu

OPA is present in most of the sheep rearing areas of the globe and, in affected flocks, tumours are present in a high proportion of sheep. OPA is associated with the ovine retrovirus, jaagsiekte sheep retrovirus (JSRV), and is transmissible only with inocula that contain JSRV. All sheep contain JSRV-related endogenous viruses, but JSRV is an exogenous virus that is associated exclusively with OPA. JSRV is detected consistently in the lung fluid, tumour and lymphoid tissues of sheep affected by both natural and experimental OPA or unaffected in-contact flockmates and never in sheep from unaffected flocks with no history of the tumour. JSRV replicates principally in the epithelial tumour cells, but also establishes a disseminated infection of several lymphoid cell types, including peripheral blood leukocytes (PBLs). Longitudinal studies in flocks with endemic OPA have revealed JSRV in PBLs before the onset of clinical OPA and even in the absence of discernible lung tumour. The prevalence of JSRV infection is 40%–80%, although only 30% of sheep appear to develop OPA lesions. A unique feature of OPA is the absence of a specific humoral immune response to JSRV, despite the highly productive infection in the lungs and the disseminated lymphoid infection. This feature is associated with reduced responsiveness to some mitogens, although the phenotypic profile of the peripheral blood remains unaltered. The reduced response is an early and sustained event during infection and may indicate that the failure of infected sheep to produce specific antibodies to JSRV is a direct consequence of infection.

1 Introduction

Ovine pulmonary adenocarcinoma (OPA, previously referred to as jaagsiekte, sheep pulmonary adenomatosis or ovine pulmonary carcinoma) has been known for many years as an infectious tumour and there is clear evidence from inter-flock and international trade of live animals that the disease is contagious (HOUWERS and TERPSTRA 1984; STEVENSON et al. 1982). OPA is a naturally occurring retrovirus-induced tumour involving two types of secretory epithelial cell of the lung, alveolar type II (ATII) cells and non-ciliated bronchiolar epithelial cells (DEMARTINI et al. 1988; DEMARTINI and YORK 1997; SHARP 1987; VERWOERD et al. 1985). OPA-affected sheep typically produce a watery nasal fluid that originates in the lungs as a product of the tumour cells. OPA has been recognized by international regulatory bodies, such as the Office International des Epizooties, as an important infectious disease of sheep because of its high prevalence in several countries, where it can cause large economic losses arising from its direct effects, as well as a loss of trade in live sheep.

OPA exhibits similar pathological and epidemiological features to bronchioloalveolar carcinoma (BAC) of man and is considered a useful model for the study of carcinogenesis in lung cancer. Both OPA and human BAC are relatively well-differentiated tumours of ATII and/or Clara cells, both exhibit multifocal growth, and both have a peak occurrence in adults (De las Heras et al. 2000; Palmarini and Fan 2001; Perk and Hod 1982).

Until recently, our understanding of the epidemiology and natural history of jaagsiekte sheep retrovirus (JSRV) infection has been incomplete, as it was based on retrospective extrapolation from clinical and pathological observations, rather than specific diagnostic tests for JSRV infection during the preclinical stages. The recent elucidation of the genomic sequence of JSRV (see the chapter by York and Querat, this volume) has led to the development of new reagents and techniques that have provided a more complete appreciation of the pathogenesis of OPA and facilitated new approaches to investigating the epidemiology of JSRV infections in affected flocks. The advances in the molecular biology of JSRV resulted in the identification of JSRV as an exogenous virus and, ultimately, led to the isolation of pathogenic molecular clones of JSRV that confirmed JSRV as the cause of OPA (see the chapter by Palmarini and Fan, this volume). In this chapter, we will review the features of naturally-occurring and experimentally-induced OPA, the natural history of JSRV in sheep flocks affected by OPA, and the virus–host interactions that lead to neoplasia.

2 Geographic Distribution and Flock Prevalence of OPA

OPA has been recognized for more than 100 years since the first descriptions in South Africa and Britain (Dykes and McFadyean 1888; Tustin 1969), since when it has been reported in many of the sheep-rearing countries of Europe, Africa, Asia, and the Americas (Fig. 1). The disease was eradicated in Iceland by a rigorous slaughter policy (Palsson 1985) and, apart from two isolated reports (Hooper and Ellard 1995; Dalefield and Alley 1988), does not appear to affect the large sheep populations of Australia and New Zealand.

Clinically apparent OPA is most often detected in sheep that are 2–4 years of age, but the disease may occur in lambs as young as 2 months

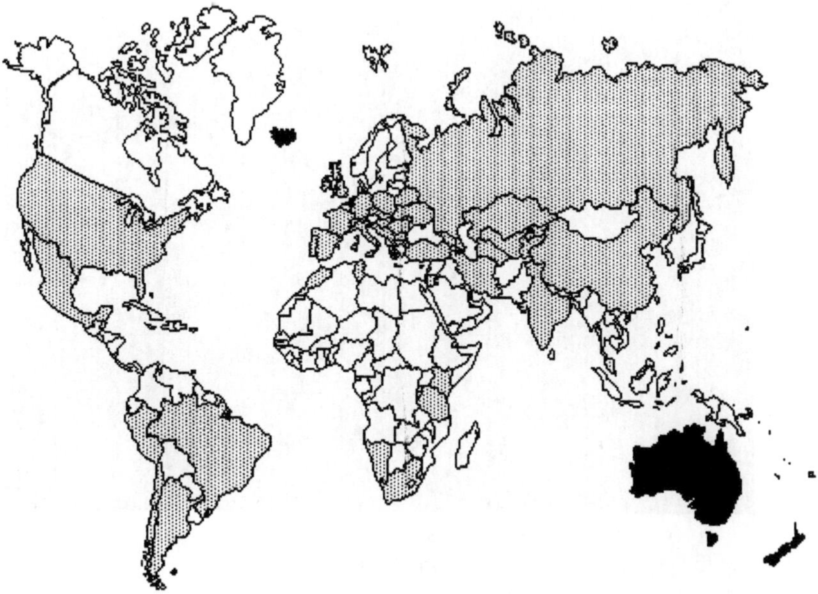

Fig. 1. World distribution of OPA. Countries in which OPA has been reported are shaded *grey*. Countries from which OPA has been eradicated (Iceland, Netherlands) or has not been reported (Australia, New Zealand) are shaded *black*. The status of OPA is unknown in countries that are *unshaded*

of age (Fig. 2). These data have been interpreted to indicate a protracted incubation period for development of clinical disease following natural infection by JSRV. However, detailed epidemiological studies, based on assays to detect subclinically infected sheep, are required to resolve this feature (see Sect. 10).

The prevalence of OPA appears to vary between countries and is endemic in some, for example Peru, Scotland and South Africa. OPA is the most common neoplasm of sheep in Britain and South Africa (Bastianello 1982; Hunter and Munro 1983) and the most common disease of adult sheep in the Central Sierra region of Peru, where it causes 2% annual mortality in adult sheep (Snyder et al. 1983). Longitudinal surveys lasting 5 years in two OPA-affected flocks in Scotland revealed that approximately 30% of the sheep had histologically confirmed lesions in their lungs and that the annual losses attributable to OPA varied between 2% and 10%. Between 12% and 21% of the sheep had subclinical tumours (J.M. Sharp and P. Dewar, unpublished results). In contrast, OPA is infrequently diag-

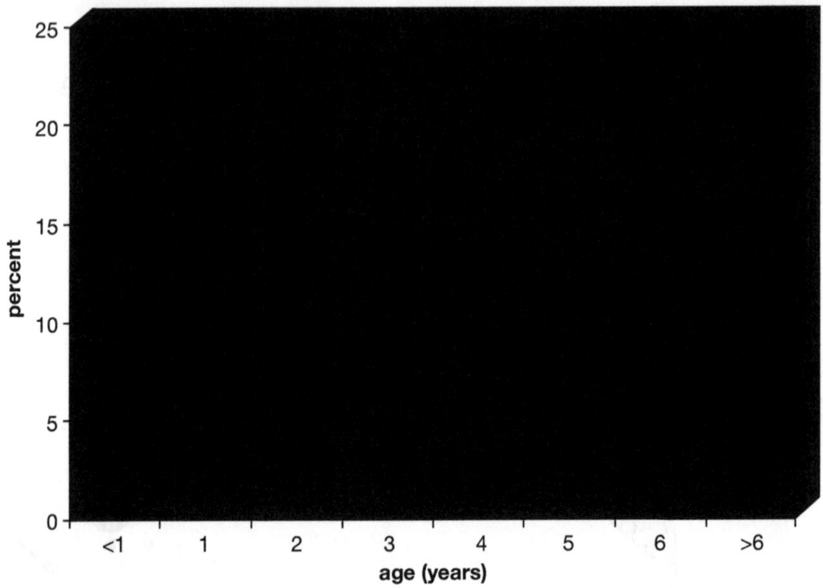

Fig. 2. Age distribution: *light bars,* natural cases of OPA; *dark bars,* natural cases of maedi

nosed in the USA or Canada, where only 11 and 43 cases have been reported (CUTLIP and YOUNG 1982; ROSADIO et al. 1987; STEVENSON et al. 1982). These apparent differences in prevalence and incidence of OPA may reflect variations in viral strains, host susceptibility, or reflect different levels of exposure to JSRV resulting from different management systems.

It is worth noting that in many countries, although not Scotland, OPA often coexists with the ovine lentivirus lung disease maedi (DEMARTINI et al. 1987, 1988; PAYNE et al. 1986). Interactions between JSRV and the ovine lentivirus have been speculated, but firm evidence is lacking (DAWSON et al. 1990; MYER et al. 1988).

3 Experimental Models of OPA

Epidemiological observations and experience have shown consistently that OPA is transmitted between flocks and even countries by movement of apparently unaffected sheep, suggesting that the disease is contagious. To determine whether OPA was a contagious and infectious disease, early transmission studies involved several routes of inoculation or aerosol exposure, using tumour cells or crude tumour extracts, and even contact-exposure to OPA-affected sheep (Dungal 1949; Wandera 1968). In these experiments, mature sheep were used and the incubation periods were long, usually more than 1 year, clinical disease infrequent, and lesions usually small. The results from these experiments supported the view that the incubation period of OPA is protracted, probably more than 6–12 months, which suggested that oncogenesis might be occurring by means of a slowly transforming virus.

More recently, following the demonstration of oncogenicity of reverse transcriptase-containing material in adult sheep (Martin et al. 1976), it was demonstrated that OPA can be induced within a much shorter time by intratracheal inoculation of newborn lambs with JSRV extracted from lung fluid or tumour (Sharp et al. 1983; DeMartini et al. 1987). In these studies, histologically confirmed OPA was induced successfully in 70%–100% of inoculated animals. Generally, JSRV prepared from lung fluid induces OPA in a higher proportion of the inoculated lambs. Most lambs develop clinical signs of OPA within 3–6 weeks of inoculation (Fig. 3) but some have developed OPA within 4–10 days (Sharp et al. 1983). The pulmonary lesions in experimentally induced OPA consist of scattered grey–red nodules throughout the lungs (see the chapter by De las Heras, González and Sharp, this volume). Very few lambs produce the watery lung fluid that is characteristic of classic OPA in naturally affected adult sheep, but copious fluid exuding from the cut surface of affected lungs is a

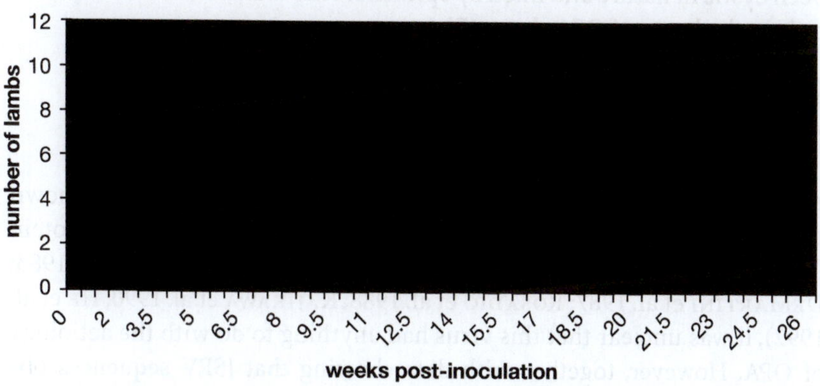

Fig. 3. Incubation period for experimentally-induced OPA in newborn lambs. Lambs were inoculated intratracheally with 5 ml of 15× JSRV extracted from 75 ml OPA lung fluid. (From Sharp et al. 1983 and unpublished)

consistent finding. The short incubation period following experimental inoculation of young lambs is in marked contrast with that obtained in the early experimental studies using mature sheep and raises the possibility that oncogenesis in these animals may be occurring by a different mechanism.

In studies where the JSRV inoculum has contained the ovine lentivirus, maedi-visna virus (MVV), inoculated lambs developed lymphoid interstitial pneumonia in addition to OPA and also developed antibodies to MVV (DeMartini et al. 1987; Rosadio et al. 1988). Inoculation of newborn lambs with MVV alone consistently induces a lymphoid interstitial pneumonia but never causes the characteristic OPA histology (DeMartini et al. 1988).

Several attempts have been made to produce models of OPA in other species, but all appear to be resistant except for goats. Reports of a very low prevalence in goats in India and Greece (Rajya and Singh 1964; Stefenou et al. 1975) have been supported by experimental transmission to young goats. Intratracheal inoculation of newborn goats with JSRV extracted from lung fluid obtained from sheep with OPA induces lesions similar to those in lambs. However, in contrast with lambs, none of the inoculated goats developed clinical disease and only 30% had small circumscribed lesions after many months (Sharp et al. 1986; Tustin et al. 1988). Thus, goats appear to be more resistant than sheep to JSRV.

Tumour cells or small pieces of tumour OPA cases have been successfully transplanted subcutaneously in athymic nude mice (Jassim 1988; Verwoerd et al. 1977; Zimber et al. 1984). The resulting tumours have been cystic in nature and lined by epithelial cells that resemble the tumour cells in the lungs of OPA sheep (Fig. 4).

4 Association of JSRV with OPA

Although lung fluid and tumour from OPA-affected sheep had been shown consistently to contain JSRV particles and the major capsid (CA) protein in a number of studies (Herring et al. 1983; Sharp and Herring 1983; DeMartini et al. 1987; Rosadio et al. 1988; Kajikawa et al. 1990; He et al. 1992), it was unclear that this virus had anything to do with the aetiology of OPA. However, together with data showing that JSRV sequences obtained from geographically-separated OPA cases show remarkable conservation and that successful transmission of OPA could be achieved only with inocula that contain JSRV, these observations made JSRV the best

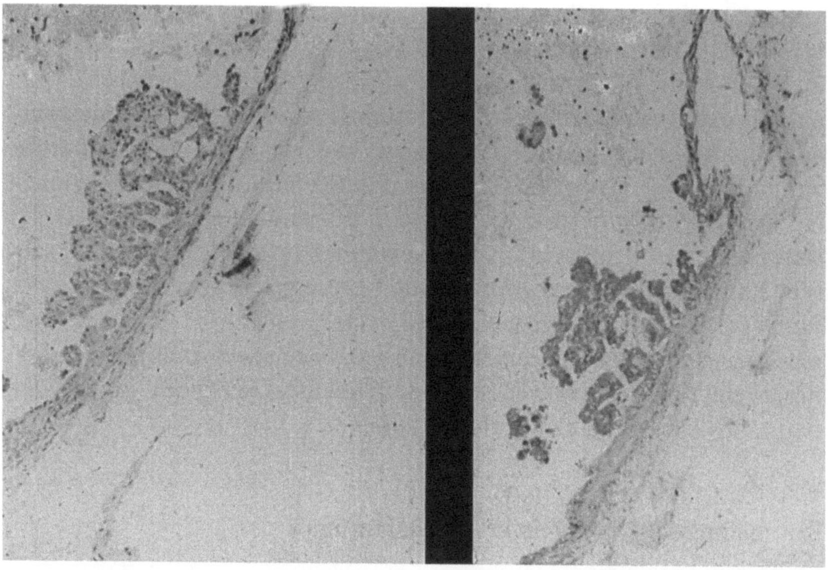

Fig. 4. OPA tumour heterotransplant in an athymic nude mouse. *Left panel:* H and E-stained section of intradermal transplant of OPA tumour cells in nude mice, *Right panel:* PAS stained section of intradermal transplant; dark staining demonstrates the presence of abundant glycogen in the OPA tumour cells

candidate as the aetiologic agent of OPA. However, definition of the pathogenic role of JSRV in OPA was complicated by the presence of 15–20 copies per sheep genome of JSRV-related endogenous sequences (see the chapter by DEMARTINI et al., this volume). These sequences increased the difficulty of detecting integrated exogenous JSRV, and raised the possibility that the expression of JSRV in OPA may represent the activation of endogenous virus(es) in the host cells as a 'downstream' event of carcinogenesis. The identification of reagents and techniques to distinguish exogenous and endogenous ovine retroviruses, therefore, was a critical development. The availability of a genomic sequence for JSRV (YORK et al. 1992) opened the way to develop reagents for JSRV.

Genetic variation between isolates of JSRV and endogenous sheep retroviruses enJSRV indicated that JSRV consists of a family of endogenous and exogenous viruses. Comparison of deduced Env amino acid sequences of JSRV strains from three continents identified residues that defined two distinct genotypes of JSRV; African (Type 1) and North American/European (Type 2) (BAI et al. 1996). Analysis of the proviral sequences confirmed the novel open reading frame, *orf X*, and showed that it was well-conserved among enJSRV and JSRV (Bai et al. 1999; Rosati et al. 2000). At the nucleotide level, the greatest differences between Type I and II isolates were seen in the long terminal repeat (LTR) (87% identical) and to a lesser degree (91%) in the *gag* and *env* genes, although at the amino acid level, Type I and II isolates differed most in the Orf X region. It is interesting to note that the deduced amino acid sequence of Orf X showed similarity to a portion of the mammalian adenosine receptor subtype 3, a member of the G protein-coupled receptor family (Bai et al. 1999; Rosati et al. 2000). The closely related ovine nasal adenocarcinoma virus, which causes nasal adenocarcinoma in goats and sheep, is only 60% identical to the JSRV LTRs and 83% and 88% identical to *env* and *gag* (see the chapter by De las Heras et al., this volume).

5
Expression of JSRV Protein in Affected Animals

To improve understanding of the aetio-pathogenesis of OPA, the distribution and sites of JSRV expression were identified, using two immunological approaches. Recombinant JSRV CA protein was expressed in *Escherichia coli* as β-galactosidase and glutathione-*S*-transferase (GST) fusion proteins and used to generate specific antisera in rabbits (Palmarini et al. 1995). These new reagents were used to develop a blocking-ELISA to detect JSRV in tissues from naturally and experimentally OPA-affected sheep, as well as age- and breed-matched controls and to determine the anatomic distribution of JSRV in sheep. JSRV was detected only in lung fluid and tumour from affected animals and not in any other tissue from affected or unaffected sheep. Immunohistochemical examination of the same tissues, using the same specific rabbit antiserum to JSRV CA, detected JSRV protein in the cytoplasm of recognizable neoplastic epithelial cells in the pulmonary alveoli of naturally and experimentally OPA-affected sheep. Although JSRV CA was observed in all OPA tumours, not all tumour cells were positive, even in the same nodule (Fig. 5a). There was no staining of non-transformed epithelial cells nor of the stromal cells and specific staining was observed only rarely in the extracellular alveolar spaces (Palmarini et al. 1995; Garcia Goti et al. 2000; Sanna et al. 2001; Platt et al. 2002). Specific staining was not demonstrated in any other tissue, with the exception of a few large lymphoblastoid cells in the paracortical zones and medullary sinuses of the tracheobronchial and medi-

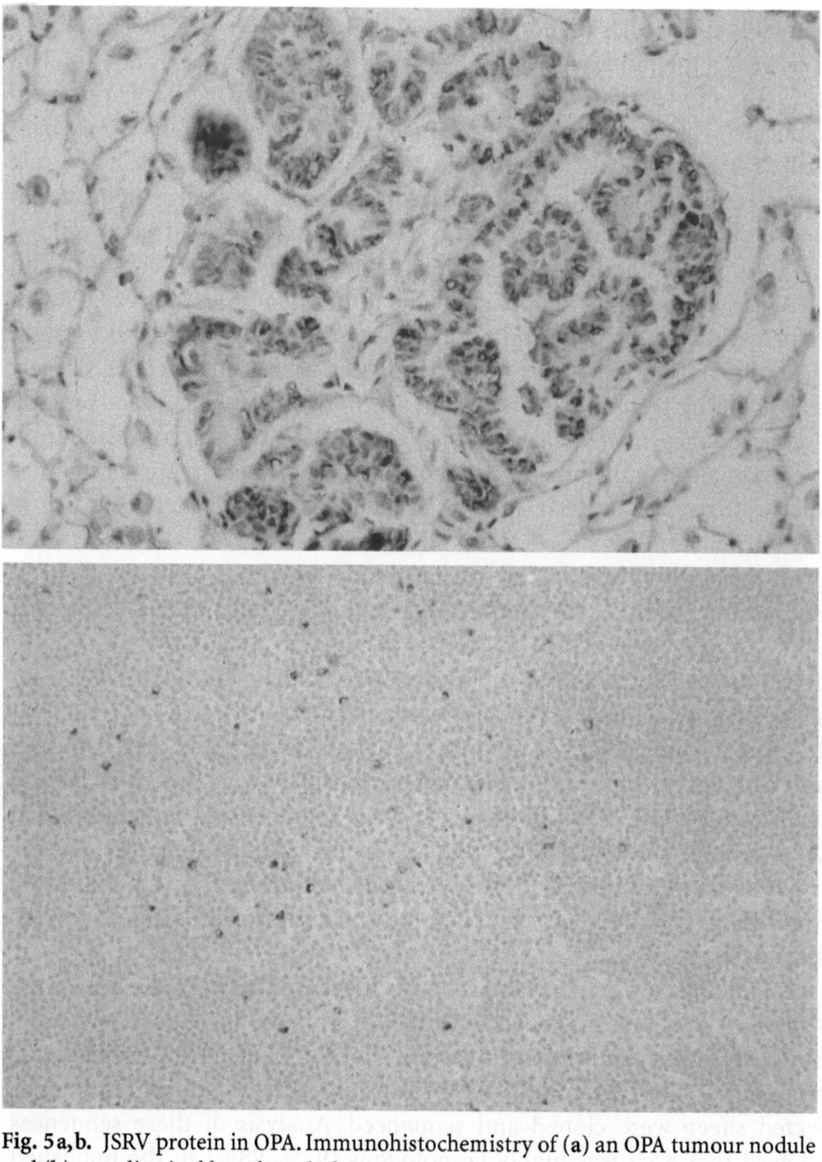

a

b

Fig. 5 a, b. JSRV protein in OPA. Immunohistochemistry of (**a**) an OPA tumour nodule and (**b**) a mediastinal lymph node from an animal with OPA, using a polyclonal rabbit antibody to JSRV CA. The *dark staining* represents expression of JSRV CA in the cytoplasm

astinal lymph nodes that drain the lung (Fig. 5b) and some macrophages free within the pulmonary alveoli (Palmarini et al. 1995; Holland et al. 1999; Platt et al. 2002). These observations indicated that JSRV appears to replicate principally in the transformed epithelial cells in OPA tumours as well as a minor subset of lymphoid cells. Furthermore, if JSRV had a role in the aetiology of OPA, continuous expression of the viral proteins might not be required to maintain the transformed state of the alveolar epithelial cells. Whilst these results highlighted the strict link between JSRV and OPA, providing further support for the notion that JSRV causes OPA, the information at this point did not rule out other possibilities. JSRV might be acting as a helper virus for some other replication defective, acutely transforming retrovirus or represent an endogenous virus, reactivated as a consequence of neoplasia (Weiss et al. 1985) or, indeed, result from recombination with endogenous sequences as has been reported in other species (Stoye et al. 1991; Golovkina et al. 1994; Bai et al. 1995).

6
Expression of JSRV RNA in Affected Animals

To resolve the above issues, the expression and distribution of JSRV transcripts was examined in various tissues from OPA-affected and unaffected animals. Both JSRV CA (Palmarini et al. 1996a) and SU (C. Cousens and J.M. Sharp, unpublished results) transcripts were detected by reverse transcription (RT)–PCR in 100% of a wide range of tissues that included OPA tumour and unaffected lung, as well as lymphoreticular tissues. Transcripts were demonstrated in both OPA-affected and -unaffected animals, providing clear evidence for transcription of related enJSRV sequences.

To determine whether there are any differences between the RNA transcripts in lung fluid (exogenous JSRV) and the endogenous transcripts obtained from non-tumour tissues, the CA and SU RT-PCR products were analysed further. CA and SU RT–PCR products from kidney of an unaffected sheep were cloned and sequenced. Analysis of these sequences showed that each was highly homologous (95%–98%) to the published JSRV sequences, but a *Sca*I restriction site in CA and an *Nde*I site in SU was absent in the endogenous derived clones (Palmarini et al. 1996a; C. Cousens and J.M. Sharp, unpublished results). The absence of these restriction sites in endogenous sequences was confirmed in a separate molecular characterization of endogenous sheep retroviruses (Bai et al.

1996). Restriction enzyme digestion of CA RT–PCR products obtained from tissues collected from OPA-affected and unaffected control sheep revealed that the *Sca*I-sensitive form of JSRV was exclusively and consistently present in tumour tissues, draining lymph nodes and lung secretions of the affected animals (PALMARINI et al. 1996a). Analysis of SU RT–PCR products revealed the presence of an *Nde*I site only in transcripts from tumour or lung secretions from affected animals. Thus, for the first time, exogenous JSRV could be distinguished from endogenous transcripts by the presence of *Sca*I and *Nde*I restriction sites that are conserved in viruses from geographically dispersed areas.

7
JSRV Proviral DNA in Affected Animals

The above results (Sect. 6), however, did not rule out the possibility that the exogenous form of JSRV may have represented an enJSRV that normally was transcriptionally silent but reactivated as a downstream event of neoplasia. The molecular characterization of a number of endogenous sheep retrovirus loci (BAI et al. 1996) and the identification, in particular, of major deletions and point mutations in the U3 region of the LTR of exogenous JSRV (BAI et al. 1996; PALMARINI et al. 1996b) allowed the development of PCRs specific for exogenous JSRV proviral DNA. Analysis of OPA tumour and paired non-tumour tissues of the same animals with the JSRV U3 specific PCR demonstrated JSRV proviral DNA in 100% of tumour and draining lymph nodes, but also in a high proportion of other lymphoreticular tissues and peripheral blood mononuclear cells (PBMCs; PALMARINI et al. 1996b; GARCIA GOTI et al. 2000; GONZALEZ et al. 2000). More recently, SANNA et al. (2001) used in situ PCR to demonstrate JSRV proviral DNA in transformed and a few untransformed type II pneumocytes, alveolar macrophages and cortical follicles and paracortex of the draining lymph nodes. Thus, JSRV provirus was found more frequently in OPA tumours than in the lymphoid tissues, particularly those associated with the extrathoracic tissues. A separate study confirmed the association of JSRV proviral DNA with OPA tumour and its discrimination from endogenous sequences (BAI et al. 1996). These results, therefore, confirmed that the exogenous form of JSRV was associated exclusively with OPA and that it could play a major role in its aetiology. Although these experiments provided strong support for the hypothesis that exogenous JSRV was the probable cause of OPA, final confirmation of this role required the isolation of the pathogenic molecular clones of JSRV, described in the chapter by PALMARINI and FAN (this volume).

8
Characteristics of Neoplastic Cells in OPA

To better understand the interaction of JSRV with pulmonary epithelial cells in the neoplastic process, PLATT et al. (2002) investigated the distribution of JSRV-CA protein in neoplastic and normal cells of the lung and correlated this with tumour cell phenotype and differentiation state in naturally occurring (OPA-N) and experimentally induced (OPA-E) cases of OPA. Overall, 82% of tumour cells had ultrastructural features consistent with ATII cells, 7% of tumour cells had features of Clara cells, and 11% of tumour cells were insufficiently differentiated to classify. The proportion of the neoplastic cell phenotypes varied within tumours and no tumour consisted of a morphologically uniform cell population. To further characterize the neoplastic cell population, sections of tumours were immunostained with antibodies to surfactant protein A (SP-A), surfactant protein C (SP-C), and Clara cell 10 kDa protein (CC-10). Overall, SP-A and SP-C were expressed in 70% and 80% of tumour cells, respectively, whereas CC-10 was expressed in 17% of tumour cells. JSRV CA protein was detected in 71% of tumour cells, in macrophages (5 of 21 tumours examined) and in non-neoplastic alveolar and bronchiolar cells (6 of 14 tumours) (Fig. 6).

Differences between the OPA-N and OPA-E were few. The experimental cases tended to have a large number of small, widely disseminated tumour nodules, whereas the tumours in naturally occurring cases consisted of large, monocentric coalescing masses involving one or more lung lobes. This latter pattern, observed in OPA-N, more closely resembled human BAC. There was no difference in the proportion of specific cell types found in OPA-E and OPA-N nor was there a difference in JSRV CA expression observed in this study. These findings help to validate the experimentally induced disease as a model for the natural disease. In general, the major difference between OPA-N and OPA-E involved the observed rapid onset of clinical disease in experimentally inoculated animals. This may reflect increased proliferation of ATII cells in neonatal lambs relative to adult sheep, and thus, an increased capacity of the inoculated virus to integrate, or it could simply be the route and/or dose of virus used to infect lambs. A preliminary study using bromodeoxyuridine labelling to estimate the

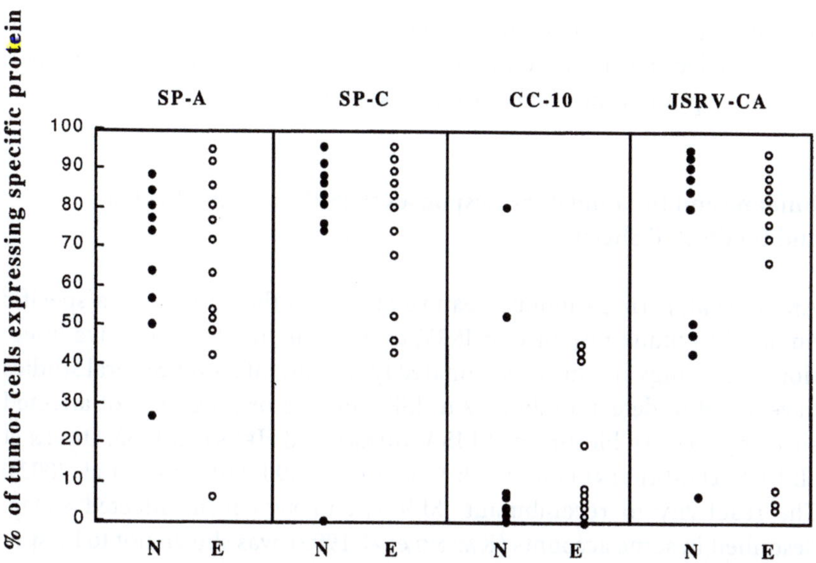

Fig. 6. Phenotyping of OPA tumour cells. Results of immunostaining of OPA tumour cells expressing surfactant protein A (*SP-A*), surfactant protein C (*SP-C*), Clara cell secretory protein (*CC-10*) and JSRV capsid protein (*JSRV CA*) in 9 cases of OPA-N (*N*) and 12 cases of OPA-E (*E*). (From PLATT et al. 2002)

replication rates of ATII cells and Clara cells in lambs (N. Kraipowich, unpublished results) demonstrated that there were more labelled ATII cells per high-power field in the 2- to 3-week-old lambs (8.5 ± 4.4) than in 16- to 18-week-old lambs (2.6 ± 1.6; $P = 0.005$). However, there was no difference in labelled Clara cells in bronchioles in the two groups (3.6 ± 3.2 vs. 2.7 ± 1.8; $P = 0.56$).

The use of immunophenotyping with SP-A, SP-C, and CC-10 together with ultrastructural findings clearly indicate that ATII cells are the predominant neoplastic cell type in OPA and that the tumour cells exhibit variable degrees of differentiation. Although JSRV CA was observed in all tumours, it was not observed in all tumour cells, even in the same nodule, and the presence of this viral protein was not limited to neoplastic cells. We propose that there are specific interactions between JSRV and its target cell that limit carcinogenesis to cells of the ATII/Clara cell lineage and that continuous production of intracellular viral protein may not be required for continuous tumour cell proliferation. Possible mechanisms for limiting neoplasia to ATII/Clara cells include a cell type specific interaction with viral protein components, such as the envelope protein, an insertional mutagenesis event that specifically transforms these cell types, or transcriptional specificity of JSRV expression.

9 Immune and Inflammatory Responses to JSRV in OPA-Affected and Unaffected Sheep

An unusual, perhaps unique, feature of OPA is the absence of a specific humoral immune response to JSRV, despite the highly productive infection in the lungs and the disseminated lymphoid infection. Several studies have failed to detect antibodies to JSRV in sera or lung fluid of affected sheep by Western blotting or ELISA (Sharp and Herring 1983; Ortin et al. 1998; DeMartini et al. 1988; Verwoerd 1990; Summers et al. 2002). The reactivity to recombinant JSRV CA in sera from affected sheep, described in some accounts (Kwang et al. 1995), was shown not to be specific and reflected the presence of antibodies to the GST fusion partner of the recombinant antigen used in the assays (Ortin et al. 1998). However, antibodies can be detected readily in the serum and lung lavage of sheep immunized with recombinant JSRV CA or SU in adjuvant (J.M. Sharp, P. Dewar and R. van der Molen, unpublished results). These results indicate that sheep are not inherently unresponsive to JSRV antigens and that the apparent tolerance, perhaps as a consequence of endogenous sequences expressed in utero (Palmarini et al. 2001) or as a direct consequence of infection by JSRV, can be broken. To date, specific cellular immune responses have not been investigated. Another prominent feature of naturally OPA-affected sheep is the marked peripheral neutrophilia and lymphopaenia, particularly affecting $CD4^{+}$ T lymphocytes although other subsets remained unaffected (Rosadio and Sharp 1992; Holland et al. 1999; Summers et al. 2002).

The persistent and disseminated infection of the lymphoreticular system by JSRV and dysregulation of the immune and inflammatory responses of infected sheep suggest that JSRV may interfere with the host immune responses. This notion is supported by studies demonstrating JSRV infection of a wide range of lymphoid cells (Holland et al. 1999). These workers showed that in naturally infected sheep, JSRV proviral burden was greatest in the macrophage/monocytic cell population (1/2500 cells), followed by B cells (1/3800 cells), $CD4^{+}$ T lymphocytes (1/6800 cells) and $CD8^{+}$ T lymphocytes (1/16700 cells). Furthermore, dissemination of

JSRV was an early event following experimental infection of young lambs. The virus was present in $CD4^+$ and $CD8^+$ T lymphocytes, B lymphocytes and adherent mononuclear cells in the pulmonary lymph nodes as early as 7 days post-inoculation, and was detected in PBMCs by 14 days post-inoculation. Although the levels of proviral DNA were too low to allow quantitative estimates of the JSRV burden, these observations are particularly significant as they indicated that dissemination of JSRV preceded detectable neoplastic transformation and, for the first time, suggested that infected animals might be detected in OPA-affected flocks before the appearance of clinical signs.

In contrast with the changes in phenotypic frequencies observed in naturally-occurring OPA, no significant alterations were identified at any time during the first 20 weeks after inoculation in the blood of young lambs experimentally inoculated with JSRV, compared to age- and breed-matched controls (Summers et al. 2002; Fig. 7). These observations indi-

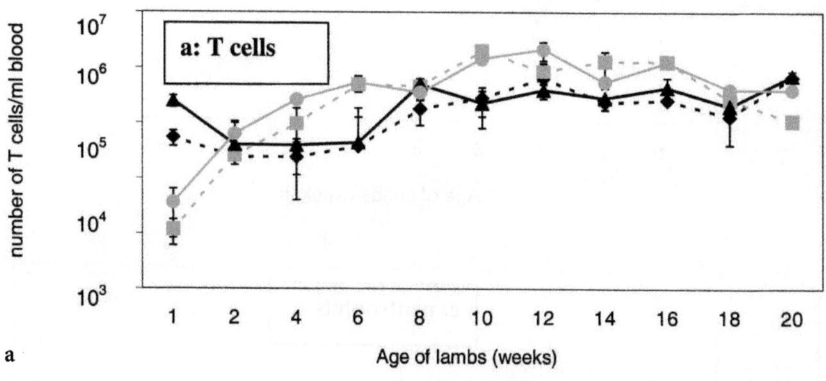

Fig. 7. a–c. Leukocyte subset frequencies in the peripheral blood during experimental JSRV infection. The peripheral blood mononuclear cells (PBMCs) were purified from the blood of lambs during experimental JSRV infection: conventionally housed control lambs, 1–16 weeks $n = 4$, 18–20 weeks $n = 2$; conventionally housed JSRV-infected lambs, 1–12 weeks $n = 6$, 14–16 weeks $n = 3$, 18 weeks $n = 2$ 20 weeks $n = 1$; specific pathogen free control lambs, 1–16 weeks $n = 4$, 18–20 weeks $n = 2$; specific pathogen free JSRV-infected lambs, 1 week $n = 6$, 2–9 weeks $n = 5$, 10–12 weeks $n = 4$, 14–16 weeks $n = 3$, 18–20 weeks $n = 2$. The PBMCs were phenotyped with monoclonal antibodies specific for CD2, CD4, $\gamma\delta$ T cells and B cells, with the results determined by flow cytometry. The results are shown in **a** and **b**, respectively, as the number of positively immunolabelled cells/lymphocyte region of the PBMCs. **c** Neutrophil counts shown as percentage neutrophils in the blood. Each time point represents the group mean±SEM. (From Summers et al. 2002)

cate that the alterations in peripheral blood lymphoid cell subsets are not an early event. Probably they do not occur in direct response to JSRV infection, but are a consequence of the superimposed bacterial infections that are common in natural cases of OPA.

By comparing the proliferative responses to mitogens of lymphocytes obtained from both natural cases of OPA and lambs inoculated experimentally with JSRV with those obtained from control animals, functional defects have been identified in the immune responses of JSRV infected sheep (SUMMERS et al. 2002). No significant differences were shown for the

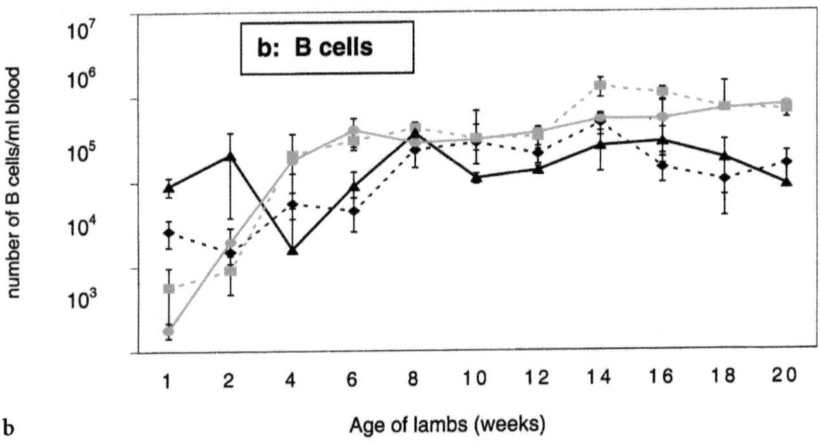

b

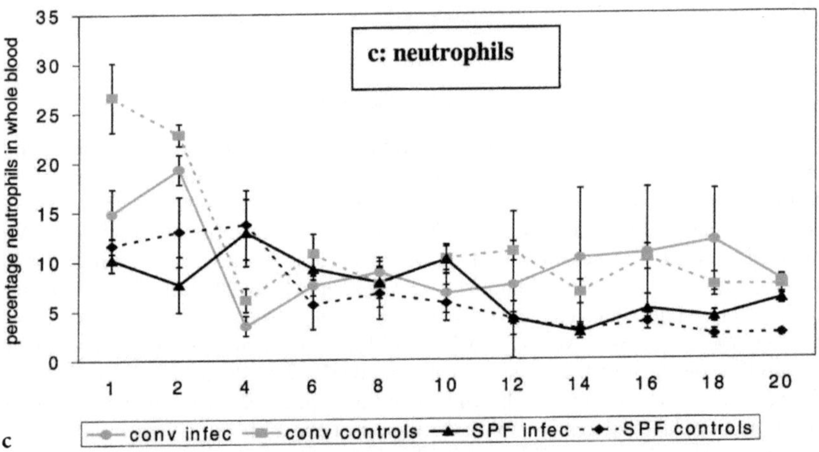

c

Fig. 7b,c

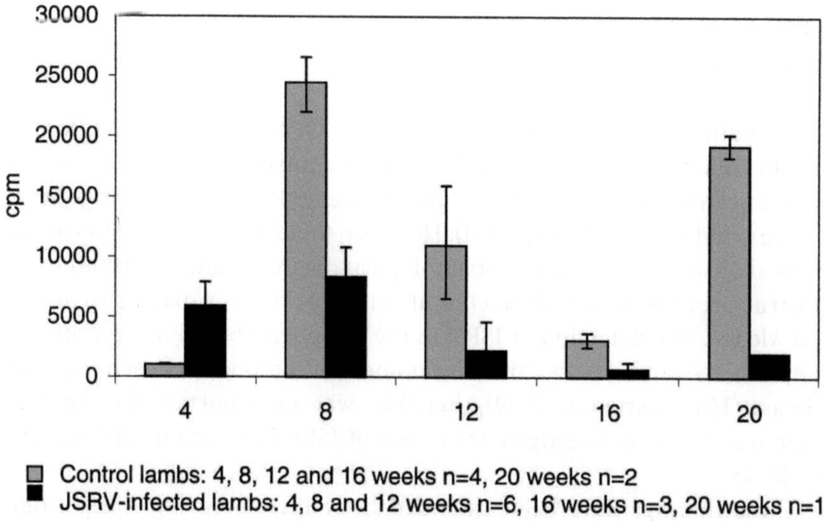

Fig. 8. Lymphoproliferative response to ConA stimulation during experimental JSRV-induced OPA. PBMCs purified from blood samples of conventionally housed JSRV-infected and control lambs were stimulated with ConA. *Histogram bars* represent group means±SEM for each mitogen. ^{3}H-thymidine uptake is shown as counts per minute (c.p.m.). Statistical significance, $P = 0.002–0.008$. (Adapted from Summers et al. 2002)

proliferative responses to stimulation with either phytohaemagglutinin or pokeweed mitogen in both natural cases of OPA and lambs inoculated experimentally with JSRV. In contrast, significantly reduced responses to concanavalin A (ConA) were evident in these animals. A reduced response was detected in the lambs as early as 8 weeks after inoculation prior to clinical disease, when the response was only 33% that of the control lambs (Fig. 8). A reduced response was maintained until the end of the experiment at 20 weeks. A reduced ConA response also was observed in the natural cases of OPA, which was only 58% that of the matched control sheep. These observations have provided the first indications that JSRV can alter the functional activity of immune cells in infected animals, although the virus apparently does not alter the phenotypic profile of the peripheral blood. Furthermore, the reduced response to ConA is an early and sustained event during infection and may indicate that the failure of infected sheep to produce specific antibodies to JSRV is a direct consequence of infection by this exogenous retrovirus.

10
Epidemiology

Longitudinal studies in flocks with endemic OPA have shown that losses due to OPA can vary between 2% and 10% annually and at necropsy the tumour can be demonstrated in approximately 30% of the breeding sheep (SHARP and DE LAS HERAS 2000). However, these findings were based on clinical observations and histopathological diagnosis and information of the true prevalence of JSRV in OPA-affected flocks was a major gap in our knowledge. The detection of JSRV in the peripheral blood of experimentally infected lambs, before the development of tumour or onset of clinical disease (HOLLAND et al. 1999), therefore, was an important finding that offered a means to investigate the extent of JSRV infection in OPA-affected flocks.

These findings have been extended in studies involving sheep from OPA-affected commercial flocks (GONZALEZ et al. 2001). Peripheral blood leukocytes and tissue samples from sheep were examined for JSRV by PCR. The animals were classified into four groups according to their OPA status as confirmed by histopathological examination: (1) sheep with classic OPA; (2) sheep with atypical OPA; (3) in-contact sheep with no evidence of OPA lesions from OPA-affected flocks; and (4) unaffected control sheep from OPA-free flocks. Overall, JSRV was detected in all OPA sheep, with either the classic or atypical form of OPA as well as in 80% of OPA in-contact sheep (Fig. 9). None of 71 samples from the control sheep was positive. Although an earlier report had indicated that as few as 1/250,000 PBLs might be positive for JSRV (HOLLAND et al. 1999), in this study JSRV was detected in the PBLs of all classic OPA sheep, 83% of the sheep with atypical OPA, as well as 40% of lesion-free in-contact sheep. These results clearly demonstrated, for the first time, that JSRV can be detected in naturally infected live sheep before the onset of clinical OPA and even in the absence of discernible lung tumours. Although only a small number of in-contact sheep were examined, it seems clear that subclinical JSRV infection can reach a high prevalence in OPA-affected flocks.

To provide further information of the dynamics of JSRV infection in OPA-affected flocks, GARCIA GOTI et al. (1999) conducted a small prospective longitudinal survey in an OPA-affected flock. JSRV was detected by PCR in PBLs from 28% of the flock. Fifteen positive sheep and five negative sheep were selected and PBLs were examined by JSRV PCR at monthly intervals for the next 4 months. JSRV was detected in only 9 of the original

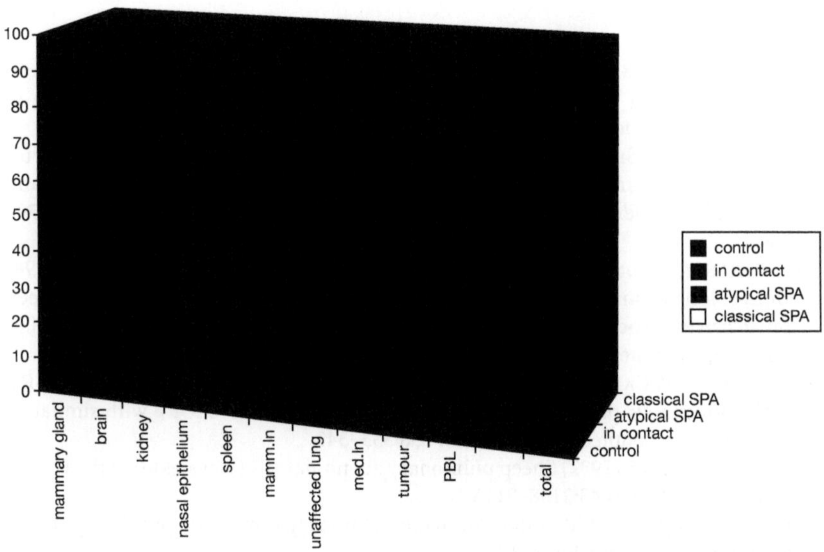

Fig. 9. Distribution of JSRV in tissues from OPA-affected sheep and their in-contact flockmates. The percentage of tissues positive for JSRV DNA by nested PCR are shown for control animals from a non-endemic OPA flock, animals with classic or atypical OPA (*SPA* in the figure), and in-contact animals from high OPA flocks

15 positive sheep during this period and in 4 of those that were negative (Table 1). These results demonstrated the fluctuation in detection of JSRV in blood and, more importantly, confirmed the high prevalence of JSRV infection in OPA-affected flocks. It is notable that despite this high prevalence, only approximately 30 % of sheep in affected flocks develop OPA lesions.

Acknowledgements. This work was supported by funds from the Scottish Executive, Environment and Rural Affairs Department, the Commission of the European Communities and the National Cancer Institute, National Institutes of Health.

Table 1. Longitudinal survey of sheep in an OPA-affected flock to detect JSRV in blood (adapted from Garcia Goti 1999)

JSRV status	Number of sheep	Time (months)					Necropsy	Total
		0	1	2	3	4		
PCR$^+$	15	15	9	5	1	4	6	11
PCR$^-$	5	5	2	3	1	1	2	4

References

Bastianello S (1982) A survey on neoplasia in domestic species over a 40 year period from 1935 to 1974 in the Republic of South Africa. II. Tumours occurring in sheep. Onderstepoort J Vet Res 49:205–209

Bai J, Payne LN, Skinner M (1995) HPRS-103 (exogenous avian leukosis virus, subgoup J) has an *env* gene related to those of endogenous elements EAV-0 and E51 and an E element found previously only in sarcoma viruses. J Virol 69:779–784

Bai J, Zhu R-Y, Stedman K, Cousens C, Carlson J, Sharp JM, DeMartini JC (1996) Unique long terminal repeat U3 sequences distinguish exogenous jaagsiekte sheep retroviruses associated with ovine pulmonary carcinoma from endogenous loci in the sheep genome. J Virol 70:3159–3168

Bai J, Bishop JV, Carlson JO, DeMartini JC (1999) Sequence comparison of JSRV with endogenous proviruses: envelope genotypes and a novel ORF with similarity to a G protein coupled receptor. Virol 258:333–343

Cutlip RC, Young S (1982) Sheep pulmonary adenomatosis (Jaagsiekte) in the United States. Am J Vet Res 43:2108–2113

Dalefield RR, Alley MR (1988) An ovine pulmonary tumour of alveolar epithelial type II cells. N Z Vet J 36:25–27

Dawson M, Done SH, Venables C, Jenkins CE (1990) Maedi-visna and sheep pulmonary adenomatosis: a study of concurrent infection. Br Vet J 146:531–533

De las Heras M, Hasleton P, Larson E, Wagner M, Minguijon E, Egan J, Dewar P, Ortin A, Gimenez-Mas JA, Gonzalez L, Palmarini M, Dalziel R, Sharp JM (2000) Human bronchioloalveolar cell carcinoma contains a protein related immunologically to the capsid protein of jaasiekte sheep retrovirus. Eur Resp J 15: 330–332

DeMartini JC, Rosadio RH, Lairmore MD (1988) The etiology and pathogenesis of ovine pulmonary carcinoma (sheep pulmonary adenomatosis). Vet Microbiol 17: 219–236

DeMartini JC, Rosadio RH, Sharp JM, Russell HI, Lairmore MD (1987) Experimental coinduction of type D retrovirus-associated pulmonary carcinoma and lentivirus-associated lymphoid interstitial pneumonia. J Natl Cancer Inst 79: 167–177

DeMartini JC, York DF (1997) Retrovirus-associated neoplasms of the respiratory system of sheep and goats. Ovine pulmonary carcinoma and enzootic nasal tumor. Vet Clin North Am Food Anim Pract 13:55–70

Dungal N (1949) Experiments with jaagsiekte. Am J Pathol 22:737–759

Dykes JR, M'Fadyean J (1888) Lung disease in sheep caused by Strongylus rufescens. J Comp Path Therap 1:139–144

García-Goti M (1999) Estudios sobre la patogenia y diagnóstico de la adenomatosis pulmonar ovina. PhD Thesis. University of Zaragoza

García-Goti M, González L, Cousens C, Cortabarria N, Extramiana AB, Minguijón E, Ortín A, de las Heras M, Sharp JM (2000) Sheep pulmonary adenomatosis: characterization of two pathological forms associated with jaagsiekte retrovirus. J Comp Pathol 122:55–65

Golovkina TV, Jaffe AB, Ross SR (1994) Coexpression of exogenous and endogenous mouse mammary tumour virus RNA in vivo results in viral recombination and broadens the virus host range. J Virol 68:5019–5026

González L, García-Goti M, Cousens C, Dewar P, Cortabarría N, Extramiana B, Ortín A, De las Heras M, Sharp JM (2001) Jaagsiekte sheep retrovirus can be detected in the peripheral blood during the preclinical period of sheep pulmonary adenomatosis, J Gen Virol 82:1355–1358

He Y, Hecht SJ, DeMartini JC (1992) Evidence for retroviral capsid and nucleocapsid antigens in ovine pulmonary carcinoma. Virus Res 25:159–167

Hecht S, Stedman K, Carlson J, DeMartini JC (1996) Distribution of endogenous type B and D sequences in ungulates and other mammals. Proc Natl Acad of Science USA 93:3297–3302

Herring AJ, Sharp JM, Scott FMM, Angus KW (1983) Further evidence for a retrovirus as the aetiological agent of sheep pulmonary adenomatosis (jaagsiekte). Vet Microbiol 8:237–249

Holland M, Palmarini M, Garcia-Goti M, Gonzalez L, McKendrick L, De las Heras M, Sharp JM (1999) Jaagsiekte retrovirus is widely distributed both in T and B lymphocytes and in mononuclear phagocytes of sheep with naturally and experimentally acquired pulmonary adenomatosis. J Virol 73:4004–4008

Hooper PT, Ellard KA (1995) A pulmonary adenoma in an Australian sheep. Aust Vet J 72:477

Houwers DJ, Terpstra C (1984) Sheep pulmonary adenomatosis. Vet Rec 114:23

Hunter AR, Munro R (1983) The diagnosis, occurrence and distribution of sheep pulmonary adenomatosis in Scotland 1975 to 1981. Brit Vet J 139:153–164

Jassim FA (1988) Identification and characterisation of transformed cells in jaagsiekte, a contagious lung tumour of sheep. PhD Thesis. University of Edinburgh

Kajikawa O, Dahlberg JE, Rosadio RH, DeMartini JC (1990) Detection and quantitation of a type D retrovirus *gag* protein in ovine pulmonary carcinoma (sheep pulmonary adenomatosis) by means of a competition radioimmunoassay. Vet Microbiol 25: 17–28

Kwang J, Keen J, Rosati S, Tolari F (1995) Development and application of an antibody ELISA for the marker protein of ovine pulmonary carcinoma. Vet Immunol Immunopathol 47:323–331

Maeda N, Palmarini M, Murgia C, Fan H (2001) Direct transformation of rodent fibroblasts by jaagsiekte sheep retrovirus DNA. Proc Natl Acad Science USA 98: 4449–4454

Martin WB, Scott FMM, Sharp JM, Angus KW, Norval M (1976) The experimental production of sheep pulmonary adenomatosis (Jaagsiekte). Nature, London 264:183–185

Myer MS, Huchzermeyer HFAK, York DF, Hunter P, Verwoerd DW, Garnett HM (1988) The possible involvement of immunosuppression caused by a lentivirus in the aetiology of jaagsiekte and pasteurellosis in sheep. Onderstepoort J Vet Res 55:127–133

Ortin A, Minguijon E, Dewar P, Garcia M, Ferrer LM, Palmarini M, Gonzalez L, Sharp JM, De las Heras M (1998) Lack of a specific immune response against a recombinant capsid protein of Jaagsiekte sheep retrovirus in sheep and goats naturally affected by enzootic nasal tumour or sheep pulmonary adenomatosis. Vet Immunol Immunopathol 61:229–237

Palmarini M, Dewar P, De las Heras M, Inglis NF, Dalziel RG, Sharp JM (1995) Epithelial tumour cells in the lungs of sheep with pulmonary adenomatosis are major sites of replication for Jaagsiekte retrovirus. J Gen Virol 76:2731–2737

Palmarini M, Cousens C, Dalziel RG, Bai J, Stedman K, DeMartini JC, Sharp JM (1996a) The exogenous form of Jaagsiekte retrovirus (JSRV) is specifically associated with a contagious lung cancer of sheep. J Virol 70:1618–1623

Palmarini M, Holland MJ, Cousens C, Dalziel RG, Sharp JM (1996b) Jaagsiekte retrovirus (JSRV) establishes a disseminated infection of the lymphoid tissues in sheep affected by pulmonary adenomatosis. J Gen Virol 77:2991–2998

Palmarini M, Maeda N, Murgia C, De-Fraja C, Hofacre A, Fan H (2001) A phosphatidylinositol-3-kinase (PI3-K) docking site in the cytoplasmic tail of the jaagsiekte sheep retrovirus transmembrane protein is essential for envelope-induced transformation of NIH 3T3 cells. J Virol 75:11002–11009

Palmarini M, Fan H (2001) Retrovirus-induced ovine pulmonary adenocarcinoma, an animal model for lung cancer. J Natl Cancer Inst 93:1603–1614

Pálsson PA: Maedi/visna of sheep in Iceland: Introduction of the disease to Iceland, clinical features, control measures, and eradication. In Sharp JM, Hoff-Jorgensen R (eds): Slow Viruses in Sheep, Goats, and Cattle. Brussels, Commission of the European Communities 1985, pp 3–19

Payne A, York DF, de Villiers EM, Verwoerd DW, Querat G, Barban V, Sauze N, Vigne R (1986) Isolation and identification of a South African lentivirus from jaagsiekte lungs. Onderstepoort J Vet Res 53:55–62

Perk K, Hod I (1982) Sheep lung carcinoma: an epidemic analogue of a human neoplasm. J Natl Cancer Inst 69:747–750

Platt J, Villafane F, Kraipowich N, DeMartini JC (2002) Alveolar type II cells expressing jaagsiekte sheep retrovirus capsid protein and surfactant proteins are the predominant neoplastic cell type in ovine pulmonary adenocarcinoma. Vet Pathol 39:341–352

Rai SK, Duh FM, Vigdorovich V, Danilkovitch-Miagkova A, Lerman MI Miller AD (2001) Candidate tumour suppressor HYAL2 is a glycosylphospatidylinositol (GPI)-anchored cell-surface receptor for Jaagsiekte sheep retrovirus, the envelope protein of which mediates oncogenic transformation. Proc Natl Acad Science USA 98:4443–4448

Rajya BS, Singh CM (1964) The pathology of pneumonia and associated respiratory disease of sheep and goats. 1. Occurrence of jaagsiekte and maedi in sheep and goats in India. Am J Vet Res 25:61–67

Roche S, Koegl M, Courtneidge SA (1994) The phosphatidylinositol 3-kinase α is required for DNA synthesis induced by some, but not all, growth factors. Proc Natl Acad Science USA 91:9185–9189

Rosadio R, Sharp JM (1992) Leukocyte frequency alterations in sheep with naturally and experimentally-induced lung cancer. Vet Med 9:49–51

Rosadio RH, Lairmore MD, Russell HI, DeMartini JC (1988) Retrovirus-associated ovine pulmonary carcinoma (sheep pulmonary adenomatosis) and lymphoid interstitial pneumonia. I. Lesion development and age susceptibility. Vet Pathol 25: 475–483

Rosadio RH, Sharp JM, Lairmore MD, Dahlberg JE, DeMartini JC (1987) Lesions and retroviruses associated with naturally occurring ovine pulmonary carcinoma (sheep pulmonary adenomatosis). Vet Pathol 25:58–66

Rosati S, Pittau M, Alberti A, Pozzi S, York DF, Sharp JM, Palmarini M (2000) An accessory open reading frame (orf-x) of jaagsiekte sheep retrovirus is conserved between different virus isolates. Virus Res 66:109–116

Sanna MP, Sanna E, De las Heras M, Leoni A, Nieddu AM, Pirino S, Sharp JM, Palmarini M (2001) Jaagsiekte sheep retrovirus is associated with ovine pulmonary carcinoma in Sardinian moufflons. J Comp Pathol 125:145–152

Sharp JM (1987) Sheep pulmonary adenomatosis: a contagious tumour and its cause. Cancer Surv 6:73–83

Sharp JM, De las Heras M (2000) Contagious respiratory tumours. In: Diseases of Sheep, 3rd edition, ed. Martin and Aitken, pp 181–186

Sharp JM, Herring AJ (1983) Sheep pulmonary adenomatosis: demonstration of a protein which cross-reacts with the major core proteins of Mason-Pfizer monkey virus and mouse mammary tumour virus. J Gen Virol 64:2323–2327

Sharp JM, Angus KW, Gray EW, Scott FMM (1983) Rapid transmission of sheep pulmonary adenomatosis (jaagsiekte) in young lambs. Arch Virol 78:89–95

Sharp JM, Angus KW, Jassim FA, et al (1986) Experimental transmission of sheep pulmonary adenomatosis to a goat. Vet Rec 19:245

Snyder SP, DeMartini JC, Ameghino E, Caletti E (1983) Coexistence of pulmonary adenomatosis and progressive pneumonia in sheep in the Central Sierra of Peru. Am J Vet Res 44:1334–1338

Stefanou D, Tsangaris TH, Lekkas ST (1975) Pulmonary adenomatosis in goats in the district of Pieria (Greece). In: 20th World Vterinary Congress, Thessaloniki, Greece. Summaries, pp 495–496

Stevenson RG, Finley GG, Long JR, Rehmtulla AJ (1982) Pulmonary adenomatosis (jaagsiekte) of sheep in Canada. Can Vet J 23:147–152

Stoye JP, Moroni C, Coffin JM (1991) Virological events leading to spontaneous AKR thymomas. J Virol 65:1273–1285

Summers C, Neill W, Dewar P, Gonzalez L, van der Molen R, Norval M, Sharp JM (2002) Systemic immune responses following infection with jaagsiekte sheep retrovirus and in the terminal stages of ovine pulmonary adenocarcinoma. J Gen Virol 83:1753–1757

Tustin RC (1969) Ovine jaagsiekte. J S Afr Vet Med Assoc 40:3–23

Tustin RC, Williamson A, York DF et al. (1988) Experimental transmission of jaagsiekte (ovine pulmonary adenomatosis) to goats. Onderstepoort J Vet Res 55: 27–32

Verwoerd DW, Tustin RC, Payne AL (1985) Jaagsiekte: An infectious pulmonary adenomatosis of sheep. In: Comparative Pathobiology of Viral Diseases, Olsen RG, Krakowka S, Blackslee JR (eds) pp 53–76. CRC Press, Inc.: Boca Raton

Verwoerd DW, Mayer-Scharrer E, Du Plessis JL (1977) Transplantation of cultured jaagsiekte (sheep pulmonary adenomatosis) cells into athymic nude mice. Onderstepoort J Vet Res 44:271–274

Verwoerd DW (1990) Jaagsiekte (ovine pulmonary adenomatosis) virus. In: Dinta Z and Morein B (eds), Virus infections of ruminants. Elsevier, pp 453–462

Wandera JG (1968) Experimental transmission of sheep pulmonary adenomatosis (jaagsiekte). Vet Rec 83:478–482
Weiss R, Teich N, Varmus H, Coffin J (eds) (1985) RNA Tumour Viruses. Cold Spring Harbor Laboratory
York DF, Vigne R, Verwoerd DW and Querat G (1992) Nucleotide sequence of the jaagsiekte retrovirus, an exogenous and endogenous type D and B retrovirus of sheep and goats. J Virol 66:4930–4939
Zimber A, Perk K, Hod I et al (1984) Heterotransplantation of experimentally - induced sheep lung adenomatosis into nude mice. Res Vet Sci 36:122–124

CHAPTER 4

Molecular Biology of Jaagsiekte Sheep Retrovirus

M. PALMARINI, H. FAN

Abstract. Jaagsiekte sheep retrovirus (JSRV) is the causative agent of ovine pulmonary adenocarcinoma (OPA), a contagious lung cancer of sheep. Until recently, research on JSRV/OPA was hampered by the lack of a tissue culture system for the propagation of the virus. Historically, patho-

M. PALMARINI
Department of Medical Microbiology and Parasitology,
College of Veterinary Medicine, University of Georgia, Athens, GA 30602-7386, USA
e-mail: mpalmari@vet.uga.edu

H. FAN
Cancer Research Institute and Department of Molecular Biology and Biochemistry,
University of California, Irvine, CA 92697, USA
e-mail: hyfan@uci.edu

logical samples (lung fluid) collected from sheep affected by OPA were the only source of infectious JSRV. Thus studies on the JSRV/OPA system were conducted only where field isolates of OPA cases were readily available. In the past 10 years, the deduction of the JSRV sequence (YORK et al. 1991; YORK 1992), the isolation of an infectious and oncogenic JSRV molecular clone ($JSRV_{21}$) (PALMARINI et al. 1999a) and the establishment of a rapid method to produce infectious virus in vitro (PALMARINI et al. 1999a) sparked many studies at the molecular level that strengthened past observations and revealed new properties of this unique virus. Here, we will review the data accumulated so far on the molecular biology of JSRV using the infectious and oncogenic $JSRV_{21}$ molecular clone as virus of reference.

1 Retroviruses

Jaagsiekte sheep retrovirus (JSRV) is a member of the family *Retroviridae*. Retroviruses are a diverse family of RNA viruses that are widespread in nature; retroviruses infecting most vertebrate species and many non-vertebrates have been identified. The initial discoveries of retroviruses resulted predominantly from studies on tumor-causing viruses (BITTNER 1936; ELLERMAN and BANG 1908; LABORATORY 1933; ROUS 1911a,b). However, retroviruses are capable of inducing many different diseases including cancers, immunodeficiencies, neuropathies, pneumonia, and bone and joint diseases; moreover, many retroviruses (perhaps the best-adapted to their host species) replicate in their hosts in the absence of any clinical symptoms. Despite the diversity of species infected, routes of transmission and disease manifestations, retroviruses are all quite similar in terms of virion structure, genomic organization and replication cycle.

Retroviral virions are spherical with a diameter of about 100–150 nm. The viral particles are surrounded by an envelope, often with prominent spikes composed of virus-encoded glycoproteins (RIFKIN and COMPANS 1971). The internal core is a spherical to rod-shaped capsid (depending on the virus). The core contains the viral RNA genome, as well as virus-encoded enzymes. The genome consists of two identical molecules of positive sense single-stranded RNA (7–10 kb in length) (BEEMON et al. 1974; BILLETER et al. 1974; QUADE et al. 1974), held together by hydrogen bonds. The genomic RNA has a structure reminiscent of cellular mRNAs, with a 5′ methylated cap structure and a poly(A) tract at the 3′ end (BENDER et al. 1978; Bender and Davidson 1976; Kung et al. 1975). In the virion, each genomic RNA molecule is hydrogen-bonded with a specific cellular tRNA molecule at a region termed the primer binding site (PBS) near the 5′ end of the RNA (Taylor and Illmensee 1975). The tRNA is important in initiating reverse transcription of viral RNA into DNA.

All retroviral genomes contain four main genes encoding virion proteins invariably in the same order: 5′-*gag-pro-pol-env*-3′ (Duesberg et al. 1976; Wang et al. 1976a–c). The *gag* gene encodes at least three proteins: the MA (matrix) protein, the major capsid protein (CA) and the NC (nucleocapsid) protein. The *pro* gene encodes a protease (PR) while the *pol* gene encodes the enzymes reverse transcriptase (RT) and integrase (IN). (For some classes of retroviruses, e.g., alpharetroviruses, PR is actually encoded by the *gag* gene, and in other retroviruses, e.g., gammaretroviruses and lentiviruses, it is encoded by the *pol* gene; for these retroviruses there is no separate *pro* gene.) The *env* gene encodes the envelope glycoproteins SU (surface) and TM (transmembrane) (Vogt 1997).

All retroviruses contain non-coding regions at the ends of the genome which are essential for the replicative strategy of these viruses. These regions are R, repeated at both ends of the viral genome; U5, a region unique to the 5′ end of the genome; and U3, unique to the 3′ end (Vogt 1997).

Some retroviruses ('complex' retroviruses) encode additional virion and non-virion proteins that are important for the regulation of viral expression. In addition, some oncogenic retroviruses (acute transforming retroviruses) have incorporated fragments of certain cellular genes (c-*onc* or proto-oncogenes) into their genomes – viral oncogenes (v-*onc*). During this transduction, the viral oncogenes are frequently altered in structure and expression; as a result, the acute transforming retroviruses rapidly induce neoplasms and can frequently transform cells in culture (Hanafusa et al. 1977; Stehelin et al. 1976).

Two major molecular events are characteristic of retroviral replication: reverse transcription of the single-stranded RNA genome into double-stranded DNA and the integration of this DNA copy into the cell genome. After viral entry the retroviral genome is reverse transcribed into DNA by the viral reverse transcriptase (Baltimore 1970; Mizutani et al. 1970; Temin and Mizutani 1970) in the cytoplasm. The viral DNA is then transferred to the nucleus and integrated via viral IN protein into the chromosomal DNA of the host to form a provirus (Brown 1990, 1997). Once integrated the provirus uses the cellular machinery for its transcription and translation (see below).

For most retroviruses, the Env proteins are translated from a singly spliced mRNA, while the Gag and Pol proteins are translated from unspliced viral RNA. Translation of Pol proteins results from read-through translation from the *gag* (and/or *pro*) gene by way of ribosomal frame shifting or amber suppression (Jamjoom 1977; Yoshinaka 1985; Jacks; Swanstrom and Wills 1997). As for many other viruses, retroviral proteins are initially translated as polyproteins - one for each of the genes. The polyproteins are then proteolytically cleaved during virion morphogenesis. For Gag and Pol polyproteins, the viral PR carries out the cleavage; a cellular protease is responsible for the cleavage of the Env polyprotein into SU and TM.

During the reverse transcription process, the resulting viral DNA acquires long terminal repeats (LTRs) at either end (Hughes et al. 1978). Each LTR contains sequences from the ends of the viral RNA in the order U3–R–U5. The LTRs are important because they contain the sequences recognized by the cellular transcription machinery for initiation of transcription and cleavage/polyadenylation. Transcription begins at the U3–R junction in the upstream LTR, and cleavage/polyadenylation occurs at the R–U5 junction in the downstream LTR. This is because the DNA sequences of the U3 region contain basal RNA pol II promoter and enhancer sequences, and usually the R region contains a cleavage/polyadenylation site.

2
JSRV Taxonomy

The *Retroviridae* are divided into seven genera: alpharetroviruses, betaretroviruses, gammaretroviruses, deltaretroviruses, epsilonretroviruses, lentiviruses and spumaviruses (Hunter et al. 2000). JSRV is a betaretrovirus. This genus includes viruses of B-type [e.g., mouse mammary tumor virus (MMTV)] and D-type morphology [e.g., Mason–Pfizer monkey virus (MPMV)] with prominent surface spikes and eccentric condensed core, or less dense surface spikes and a cylindrical core respectively. In contrast to most retroviruses where particle assembly involves budding from the plasma membrane, capsid assembly for betaretroviruses occurs in the cytoplasm in structures previously described as A-type particles (Coffin 1992). A-type particles also have been observed in ovine pulmonary adenocarcinoma (OPA) tumor cells (Bucciarelli 1973; Hod et al. 1976; Perk and Hod 1971; Perk et al. 1971). The relationship of JSRV with betaretroviruses was initially evident from serological assays. Extracts

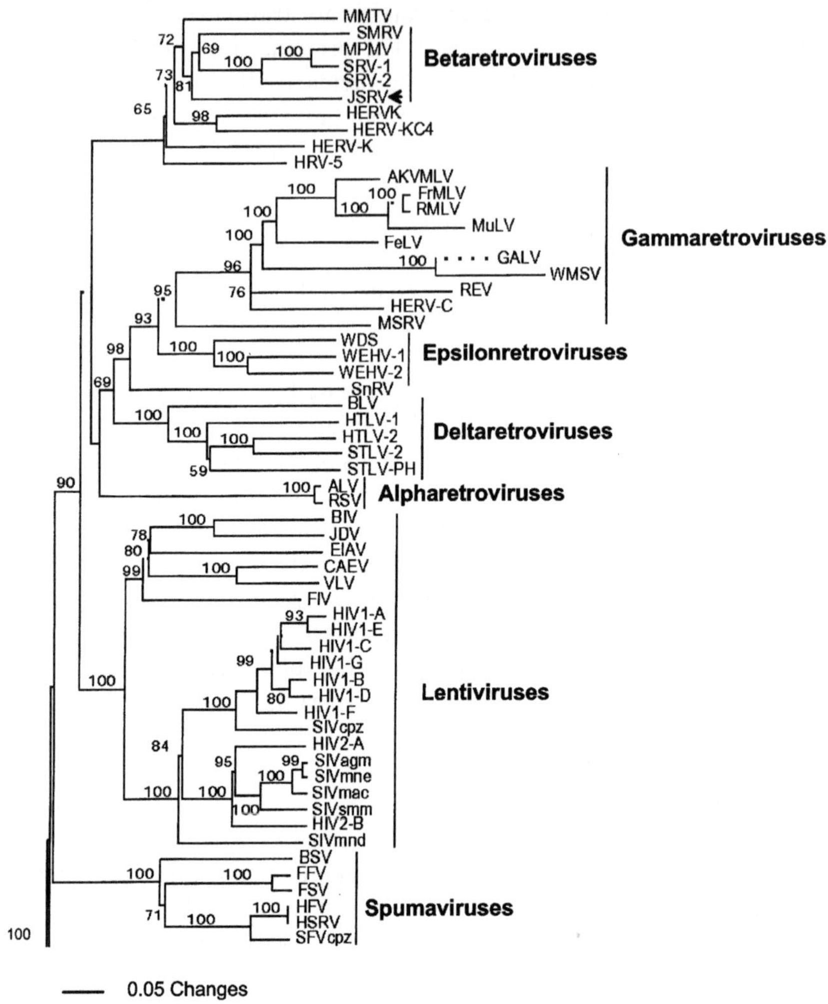

Fig. 1. Phylogeny of JSRV. Maximum-parsimony tree (random addition, tree bisection-reconnection, 1,000 replicates) based on *pol* sequences of members of the family of the *Retroviridae*. Retroviruses are divided into seven genera and JSRV (indicated by an *arrow*) is a betaretrovirus, a genus that contains exogenous and endogenous viruses of monkeys, mice, small ruminants and humans. Support of the nodes was evaluated using the bootstrap technique (Felsenstein 1985) with 1,000 replicates

from lung tumors and secretions from OPA-affected sheep cross-reacted with antisera raised against the CAs of MMTV and MPMV (SHARP and HERRING 1983), and other betaretroviruses – Squirrel monkey retrovirus and Langur virus (KAJIKAWA et al. 1990). The position of JSRV in the phylogenetic tree of retroviruses (based on RT protein sequence similarities) is illustrated in Fig. 1.

3 Isolation of the $JSRV_{21}$ Molecular Clone

The JSRV nucleotide sequence was first deduced by YORK and collaborators (YORK et al. 1991, 1992). Starting with lung fluid from a South African OPA case, these investigators carried out partial purification of the virus, followed by extraction of the viral RNA and oligo(dT)-primed in vitro reverse transcription. A series of overlapping cDNA clones was obtained, and sequencing allowed these investigators to piece together the complete genome sequence of a novel retrovirus, JSRV (reviewed in the chapter by YORK and QUERAT, this volume). The deduction of the JSRV sequence was a major advance that opened the way to molecular studies of JSRV/OPA. However a reconstructed full-length clone did not show infectivity in vitro nor in vivo (G. Querat, personal communication). Thus, it was still unclear if JSRV was necessary and sufficient to induce OPA. In addition, hybridization of JSRV DNA with unaffected (normal) sheep tissues revealed the presence in the sheep genome of 15–20 copies of endogenous JRSV-related retroviruses (enJSRVs) (HECHT et al. 1994, 1996; YORK 1992). The high degree of homology between the exogenous JSRV and enJSRVs (see the chapter by DEMARTINI et al., this volume) meant that most probes obtained from the exogenous JSRV sequence cross-reacted with the enJSRVs, making it difficult to detect the exogenous virus above the enJSRV background.

The identification of exogenous JSRV-specific restriction endonuclease polymorphisms (BAI et al. 1996; PALMARINI et al. 1996a) and the resulting development of PCR-based diagnostic tools for the presence of exogenous JSRV-specific DNA (PALMARINI et al. 1996b) paved the way to finally obtaining an infectious molecular clone of JSRV.

In order to obtain an infectious JSRV molecular clone, we chose the strategy of cloning a full-length JSRV provirus. A lambda phage library (using a restriction endonuclease known not to cut within exogenous JSRV DNA) was constructed from tumor DNA from a natural OPA case from the UK (Palmarini et al. 1999a). The tumor library was screened by sib-selection (Goubin et al. 1983) followed by plaque hybridizations using JSRV DNA probes. The sib-selection used PCR amplifications followed by restriction endonuclease digestion that could distinguish exogenous JSRV from enJSRVs (Palmarini et al. 1996a, b). This combination of techniques was important because the currently available JSRV hybridization probes cross-hybridized with the endogenous JSRV-related sequences. The cloning strategy resulted in the isolation of a full-length exogenous JSRV proviral clone ($\lambda JSRV_{21}$). The insert from this clone was subcloned into a plasmid to give $pJSRV_{21}$. Sequence analysis showed that $JSRV_{21}$ possesses the hallmarks of integrated retroviral proviruses (Dhar et al. 1980; Hughes et al. 1981; Van Beveren et al. 1980), such as the presence of a CA/TG dinucleotide pair present at the termini of the upstream and downstream viral long terminal repeats (LTRs), the loss of two nucleotides from the termini of the LTRs during integration and an apparent duplication of six nucleotides of cellular flanking sequences (TGTGTC) at the integration site. Thus $JSRV_{21}$ was indeed an integrated provirus and sequence analysis (see below) demonstrated that it was full-length and with the expected open reading frames (ORFs).

Recently another integrated JSRV proviral clone has been isolated, in which the starting material was an OPA-derived cell line (JS7), also from an OPA tumor from the UK (DeMartini et al 2001).

3.1 Genetic Structure of JSRV

So far, three strains of JSRV have been completely sequenced or deduced: the original South-African strain (JSRV-SA) (York et al. 1991; York 1992), $JSRV_{21}$ (Palmarini et al. 1999a) and $JSRV_{JS7}$ (DeMartini et al. 2001). Partial sequences of JSRV isolates have been obtained also from different strains originating from various geographical locations (Bai et al. 1996, 1999; Hecht et al. 1994; Palmarini et al. 1996a, b; Rosati et al. 2000). Below, we will discuss the genetic structure of JSRV taking as reference strain $JSRV_{21}$ (the first isolated infectious full-length molecular clone) and compare it to the other full or partial JSRV sequences available.

The sequence of $JSRV_{21}$ confirmed the genomic structure deduced for JSRV-SA (York et al. 1991; York 1992) (Fig. 2a). By analogy to other retroviruses, the $JSRV_{21}$ RNA genome is 7455 nucleotides (nt) whereas the DNA integrated proviral form is 7,834 nt. The $JSRV_{21}$ U3 region is 266 base pairs

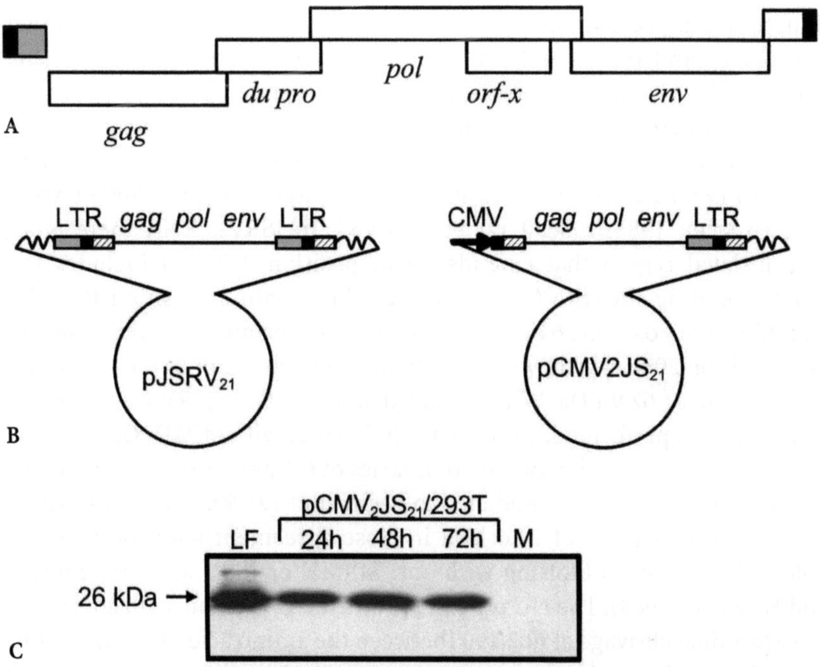

Fig. 2 A–C. JSRV genetic structure, $JSRV_{21}$-based plasmid constructs and in vitro synthesis of viral particles. **A** Schematic representation of the genomic organization of JSRV is shown using standard retroviral notation. Note the presence of an accessory ORF (*orf-x*) overlapping *pol*. **B** The $pJSRV_{21}$ and $pCMV2JS_{21}$ plasmid constructs. In $pCMV2JS_{21}$ the U3 region of the proximal LTR was replaced by the human CMV promoter. **C** Western blotting of concentrated supernatant (300×) from 293T cells transiently transfected with $pCMV2JS_{21}$ and collected 24, 48 and 72 h post-transfection. The filters were probed with a rabbit polyclonal antiserum against the major capsid protein (CA) of JSRV (PALMARINI et al. 1995). Lung secretions collected from an OPA-affected animal and concentrated the same way as the 293T supernatant was used as a positive control (*LF*). Concentrated supernatant from mock-transfected 293T cells was used as a negative control (*M*). The 26-kDa CA protein is indicated. (Modified from PALMARINI et al. 1999a, with permission from the American Society for Microbiology)

(bp) long and contains putative enhancer binding motifs, a TATA box at around –30 bp and the polyA signal. The R region is 13 bp long and the U5 is 115 bp. The determinants of transcriptional tissue specificity in the JSRV LTR will be discussed in Sect. 5.

The $JSRV_{21}$ genome has the classical retroviral genes *gag*, *pro*, *pol* and *env* with an additional ORF, known as *orf-x*, that overlaps *pol* and whose function is at present unknown. The genetic organization of JSRV is that one of a betaretrovirus with *pro* in a different ORF from *pol*. The *gag* gene encodes the viral core proteins, as mentioned above. In all retroviruses the Gag polyprotein is cleaved into at least three proteins (always located in the same relative positions 5′ to 3′), MA, the CA and NC (WILLS and CRAVEN 1991). Betaretroviruses also have additional Gag proteins (BRADAC and HUNTER 1984, 1986; HIZI et al. 1989). In $JSRV_{21}$ RNA there is an untranslated region that extends up to position 263 that includes the primer binding site ($tRNA_{lys1,2}$) and the splice donor at position 193. The first AUG is at position 264 and the first in-frame termination codon for *gag* is at position 2096, which would result in a polypeptide of 611 amino acids with a mass of 67.9 kDa. While detailed studies of the JSRV Gag proteins have not been performed, amino acid similarities with MPMV Gag protein allow likely identification of the boundaries of CA and possibly also of NC. CA coding sequences are predicted to begin at nt 1,029 and end at nt 1,686 to give a polypeptide of 24.6 kDa in close agreement with the p25/p26 detected by Western blotting with anti-MPMV or JSRV antisera (SHARP and HERRING 1983). The NC protein follows the CA region, with a predicted C-terminal cleavage at nt 2,190 (between the amino acids Y–GA). On the other hand, the amino acid sequence similarities between JSRV and MPMV/MMTV Gag do not allow prediction of the boundary of the MA protein, or if there might be an additional protein between MA and CA.

Amino acid sequence comparisons between JSRV and MPMV/MMTV Gag proteins also indicate that JSRV Gag probably has several functional motifs important for betaretrovirus replication. A consensus sequence for myristilation conserved in other betaretroviruses is present at the N terminus of the Gag protein. Myristilation is necessary for the transport of completed capsid to the plasma membrane for MPMV (RHEE and HUNTER 1987). The predicted JSRV MA protein also contains sequences very similar to the putative cytoplasmic targeting/retention signal present in MPMV and MMTV (RHEE and HUNTER 1990), although R55 that is absolutely required for MPMV morphogenesis is not conserved in $JSRV_{21}$. The predicted NC protein contains two Cys–His motifs critical for viral RNA packaging in other retroviruses (BERKOWITZ et al. 1996). Finally, there is a sequence C-terminal to the predicted end of JSRV NC that might comprise a small protein analogous to MPMV p4. MPMV p4 has been shown to interact with a cytosolic chaperonin TriC, suggesting a role in assisting the proper folding and assembly of the *gag* gene products (HONG et al. 2001).

The JSRV *pro* gene overlaps *gag* and encodes most probably two proteins, a deoxyuridine triphosphatase (dUTPase) in the 5′ portion of the gene and the retroviral protease that is important in virion morphogenesis (DOUGHERTY and SEMLER 1993). In $JSRV_{21}$ the *pro* ORF starts 53 nt downstream from the previously predicted putative *pro* start in JSRV(SA); in particular there are two stop codons in $JSRV_{21}$ at position 1919 and 1931 that are not present in JSRV(SA). The *pro* ORF in $JSRV_{JS7}$, another isolate from the UK, starts in the analogous position of $JSRV_{21}$ (DEMARTINI et al. 2001). Thus, for $JSRV_{21}$ (and for $JSRV_{JS7}$) the −1 translational frameshift that presumably occurs during synthesis of the Gag–Pro–Pol polyprotein precursor must occur upstream of the stop codon at nt 1919 (ELDER et al. 1992). The 5′ portion of the *pro* has homology to the dUTPase gene which has been identified in other betaretroviruses and lentiviruses (ELDER et al. 1992; KOPPE et al. 1994). dUTPase catalyzes hydrolysis of deoxyuridine triphosphate (dUTP) into deoxyuridine monophosphate and inorganic monophosphate and its function is primarily to inhibit the misincorporation of uracil into DNA (ELDER et al. 1992). We assessed the dUTPase activity of $JSRV_{21}$ by deleting the conserved IAQLL motif in the putative dUTPase region and comparing the capacity to incorporate dUTP of the deleted mutant versus the wild-type virus (M. Palmarini, G. Querat and H. Fan, unpublished results). The dUTPase mutant incorporated uracil approximately five times more efficiently than the wild type, supporting the idea that JSRV encodes a functional dUTPase.

The 3′ half of the *pro* gene is predicted to encode PR. The DTG active site of the viral protease, that is conserved in most retroviruses, is also present in the $JSRV_{21}$ *pro* at position 2474–2482.

The *pol* ORF is predicted to encode the viral RT and IN; in $JSRV_{21}$ the *pol* gene is between nt 2,827 and nt 5,451. The conserved active RT site YMDD is present in $JSRV_{21}$ at position 3,403–3,415. The RT includes the RNAseH domain that serves to degrade viral genomic RNA once it has been copied into DNA.

Of particular interest is the presence in the JSRV genome of *orf-x*, an accessory ORF of unknown function that overlaps *pol* (nt 4,463–5,101) (YORK 1992). ORFs homologous to *orf-x* are not present in other betaretroviruses and it has been debated if this ORF truly represents an accessory gene for JSRV (BAI et al. 1999; ROSATI et al. 2000; YORK 1992). For instance, the codon usage of the *orf-x* is very different from the codon usage of *pol* (YORK 1992). However, *orf-x* is conserved in all the JSRV isolates (originating from different geographical locations) known to date and in most of the endogenous JSRV-related loci identified so far (BAI et al. 1999; ROSATI et al. 2000; YORK 1992). Moreover, we identified two subgenomic spliced mRNAs with splice acceptors in or in the vicinity of the *orf-x* reading frame (see below) (PALMARINI et al. 2002). The first AUG for the *orf-x* of $JSRV_{21}$ is at position 4560 and the predicted protein would be 179 amino acids (aa) long with an approximate molecular weight of 20.9 kDa. The Orf-x is very hydrophobic with four putative transmembrane domains. The only homology found between Orf-x and proteins in the database is with the adenosine A3 receptor (22% identity and 48% similarity in a stretch of 76 aa) (AUCHAMPACH et al. 1997) but the relevance of this is unknown and the potential function of *orf-x* remains the object of speculation. A $JSRV_{21}$-derived mutant with two stop codons inserted in the *orf-x* reading frame was still able to infect cells in vitro and induce tumor nodules in vivo although no quantitative data on the in vivo experiments are yet available (M. Palmarini, J.M. Sharp and H. Fan, unpublished results). In addition, mutation of the *orf-x* reading frame does not affect the ability of JSRV to morphologically transform NIH-3T3 cells in culture (MAEDA et al 2001).

The *env* gene (nt 5,348–7,195) encodes for the glycoproteins of the Envelope, and their function is to mediate viral entry into the cells. The SU glycoprotein presumably interacts with the cellular receptor. The JSRV receptor has recently been identified as the glycosylphosphatidylinositol-anchored protein hyaluronidase-2 (Hyal-2) (RAI et al. 2001; see the chapter by MILLER, this volume). The TM glycoprotein presumably anchors SU (by covalent or non-covalent bonding) to the lipid bilayer of the viral envelope through a membrane-spanning domain. A hydrophobic stretch of amino acids is present in JSRV TM, which probably comprises the membrane-spanning domain; it is followed by a cytoplasmic tail (44 aa) that presumably remains on the internal side of the viral membrane (or in the cytoplasmic face of the infected cell plasma membrane). The JSRV envelope is a primary determinant of viral-induced transformation. Expression of JSRV Env is sufficient to induce transformation of rodent fibroblasts in vitro (MAEDA et al. 2001; RAI et al. 2001). The cytoplasmic tail mediates the transformation process possibly via signal transduction involving the activation of the PI-3 K/Akt pathway (PALMARINI and FAN 2001; PALMARINI et al. 2001b; see the chapter by FAN et al., this volume).

Like all secreted cellular proteins, the N-terminal portion of the env polyprotein contains a hydrophobic signal peptide that is presumably responsible for coupling the nascent chain to a signal recognition particle followed by transport into the endoplasmic reticulum. The most likely signal peptide cleavage site is between residues 81 and 82 (AAA–AF) of the Env polypeptide. Subsequently, Env is presumably transported through the Golgi where it is glycosylated. In $JSRV_{21}$ there are six potential N-glycosylation sites in SU and three in TM in which there are also two putative O-glycosylation sites. The Env precursor is then cleaved by cellular protease to give rise to SU and TM. The boundary between SU and TM is probably between residues 385 and 386 (RPKR–GLS). The predicted molecular weight for the Env precursor (unglycosylated) is 69.4 kDa (33 kDa for SU and 26.9 for TM) while the apparent molecular weight is going to be influenced by the glycosylation events. In NIH-3T3 fibroblasts expressing a JSRV envelope protein tagged with a C-terminal influenza HA epitope, a TM protein of around 35 kDa has been detected (N. Maeda and H. Fan, unpublished results).

The JSRV Envelope appears to be quite resistant to procedures that generally inactivate most retroviruses (Burns et al. 1993; Zsengeller et al. 1999). For instance, the titer of retroviral vectors bearing JSRV envelope remain unaltered after incubation at room temperature for 30 min or after six freeze–thaw cycles. The same procedures decreased the titer of the same vectors bearing the amphotropic murine leukemia virus (MuLV) 4070A envelope by 30% and 40% respectively (Coil et al. 2001). MuLV vectors containing JSRV envelope are also resistant to treatment with Survanta, a clinical-grade bovine-derived surfactant that inactivates the same vectors with amphotropic MuLV envelope (Zsengeller et al. 1999). JSRV appears to be transmitted through the respiratory route, and the resistance of the virus to the lung surfactant might explain why this virus is one of the few retroviruses transmitted via aerosol. In addition, the resistance to lung surfactant of the JSRV envelope may have application to the design of retroviral vectors for gene therapy in the lung (Coil et al. 2001).

3.2 Sequence Homology Among JSRV Isolates

There is very low sequence diversity between JSRV isolates. For instance, the $JSRV_{21}$ (Palmarini et al. 1999a) and the $JSRV_{JS7}$ (DeMartini et al. 2001) molecular clones are virtually identical (99.2% identity); they originate from OPA tumors from the UK collected approximately 10 years apart. $JSRV_{21}$ and $JSRV_{JS7}$ are 93.3% identical to the deduced South African

$JSRV_{SA}$ sequence. Amino acid sequences of the *env* genes of isolates from Europe, North America, and Africa are 95.6% to 100% identical. Isolates from different geographical locations are phylogenetically distinguishable (BAI et al. 1996, 1999; ROSATI et al. 2000) into a type I from UK and North America and a type II from Africa, although these differences may not have biological significance. On the other hand, differences in U3, *gag* and *env* between the exogenous JSRV and the JSRV-related endogenous sequences (BAI et al. 1996; PALMARINI et al. 1996a,b) have been shown to be major determinants in cell tropism and cell transformation (PALMARINI et al. 2000a, b, 2001b). The biology of enJSRVs will be discussed elsewhere in this volume (see the chapter by DEMARTINI et al. this volume).

4
Synthesis of $JSRV_{21}$ Particles In Vitro

The isolation of the $JSRV_{21}$ molecular clone allowed synthesis of $JSRV_{21}$ viral particles for the first time. Initially, we attempted to recover virus particles by transient transfection of $pJSRV_{21}$ into human 293T cells, as these cells are highly transfectable and they have been used for production of other retroviruses by transient transfection. This approach did not yield detectable virus in the culture supernatants. One possible explanation was that the JSRV LTR might not be active in 293T cells. Therefore, the U3 region of the upstream LTR in $pJSRV_{21}$ was replaced with the human cytomegalovirus (CMV) immediate early enhancer/promoter (Fig. 2b). In this plasmid, termed pCMV2JS21, the CMV promoter was placed so that the resulting RNA transcript would be very similar to wild-type JSRV RNA. When pCMV2JS21 was transfected into 293T cells, substantial amounts of $JSRV_{21}$ virus were released into the supernatant. Western blot analysis for JSRV CA protein indicated that the amount of virions produced from pCMV2JS21-transfected 293T cells was comparable to that present in lung fluid from OPA-affected sheep (Fig. 2c). Moreover, the supernatants from transfected 293T cells showed CA protein of the mature (cleaved) size; because gag polyprotein cleavage only occurs after particles have formed and budded from the cell, this strongly suggested that the transfected 293T cells were producing complete virus particles. Enzymatically active RT could also be detected in the transfected 293T cell supernatants by standard exogenous assays.

Supernatants from $pCMV2JS_{21}$-transfected 293T cells were also analyzed by isopycnic centrifugation in sucrose density gradients. The peak of

RT activity was detected at approximately 1.15 g/ml. This was consistent the buoyant density of retroviruses in general (Vogt 1997) although it was slightly lower than that reported for JSRV isolated directly from OPA lung secretions (1.16–1.18 g/ml) (Herring et al. 1983; Palmarini et al. 1995; Sharp and Herring 1983; York et al. 1991). Treatment of transfected 293T supernatants with 0.1 % Triton X-100 prior to centrifugation shifted the RT peak to 1.218–1.238 g/ml, consistent with the release of viral cores. The apparent discrepancy in buoyant density between JSRV virions isolated from lung fluid and those obtained by in vitro by transfection is interesting, and the basis for this remains to be determined.

4.1 JSRV Is the Etiological Agent of OPA

Four newborn black-face lambs were inoculated intratracheally with concentrated $JSRV_{21}$ viral particles obtained from transfected 293T cells, in order to test if JSRV is alone necessary and sufficient to induce OPA (Palmarini et al. 1999a). All four animals were sacrificed at 4 months post-inoculation when one of the animals showed respiratory distress suggestive of OPA. At post-mortem, the lungs of the clinically affected lamb where enlarged and showed OPA lesions. Small foci with similar features also were observed in the caudal lobes of one other animal. Histological examination revealed the presence of multifocal neoplastic foci (Fig. 3). Lesions comprised many small intra-alveolar and bronchiolar papilliform proliferations of cuboidal or prismatic epithelial cells. Some of these neoplastic nodules had an interstitial myxoid or fibrotic appearance. Alveoli adjacent to tumor nodules contained a small number of alveolar macrophages. The above lesions were consistent with previously described features of OPA. The presence of JSRV in the tumor lesions was confirmed by immunohistochemistry for viral CA antigen and PCR for viral DNA. One of the two inoculated lambs that did not show any sign of disease was also infected by JSRV, as shown by PCR. As expected, two uninoculated control lambs showed no signs of disease, and at necropsy their lungs showed no signs of macroscopic or histologic OPA lesions, or JSRV immuno-reactive material. Similar results have been obtained recently for another JSRV isolate, $JSRV_{JS7}$ (DeMartini et al. 2001; Jassim 1988). $JSRV_{JS7}$ induced histological OPA foci in one out of four inoculated lambs.

These experiments conclusively demonstrated that JSRV is the etiological agent of OPA.

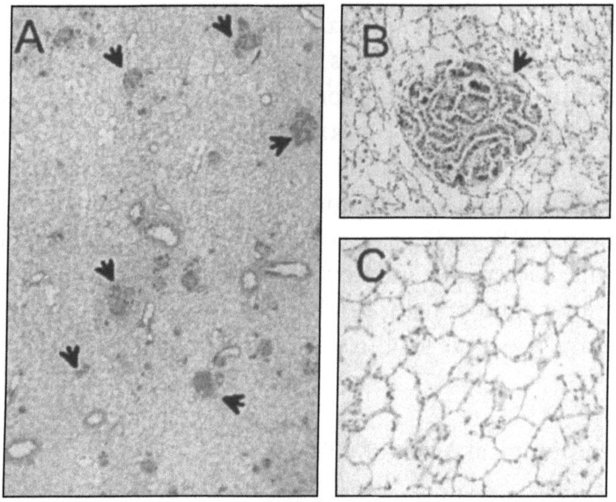

Fig. 3 A–C. Induction of OPA in $JSRV_{21}$-infected lambs. Lung sections from $JSRV_{21}$-infected lambs and uninoculated controls were fixed in neutral 10% formalin, embedded in paraffin, and sectioned by routine procedures. Hematoxylin and eosin-stained lung tumor sections (**A, C**) and immunohistochemistry (**B**) are shown. **A** Low-magnification micrograph showing many neoplastic foci in the microscopic field (some indicated by *arrows*). Immunohistochemistry for JSRV CA antigen developed with an avidin–biotin peroxidase complex kit (ABC; Vector laboratories) is shown in **B**, where most of the neoplastic cells have a brownish cytoplasmic stain indicative of positive reaction. **C** Lung section from an uninoculated control animal. (Partially derived from PALMARINI et al. 1999a, with permission from the American Society for Microbiology)

4.2
In Vitro Infection of Sheep Cell Lines

The production of JSRV particles by transient transfection of 293T cells allowed in vitro production of infectious and oncogenic virus particles for the first time. Thus, we re-examined the possibility of establishing an in vitro infection system for JSRV by infecting a variety of sheep cell lines (PALMARINI et al. 1999b). Infection of concentrated $JSRV_{21}$ into ovine choroid plexus, testes (OAT-T3), turbinate (FLT) and intestinal carcinoma (ST6) (NORVAL et al. 1981) cell lines resulted in establishment of infection as measured by PCR amplification. This represented genuine infection as: (1) heat inactivation of the virus eliminated PCR amplification; (2) the levels of viral DNA increased with passage of the infected cells; (3) infected cells released active RT as measured by the sensitive product enhancement reverse transcriptase assay (Pyra et al. 1994); and (4) infection could be transferred from the in vitro infected cells to other uninfected cells by co-cultivation. Interestingly, the RT activity released from the in vitro infected cells banded at 1.15 g/ml, similar to that observed for virions obtained by transfection of 293T with pCMV2JS21. However, the amount of virus released from infected cells was low and none of the cell lines tested was a convenient source for infectious virus.

The fact that all ovine cell lines tested could be productively infected with $JSRV_{21}$ suggested that the JSRV receptor is present on many ovine cell types. This was later confirmed by Rai et al. (2001), who identified Hyal-2 as the cellular receptor for JSRV (Rai et al. 2000, 2001). Hyal-2 is indeed expressed in many different cell types (Lepperdinger et al., 1998). Also in vivo, JSRV DNA and RNA are detected in both the differentiated epithelial cells of the lungs and various lymphoid cells (Gonzalez et al. 2001; Holland et al. 1999; Palmarini et al. 1996b) further reinforcing the idea that JSRV can enter several cell types.

5
JSRV Expression

A feature of JSRV is its tight cell type specificity for viral expression. For most other retroviral systems, there is substantial infection of numerous cell types within the infected host above and beyond the resulting tumor cells (Rosenberg and Jolicoeur 1997). In sheep with natural or experimentally-induced OPA, JSRV is abundantly expressed only in the transformed epithelial cells of the lungs (Palmarini et al. 1995). However, as mentioned above, JSRV DNA and (in some cases) RNA can be detected by sensitive assays in a variety of lymphoid tissues (Palmarini et al. 1996b). Therefore, JSRV infects many different cell types (enters and reverse transcribes), but it is highly expressed only in the epithelial tumor cells of the lungs. In particular, the restriction of viral expression to epithelial tumor cells in the lungs is not due to the restricted expression of the JSRV receptor to those cells.

Besides the viral envelope, another major determinant of retroviral tissue tropism are the enhancer sequences in the LTR (Athas et al. 1994; Fan 1990, 1994; Rosenberg and Jolicoeur 1997). We will describe below the preferential activity of the JSRV LTRs in type II pneumocytes and Clara cells and how they interact with transcription factors specific for the differentiated lung epithelial cells.

5.1 The JSRV LTR as a Determinant of Tissue Specificity

As described above, the viral promoter and enhancer sequences in the U3 region of the upstream LTR drive the expression of retroviral genes. Promoters are DNA sequence elements that contain binding sites for various protein factors that initiate and regulate transcription (LEWIN 1990). All known retroviruses, in common with many viral and cellular genes, have a TATA sequence at around –30 nt in U3. The TATA box binds a cellular factor (TATA-binding protein) that binds other cellular factors and RNA polymerase II to initiate transcription from the DNA template (GREENBLATT 1991). Enhancers are cis-acting sequences that enhance transcription from an adjacent promoter. For retroviruses, LTR enhancers are usually located 100–200 bp upstream of the TATA box, and they influence tissue-specific expression from the viral LTR and/or tumor induction (ATHAS et al. 1994; FAN 1990, 1994).

One obstacle in the study of JSRV expression has been that the target cells for JSRV transformation, type II pneumocytes and Clara cells, are difficult to isolate and they lose their differentiated phenotypes quite rapidly when cultured in vitro (MASON and SHANNON 1997; PLOPPER et al. 1997). Recently however, mouse cell lines derived from differentiated epithelial cells of the lungs that retain their differentiated state have been established (MALKINSON et al. 1997). The MLE-15 line was derived from a lung tumor arising in transgenic mice harboring the SV40 large T antigen under transcriptional control of the promoter/enhancer region from the human surfactant protein C (SP-C) gene (WIKENHEISER et al. 1993). Likewise, the mtCC1–2 line was derived from Clara cells from transgenic mice expressing SV40 T antigen under the transcriptional control of the CC10 promoter/enhancer (MAGDALENO et al. 1997). Both cell lines have been useful in studying expression of cellular lung epithelial-specific proteins such as the surfactant proteins (SP-A, -B and -C) and of the Clara cell-specific CC10 (BACHURSKI et al. 1997; BINGLE et al. 1995; BOHINSKI et al. 1994; BRAUN and SUSKE 1998; BRUNO et al. 1995; CLEVIDENCE et al. 1994; MARGANA and BOGGARAM 1997). We used the MLE-15 and the mtCC1–2 cell lines as tools to understand the transcriptional regulation of the JSRV LTR.

We performed luciferase reporter assays of the activity of the JSRV LTR in lung epithelial cell lines vs. non-lung cell lines, including those of testicular carcinoma, mammary carcinoma, mouse kidney cells, myoblasts and lymphoid origins. The relative activity of a reporter plasmid driven by the

JSRV LTR (pJS21-luc) was higher in MLE-15 and mtCC1-2 compared to any other mouse cell line tested ($n = 13$) (PALMARINI et al. 2000a). This suggested that the JSRV LTR is preferentially expressed in differentiated lung epithelial cells. Furthermore, a series of overlapping deletions in the U3 region of pJS21-luc, indicated the presence of transcriptional enhancers active in MLE-15 cells (and at a lower level in mtCC1-2). Analysis of the same set of deletions in other mouse cell lines where the JSRV LTR has a relatively low activity (e.g., NIH-3T3) indicated that the JSRV LTR enhancers are not active in those cells. In MLE-15 cells, approximately half of the enhancer activity was associated with elements between positions -51 and -240, while mutation or deletion of distal elements between -240 and -267 also substantially reduced activity. An internal deletion of the JSRV LTR of the central distal elements (pJS21 [Δ-209/-166]) also eliminated approximately half of the transcriptional activity in MLE-15 cells. This supported the importance of the central distal elements of the JSRV LTR for enhancer activity in MLE-15 cells. In summary, central distal and upstream distal elements both contribute to the enhancer activity of the JSRV LTR in MLE-15 cells (and presumably primary type II pneumocytes).

Elements of the JSRV LTR located both upstream and downstream from the TATA box are important for optimal function and cell-specificity of the JSRV LTR. For instance, the JSRV enhancers are able to activate heterologous promoters (from SV40 and M-MLV) in MLE-15 cells, but optimal activation was achieved only with the homologous JSRV promoter (PALMARINI et al. 2000a). Moreover, deletion of the whole U5 or a portion of it reduced the activity of the JSRV LTR in MLE-15 cell reporter assays, suggesting that sequences downstream of the transcriptional start site may also influence JSRV transcription. Further studies are necessary to establish the role of R and U5 in the JSRV LTR expression. The R region has been shown to be important for transcription for human and primate lentiviruses (RABSON and GRAVES 1997) and for other simple retroviruses such as MuLV, MMTV, bovine leukemia viruses and chicken reticuloendotheliosis virus (CUPELLI et al. 1998; KISS-TOTH and UNK 1994; PIERCE et al. 1993; RIDGWAY et al. 1989).

The $JSRV_{21}$ LTR did show particularly high activity in other type II pneumocyte/Clara cell-derived cell lines such as the human H441, H358 and A549 or the JS-7 cell line (derived from the tumor cells of a sheep with OPA) (JASSIM 1988). To a large degree, this can be explained by the fact that these cell lines have lost some of the characteristics of differentiated lung epithelial cells. For instance, Clara cell-derived H441 cells express only SP-A but not the other pulmonary surfactant proteins, or CC10 or HNF-3β (BINGLE et al. 1995). These latter proteins are all expressed in normal Clara cells. However, A441 cells maintain the ability to express reporter genes driven by the CC10 promoter/enhancer. Similarly, H358 also expresses only SP-A, and A549 expresses neither SP-A nor SP-B. No data are available for the surfactant production of the JS-7 cell line but they have lost the morphological phenotype of type II pneumocytes after passage in vitro (JASSIM 1988). It is noteworthy that the lack of JSRV LTR activity in different lung epithelial cell lines correlates with the absence expression of the endogenous SP-B gene. This in turn suggests that the tissue-specific transcriptional activity of the JSRV LTR might be influenced by transcription factors that specifically activate the SP-B promoter. It should be noted that the LTR deletion studies implicate central distal and upstream sequences for efficient enhancer function in MLE-15 cells, but these experiments have not distinguished between motifs/factors important for efficient LTR expression in MLE-15 cells, vs. motifs/factors that are important for the tissue specific expression in these cells. For instance, while studies have indicated the importance of the JSRV R and U5 regions for LTR expression in MLE-15 cells, they are not by themselves capable of conferring tissue-specificity. A chimeric construct containing just the JSRV promoter proximal elements, R and U5 was activated by SV40 enhancers for expression in TCMK cells, where the native JSRV LTR is poorly active.

5.2
HNF-3β Activates the JSRV LTR In Vitro

Since the JSRV LTR is specifically active in differentiated lung epithelial cells we surveyed the JSRV U3 region for potential binding sites for transcription factors known to be important in lung-specific gene expression. It was noteworthy that the JSRV LTR contains two putative binding sites for HNF-3, 'upstream' at position −147 to −128 and 'downstream' at position −48 to −29 (Fig. 4). HNF-3, a transcription factor of the hepatocyte nuclear factor −3/forkhead homology protein family, is expressed in liver and lungs and is a critical factor in development of these tissues (COSTA et al. 2001). The two isoforms, HNF-3α and HNF-3β, are highly expressed in both type II pneumocytes and Clara cells (CLEVIDENCE et al. 1994; KAESTNER et al. 1994) where they are important in the expression of SP-B, CC-10 and other lung-specific genes (BINGLE et al. 1995; BOHINSKI et al.

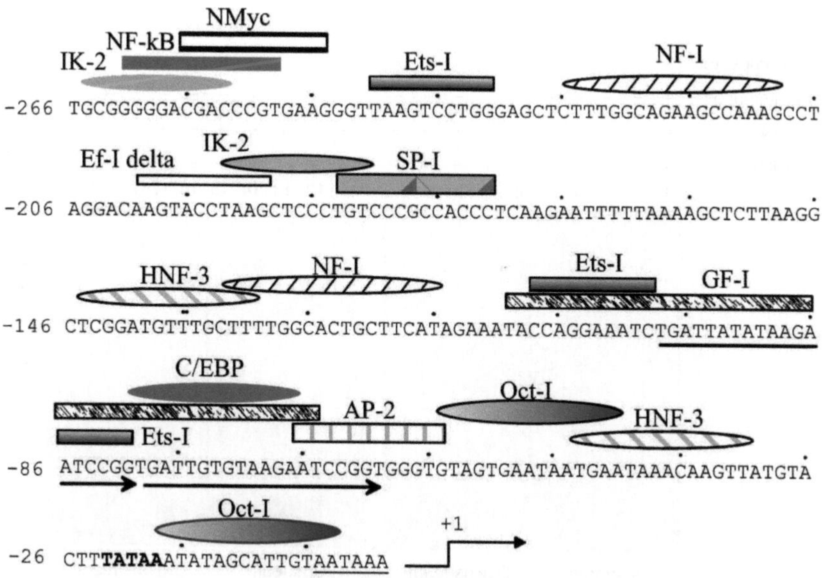

Fig. 4. Potential transcriptional factor binding sites in the JSRV LTR. The U3 sequences in the JSRV LTR were analyzed for potential factor binding sites by the program MatInspector v2.2 (Genomatix) (Quandt et al. 1995). The sites with the best matches to consensus sequences are shown. Direct repeats are *underlined by arrows*, the TATA box is in *bold* and the polyA site is *underlined*

1994; Braun and Suske 1998; Bruno et al. 1995; Clevidence et al. 1994; Margana and Boggaram 1997; Sawaya et al. 1993).

In co-transfection experiments in NIH-3T3 cells that normally show low enhancer activity for the JSRV LTR, the JSRV LTR could be activated in a dose-dependent manner by HNF-3α and HNF-3β expression plasmids. This indicated that (as previously reported), NIH-3T3 cells do not express either HNF-3 isoform, and it was consistent with a role for these factors in JSRV expression in lung epithelial cells (Palmarini et al. 2000a).

Mutation or deletion of the upstream HNF-3 binding site in the JSRV LTR resulted in a four- to fivefold reduction in the luciferase activity in MLE-15 cells (McGee-Estrada et al. 2001). In contrast, mutation or deletion of the downstream HNF-3 binding site did not affect the JSRV LTR activity in these cells. Electrophoretic mobility shift assays (EMSAs) with radiolabeled double-stranded oligonucleotides corresponding again to the upstream or downstream HNF-3 binding site combined with antibody super-shifts indicated that: (1) HNF-3β but not HNF-3α is important for

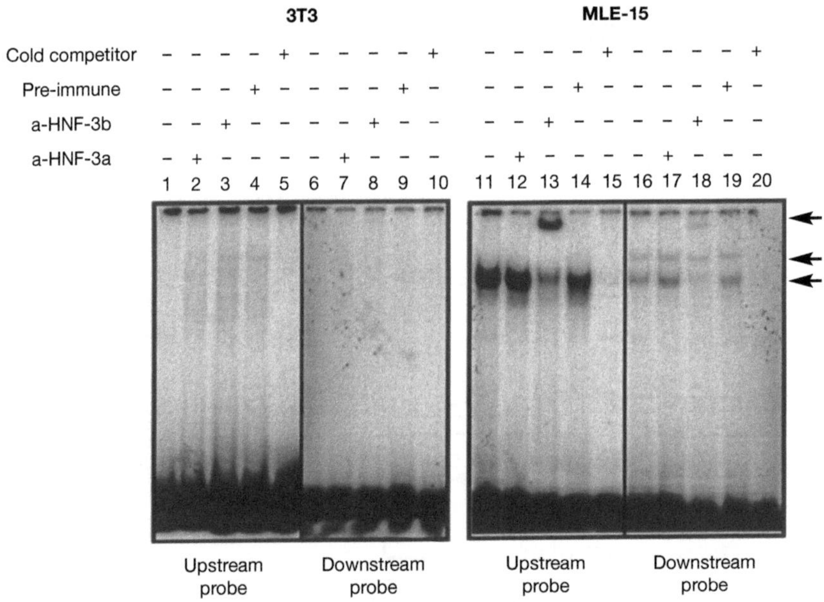

Fig. 5. Detection of cellular factors binding to the JSRV HNF-3 consensus elements. Nuclear extracts from NIH3T3 (*left*) or MLE-15 (*right*) cells were incubated with ^{32}P-labeled oligonucleotides encompassing the upstream or downstream HNF-3 binding sites. Some reactions were preincubated with rabbit anti-mouse HNF-3α (*lanes 2, 7, 12,* and *17*), HNF-3β (*lanes 3, 8, 13,* and *18*), or preimmune (*lanes 4, 9, 14,* and *19*) serum. Competition with excess unlabeled homologous oligonucleotide (*lanes 5, 10, 15,* and *20*) confirmed specific protein–DNA interactions. (From McGee-Estrada et al. 2001, with permission from the American Society for Virology)

JSRV LTR activity in MLE-15 cells; and (2) that the affinity of the upstream binding site for HNF-3β is higher than for the downstream site (Fig. 5) (McGee-Estrada et al.).

Interestingly, when analogous experiments on the HNF-3 site mutant LTRs were performed in NIH-3T3 cells co-transfected with an HNF-3β expression plasmid, different results were obtained. In the co-transfected NIH3T3 cells, mutation or deletion of the upstream HNF-3 site had no effect on JSRV LTR activity, while mutation or deletion of the downstream site substantially inhibited activity. Nevertheless, as for MLE-15 cells, the affinity of the upstream HNF-3 site for HNF-3 was higher than for the downstream site.

One possible explanation for the differential importance of the upstream and downstream HNF-3 sites in MLE-15 vs. NIH3T3 cells could be that HNF-3 might act cooperatively with other transcription factors in activating LTR transcription. HNF-3 has been found to cooperate with other transcription factors such as thyroid transcription factor-1 and nuclear factor (NF)-1 in transcription of cellular genes (ARENZANA and RODRIGUEZ DE CORDOBA 1996; BOHINSKI et al. 1994; NORD et al. 2000; TSAY et al. 1997). Interestingly in the JSRV LTR, the upstream HNF-3 site is immediately adjacent to an NF-1 site. It is possible that the simultaneous binding of NF-1 and HNF-3 in MLE-15 cells leads to efficient activation of transcription but this has still to be addressed experimentally. It seems possible that MLE-15 cells (and by analogy normal type II pneumocytes) contain an NF-1 factor that can cooperate with HNF-3β, while NIH3T3 cells do not. Numerous NF-1 isoforms have been described, some of which show differential expression in different tissues (APT et al. 1993; GRAVES et al. 1991; MINK et al. 1992). Conversely, NIH3T3 cells might express factors that can cooperate with HNF3 bound to the downstream site in the JSRV LTR, while MLE-15 cells might not. Investigation of other cellular factors that bind to the JSRV LTR is in progress.

It was also noteworthy that when the HNF-3 site mutants were tested in the mtCC1–2 cells, neither the upstream nor the downstream sites appeared to be important for expression (MCGEE-ESTRADA et al.). This suggests that other transcription factors and LTR motifs may be important for JSRV expression in Clara cells.

In summary, HNF-3β is important for expression of the JSRV LTR in MLE-15 cells, and presumably primary type II pneumocytes. The upstream HNF-3 site is the target of HNF-3β binding, and it is likely that this interaction reflects the importance of the central–distal elements identified in the LTR deletional analysis. On the other hand, activation of the JSRV in Clara cells apparently involves binding of other transcription factors in the central–distal region.

5.3
An NFκB-Like Binding Site Is Important for Expression of the JSRV LTR

As described above, deletion mutants of the JSRV LTR indicated that approximately one-half of the JSRV LTR enhancer activity in MLE-15 and mtCC1–2 cells could be attributed to an element(s) in the distal region (–239 to –266) (PALMARINI et al. 2000a). This region contains an NFκB-like binding site with one mismatch (5′-GGGACGACC-3′) from the canonical NFκB consensus binding sequence (5′-GGGPuNNPyPyCC-3′).

Mutation of the NFκB-like site from the JSRV LTR reduced transcriptional activity in MLE-15 and mtCC1–2 cells but it did not affect the level of expression in the other murine cell lines. Thus, these data supported the idea that the NFκB-like element is important for the high-level expression of the JSRV LTR in lung epithelial-derived cells, but that it is not important for the low-level expression in non-lung epithelial cells. We tested for the presence of nuclear factors that could bind to this sequence by EMSA and found that several complexes of different mobility could be detected in extracts from both the MLE-15 cells and from cells that do not support high-level expression of the JSRV LTR (e.g., 3T3 and TCMK). Thus the factor or factors that bind to the JSRV NFκB-like sequences may be ubiquitously expressed and are not expressed specifically in MLE-15 cells.

We used antibodies to the p50 and p52 members of the NFκB protein complex in attempts to supershift or inhibit complex formation in nuclear extracts from MLE 15 cells. However neither antibody showed inhibition or a supershift of any of the complexes. Thus the protein(s) that bind to the NFκB-like site in the JSRV LTR may be previously unidentified NFκB-like proteins, or unrelated factors. It should be noted that the NFκB-like site also overlaps with an Ik-2-like binding site for Ikaros-related proteins. The *Ikaros* gene is expressed only in hematopoietic cells (Molnar and Georgopoulos 1994) and there have been no reports of expression in differentiated lung cells.

5.4
JSRV mRNA Splicing Pattern

As discussed in Sect. 1, retroviral proviruses contain a single transcriptional unit that is transcribed in the nucleus to produce full-length genomic RNA. The viral RNA is then transported to the cytoplasm with and without mRNA splicing to yield the viral mRNAs. All retroviruses use full-length genomic RNA as mRNA for the Gag, Pro and Pol products. A singly spliced mRNA consisting of the 5′ end of the genome, the *env* gene, is used for translation of the envelope proteins (Hayward 1977; Weiss et al. 1977). Retroviruses that use only these two mRNAs are generally termed ‘simple’ retroviruses. In contrast, ‘complex’ retroviruses additionally encode other subgenomic spliced mRNAs (singly or doubly spliced) for regulatory proteins such as the Tat and Tax transactivators of human immunodeficiency virus and human T cell lymphotropic virus, respectively (Rabson and Graves 1997). All subgenomic mRNAs (*env* included) are made by splicing a leader sequence from the 5′ end of viral RNA to an acceptor site(s) within the viral genome. All spliced viral mRNAs typically result from the same 5′ splice donor site; as a result the subgenomic mRNAs share 5′ and 3′ ends.

In contrast to full-length viral mRNA which is translated on polyribosomes that are free in the cytoplasm, the subgenomic *env* mRNA is translated on polyribosomes associated with the rough endoplasmic reticulum. Cleavage of the Env polyprotein takes place in the Golgi apparatus, yielding the SU and TM proteins. As described above, these proteins remain associated with each other at the surface of the infected cell, and on the surface of virions.

The mRNAs encoded by JSRV were investigated in a recent study (Palmarini et al. 2002). Initially, RNA was obtained from 293T cells transiently transfected with pCMV2JS21, as these cells express the transfected JSRV genome at high levels and they produce infectious virus (see above). This facilitated initial Northern blot analyses. Subsequently, a productively infected deer lung cell line infected was studied: because deer cells lack endogenous JSRV-related proviruses this facilitated analysis of exogenous JSRV transcripts in a productively infected cell line. Finally, tumor tissues from OPA-affected sheep were also analyzed (Palmarini et al. 2002). Figure 6 show the major JSRV mRNA species present in productively infected and transfected cells. These were deduced from a combination of Northern blot hybridization and reverse transcription–PCR cloning followed by sequencing. Major viral transcripts in JSRV-expressing cells included the 7.5 kb full-length genome and a singly spliced mRNA of 2.4 kb. The splice donor in $JSRV_{21}$ is located at position 193 upstream of *gag* while the splice acceptor is located in *env* at position 5347. However, other additional spliced subgenomic viral RNAs were also observed in pCMV2JS21-transfected cells and some of these mRNAs were also observed in JSRV-infected cells and in OPA tumors, suggesting that they are biologically relevant.

One of the most interesting transcripts was a 3.2-kb mRNA detected in 293T cells transfected with pCMV2JS21, and also in OPA tumors. This 3.2-kb RNA (actually two related RNAs) corresponds to singly spliced transcripts employing the standard splice donor at position 193, and two alternative splice acceptors immediately upstream of *orf-x* at position 4,471 and 4,369. Ribosome scanning from the 5′ end of these mRNAs would in both cases result in translation of the *orf-x* reading frame, yielding a putative protein of 20.9 kDa. The existence of the 3.2 kb mRNA is the first indication that the *orf-x* reading frame is actually expressed.

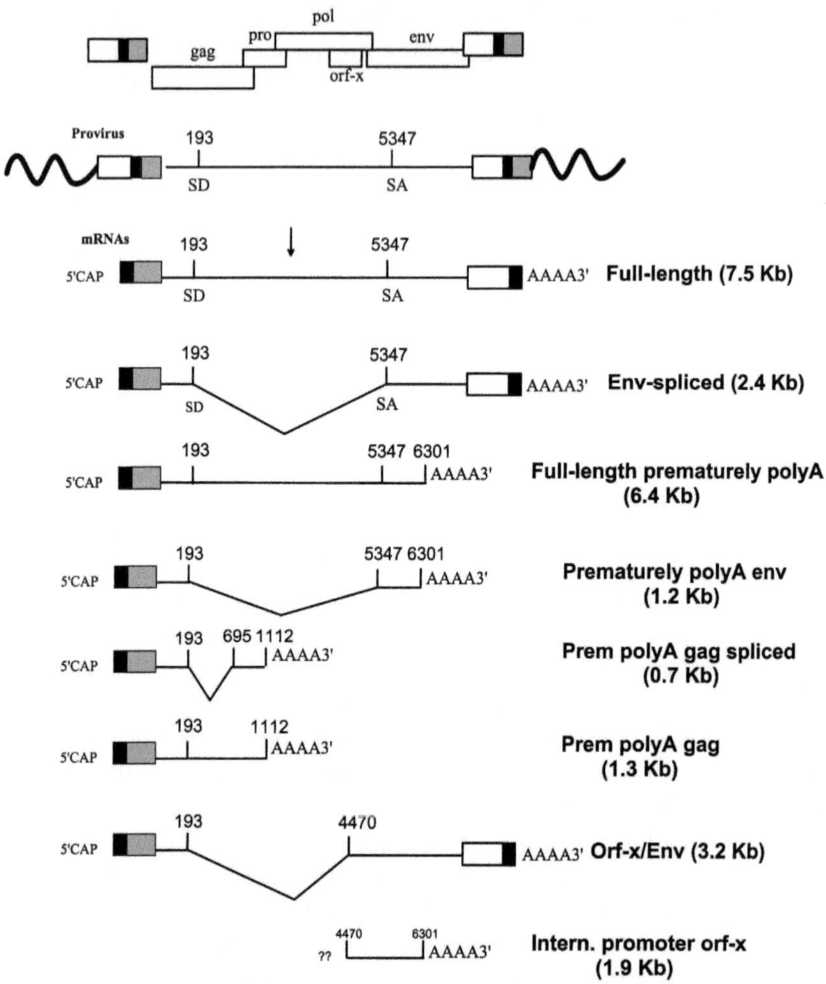

Fig. 6. Major JSRV mRNA species. The organization of the major JSRV RNA species is shown. The position of the splice donor, splice acceptor and premature termination are indicated. The molecular weights of the transcripts are indicated *in parentheses*. For more details, please refer to the text

Another interesting phenomenon was observed in pCMV2JS21-transfected cells: two internal non-canonical polyadenylation sites in *gag* and *env* result in the presence of truncated transcripts. A non-canonical polyA site in *env* (ATTAAA, position 6,283–6,288) results in prematurely polyadenylated transcripts corresponding to singly spliced *env* (1.2 kb, encoding only SU) (MAEDA et al. 2001) and unspliced viral RNA (6.4 kb). While the relevance of these truncated RNAs is still unclear, at least the prematurely polyadenylated RNA of 1.2 kb is also present in cells infected by JSRV in vitro and possibly in tumor tissues. One additional transcript truncated in *env* was also detected as an *orf-x*/*env* transcript (1.9 kb). This latter transcript may be derived from an internal promoter as it did not hybridize with a probe for the 5′ end of JSRV RNA.

In addition to the transcripts truncated in *env*, two prematurely polyadenylated *gag* transcripts in transfected 293T cells were also detected, resulting from a non-canonical polyA signal (ATTAAA) at position 1,091. One transcript is spliced and uses the splice donor at position 193 and the splice acceptor at position 694. We did not detect prematurely polyadenylated *gag* transcripts in JSRV infected deer cells or in tumor tissues.

While the biological significance of the prematurely polyadenylated viral RNAs is not yet known, they might be relevant for the design of JSRV-based retroviral vectors. The polyadenylation sites in *gag* and *env* seem to be highly functional in transfected cells and they might thus decrease the titer of JSRV-based retroviral vectors that are generated by transient transfection of cells.

6 Conclusions

Major advances have been made recently in understanding the molecular biology of JSRV. Most notably, important insights into the molecular basis for the viral tropism and for the mechanisms of virus-induced transformation have emerged (see the chapters by FAN et al. and MILLER, this volume). The importance of the LTR for the virus-specific expression in the differentiated epithelial lung cells and of the viral envelope in oncogenesis has been deduced by experiments carried out entirely in vitro. A major goal of future research will be to test the relevance of results from experiments in vitro to infection and oncogenicity in vivo.

Another point of further investigation is the role of *orf-x* in JSRV biology and OPA pathogenesis. The *orf-x* reading frame is conserved among several JSRV isolates, as well as some endogenous proviruses. As described above, a spliced *orf-x* transcript has been detected, supporting the notion that this ORF is indeed expressed. There are currently no clues as to the possible function of this putative gene, except that it is not necessary for particle formation or transformation in vitro.

Infectious JSRV particles can now be readily obtained in vitro by transient transfection of 293T cells. As illustrated throughout the article this was a major milestone for JSRV research. However, no tissue culture system for high level propagation of JSRV is available yet, which has hindered quantification of viral infectivity. Current research in type II pneumocyte and Clara cell biology might facilitate new attempts to establish a highly efficient JSRV cell culture system.

Finally, the unique properties of JSRV make this virus an ideal candidate for developing viral vectors for gene therapy of lung diseases.

Acknowledgements. We are grateful to David Posada (Brigham Young University) for the phylogenetic tree of the Retroviridae. We thank our coworkers, Naoyoshi Maeda, Kathy McGee Estrada, Andy Hofacre at UCI and Claudio Murgia, Alberto Alberti, Manuela Mura and Elena Grego at UGA for their support and enthusiasm in this area. Funding was provided by the Georgia Cancer Coalition and Grant CA95706-01 from the National Institutes of Health.

Abbreviations

CA	major capsid protein
CMV	cytomegalovirus
dUTPase	deoxyuridine triphosphatase
ENV	envelope
PI3-K	phosphatidylinositol 3-kinase
HNF-3	hepatocyte nuclear factor
IN	integrase
JSRV	Jaagsiekte sheep retrovirus
LTR	long terminal repeat
MA	matrix
MMTV	mouse mammary tumor virus
MPMV	Mason-Pfizer monkey virus
MuLV	murine leukemia virus
NC	nucleocapsid
OPA	ovine pulmonary adenocarcinoma
PERT	product enhancement reverse transcriptase
PR	protease
RT	reverse transcriptase
SU	surface
TM	transmembrane

References

Apt D, Chong T, Liu Y, Bernard HU (1993) Nuclear factor I and epithelial cell-specific transcription of human papillomavirus type 16. J Virol 67(8):4455–63

Arenzana N, Rodriguez de Cordoba S (1996) Promoter region of the human gene coding for beta-chain of C4b binding protein. Hepatocyte nuclear factor-3 and nuclear factor-I/CTF transcription factors are required for efficient expression of C4BPB in HepG2 cells. J Immunol 156(1):68–75

Athas GB, Starkey CR, Levy LS (1994) Retroviral determinants of leukemogenesis. Crit Rev Oncog 5(2–3):169–99

Auchampach JA, Jin X, Wan TC, Caughey GH, Linden J (1997) Canine mast cell adenosine receptors: cloning and expression of the A3 receptor and evidence that degranulation is mediated by the A2B receptor. Mol Pharmacol 52(5):846–60

Bachurski CJ, Kelly SE, Glasser SW, Currier TA (1997) Nuclear factor I family members regulate the transcription of surfactant protein-C. J Biol Chem 272(52): 32759–66

Bai J, Bishop JV, Carlson JO, DeMartini JC (1999) Sequence comparison of JSRV with endogenous proviruses: envelope genotypes and a novel ORF with similarity to a G-protein-coupled receptor. Virology 258(2):333–43

Bai J, Zhu RY, Stedman K, Cousens C, Carlson J, Sharp JM, DeMartini JC (1996) Unique long terminal repeat U3 sequences distinguish exogenous jaagsiekte sheep retroviruses associated with ovine pulmonary carcinoma from endogenous loci in the sheep genome. J Virol 70(5):3159–68

Baltimore D (1970) Viral RNA-dependent DNA polymerase. London 226:1209–1211

Beemon K, Duesberg P, Vogt P (1974) Evidence for crossing-over between avian tumor viruses based on analysis of viral RNAs. Proc Natl Acad Sci USA 71(10):4254–8

Bender W, Chien YH, Chattopadhyay S, Vogt PK, Gardner MB, Davidson N (1978) High-molecular-weight RNAs of AKR, NZB, and wild mouse viruses and avian reticuloendotheliosis virus all have similar dimer structures. J Virol 25(3): 888–96

Bender W, Davidson N (1976) Mapping of poly(A) sequences in the electron microscope reveals unusual structure of type C oncornavirus RNA molecules. Cell 7(4): 595–607

Berkowitz R, Fisher J, Goff SP (1996) RNA packaging. Curr Top Microbiol Immunol 214:177–218

Billeter MA, Parsons JT, Coffin JM (1974) The nucleotide sequence complexity of avian tumor virus RNA. Proc Natl Acad Sci USA 71(9):3560–4

Bingle CD, Hackett BP, Moxley M, Longmore W, Gitlin JD (1995) Role of hepatocyte nuclear factor-3 alpha and hepatocyte nuclear factor-3 beta in Clara cell secretory protein gene expression in the bronchiolar epithelium. Biochem J 308(Pt 1) 197–202

Bittner JJ (1936) Some possible effects of nursing on the mammary gland tumour incidence in mice. Science 84:162

Bohinski RJ, Di Lauro R, Whitsett JA (1994) The lung-specific surfactant protein B gene promoter is a target for thyroid transcription factor 1 and hepatocyte nuclear factor 3, indicating common factors for organ-specific gene expression along the foregut axis. Mol Cell Biol 14(9):5671–81

Bradac J, Hunter E (1984) Polypeptides of Mason-Pfizer monkey virus. I. Synthesis and processing of the gag-gene products. Virology 138(2):260–75

Bradac JA, Hunter E (1986) Polypeptides of Mason-Pfizer monkey virus. III. Translational order of proteins on the gag and env gene specified precursor polypeptides. Virology 150(2):503–8

Braun H, Suske G (1998) Combinatorial action of HNF3 and Sp family transcription factors in the activation of the rabbit uteroglobin/CC10 promoter. J Biol Chem 273(16):9821–8

Brown PO (1990) Integration of retroviral DNA. Curr Top Microbiol Immunol 157: 19–48

Brown PO (1997) Integration. In: Retroviruses (JM Coffin, SH Hughes and HE Varmus, Eds.), pp 161–203. Cold Spring Harbor Laboratory Press, New York

Bruno MD, Bohinski RJ, Huelsman KM, Whitsett JA, Korfhagen TR (1995) Lung cell-specific expression of the murine surfactant protein A (SP-A) gene is mediated by interactions between the SP-A promoter and thyroid transcription factor-1 [published erratum appears in J Biol Chem 1995 Jul 7; 270(27):16482]. J Biol Chem 270(12):6531–6

Bucciarelli E (1973) [Pulmonary adenomatosis of the sheep (Jaagsiekte). Ultrastructural study]. Lav Ist Anat Istol Patol Perugia 33(3):99–117

Burns JC, Friedmann T, Driever W, Burrascano M, Yee JK (1993) Vesicular stomatitis virus G glycoprotein pseudotyped retroviral vectors: concentration to very high titer and efficient gene transfer into mammalian and nonmammalian cells. Proc Natl Acad Sci USA 90(17):8033–7

Clevidence DE, Overdier DG, Peterson RS, Porcella A, Ye H, Paulson KE, Costa RH (1994) Members of the HNF-3/forkhead family of transcription factors exhibit distinct cellular expression patterns in lung and regulate the surfactant protein B promoter. Dev Biol 166(1):195–209

Coffin JM (1992) Structure and classification of retroviruses. In: The Retroviridae (JA Levy, Ed.), Vol 1, pp 19–49. Plenum Press, New York

Coil DA, Strickler JH, Rai SK, Miller AD (2001) Jaagsiekte Sheep Retrovirus Env Protein Stabilizes Retrovirus Vectors against Inactivation by Lung Surfactant, Centrifugation, and Freeze- Thaw Cycling. J Virol 75(18):8864–7

Costa RH, Kalinichenko VV, Lim L (2001) Transcription factors in mouse lung development and function. Am J Physiol Lung Cell Mol Physiol 280(5):L823–38

Cupelli L, Okenquist SA, Trubetskoy A, Lenz J (1998) The secondary structure of the R region of a murine leukemia virus is important for stimulation of long terminal repeat-driven gene expression. J Virol 72(10):7807–14

DeMartini JC, Bishop JV, Allen TE, Jassim FA, Sharp JM, de Las Heras M, Voelker DR, Carlson JO (2001) Jaagsiekte Sheep Retrovirus Proviral Clone JSRV(JS7), Derived from the JS7 Lung Tumor Cell Line, Induces Ovine Pulmonary Carcinoma and Is Integrated into the Surfactant Protein A Gene. J Virol 75(9):4239–46

Dhar R, McClements WL, Enquist LW, Vande Woude GF (1980) Nucleotide sequences of integrated Moloney sarcoma provirus long terminal repeats and their host and viral junctions. Proc Natl Acad Sci USA 77(7):3937–41

Dougherty WG, Semler BL (1993) Expression of virus-encoded proteinases: functional and structural similarities with cellular enzymes. Microbiol Rev 57(4):781–822

DUESBERG PH, WANG LH, BEEMON K, KAWAI S, HANAFUSA H (1976) Sequences and functions of Rous sarcoma virus RNA. Hamatol Bluttransfus 19:327–40

ELDER JH, LERNER DL, HASSELKUS-LIGHT CS, FONTENOT DJ, HUNTER E, LUCIW PA, MONTELARO RC, PHILLIPS TR (1992) Distinct subsets of retroviruses encode dUTPase. J Virol 66(3):1791–4

ELLERMAN V, BANG O (1908) Experimentelle leukamie bei Huhnern. Fizentralblatt Bakteriologie. Zentralblatt der bakteriologie 46:595–609

FAN H (1990) Influences of the long terminal repeats on retrovirus pathogenicity. Seminars in Virology 1:165–174

FAN H (1994) Retroviruses and their role in cancer. In: The Retroviridae (JA Levy, Ed.), Vol 3, pp 313–362. 4 vols. Plenum Press, New York

FELSENSTEIN J (1985) Confidence limits on phylogenies: an approach using the bootstrap. Evolution 39:783–791

GONZALEZ L, GARCIA-GOTI M, COUSENS C, DEWAR P, CORTABARRIA N, EXTRAMIANA AB, ORTIN A, DE LAS HERAS M, SHARP JM (2001) Jaagsiekte sheep retrovirus can be detected in the peripheral blood during the pre-clinical period of sheep pulmonary adenomatosis. J Gen Virol 82(Pt 6):1355–8

GOUBIN G, GOLDMAN DS, LUCE J, NEIMAN PE, COOPER GM (1983) Molecular cloning and nucleotide sequence of a transforming gene detected by transfection of chicken B-cell lymphoma DNA. Nature 302(5904):114–9

GRAVES RA, TONTONOZ P, ROSS SR, SPIEGELMAN BM (1991) Identification of a potent adipocyte-specific enhancer: involvement of an NF-1-like factor. Genes Dev 5(3): 428–37

GREENBLATT J (1991) Roles of TFIID in transcriptional initiation by RNA polymerase II. Cell 66(6):1067–70

HANAFUSA H, HALPERN CC, BUCHHAGEN DL, KAWAI S (1977) Recovery of avian sarcoma virus from tumors induced by transformation- defective mutants. J Exp Med 146(6):1735–47

HAYWARD WS (1977) Size and genetic content of viral RNAs in avian oncovirus-infected cells. J Virol 24(1):47–63

HECHT SJ, CARLSON JO, DEMARTINI JC (1994) Analysis of a type D retroviral capsid gene expressed in ovine pulmonary carcinoma and present in both affected and uneffected sheep genomes. Virology 202:480–484

HECHT SJ, STEDMAN KE, CARLSON JO, DEMARTINI JC (1996) Distribution of endogenous type B and type D sheep retrovirus sequences in ungulates and other mammals. Proc Natl Acad Sci USA 93(8):3297–302

HERRING AJ, SHARP JM, SCOTT FM, ANGUS KW (1983) Further evidence for a retrovirus as the aetiological agent of sheep pulmonary adenomatosis (jaagsiekte). Vet Microbiol 8(3):237–49

HIZI A, HENDERSON LE, COPELAND TD, SOWDER RC, KRUTZSCH HC, OROSZLAN S (1989) Analysis of gag proteins from mouse mammary tumor virus. J Virol 63(6): 2543–9

HOD I, HERZ A, ZIMBER A (1976) Pulmonary carcinoma (jaagsiekte) of sheep. American Journal of Pathology 86:545–558

HOLLAND MJ, PALMARINI M, GARCIA-GOTI M, GONZALEZ L, MCKENDRICK I, DE LAS HERAS M, SHARP JM (1999) Jaagsiekte retrovirus is widely distributed in both T and B lymphocytes and mononuclear phagocytes of sheep with naturally and experimentally acquired pulmonary adenomatosis. Journal of Virology 73:4004–4008

Hong S, Choi G, Park S, Chung AS, Hunter E, Rhee SS (2001) Type D retrovirus Gag polyprotein interacts with the cytosolic chaperonin TRiC. J Virol 75(6):2526–34

Hughes SH, Mutschler A, Bishop JM, Varmus HE (1981) A Rous sarcoma virus provirus is flanked by short direct repeats of a cellular DNA sequence present in only one copy prior to integration. Proc Natl Acad Sci USA 78(7):4299–303

Hughes SH, Shank PR, Spector DH, Kung HJ, Bishop JM, Varmus HE, Vogt PK, Breitman ML (1978) Proviruses of avian sarcoma virus are terminally redundant, co- extensive with unintegrated linear DNA and integrated at many sites. Cell 15(4): 1397–410

Hunter E, Casey J, Hahn B, hayami M, Korber B, Kurth R, Neil J, Rethwilm A, Sonigo P, Stoye J (2000) The Retroviridae. In: Virus Taxonomy. Seven report of the international committee on taxonomy of viruses. (MHV van Regenmortel, CM Fauquet, DHL Bishop, EB Carstens, MK Estes, SM Lemon, J Maniloff, MA Mayo, DJ McGeoch, CR Pringle and RB Wickener, eds), pp 369–387. Academic Press, San Diego

Jassim FA (1988) PhD Thesis. University of Edinburgh, Edinburgh, UK

Kaestner KH, Hiemisch H, Luckow B, Schutz G (1994) The HNF-3 gene family of transcription factors in mice: gene structure, cDNA sequence, and mRNA distribution. Genomics 20(3):377–85

Kajikawa O, Dahlberg JE, Rosadio RH, De Martini JC (1990) Detection and quantitation of a type D retrovirus gag protein in ovine pulmonary carcinoma (sheep pulmonary adenomatosis) by means of a competition radioimmunoassay. Vet Microbiol 25(1):17–28

Kiss-Toth E, Unk I (1994) A downstream regulatory element activates the bovine leukemia virus promoter. Biochem Biophys Res Commun 202(3):1553–61

Koppe B, Menendez-Arias L, Oroszlan S (1994) Expression and purification of the mouse mammary tumor virus gag-pro transframe protein p30 and characterization of its dUTPase activity. J Virol 68(4):2313–9

Kung HJ, Bailey JM, Davidson N, Vogt PK, Nicolson MO, McAllister RM (1975) Electron microscope studies of tumor virus RNA. Cold Spring Harb Symp Quant Biol 39(Pt 2):827–34

Laboratory S. o. t. RBJM (1933) Theexistence of non-chromosomal influence of mammary tumors in mice. Science 78:465–466

Lepperdinger G, Strobl B, Kreil G (1998) HYAL2, a human gene expressed in many cells, encodes a lysosomal hyaluronidase with a novel type of specificity. J Biol Chem 273(35):22466–70

Lewin B (1990) Commitment and activation at pol II promoters: a tail of protein- protein interactions. Cell 61(7):1161–4

Maeda N, Palmarini M, Murgia C, Fan H (2001) Direct transformation of rodent fibroblasts by jaagsiekte sheep retrovirus DNA. Proc Natl Acad Sci USA 98(8): 4449–4454

Magdaleno SM, Wang G, Jackson KJ, Ray MK, Welty S, Costa RH, DeMayo FJ (1997) Interferon-gamma regulation of Clara cell gene expression: in vivo and in vitro. Am J Physiol 272(6 Pt 1):L1142–51

Malkinson AM, Dwyer-Nield LD, Rice PL, Dinsdale D (1997) Mouse lung epithelial cell lines – tools for the study of differentiation and the neoplastic phenotype. Toxicology 123:53–100

Margana RK, Boggaram V (1997) Functional analysis of surfactant protein B (SP-B) promoter. Sp1, Sp3, TTF-1, and HNF-3alpha transcription factors are necessary for lung cell- specific activation of SP-B gene transcription. J Biol Chem 272(5): 3083–90.

Mason RJ, Shannon JM (1997) Alveolar type II pneumocytes. In: The Lung: Scientific Foundations (RG Crystal, JB West, ER Weibel and PJ Barnes, eds), Vol 1, pp 543–555. 2 vols. Lippincott-Raven, Philadelphia

McGee-Estrada K, Palmarini M, Fan H (2002) HNF-3β is a critical factor for the expression of the Jaagsiekte sheep retrovirus (JSRV) long terminal repeat in type II pneumocytes but not in clara cells. Virology 292:87–97

Mink S, Hartig E, Jennewein P, Doppler W, Cato AC (1992) A mammary cell-specific enhancer in mouse mammary tumor virus DNA is composed of multiple regulatory elements including binding sites for CTF/NFI and a novel transcription factor, mammary cell-activating factor. Mol Cell Biol 12(11):4906–18

Mizutani S, Boettiger D, Temin HM (1970) A DNA-depenent DNA polymerase and a DNA endonuclease in virions of Rous sarcoma virus. Nature 228(270):424–7

Molnar A, Georgopoulos K (1994) The Ikaros gene encodes a family of functionally diverse zinc finger DNA-binding proteins. Mol Cell Biol 14(12):8292–303

Nord M, Cassel TN, Braun H, Suske G (2000) Regulation of the Clara cell secretory protein/uteroglobin promoter in lung. Ann NY Acad Sci 923:154–65

Norval M, Head KW, Else RW, Hart H, Neill WA (1981) Growth in culture of adenocarcinoma cells from the small intestine of sheep. Br J Exp Pathol 62(3):270–82

Palmarini M, Cousens C, Dalziel RG, Bai J, Stedman K, DeMartini JC, Sharp JM (1996a) The exogenous form of Jaagsiekte retrovirus is specifically associated with a contagious lung cancer of sheep. J Virol 70(3):1618–23

Palmarini M, Datta S, Omid R, Murgia C, Fan H (2000a) The long terminal repeats of Jaagsiekte sheep retrovirus (JSRV) are preferentially active in type II pneumocytes. Journal of Virology 74:5776–5787

Palmarini M, Dewar P, De las Heras M, Inglis NF, Dalziel RG, Sharp JM (1995) Epithelial tumour cells in the lungs of sheep with pulmonary adenomatosis are major sites of replication for Jaagsiekte retrovirus. J Gen Virol 76(Pt 11):2731–7

Palmarini M, Fan H (2001) Retrovirus-induced ovine pulmonary adenocarcinoma, an animal model for lung cancer. Journal of the National Cancer Institute 93:1603–1614

Palmarini M, Gray CA, Carpenter K, Fan H, Bazer FW, Spencer T (2001a) Expression of Endogenous Betaretroviruses in the Ovine Uterus: Effectes of Neonatal Age, Estrous Cycle, Pregnancy and Progesterone. Journal of Virology 75: 11319–11327

Palmarini M, Hallwirth C, York D, Murgia C, de Oliveira T, Spencer T, Fan H (2000b) Molecular cloning and functional analysis of three type D endogenous retroviruses of sheep reveals a different cell tropism from that of the highly related exogenous jaagsiekte sheep retrovirus. Journal of Virology 74:8065–8076

Palmarini M, Holland MJ, Cousens C, Dalziel RG, Sharp JM (1996b) Jaagsiekte retrovirus establishes a disseminated infection of the lymphoid tissues of sheep affected by pulmonary adenomatosis. J Gen Virol 77(Pt 12):2991–8

Palmarini M, Maeda N, Murgia C, De-Fraja C, Hofacre A, Fan H (2001b) A phosphatidylinositol-3-kinase (PI-3 K) docking site in the cytoplasmic tail of the Jaagsiekte sheep retrovirus transmembrane protein is essential for envelope-induced transformation of NIH3T3 cells. Journal of Virology 75:1002–1009

Palmarini M, Murgia C, Fan H (2002) Spliced and prematurely polyadenylated Jaagsiekte sheep retrovirus (JSRV)-specific RNAs from infected or transfected cells. Virology 294:180–188

Palmarini M, Sharp JM, De las Heras M, Fan H (1999a) Jaagsiekte sheep retrovirus is necessary and sufficient to induce a contagious lung cancer in sheep. Journal of Virology 73:6964–6972

Palmarini M, Sharp JM, Lee C, Fan C (1999b) In vitro infection of ovine cell lines by jaagsiekte sheep retrovirus (JSRV). Journal of Virology 73:10070–10078

Perk K, Hod I (1971) Demonstration of virus particles in ovine pulmonary adenomata. Experientia 27(4):437–8

Perk K, Hod I, Nobel TA (1971) Pulmonary adenomatosis of sheep (jaagsiekte). I. Ultrastructure of the tumor. J Natl Cancer Inst 46(3):525–37

Pierce J, Fee BE, Toohey MG, Peterson DO (1993) A mouse mammary tumor virus promoter element near the transcription initiation site. J Virol 67(1):415–24

Plopper CG, Dallas MH, Buckpitt AR (1997) Clara cells. In: The Lung: Scientific Foundations (RG Crystal, JB West, ER Weibel and PJ Barnes, eds), Vol 1, pp 517–533. 2 vols. Lippincott-Raven, Philadelphia

Pyra H, Boni J, Schupbach J (1994) Ultrasensitive retrovirus detection by a reverse transcriptase assay based on product enhancement. Proc Natl Acad Sci USA 91(4):1544–8

Quade K, Smith RE, Nichols JL (1974) Evidence for common nucleotide sequences in the RNA subunits comprising Rous sarcoma virus 70 S RNA. Virology 61(1):287–91

Quandt K, Frech K, Karas H, Wingender E, Werner T (1995) MatInd and MatInspector: new fast and versatile tools for detection of consensus matches in nucleotide sequence data. Nucleic Acids Res 23(23):4878–84

Rabson AB, Graves BJ (1997) Synthesis and processing of viral RNA. In: Retroviruses (JM Coffin, SH Hughes and H Varmus, eds), pp 205–261. Cold Spring Harbor laboratory press, Cold Spring Harbor (NY)

Rai SK, DeMartini JC, Miller AD (2000) Retrovirus vectors bearing jaagsiekte sheep retrovirus Env transduce human cells by using a new receptor localized to chromosome 3p21.3. J Virol 74(10):4698–704

Rai SK, Duh FM, Vigdorovich V, Danilkovitch-Miagkova A, Lerman MI, Miller AD (2001) Candidate tumor suppressor HYAL2 is a glycosylphosphatidylinositol (GPI)-anchored cell-surface receptor for jaagsiekte sheep retrovirus, the envelope protein of which mediates oncogenic transformation. Proc Natl Acad Sci USA 98(8):4443–4448

Rhee SS, Hunter E (1987) Myristylation is required for intracellular transport but not for assembly of D-type retrovirus capsids. J Virol 61(4):1045–53

Rhee SS, Hunter E (1990) A single amino acid substitution within the matrix protein of a type D retrovirus converts its morphogenesis to that of a type C retrovirus. Cell 63(1):77–86

Ridgway AA, Kung HJ, Fujita DJ (1989) Transient expression analysis of the reticuloendotheliosis virus long terminal repeat element. Nucleic Acids Res 17(8):3199–215

RIFKIN D, COMPANS RW (1971) Identification of the spike proteins of Rous sarcoma virus. Virology 46(2):485–9

ROSATI S, PITTAU M, ALBERTI A, POZZI S, YORK DF, SHARP JM, PALMARINI M (2000) An accessory open reading frame (orf-x) of jaagsiekte sheep retrovirus is conserved between different virus isolates. Virus Research 66:109–116

ROSENBERG N, JOLICOEUR P (1997) Retroviral pathogenesis. In: Retroviruses (JM Coffin, S Hughes and HE Varmus, eds), pp 475585. 1 vols. Cold Spring Harbor Laboratory Press, New York

ROUS P (1911 a) A sarcoma of the fowl transmissible by an agent separable from the tumour cells. Journal of Experimental Medicine 13:397–411

ROUS P (1911 b) Transmission of a malignant new growth by means of a cell-free filtrate. Journal of the American Medical Association 56 198

SAWAYA PL, STRIPP BR, WHITSETT JA, LUSE DS (1993) The lung-specific CC10 gene is regulated by transcription factors from the AP-1, octamer, and hepatocyte nuclear factor 3 families. Mol Cell Biol 13(7):3860–71

SHARP JM, HERRING AJ (1983) Sheep pulmonary adenomatosis: demonstration of a protein which cross-reacts with the major core proteins of Mason-Pfizer monkey virus and mouse mammary tumour virus. J Gen Virol 64(Pt 10):2323–7

STEHELIN D, VARMUS HE, BISHOP JM, VOGT PK (1976) DNA related to the transforming gene(s) of avian sarcoma viruses is present in normal avian DNA. Nature 260(5547):170–3

SWANSTROM R, WILLS JW (1997) Synthesis, assembly, and processing of viral proteins. In: Retroviruses (JM Coffin, SH Hughes and HE Varmus, eds), pp 263–364. Cold Spring Harbor Laboratory Press, New York

TAYLOR JM, ILLMENSEE R (1975) Site on the RNA of an avian sarcoma virus at which primer is bound. J Virol 16(3):553–8

TEMIN HM, MIZUTANI S (1970) RNA-dependent DNA polymerase in virions of Rous sarcoma virus. Nature 226(252):1211–3

TSAY W, LEE YM, LEE SC, SHEN MC, CHEN PJ (1997) Synergistic transactivation of HNF-1alpha, HNF-3, and NF-I contributes to the activation of the liver-specific protein C gene. DNA Cell Biol 16(5):569–77

VAN BEVEREN C, GODDARD JG, BERNS A, VERMA IM (1980) Structure of Moloney murine leukemia viral DNA: nucleotide sequence of the 5′ long terminal repeat and adjacent cellular sequences. Proc Natl Acad Sci USA 77(6):3307–11

VOGT VM (1997) Retroviral virions and genomes. In: Retroviruses (JM Coffin, SH Hughes, and HE Varmus, eds.), pp 27–69. Cold Spring Harbor laboratory Press, New York

WANG L, GALEHOUSE D, MELLON P, DUESBERG P, MASON WS, VOGT PK (1976 a) Mapping oligonucleotides of Rous sarcoma virus RNA that segregate with polymerase and group-specific antigen markers in recombinants. Proc Natl Acad Sci USA 73(11):3952–6

WANG LH, DUESBERG P, MELLON P, VOGT PK (1976 b) Distribution of envelope-specific and sarcoma-specific nucleotide sequences from different parents in the RNAs of avian tumor virus recombinants. Proc Natl Acad Sci USA 73(4):1073–7

WANG LH, DUESBERG PH, KAWAI S, HANAFUSA H (1976 c) Location of envelope-specific and sarcoma-specific oligonucleotides on RNA of Schmidt-Ruppin Rous sarcoma virus. Proc Natl Acad Sci USA 73(2):447–51

Weiss SR, Varmus HE, Bishop JM (1977) The size and genetic composition of virus-specific RNAs in the cytoplasm of cells producing avian sarcoma-leukosis viruses. Cell 12(4):983–92

Wikenheiser KA, Vorbroker DK, Rice WR, Clark JC, Bachurski CJ, Oie HK, Whitsett JA (1993) Production of immortalized distal respiratory epithelial cell lines from surfactant protein C/simian virus 40 large tumor antigen transgenic mice. Proc Natl Acad Sci USA 90(23):11029–33

Wills JW, Craven RC (1991) Form, function, and use of retroviral gag proteins. *Aids* 5(6):639–54

York DF, Vigne R, Verwoerd DW, Querat G (1991) Isolation, identification, and partial cDNA cloning of genomic RNA of jaagsiekte retrovirus, the etiological agent of sheep pulmonary adenomatosis. J Virol 65(9):5061–7

York DF, Vigne R, Verwoerd DW, Querat G (1992) Nucleotide sequence of the Jaaksiekte retrovirus, an exogenous and endogenous type D and B retrovirus of sheep and goats. Journal of Virology 66:4930–4939

Zsengeller ZK, Halbert C, Miller AD, Wert SE, Whitsett JA, Bachurski CJ (1999) Keratinocyte growth factor stimulates transduction of the respiratory epithelium by retroviral vectors. Hum Gene Ther 10(3):341–53

Chapter 5

Endogenous Retroviruses Related to Jaagsiekte Sheep Retrovirus

J. C. DeMartini, J. O. Carlson, C. Leroux, T. Spencer, M. Palmarini

J. C. DeMartini, J. O. Carlson
Department of Microbiology, Immunology and Pathology,
College of Veterinary Medicine and Biomedical Sciences, Colorado State University,
Fort Collins, CO 80523-1619, USA
e-mail: james.demartini@colostate.edu

C. Leroux
Université Claude Bernard, UMR754 INRA-UCB-ENVL, 50 Avenue Tony Garnier,
69366 Lyon Cedex, France
e-mail: cleroux@rockefeller.univ.lyon1.fr

T. Spencer
Center for Animal Biotechnology and Genomics, Department of Animal Science,
442 Kleberg Center, 2471 TAMU, Texas A&M University, College Station,
TX 77843-2471, USA
e-mail: tspencer@ansc.famu.edu

M. Palmarini
Department of Medical Microbiology and Parasitology,
College of Veterinary Medicine, University of Georgia, Athens, GA 30607, USA
e-mail: mpalmari@vet.uga.edu

Abstract. Ovine betaretroviruses consist of exogenous viruses [jaagsiekte sheep retrovirus (JSRV) and enzootic nasal tumor virus, (ENTV)] associated with neoplastic diseases of the respiratory tract and 15–20 endogenous viruses (enJSRV) stably integrated in the ovine and caprine genome. Phylogenetic analysis of this group of retroviruses suggests that the enJSRV can be considered as 'modern' endogenous retroviruses with active, exogenous counterparts. Sequence analysis of JSRV, ENTV and enJSRV suggests that enJSRV do not directly contribute to the pathogenesis of ovine pulmonary adenocarcinoma (OPA) or enzootic nasal tumor through large-scale recombination events, but small-scale recombination or complementation of gene function cannot be excluded; experiments involving enJSRV-free sheep, which have not been found, would be needed to investigate this possibility. Evidence of expression of enJSRV structural proteins in tissues of the reproductive tract and lung implies that they do not have a primary role in disease. However, experimental exploitation of exogenous/endogenous retrovirus sequence differences by producing chimeras has been useful in establishing the determinants of JSRV Env-induced transformation. Even if enJSRV do not have a direct role in OPA, their expression during ontogeny or in neonatal life may impact the likelihood of exogenous JSRV infection and disease outcome via the induction of immunological tolerance. Aside from any role in disease, enJSRV loci may serve as useful genetic markers in the sheep and their frequent expression in the reproductive tract of the ewe may portend an important physiologic role in sheep.

1 Introduction

Ovine pulmonary adenocarcinoma (OPA) is a contagious neoplasm caused by jaagsiekte sheep retrovirus (JSRV) (Palmarini et al. 1999; DeMartini et al. 2001). When the first clones of JSRV were isolated and sequenced, they were soon found to hybridize to proviral sequences in the sheep and goat genomes (York et al. 1991). Using JSRV capsid (CA) and envelope (SU) region probes, these endogenous retrovirus sequences related to JSRV (enJSRV) were found to be widely distributed among the *Ovis* and *Capra* genera (Hecht et al. 1996). For several years, progress in clarifying the etiologic agent of OPA was impeded by the inability to distinguish enJSRV from exogenous JSRV sequences. Then subtle differences in a *gag* restriction site (Palmarini et al. 1996a) and long terminal repeat

(LTR) U3 and envelope transmembrane (TM) nucleotide sequences (Bai et al. 1996) were found to be useful in differentiating between enJSRV and the pathogenic, exogenous form of JSRV which was consistently associated with OPA (Bai et al. 1996; Palmarini et al. 1996a).

In addition to their potential role in the host immune or pathogenic response to JSRV, enJSRV are of more basic biological interest. In particular, examination of gene arrangement and coding capacity in comparison to JSRV and other endogenous members of the *Betaretroviridae* provides insight into the evolutionary phylogeny of these viral sequences. Moreover, analysis of integration sites and tissue expression patterns is useful in assessing the functional significance of enJSRV. When the family of enJSRV is fully described and mapped to the genome of sheep and other ruminants, these endogenous retroviral sequences also may be useful as genetic markers for livestock productivity or disease traits.

Endogenous retroviruses (ERV) are vertically transmitted as stable mendelian genes in the germline of most eukaryotes (Boeke and Stoye 1997), and it is believed that they derive from integration of exogenous (horizontally transmitted) retroviruses in the germ cells of a specific host. During evolutionary time, ERV have shaped the genome of virtually all animal species. Depending on the length of time since they are presumed to have entered the genome, ERV are classified as 'ancient' or 'modern'. Ancient viruses were acquired in evolutionarily 'ancient' times as indicated by their identical integration in more than one species. These ERV also have accumulated numerous point mutations or deletions in their coding or regulatory regions that render them replication incompetent and transcriptionally inactive. Conversely, modern ERV are less likely to have accumulated defective genes and sequence variability between the two LTRs, and they are more likely to have exogenous, infectious, horizontally transmitted counterparts (i.e., they are still 'active').

The biological effects of ERV are largely unknown. Given the longstanding presence of ERV in the genome, evolutionary theory would predict they might have more beneficial effects than detrimental effects and this may indeed be the case. Expression of ERV genes may prevent infection by related exogenous retroviruses and consequent disease by receptor interference or other post-entry mechanisms (Best et al. 1997; Ruscetti et al. 1981), or by clonal deletion of T cells required for infection (Golovkina et al. 1992). Another interesting example of a beneficial ERV effect is expression of salivary amylase in primates due to integration of ERV containing a parotid-specific enhancer element in the flanking region of a pancreatic amylase gene (Ting et al. 1992). In contrast, as described in mice, cats, and chickens, more pathogenic retroviruses can result from reactivation or recombination among different endogenous loci (Golovkina et al. 1996; Stoye et al. 1991; Stoye and Coffin 1987) or recombination of exogenous viruses with ERV (Benson et al. 1998; Golovkina et al. 1994, 1997; Neil et al. 1991; Ruscetti et al. 1981). More speculatively, ERV have been suggested to have a role in human diseases such as autoimmunity (Nakagawa and Harrison 1996), testicular tumors (Lower et al. 1996; Sauter et al. 1995), and multiple sclerosis (Perron et al. 1997).

In this chapter, we summarize what is known regarding the distribution of enJSRV among vertebrates, their genome structure and organization as well as phylogenetic relationships with JSRV and similar retroviruses, and we conclude with recent data on expression of enJSRV in vivo and its possible significance.

2 Vertebrate Distribution of enJSRV

The distribution of enJSRV within sheep families and among small ruminants and other vertebrate species have been extensively studied by Southern blot hybridization. Probes derived from the JSRV *gag*, *pol*, and *env* genes hybridize to sheep genomic DNA that has been cut by different restriction enzymes consistently yield 15–20 bands of varying intensity (Fig. 1; Hecht et al. 1994, 1996). When a JSRV LTR probe is used to hybridize to sheep genomic DNA that has been cut by *Sac*I, approximately twice as many bands are observed (Bai et al. 1999) as when using a *gag*, *pol*, or *env* probe (Fig. 1). Determining whether there are single LTR integrations, as found with ERV in other animal species (Boeke and Stoye 1997), will require further study. In Fig. 1, an intense *Sac*I band of approximately 7.1 kb detected by all three probes is consistent with a *Sac*I internal fragment predicted for the enJSRV sequence designated enJS56A1 (Palmarini et al. 2000b). In previous work, DNA fragment polymorphisms were observed between parents and offspring, indicating that the enJSRV loci segregate according to mendelian genetics (Hecht et al. 1994).

When genomic DNA from 13 breeds of domestic sheep (*Ovis aries*) exhibiting diverse breeding histories was hybridized to CA and *env* SU probes, many of the larger bands were shared among all sheep tested

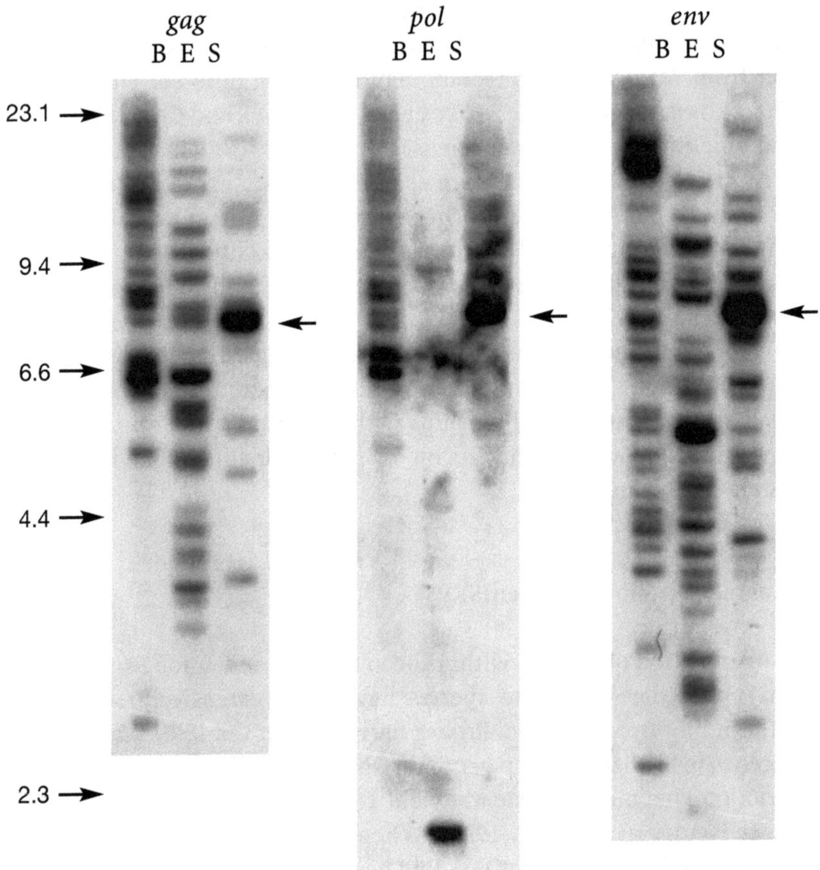

Fig. 1. Endogenous JSRV-related sequences detected by Southern hybridization in genomic sheep DNA digested by *Bam*H1 (*B*), *Eco*R1 (*E*) and *Sac*I (*S*) using probes derived from the *gag*, *pol*, and *env* regions of JSRV (Hecht et al. 1994, 1996). Hybridization conditions were low stringency (55 °C; T_m −30 °C)

(Hecht et al. 1996). This indicates that these enJSRV sequences were fixed in the genome early in the evolutionary history of domestic sheep. In addition, similar hybridization patterns were observed in animals of three wild *Ovis* species, including the moufflon, a European wild sheep breed thought to be the predecessor of domestic sheep (Table 1; Hecht et al. 1996). This suggests that the endogenous sequences became fixed in the ancestral *Ovis* genome prior to the divergence of these species, and that they have been relatively inactive since then. The conservation of endogenous viral numbers and presumed chromosomal sites among sheep is quite different from the more extensive variation seen in endogenous viruses of laboratory and wild mice (Stoye and Coffin 1988; Frankel et al. 1990; Tomonaga and Coffin 1998) and in avian viruses of chickens, jungle fowl, and pheasants (Hughes et al. 1981).

The intriguing data from sheep prompted examination of DNA from other members of the subfamily Caprinae and more distantly related mammalian species (Table 1; Hecht et al. 1996). Closely related viruses were found in six breeds of domestic goats (*Capra hircus*) in similar copy numbers as in sheep; the goat hybridization pattern was different from that of sheep but was generally conserved among goats. Southern blot profiles (15–20 bands) were unchanged between low and high stringency conditions, indicating relatively high similarity between the sheep and goat viruses (Hecht et al. 1996). However, the differences in restriction enzyme profiles between sheep and goat lineages suggest that much of the amplification from founding viruses within the respective genomes occurred after the divergence of goats and sheep 4–10 million years ago (Irwin et al. 1991). Among seven other genera of Caprinae, only the mountain goat and aoudad

Table 1. Endogenous retroviral sequences related to JSRV CA and SU regions in representative ungulate species and other mammals detected by Southern blot hybridization at high stringency (~20°C below calculated probe T_m)

Common name	CA probe	SU probe
Domestic sheep	15–20	15–20
Bighorn sheep	15–20	15–20
Moufflon sheep	15–20	15–20
Domestic goats	15–20	15–20
Markor	17	6
Goral	6	7
Himilayan tahr	11	7
Aoudad	0	0
Mountain goat	0	0
Domestic cattle	1–2	1–3
Cape buffalo	0	1
Moose	5	2
Wapiti (American elk)	1–2	0
Zebra	1	1
Mountain lion	0	0
Mouse	0	0
Human	0	0

did not have bands detectable with either the CA or SU probes at high stringency (Table 1). These animals may therefore be useful for in vitro or in vivo experiments to investigate JSRV replication and oncogenicity in ovine/caprine species that lack JSRV-related endogenous sequences. Interestingly, mountain goat fibroblasts can be infected using a JSRV pseudotyped Moloney murine leukemia virus (Rai et al. 2000), implying that cells of this genus express competent receptors for JSRV (T. Allen, unpublished results). Bands hybridizing at high stringency to CA and SU probes were found in three of nine other families of ungulates, but not in nine families of more distantly related animals, including humans. The diversity of species with endogenous loci did not follow phylogenetic classification but rather suggested that horizontal transfer among species occurred.

3
Chromosomal Distribution, Structure, and Phylogeny of enJSRV

3.1
Chromosomal Distribution

Southern blots of sheep DNA cut with different restriction enzymes consistently show 15–20 bands when hybridized with JSRV probes (Fig. 1). Some insight into the distribution of these endogenous proviruses in the sheep genome has been gained by PCR analysis of DNA from sheep/hamster cell lines and by fluorescence in situ hybridization (FISH). Chromosomal DNA was isolated from a panel of sheep/hamster cell lines that collectively contain all of the sheep chromosomes (Burkin et al. 1998). Each cell line contains one or a few sheep chromosomes. When DNA from each of these cell lines was used as a template for PCR with primers for the enJSRV *gag* region, a product was amplified from most but not all lines (Carlson et al. 2002). This suggests that many of the sheep chromosomes have at least one enJSRV provirus. This was confirmed by Southern blot analysis on several sheep/hamster lines that contain a single chromosome. When JSRV probes are used in FISH on sheep metaphase chromosomes, sites of hybridization are usually on several chromosomes and prominent sites of hybridization are consistently seen near the centromeres of two medium-sized acrocentric chromosomes (probably chromosome 6) (Fig. 2). It seems likely that there are multiple copies of enJSRV at the chromosome 6 sites of hybridization. Similarly, two sites of hybridization also were seen in interphase nuclei (D. Knudson and S. Brown,

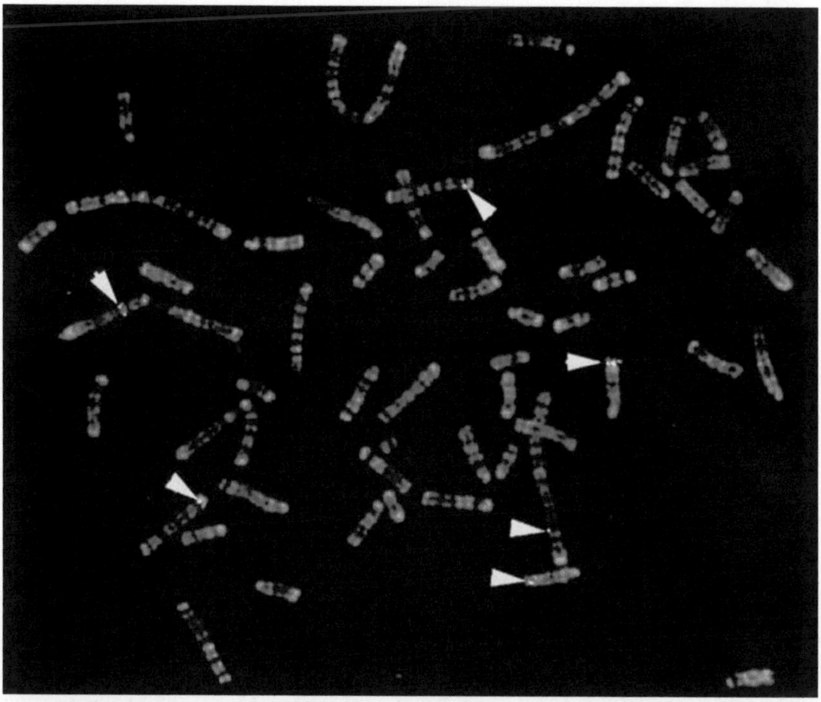

Fig. 2. Fluorescent in situ hybridization localization of enJSRV to sheep chromosomes. Integration sites of enJSRV (*arrowheads*) were analyzed by FISH using a full-length JSRV probe on an ovine tissue fibroblast metaphase chromosome spread

unpublished results). This suggests that there is a locus in the sheep genome that contains multiple copies of enJSRV within approximately 50 kb of each other. This is also consistent with Southern blot data on a hybrid cell line containing chromosome 6 (J. Bishop, unpublished results). The existence of a multicopy locus seems unusual in light of the random nature of retroviral integration.

3.2 Structure and Genomic Organization

Two approaches have been undertaken to identify and characterize the nucleotide sequence and genomic organization of enJSRV in the sheep genome: PCR amplification and molecular cloning of proviruses isolated from sheep lambda libraries. In the first demonstration of a practical method to distinguish tumor-associated, exogenous JSRV from enJSRV sequences in the sheep genome, a *Sca*I restriction site in *gag* was found to be a reliable molecular marker for the exogenous virus (PALMARINI et al. 1996a). Subsequently, a long-range PCR procedure was used to synthesize full-length endogenous proviral sequences by using DNA primers designed from JSRV sequences, with sheep DNA templates (BAI et al. 1996). Clones were derived from six different loci, including four full-length and two with internal deletions in the *pol* region. The six loci were easily distinguished from each other and from the exogenous JSRV by their unique restriction enzyme profiles. Nucleotide sequences of the entire LTR, the TM domain of *env*, the 5′ untranslated leader region (5′ LTR), and part of *gag* were determined for the six endogenous loci. The proviruses of the six endogenous loci had greater than 97% sequence identity in the regions sequenced and 95% identity to JSRV in the 5′ LTR and *gag* regions. The endogenous LTR sequences were 94%–98% identical among sheep of four diverse breeds, and similar to JSRV, particularly in R and U5. However, there was only 78% sequence identity between the endogenous loci and JSRV in the entire U3 region, and, because of deletions and point mutations in JSRV, there was only 56% identity in the 3′ 173 bp U3 region. This region of divergence in U3 has been widely exploited in distinguishing JSRV-infected cells and tissues from sheep enJSRV sequences by PCR and by Southern hybridization. Another region of divergence in the TM region of *env*, described more fully below, also has been used for this purpose. Ultimately, the differences between enJSRV and JSRV in the *gag*, LTR, and TM regions were used to isolate infectious proviral clones from tumor tissue and a tumor cell line (DEMARTINI et al. 2001; PALMARINI et al. 1999; see the chapter by SHARP and DEMARTINI, this volume).

Recently, the complete nucleotide sequence and genome structure of three enJSRV has been determined (Fig. 3; PALMARINI et al. 2000b). One of the endogenous proviruses (enJS56A1; Fig. 3) was full length and maintained open reading frames in all its structural genes but could not assemble viral particles when highly expressed in human 293T cells. By making chimeras between this provirus and $JSRV_{21}$, the defect for virus assembly was found to be in the *gag* gene (PALMARINI et al. 2000b). The other two endogenous proviruses had either premature stop codons in *gag* and *pol* or major deletions in *env* or *pol* (Fig. 3). All three endogenous proviral loci had both upstream and downstream LTRs, the hallmark of complete proviruses, and all had the previously mentioned differences from JSRV in the U3 region. The upstream and downstream LTRs of one endogenous

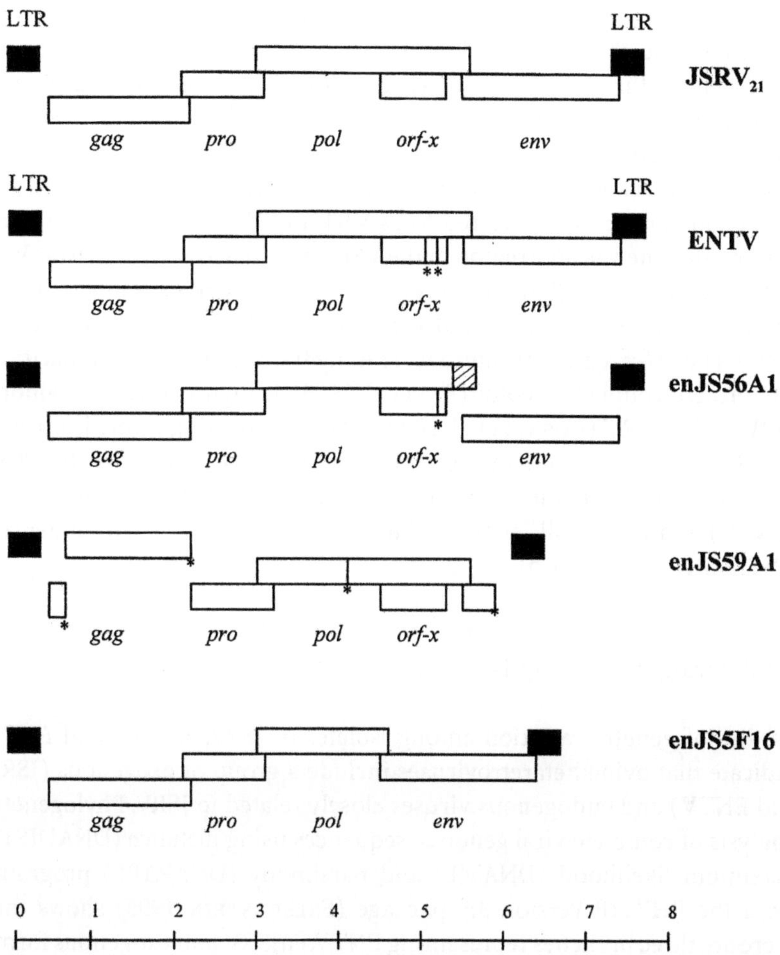

Fig. 3. Genomic structures of endogenous and exogenous JSRV-related retroviruses of sheep. Premature stop codons are indicated by a *vertical bar underlined by an asterisk*. For convenience, the *gag* open reading frame has been fixed in the same reading frame of all sequences shown. The *numbered bar* indicates distances in kilobases. The exogenous JSRV and ENTV show the canonical retrovirus *gag, pro, pol* and *env* with *pro* in a different open reading frame from *pol*, the same for alpha-, beta- and gammaretroviruses. An additional open reading frame (*orf-x*) overlapping *pol* is present in JSRV but is interrupted by two stop codons in ENTV (Cousens et al. 1999), en JS56A1 is the only one of the three endogenous proviruses cloned in this study to maintain full (or nearly full) open reading frames in all structural genes. EnJS59A1 has premature stop codons in *gag* and *pol* and a major deletion in *env*, enJSF16 has a deletion in *pol*. Different peptide sequences at the 3′ end of the *pol* gene in enJS56A1 due to a frameshift are indicated by *cross-hatching*. The LTRs are indicated by *solid boxes*

provirus were identical whereas those of two proviruses displayed two- and four-base changes, a finding useful in considering the evolutionary phylogeny of the proviruses. As expected for endogenous proviruses, other point mutations, insertions and deletions were identified in the structural genes. Alignment of deduced amino acid sequences revealed two variable regions within *gag* of polymorphism between endogenous and exogenous proviruses (referred to as VR1 and VR2 in PALMARINI et al. 2000b) and one variable region in the TM region of *env* (referred to as VR3 in PALMARINI et al. 2000b); the *env* gene is otherwise remarkably well conserved among endogenous and exogenous JSRV sequences (BAI et al. 1999). The *orf-x* region, an alternate reading frame in *pol* with similarity to a G-protein coupled receptor (BAI et al. 1999), is highly conserved among JSRV and enJSRV (ROSATI et al. 2000). The *orf-x* reading frame is open in all JSRV sequences and in most enJSRV sequences amplified by PCR, but there are premature stop codons in enzootic nasal tumor virus (ENTV) and in the full-length enJSRV provirus that has been sequenced (PALMARINI et al. 2000b).

3.3
Viral Phylogenetic Analysis

Studies of genetic variation among isolates of JSRV, enJSRV, and ENTV indicate that ovine betaretroviruses include a group of exogenous (JSRV and ENTV) and endogenous viruses closely related to JSRV. Phylogenetic analysis of complete viral genomic sequences using distance (DNADIST), maximum likelihood (DNAML) and parsimony (DNAPARS) programs from the PHYLIP Version 3.5 package (FELSENSTEIN 1995) shows that there are three branches representing ENTV, enJSRV, and exogenous forms of JSRV, confirming an initial report based on limited *gag* sequences (COUSENS et al. 1999). Furthermore, separate DNADIST analysis of U3, *env* and *gag/pol* regions of the genome gave similar results. These results are all well supported by bootstrapping. This suggests that there has been little or no recombination between the three branches since they were established.

Neighbor-joining phylogenetic trees have been generated for all available U3 (Fig. 4), *env*, and *gag/pol* sequences of endogenous and exogenous JRSV-related proviruses. The three branches described above were identified for each tree. The exogenous JSRVs could be further divided into two branches corresponding to sequences derived from Africa (type I) or from

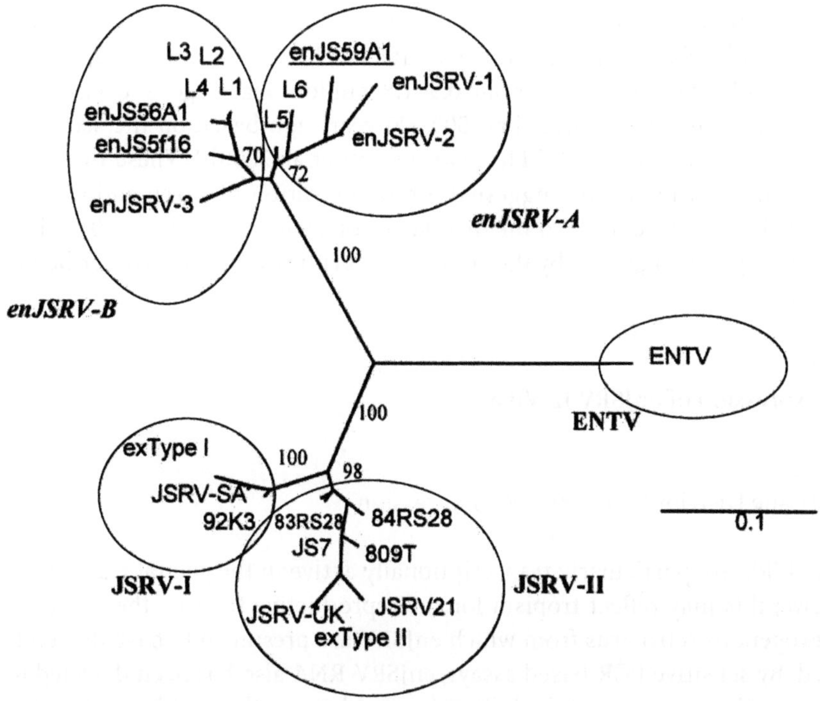

Fig. 4. Phylogenetic analysis of envelope sequences of endogenous and exogenous JSRV-related retroviruses. An unrooted tree of the LTR U3 region was derived by neighbor joining. To show consistency, all bootstrap values obtained with 1,000 replications of bootstrap sampling are shown. Sequences used for the analysis are termed as in their original references with the exception of loci 1–6, which are indicated as L1–6. GenBank accession numbers: AF105220 ($JSRV_{21}$); M80216 (JSRV-SA); X95446 (endogenous locus 2); Y16627 (ENTV); Y18301 to Y18305 (JS7, 809T, 83RS2, 84RS2 and 92K3); Z66531 to Z66533 (enJSRV1 to –3); (AF136224) enJS5F16; (AF136225) enJS59A1; AF153615 (enJS56A1)

the USA and the UK (type II) (Bai et al. 1996, 1999). Analysis of several French JSRV isolates in the *env* and LTR regions demonstrates that they are genetically closely related to type I JSRV, suggesting a common ancestor to the French and South African viruses (C. Leroux, unpublished results). In all of the generated trees, the enJSRV loci also could be divided into two groups, designated A and B (U3 tree in Fig. 4).

Based on analysis of the variability between 5′ and 3′ LTRs of single loci, the time of their integration in the sheep genome can be estimated. The LTRs were presumably identical at the time of integration and would presumably mutate at the rate of non-coding sequences and pseudogenes, as a molecular clock (Dangel et al. 1995). Thus, integration of enJS56A1 and enJS59A1 occurred about 0.9–1.8 million years ago and enJS5F16 may have integrated less than 500,000 years ago based on the sequence identity of the 5′ and 3′ LTRs (Palmarini et al. 2000b). These times are within the time frame suggested for amplification of sheep endogenous proviruses subsequent to the divergence of sheep and goats 4–10 million years ago, as suggested by the Southern hybridization studies described in Sect. 1.

4 Expression of enJSRV In Vivo

4.1 Tissue Distribution of enJSRV Expression

enJSRV are particularly transcriptionally active in the genital tract of the ewe; this may reflect tropism for the reproductive tract by the ancestral exogenous retrovirus from which enJSRV are presumed to have descended. By sensitive PCR-based assays, enJSRV RNA also has been detected in every tissue examined, including lungs, kidneys, thymus, bone marrow, spleen, mediastinal lymph nodes and leukocytes (Palmarini et al. 1996b). However, in situ hybridization indicates that enJSRV expression is mostly limited to the reproductive tract. High levels of enJSRV expression are detectable in the endometrial luminal and glandular epithelium of the uterus (Palmarini et al. 2000b; Spencer et al. 1999), and in the epithelia of the oviducts and in the cervix (Fig. 5; Palmarini et al. 2001b). In the female reproductive tract, lower levels of enJSRVs RNA are detected in the posterior and anterior regions of the vagina. Tissues outside the reproductive tract show moderate or no signal above background with the exception of cells in the lamina propria of the gut which appear strongly positive by in situ hybridization (Palmarini et al. 2000b). Low levels of RNA expression also have been observed in the bronchiolar epithelium of the lungs (Palmarini et al. 2000b).

enJSRV expression is not limited to RNA. The epithelia of the lumen and glands of the endometrium of the uterus are positive for enJSRV proteins by immunofluorescence when antisera to the highly related exogenous JSRV CA or envelope (Env) proteins are used (Palmarini et al., 2001b). It has not been established if the expression of viral proteins leads

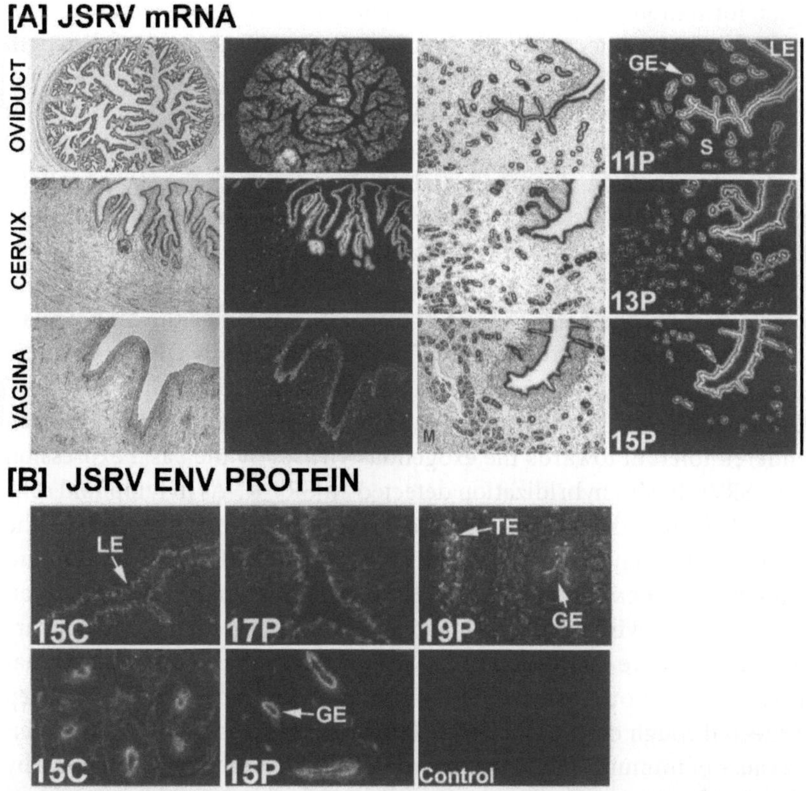

Fig. 5 A, B. Expression of enJSRV RNA and protein in the female ovine reproductive tract. A In situ hybridization analysis of enJSRV RNA expression in different tissues of the adult ovine female reproductive tract. Cross-sections of different tissues from day 9 pregnant ewes (*left*) or uterus of pregnant ewes (*P*, at day 11, 13, and 15 of pregnancy) were hybridized with radiolabeled antisense or sense ovine enJSRV cRNA probes. Protected transcripts were visualized by liquid emulsion autoradiography for 1 week and imaged under brightfield or darkfield illumination. Magnification, ×260. B Immunoreactive enJSRV Env protein expression in the ovine endometrium. Immunofluorescence staining of frozen uterine sections from cyclic (*C*, at day 15 of the estrus cycle) and pregnant (*P*, at day 15, 17, and 19 of pregnancy) ewes was conducted using specific anti-ovine enJSRV Env antibodies or irrelevant IgG as a control. Magnification, ×230. *LE*, Lumenal epithelium; *GE*, glandular epithelium; *S*, stroma; *TE*, trophectoderm

to the formation of viral particles. A defect for particle assembly and/or egress has been detected in the first half of the *gag* gene of a full-length enJSRV locus (enJS56A1), but this does not necessarily signify that all the enJSRV loci have the same defect (Palmarini et al. 2000b). Northern blot analysis of endometrial RNA detects the enJSRV-specific full length and singly spliced mRNA (Spencer et al. 1999) in line with the classic mRNA splicing pattern of betaretroviruses (Rabson and Graves 1997) and of the exogenous JSRV (Palmarini et al. 2001b).

The expression of enJSRV genes in the newborn lamb may explain some aspects of the pathogenesis of diseases induced by the highly related exogenous viruses JSRV and ENTV. Sheep affected by OPA or enzootic nasal tumor do not have circulating antibodies against JSRV or ENTV (Ortin et al. 1998; Sharp and Herring 1983). Thus, sheep might been rendered tolerant towards the exogenous viruses by the basal expression of enJSRVs. In situ hybridization detected enJSRV RNAs in lymphoid cells associated with the lamina propria of the small intestine and in the thymuses of sheep fetuses. Low level expression of enJSRV also was found in the lungs (Spencer et al. 2002). How the induction of tolerance to an exogenous retrovirus could be beneficial for its host is not readily apparent. The extensive expression of enJSRV in the genital tract suggests that ancestral exogenous forms of JSRV may have been transmitted from sheep to sheep through coitus. Ancient enJSRV-related viruses might have been the cause of immuno-mediated disorders so that induction of tolerance by related viruses integrated in the germline of the host could have resulted in an evolutionary advantage. These speculations are obviously very difficult to address experimentally. Alternatively or additionally, expression of enJSRV RNA and proteins in gut or other epithelial cells may prevent infection or inhibit replication by related exogenous JSRV via receptor blockade or interference mechanisms. Supporting this hypothesis is the fact that enJSRVs use the same receptor as the exogenous JSRV and ENTV use (Spencer et al. 2002), and expression of enJSRV Env in vitro interferes with JSRV entry (Spencer et al. 2002).

4.2 Effects of Neonatal Age, Estrus Cycle and Pregnancy on enJSRV Expression

enJSRV expression has been studied in the uteri of neonatal ewe lambs between postnatal day (PND) 1 and 56 when active uterine gland morphogenesis occurs. The endometrial glandular epithelium differentiates and buds into the underlying stroma from the lumenal epithelium between PND 1 and 7. Between PND 14 and 56 there is active branching and coiling of the nascent endometrial glands. By PND 56 the uterine wall is histologically mature. By in situ hybridization, no enJSRV expression was detected at PND 1 while it was detected in the lumenal epithelium and in the branching glandular epithelium at day 7 and in all epithelia on PND 56.

The estrous cycle and pregnancy also influence the levels of enJSRV expression in the endometrium. In cyclic ewes, endometrial enJSRV RNA is increased 12-fold between days 1 and 13 and then decreases to day 15. The increase in endometrial enJSRVs expression is correlated with the levels of progesterone in the blood and expression of the progesterone receptor (Spencer and Bazer 1995). Indeed, the initial increase of enJSRV RNAs between days 1 and 13 of the estrus cycle correlates with the formation of the corpus luteum after ovulation and consequent increase of progesterone in the peripheral blood. Continuous exposure to progesterone decreases the expression of progesterone receptor in the epithelium of the uterus and this explains the lower levels of enJSRV RNAs observed by day 11 in the cyclic ewe.

In the pregnant ewe, enJSRV RNAs also were highly expressed between days 11 and 13 and then progressively declined to almost undetectable levels. The decrease in enJSRV expression is probably influenced by the loss of expression of the progesterone receptor in the epithelium of the lumen and the glands. Interestingly, both enJSRV RNAs and proteins were detected in binucleate cells of the conceptus trophectoderm (Fig. 5; Palmarini et al. 2001b). Binucleate trophectodermal cells fuse with the endometrial luminal epithelium in both caruncular and intercaruncular areas to form a syncytium (Wooding 1982). Only the binucleate cells display invasive properties in the placenta of ruminants as well as other species. Of interest for comparative physiology is that the presence of enJSRV gene expression in the developing placenta is strikingly similar to that observed for the human endogenous retrovirus, HERV-W (Blond et al. 1999). HERV-W is specifically expressed in the syncytiotrophoblast of the human placenta that is formed by fusion of the trophoblast with the epithelium of the maternal uterus. The HERV-W envelope induces the formation of syncytia when expressed in vitro, thereby advancing the hypothesis that HERV-W is involved in human placental morphogenesis (Blond et al. 2000; Mi et al. 2000). These observations would support the theory that an ancient retroviral infection had profound consequences for mammalian evolution

(Harris 1991). Indeed, the presence of enJSRV transcripts and viral proteins in uterine epithelial cells argue for involvement of this betaretrovirus in synepitheliochorial placentation in sheep.

4.3 The enJSRV LTR as the Major Determinant of Viral Tropism

The correlation of enJSRV expression with the levels of progesterone in the blood suggests that the enJSRV LTR (or at least the LTR of some enJSRV loci), where the viral promoter and enhancers are located, is influenced by progesterone. In transient transfection assays, the LTR of the enJS59A1 locus has been activated almost 10-fold by progesterone (Palmarini et al. 2001b). The latter had minimal effect on the LTRs of the exogenous JSRV and ENTV. In contrast, the LTR of the exogenous JSRV is activated by lung specific transcription factors such as HNF-3β (McGee-Estrada et al. 2002; Palmarini et al. 2000a), while the available enJSRV LTRs are not affected by this factor (Palmarini et al. 2000b). These results support the hypothesis that the exogenous JSRV and ENTV developed their pulmonary tropism relatively recently and probably after the integration of the enJSRV loci in the sheep germline. The possibility of pneumotropism by precursors of some enJSRV cannot be excluded given that low levels of enJSRV expression is detected in the lungs (Palmarini et al. 2000b) and not all the loci have been sequenced.

Acknowledgements. This work was supported by grant 1RO1 CA 59116 from the National Cancer Institute, National Institutes of Health, the Région Rhône Alpes, Ligue Nationale contre le Contre le Cancer: comités de la Loire, de l'Ardèche, du Rhône, and Biotechnology awards from the University of Georgia.

Abbreviations

CA	viral capsid protein
ERV	endogenous retrovirus
enJSRV	endogenous jaagsiekte sheep retrovirus
FISH	fluorescent in situ hybridization
JSRV	jaagsiekte sheep retrovirus
OPA	ovine pulmonary adenocarcinoma
SU	viral envelope surface protein
Env	viral envelope protein

References

Bai J, Zhu R-Y, Stedman K, Cousens C, Carlson J, Sharp JM, DeMartini JC (1996) Unique long terminal repeat U3 sequences distinguish exogenous jaagsiekte sheep retroviruses associated with ovine pulmonary carcinoma from endogenous loci in the sheep genome. J Virol 70:3159–3168

Bai J, Bishop JV, Carlson JO, DeMartini JC (1999) Sequence comparison of JSRV with endogenous proviruses: Envelope genotypes and a novel ORF with similarity to a G-protein coupled receptor. Virology 258:333–343

Benson SJ, Ruis BL, Fadly AM, Conklin KF (1998) The unique envelope gene of the subgroup J avian leukosis virus derives from ev/J proviruses, a novel family of avian endogenous viruses. J Virol 72:10157–10164

Best S, Le Tissier PR, Stoye JP (1997) Endogenous retroviruses and the evolution of resistance to retroviral infection. Trends Microbiol 5:313–318

Blond JL, Beseme F, Duret L, Bouton O, Bedin F, Perron H, Mandrand B, Mallet F (1999) Molecular characterization and placental expression of HERV-W, a new human endogenous retrovirus family. J Virol 73(5):1175–1185

Blond JL, Lavillette D, Cheynet V, Bouton O, Oriol G, Chapel-Fernandes S, Mandrand B, Mallet F, Cosset FL (2000) An envelope glycoprotein of the human endogenous retrovirus HERV-W is expressed in the human placenta and fuses cells expressing the type D mammalian retrovirus receptor. J Virol 74:3321–3329

Boeke JD, Stoye JP (1997) Retrotransposons, endogenous retroviruses, and the evolultion of retroelements. In: Coffin JM, Hughes SH, Varmus HE (eds) Retroviruses, pp 343–436. Cold Spring Harbor Laboratory: Plainview, NY

Burkin DJ, Broad TE, Lambeth MR, Burkin HR, Jones C (1998) New gene assignments using a complete, characterized sheep - hamster somatic cell hybrid panel. Anim Genet 29:48–54

Carlson J, Bishop J, Lyon M, Vaiman A, Cribiu E, Mornex J-F, Brown S, Knudson D, Leroux C, DeMartini J (2002) Chromosomal distribution of endogenous Jaagsiekte sheep retrovirus proviral sequences in the sheep genome. Submitted

Cousens C, Minguijon E, Dalziel RG, Ortin A, Garcia M, Park J, Gonzalez L, Sharp JM, De las Heras M (1999) Complete sequence of enzootic nasal tumor virus, a retrovirus associated with transmissible intranasal tumors of sheep. J Virol 73(5):3986- 3993

Dangel AW, Baker BJ, Mendoza AR, Yu CY (1995) Complement component C4 gene intron 9 as a phylogenetic marker for primates: long terminal repeats of the endogenous retrovirus ERV-K(C4) are a molecular clock of evolution. Immunogenetics 42:41–52

DeMartini JC, Bishop JV, Allen TE, Jassim FA, Sharp JM, De las Heras M, Voelker DR, Carlson JO (2001) Jaagsiekte sheep retrovirus provirus clone JSRVJS7, derived form the JS7 lung tumor cell line, induces ovine pulmonary carcinoma and is integrated into the surfactant protein A gene. J Virol 75:4239–4246

Felsenstein J (1995) PHYLIP: phylogeny inference package, 3.5 ed. University of Washington, Seattle

Frankel WN, Stoye JP, Taylor BA, Coffin JM (1990) A linkage map of endogenous murine leukemia proviruses. Genetics 124:221–236

Golovkina TV, Chervonsky A, Dudley JP, Ross SR (1992) Transgenic mouse mammary tumor virus superantigen expression prevents viral infection. Cell 69: 637–645

Golovkina TV, Jaffe AB, Ross SR (1994) Coexpression of exogenous and endogenous mouse mammary tumor virus RNA in vivo results in viral recombination and broadens the virus host range. J Virol 68:5019–5026

Golovkina TV, Prakash O, Ross SR (1996) Endogenous mouse mammary tumor virus Mtv-17 is involved in Mtv-2- induced tumorigenesis in GR mice. Virology 218: 14–22

Golovkina TV, Piazzon I, Nepomnaschy I, Buggiano V, de OV, Ross SR (1997) Generation of a tumorigenic milk-borne mouse mammary tumor virus by recombination between endogenous and exogenous viruses. J Virol 71:3895–3903

Harris JR (1991) The evolution of placental mammals. FEBS Lett 295:3–4

Hecht SJ, Carlson JO, DeMartini JC (1994) Analysis of a type D retroviral capsid gene expressed in ovine pulmonary carcinoma and present in both affected and unaffected sheep genomes. Virology 202:480–484

Hecht SJ, Stedman K, Carlson JO, DeMartini JC (1996) Distribution of endogenous type B and D retroviral sequences in ungulates and other mammals. Proc Natl Acad Sci USA 93:3297–3302

Hughes SH, Toyoshima K, Bishop JM, Varmus HE (1981) Organization of the endogenous proviruses of chickens: implications for origin and expression. Virology 108:189–207

Irwin DM, Kocher TD, Wilson AC (1991) Evolution of the cytochrome b gene of mammals. J Mol Evol 32:128–144

Lower R, Lower J, Kurth R (1996) The viruses in all of us: characteristics and biological significance of human endogenous retrovirus sequences. Proc Natl Acad Sci USA 93:5177–5184

McGee-Estrada K, Palmarini M, Fan H (2002) HNF-3beta is a critical factor for the expression of the jaagsiekte sheep retrovirus (JSRV) long terminal repeat in type II pneumoncytes but not in Clara cells. Virology 292:87–97

Mi S, Lee X, Li X, Veldman GM, Finnerty H, Racie L, LaVallie E, Tang XY, Edouard P, Howes S, Keith JC Jr, McCoy JM (2000) Syncytin is a captive retroviral envelope protein involved in human placental morphogenesis. Nature 403: 785–789

Nakagawa K, Harrison LC (1996) The potential roles of endogenous retroviruses in autoimmunity. Immunol Rev 152:193–236

Neil JC, Fulton R, Rigby M, Stewart M (1991) Feline leukaemia virus: generation of pathogenic and oncogenic variants. Curr Top Microbiol Immunol 171:67–93

Ortin A, Minguijon E, Dewar P, Garcia M, Ferrer LM, Palmarini M, Gonzalez L, Sharp JM, De LH (1998) Lack of a specific immune response against a recombinant capsid protein of jaagsiekte sheep retrovirus in sheep and goats naturally affected by enzootic nasal tumour or sheep pulmonary adenomatosis. Vet Immunol Immunopathol 61:229–237

Palmarini M, Dewar P, De las Heras M, Inglis NF, Dalziel RG, Sharp JM (1995) Epithelial tumour cells in the lungs of sheep with pulmonary adenomatosis are major sites of replication for jaagsiekte retrovirus. J Gen Virol 76:2731–2737

Palmarini M, Cousens C, Dalziel RG, Bai J, Stedman K, DeMartini JC, Sharp JM (1996a) The exogenous form of jaagsiekte sheep retrovirus (JSRV) is specifically associated with a contagious lung cancer of sheep. J Virol 70:1618–1623

Palmarini M, Holland MJ, Cousens C, Dalziel RG, Sharp JM (1996b) Jaagsiekte retrovirus establishes a disseminated infection of the lymphoid tissues of sheep affected by pulmonary adenomatosis. J Gen Virol 77:2991–2998

Palmarini M, Sharp JM, De las Heras M, Fan H (1999) Jaagsiekte sheep retrovirus is necessary and sufficient to induce a contagious lung cancer in sheep. J Virol 73: 6964–6972

Palmarini M, Datta S, Omid R, Murgia C, Fan H (2000a) The long terminal repeat of jaagsiekte sheep retrovirus is preferentially active in differentiated epithelial cells of the lungs. J Virol 74:5776–5787

Palmarini M, Hallwirth C, York D, Murgia C, de Oliveira T, Spencer T, Fan H (2000b) Molecular cloning and functional analysis of three type D endogenous retroviruses of sheep reveal a different cell tropism from that of the highly related exogenous jaagsiekte sheep retrovirus. J Virol 74:8065–8076

Palmarini M, Maeda N, Murgia C, De-Fraja C, Hofacre A, Fan H (2001a) A phosphatidylinositol-3-kinase (PI-3 K) docking site in the cytoplasmic tail of the jaagsiekte sheep retrovirus transmembrane protein is essential for envelope-induced transformation of NIH3T3 cells. J Virol 75:11002–11009

Palmarini M, Gray CA, Carpenter K, Fan H, Bazer FW, Spencer T (2001b) Expression of endogenous betaretroviruses in the ovine uterus: effectes of neonatal age, estrous cycle, pregnancy and progesterone. J. Virol. 75:11319–11327

Palmarini M, Murgia C, Fan H (2002) Spliced and prematurely polyadenylated jaagsiekte sheep retroviurus (JSRV)-specific RNAs from infected or transfected cells. Virology 294:180–188

Perron H, Garson JA, Bedin F, Beseme F, Paranhos-Baccala G, Komurian-Pradel F, Mallet F, Tuke PW, Voisset C, Blond JL, Lalande B, Seigneurin JM, Mandrand B (1997) Molecular identification of a novel retrovirus repeatedly isolated from patients with multiple sclerosis. The Collaborative Research Group on Multiple Sclerosis. Proc Natl Acad Sci USA 94:7583–7588

Rabson AB, Graves BJ (1997) Synthesis and processing of viral RNA. In: Coffin JM, Varmus H (eds) Retroviruses, pp 205–261. Cold Spring Harbor Press: New York

Rai SK, DeMartini JC, Miller AD (2000) Retrovirus vectors bearing jaagsiekte sheep retrovirus Env transduce human cells by using a new receptor localized to chromosome 3p21.3. J Virol 74:4698–4704

Rosati S, Pittau M, Alberti A, Pozzi S, York DF, Sharp JM, Palmarini M (2000) An accessory open reading frame (orf-x) of jaagsiekte sheep retrovirus is conserved between different virus isolates. Virus Res 66:109–116

Ruscetti S, Davis L, Feild J, Oliff A (1981) Friend murine leukemia virus-induced leukemia is associated with the formation of mink cell focus-inducing viruses and is blocked in mice expressing endogenous mink cell focus-inducing xenotropic viral envelope genes. J Exp Med 154:907–920

Sauter M, Schommer S, Kremmer E, Remberger K, Dolken G, Lemm I, Buck M, Best B, Neumann-Haefelin D, Mueller-Lantzsch N (1995) Human endogenous retrovirus K10: expression of gag protein and detection of antibodies in patients with seminomas. J Virol 69:414–421

Sharp JM, Herring AJ (1983) Sheep pulmonary adenomatosis: demonstration of a protein which crossreacts with the major core proteins of Mason-Pfizer monkey virus and mouse mammary tumor virus. J Gen Virol 64:2223–2227

Spencer TE, Bazer FW (1995) Temporal and spatial alterations in uterine estrogen receptor and progesterone receptor gene expression during the estrous cycle and early pregnancy in the ewe. Biol Reprod 53:1527–1543

Spencer TE, Stagg AG, Joyce MM, Jenster G, Wood CG, Bazer FW, Wiley AA, Bartol FF (1999) Discovery and characterization of endometrial epithelial messenger ribonucleic acids using the ovine uterine gland knockout model. Endocrinology 140:4070– 4080

Spencer TE, Mura M, Gray AC, Griebel PJ, Palmarini M (2002) Receptor usage and fetal expression of orine endogenous betaretroviruses: implications for the co-evolution of endogenous and exogenous retroviruses. J Virol In Press

Stoye JP, Coffin JM (1988) Polymorphism of murine endogenous proviruses revealed by using virus class - specific oligonucleotide probes. J Virol 62:168–175

Stoye JP, Moroni C, Coffin JM (1991) Virological events leading to spontaneous AKR thymomas. J Virol 65:1273–1285

Ting CN, Rosenberg MP, Snow CM, Samuelson LC, Meisler MH (1992) Endogenous retroviral sequences are required for tissue-specific expression of a human salivary amylase gene. Genes Dev 6:1457–1465

Tomonaga K, Coffin JM (1998) Structure and distribution of endogenous nonecotropic murine leukemia viruses in wild mice. J Virol 72:8289–8300

Wooding FB (1982) The role of the binucleate cell in ruminant placental structure. J Reprod Fertil Suppl 31:31–39

York DF, Vigne R, Verwoerd DW, Querat G (1991) Isolation, identification, and partial cDNA cloning of genomic RNA of jaagsiekte retrovirus, the etiological agent of sheep pulmonary adenomatosis. J Virol 65:5061–5067

CHAPTER 6

Transformation and Oncogenesis by Jaagsiekte Sheep Retrovirus

H. FAN, M. PALMARINI, J.C. DEMARTINI

H. FAN
Department of Molecular Biology and Biochemistry, and Cancer Research Institute, University of California, Irvine, CA 92697, USA
e-mail: hyfan@uci.edu

M. PALMARINI
Department of Medical Microbiology and Parasitology, College of Veterinary Medicine, University of Georgia, Athens, GA 30607, USA
e-mail: mpalmari@vet.uga.edu

J.C. DEMARTINI
Department of Microbiology, Immunology and Pathology, College of Veterinary Medicine and Biomedical Sciences, Colorado State University, Fort Collins, CO 80523-1671, USA
e-mail: james.demartini@colostate.edu

Abstract. Jaagsiekte sheep retrovirus (JSRV) is an exogenous retrovirus of sheep that induces a contagious lung cancer, ovine pulmonary adenocarcinoma (OPA). JSRV is a potent carcinogen in the experimental setting, inducing end-stage tumors at around 6 weeks of age when newborn lambs are inoculated intratracheally. Despite this rapid oncogenesis, inspection of the JSRV genome sequence does not reveal any obvious viral oncogenes. In this review, recent advances in studies of JSRV oncogenic transformation are described. Molecular cloning of an infectious and oncogenic JSRV provirus was instrumental in the studies. DNA transfection of JSRV proviral DNA into mouse NIH3T3 cells results in morphological transformation, indicating that the JSRV genome carries an oncogene. Further experiments identified the JSRV envelope protein as the transforming gene, and a PI3 kinase docking site in the cytoplasmic tail of the transmembrane (TM) protein was shown to be necessary for transformation. Avian DF-1 cells infected with an avian retroviral vector (RCAS) expressing the JSRV envelope protein also undergo tumorigenic transformation. Possible mechanisms of transformation are discussed, and a cooperating role for insertional activation of proto-oncogenes in tumorigenesis is also considered. The transforming potential of the JSRV envelope protein may be necessary for JSRV infection and replication in vivo.

1 Introduction

Over the years, retroviruses have played important roles in investigations of oncogenesis. In particular, research on these viruses led to the discovery of viral oncogenes and cellular proto-oncogenes (Bishop and Varmus 1984; Rosenberg and Jolicoeur 1997). In addition, naturally occurring retroviruses are important in a number of animal and human cancers. Moreover, oncogenic principles elucidated by retroviruses have been found to be applicable to non-virus associated human cancers as well. In this chapter, we will review the mechanisms of oncogenesis by jaagsiekte sheep retrovirus (JSRV), and compare it to other retroviruses.

The replication cycle of retroviruses has been studied extensively, and is comprehensively reviewed in several chapters and volumes (Coffin 1996; Coffin et al. 1997). Briefly, retroviruses are positive stranded RNA-containing viruses. Each virus particle contains two identical copies of viral genomic RNA (7–12 kb, depending on the virus); the viral RNA is contained within a viral capsid or core that is made up of viral structural proteins (encoded by the *gag* gene) as well as viral enzymes (encoded by the *pol* gene). The viral core is surrounded by a viral envelope, consisting of viral proteins (encoded by the *env* gene) embedded in host cell-derived lipid bilayer. There are two envelope proteins, the surface glycoprotein (SU) and the transmembrane protein (TM). The SU protein is on the surface of the virus particle, and is responsible for binding of virus to a receptor on the surface of the infected cell. The TM protein spans the lipid bilayer of the virion, and serves to anchor the SU protein to the virus particle. During infection, virions bind to specific receptors on the surface of the infected cell; subsequently the viral cores are internalized into the cytoplasm (viral binding and entry). In the cytoplasm, virion-bound reverse transcriptase is activated, after which it reverse transcribes the viral RNA into linear double stranded viral DNA. The viral DNA is then transported to the nucleus where it is integrated into host chromosomal DNA by the viral integrase, to form the provirus. Proviral DNA is then transcribed by cellular RNA polymerase II to give full-length viral RNA molecules. These viral RNAs are then transported to the cytoplasm with and without mRNA splicing. For most retroviruses (simple retroviruses) a singly spliced mRNA encodes the envelope polyprotein; unspliced mRNA is translated into Gag and Gag–Pol polyproteins. Other unspliced viral RNA combines with the viral polyproteins to form virus particles that bud from the surface of the cell. Recently budded virus particles are immature, in that the Gag and Gag-Pol polyproteins are uncleaved. Shortly after budding, the viral protease (also encoded by the *pol* gene) becomes activated, leading to cleavage of the polyproteins. It should be noted that for most retroviruses, infection is not a lytic process. Rather, the end result is a cell that is continually producing virus. The non-lytic nature of retroviral infection favors cell transformation and oncogenesis.

As a result of the reverse transcription process, the product viral DNA is somewhat larger than the template viral RNA. In particular, direct terminal repetitions are generated at either end of the viral DNA – the long terminal repeats or LTRs (Telisnitsky and Goff 1997; see the chapter by Palmarini and Fan, this volume). The viral LTRs are important for viral expression, as they contain the transcriptional signals for RNA synthesis and cleavage/polyadenylation. In particular, the U3 region of the LTR contains viral enhancer sequences as well as a basal promoter, while a cleavage/polyadenylation signal is present in the R region. In the infected cell, initiation of transcription begins at the U3/R boundary in the upstream LTR and cleavage/polyadenylation occurs at the R/U5 boundary of the downstream LTR.

2
Oncogenesis by Previously Studied Retroviruses

Oncogenic retroviruses have been known for nearly 100 years. The first retroviral cancer was described for avian myeloblastosis virus by Ellerman and Bang (Ellerman and Bang 1908). Subsequently, other important oncogenic animal retroviruses were discovered, including Rous sarcoma virus (Rous 1911), murine mammary tumor virus (MMTV) (Bittner 1942) and murine leukemia virus (MuLV; several strains) (Gross 1951). More recently, bovine leukemia virus (BLV) (Van der Maaten et al. 1972) and the related human retrovirus human T cell lymphotropic virus (HTLV; Poiesz et al. 1980) were identified. There are several mechanisms by which these viruses induce cancers, and they will be first reviewed here. A comprehensive description of retroviral oncogenesis has been published recently (Rosenberg and Jolicoeur 1997).

2.1
Acute Transforming Retroviruses

Oncogenic retroviruses have been historically divided on the basis of the rapidity with which they induce disease. One group of retroviruses is the acute transforming retroviruses that induce neoplasms rapidly. Acute transforming retroviruses have been identified in many animal species, including birds, cats, mice and primates. Classic experiments on such viruses were carried out on avian sarcoma virus [e.g., Rous sarcoma virus (RSV); Duesberg and Vogt 1970; Stehelin et al. 1976]. Acute transforming retroviruses may induce tumors in susceptible animals within a few weeks of infection, and the tumors are generally polyclonal. These viruses frequently can morphologically transform cells when infected in culture as well. The rapid oncogenic and cell transformation properties of these viruses are associated with the presence of particular genes collectively referred to as viral oncogenes. For instance, the oncogene of RSV is v-*src* (Duesberg and Vogt 1970), which encodes a tyrosine specific protein kinase (Collett and Erikson 1978; Sefton et al. 1980). A very important discovery was that viral oncogenes are derived from normal cell genes, collectively referred to as proto-oncogenes (STEHELIN et al. 1976). As a class, proto-oncogenes are involved in various aspects of cell growth and regulation in normal cells. Acute transforming retroviruses result from viral transduction of a cellular proto-oncogene. During this process, the coding sequences for the viral oncogene typically are altered in comparison to the cellular proto-oncogene through base substitutions, domain deletions, fusion to other sequences, or a combination of these (ROSENBERG and JOLICOEUR 1997). As a result, viral oncogenes cause uncontrolled stimulation of cell growth.

The viral oncogenes in acute transforming retroviruses are not required for viral replication. Indeed, most acute transforming retroviruses are replication-defective. They are dependent on co-infection with a replication-competent retrovirus – a helper virus. For instance, some stocks of RSV are dependent on a helper virus referred to as Rous-associated virus.

2.2
Non-Acute Retroviruses

The second broad group of oncogenic retroviruses is referred to as non-acute retroviruses. These are replication-competent retroviruses, and they lack oncogenes. As a result, they induce tumors slowly (typically months to years), and they do not transform cells in culture. A common mechanism for oncogenesis by non-acute retroviruses has been identified: insertional activation of proto-oncogenes (FAN 1997). Normally during retroviral infection, integration of proviral DNA is a random process with respect to the host chromosomal DNA (STEFFEN and WEINBERG 1978; WITHERS-WARD et al. 1994). That is, the chromosomal sites of proviral insertion will be different for each infected cell. However, in tumors induced by non-acute retroviruses, it has been found that there are common chromosomal sites of insertion. For instance, in B-lymphomas induced by avian leukosis virus, proviral insertion next to the *c-myc* proto-oncogene occurs (HAYWARD et al. 1981). In the tumors, proviral insertion results in transcriptional over-expression of the adjacent proto-oncogene. This occurs by one of two general mechanisms (Fig. 1). The first mechanism is 'promoter insertion' (HAYWARD et al. 1981). In this case, transcription initiates in the downstream (or in some cases upstream) viral LTR, followed by read-through into the downstream proto-oncogene sequences. Thus the proto-oncogene is placed under transcriptional control of the very active

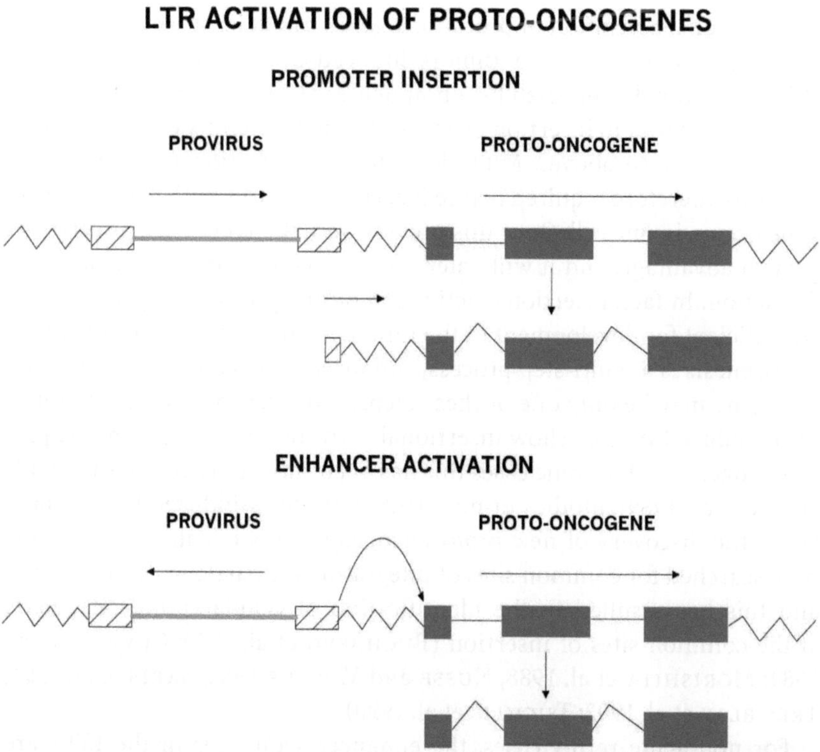

Fig. 1. LTR activation of proto-oncogenes. In tumors induced by non-acute retroviruses, LTR activation of proto-oncogenes frequently occurs. In the tumor cells, proviruses are inserted in the vicinity of a cellular proto-oncogene. For each case, the organization of the inserted viral DNA is shown in the *upper line*, with the provirus in the sequences between the *hatched boxes*, and the proto-oncogene exons indicated in the *shaded boxes*. Directions of transcription are indicated by the *arrows*. The resulting RNA transcript is shown below the DNA organization. In promoter insertion, read-through transcription from the viral LTR (*hatched box*) into the proto-oncogene places the proto-oncogene under transcriptional control of the highly active viral LTR promoter/enhancer. In enhancer activation, the viral enhancer (contained in the LTR) stimulates transcription from the proto-oncogene's endogenous promoter

viral LTR. The second mechanism of insertional activation is 'enhancer activation'. Transcriptional enhancers can function in a position-independent and orientation-independent manner. Thus in these tumors, proto-oncogene activation results from influence of an adjacent viral enhancer on the endogenous promoter of the proto-oncogene (NUSSE and VARMUS 1982; PAYNE et al. 1982).

Insertional activation of proto-oncogenes can at least partially explain the longer time course for tumors induced by non-acute retroviruses. When an animal is infected by a non-acute retrovirus, infections in most cells do not result in insertion of a provirus in the vicinity of a critical proto-oncogene (see above). Multiple rounds of infection (and additional time) are therefore required before insertion next to a critical proto-oncogene occurs in one cell. Once this process occurs, that particular cell has a growth advantage, and it will enter into the pathway of oncogenic transformation. In fact, insertional activation of one proto-oncogene may not be sufficient for development of the tumor. As for most cancers, retroviral oncogenesis is a multi-step process, and insertional activation of a proto-oncogene may be only one of these steps. Indeed, some non-acute retrovirus-induced tumors show insertional activation of more than one proto-oncogene, and in some cases this has been shown to occur sequentially (Bear et al. 1989). Studies of non-acute retrovirus-induced tumors have led to the discovery of new proto-oncogenes. In particular, investigators have searched for common sites of integration for particular tumor types, and this has resulted in the identification of cellular proto-oncogenes at the common sites of insertion (Buchberg et al. 1990; Cuypers et al. 1984; Morishita et al. 1988; Nusse and Varmus 1982; Saris et al. 1991; Tremblay et al. 1992; Tsichlis et al. 1990).

For non-acute retroviruses, the enhancer sequences in the LTRs are important determinants for disease (Chatis et al. 1984; DesGroseillers and Jolicoeur 1984; Lenz et al. 1984). This reflects the fact that in order for a retrovirus to induce tumors of a particular cell type, the LTR must have enhancer sequences that function efficiently in it. For instance the Friend strain of murine leukemia virus (MuLV) induces erythroleukemia, while the Moloney (M) strain induces T-lymphoma. It has been shown that a chimeric Friend MuLV containing only the enhancer sequences from M-MuLV induces T-lymphoma, while the reciprocal M-MuLV recombinant induces erythroleukemia (Chatis et al. 1984; Li et al. 1987). It has also been shown that the enhancer sequences of F-MuLV are preferentially active in erythroid vs. T-lymphoid cells, while the opposite is true for M-MuLV enhancers (Short et al. 1987).

2.3 HTLV and BLV

HTLV-I, HTLV-II and BLV are deltaretroviruses. These viruses induce adult T-cell leukemia (ATL) and B-lymphosarcoma, respectively. In the case of HTLV, leukemogenesis is relatively inefficient, with a small percentage of infected individuals ever developing the disease (Kondo et al. 1987; Murphy et al. 1989). Moreover, the latency for development of ATL is very long – typically decades. The common feature of the deltaretroviruses is that they have a region in their genomes that encodes several accessory proteins via multiply spliced mRNAs (Seiki et al. 1983). The best understood accessory proteins are Tax and Rex – Tax is a transcriptional transactivator (Sodroski et al. 1984), while Rex influences the efficiency of transporting unspliced or singly spliced viral mRNAs to the cytoplasm (Ahmed et al. 1990; Inoue et al. 1987). The Tax transactivator activates both the HTLV LTR itself, as well as cellular genes such as interleukin (IL)-2 (T-cell growth factor), IL-2 receptor, etc. (Greene et al. 1986). HTLV-I and HTLV-II can immortalize primary human T lymphocytes, so that they display some aspects of transformation. The likely mechanism for the T-cell immortalization is the simultaneous induction of IL-2 and IL-2 receptor in infected cells. This leads to an autocrine loop, in which the infected cell undergoes continual growth stimulation through the IL-2/IL-2R. It is noteworthy that there is a long latency between viral infection and the development of ATL in infected individuals; moreover, the resulting ATL cells frequently show relatively little expression of HTLV proteins (Reitz et al. 1983; Seiki et al. 1984). Thus it is possible that HTLV-I infection leads to initial expansion of the target cells for leukemogenesis, but that subsequent (non-viral) events are necessary for the ultimate development of leukemia.

2.4 MMTV

MMTV (the prototypic betaretrovirus) is a non-acute retrovirus. However some features of this virus merit particular mention. In addition to the standard retroviral proteins, MMTVs encode a small protein called superantigen or Sag (Rosenberg and Jolicoeur 1997; Ross et al. 1997). The MMTV Sag can bind to the T-cell receptor (TCR) complex in conjunction with CD4 protein on CD4-positive T lymphocytes, and in doing so stimulates activation of these cells. (Sags from different MMTV strains interact with different classes of TCR molecules, thus activating different sets of CD4-positive T lymphocytes.) It has become clear that the Sag protein is important for MMTV replication in vivo. The normal route of infection is through the milk – when nursing pups drink MMTV infected milk, initial infection occurs in the small intestines where B lymphocytes are the first cells infected (Golovkina et al. 1992). The infected B lymphocytes then transfer infection to T lymphocytes that amplify the infection, but in order for this to occur efficiently, activation (and multiplication) of the T lymphocytes must occur. Subsequently, infected B and T lymphocytes migrate to the mammary gland, where they infect mammary epithelial cells (Dzuris et al. 1997; Held et al. 1993). Thus MMTV encodes a viral protein that stimulates target cell multiplication as part of its normal life cycle.

3 JSRV As an Oncogenic Virus

3.1 Background

Several reviews on the biology of JSRV-induced disease have appeared (DeMartini and York 1997; Palmarini et al.1997; Palmarini and Fan 2001) (see the chapter by De las Heras and González and by Sharp and DeMartini, this volume). Briefly, JSRV is an exogenous retrovirus of sheep that is associated with development of lung cancer. The infection is broadly distributed geographically. In high endemic areas, the disease is of considerable economic importance. As many as 50% of animals in affected herds may be infected, and it has been estimated that the lifetime risk for development of the lung cancer is approximately 25% in an affected flock (see the chapter by Sharp and DeMartini, this volume). The disease induced by JSRV is referred to as ovine pulmonary adenocarcinoma or OPA [also sheep pulmonary adenomatosis (SPA) or ovine pulmonary carcinoma (OPC)]. OPA is a tumor of secretory epithelial cells in the lung – type II pneumocytes and Clara cells. Type II pneumocytes secrete lung surfactant, while Clara cells secrete other proteins for the distal airways. When infected sheep develop OPA, they exhibit respiratory distress that is directly related to the extension of the tumor mass, the accumulation of fluid in the lungs and the occurrence of secondary bacterial infection – hence the name jaagsiekte (Afrikaans for 'chasing' sickness). A field test for OPA is to elevate the body cavity over the mouth, at which point lung fluid will drain from the nose. In addition to being diagnostic for OPA, the lung fluid has been very useful experimentally. If lung fluid is filtered and concentrated, it will reproduce the disease when inoculated intratracheally into susceptible animals. The fact that filtered lung fluid could induce OPA demonstrated that there was a viral etiology.

3.2 Experimental Transmission of OPA

While the exact route of spread for JSRV infection in nature is not completely understood, it seems likely that spread of infection may involve inhalation of aerosols from an affected animal in the context of the close quarters of sheep flocks. When OPA lung fluid from field isolates is concentrated by ultracentrifugation and injected intratracheally into newborn lambs, disease occurs extremely rapidly. Under these conditions, animals develop end-stage disease from between 3 and 6 weeks post-inoculation (Sharp et al. 1983; Verwoerd et al. 1980) (see also the chapter by Sharp and DeMartini, this volume). In fact, in some cases the disease occurs as soon as between 10 days and 2 weeks (J.M. Sharp, personal communication). In contrast, if older animals are inoculated experimentally, disease occurs much more slowly. Likewise, in high endemic flocks, the rate at which infected animals progress to end-stage disease is slower as well. (In early experiments, non-concentrated lung fluid also induced OPA, although with a longer incubation time.) There are several possible explanations for the rapid onset of OPA after infection of newborns. One possible explanation is that newborn lambs are unable to mount an efficient immune response to the virus, while adults can. However, infected animals in general do not raise humoral immune responses to JSRV infection (Ortin et al. 1998; Sharp and Herring 1983), perhaps due to the early expression of closely related JSRV-like endogenous retroviruses in the sheep genome (see the chapter by DeMartini et al., this volume). (Cell-mediated immune responses to JSRV have not been extensively characterized in infected animals.) Another possible explanation is that the primary targets for JSRV infection/oncogenesis are cells that are present at high concentrations in newborn animals, but at lower concentrations in adults. It seems possible that a common precursor for type II pneumocytes and Clara cells might be the actual target for JSRV infection, and such a precursor would very likely be at higher concentrations in newborn lungs compared to adult lungs.

3.3
Progress in the Molecular Biology of JSRV

Beginning in 1990, progress has been made in understanding the molecular biology of JSRV (see the chapter by M. PALMARINI, H. FAN, this volume). Prior to that time, investigators had developed experimental evidence for the viral agent associated with OPA being a retrovirus. In particular, reverse transcriptase activity was found in OPA lung fluid (HERRING et al. 1983; MARTIN et al. 1976; PERK et al. 1974; VERWOERD et al. 1985), and immunological reagents for MMTV or the simian Mason-Pfizer monkey virus (MPMV) proteins were found to cross-react with OPA tumor tissues (SHARP and HERRING 1983; VERWOERD et al. 1985). In 1991 and 1992, the complete nucleotide sequence for a novel retrovirus present in OPA lung fluid was deduced from a series of cDNA clones prepared from reverse transcribed viral RNA (YORK et al. 1991, 1992). The putative virus was designated jaagsiekte sheep retrovirus or JSRV, the presumed causative agent of OPA. The genome organization of JSRV is shown in Fig. 2A. It has a structure typical of betaretroviruses, with *gag*, *pro* and *pol* all in different reading frames. Phylogenetically, JSRV was found to group with the betaretroviruses MMTV and MPMV, consistent with the immunological cross-reactivity with OPA tumors. Another noteworthy feature of the JSRV genome was an alternate reading frame in the pol gene – *orf-x*. The coding sequences for *orf-x* did not bear strong resemblance to other cellular or viral genes, although a weak suggestion of similarity with the adenosine A3 receptor was noted (BAI et al. 1999). Unfortunately, at the time it was not possible to demonstrate infectivity or oncogenicity with the deduced JSRV sequence, because reconstructed full-length clones were not infectious (D. YORK and G. QUERAT, unpublished results). To some extent, this may have been due to the fact that a tissue culture system for the virus did not exist at the time.

Despite the inability to demonstrate that the putative JSRV virus was the positive agent of OPA, the availability of the sequence allowed generation of important hybridization and immunological probes (BAI et al. 1996; PALMARINI et al. 1996a; PALMARINI et al. 1996b). The early experiments also made it clear that highly related endogenous JSRV-like proviruses exist in multiple copies in the sheep genome (HECHT et al. 1996; YORK et al. 1992). Therefore, molecular studies had to differentiate between endogenous JSRV-like proviruses and exogenous JSRV. Once such procedures were established (PALMARINI et al. 1996a), it was possible to show that all OPA

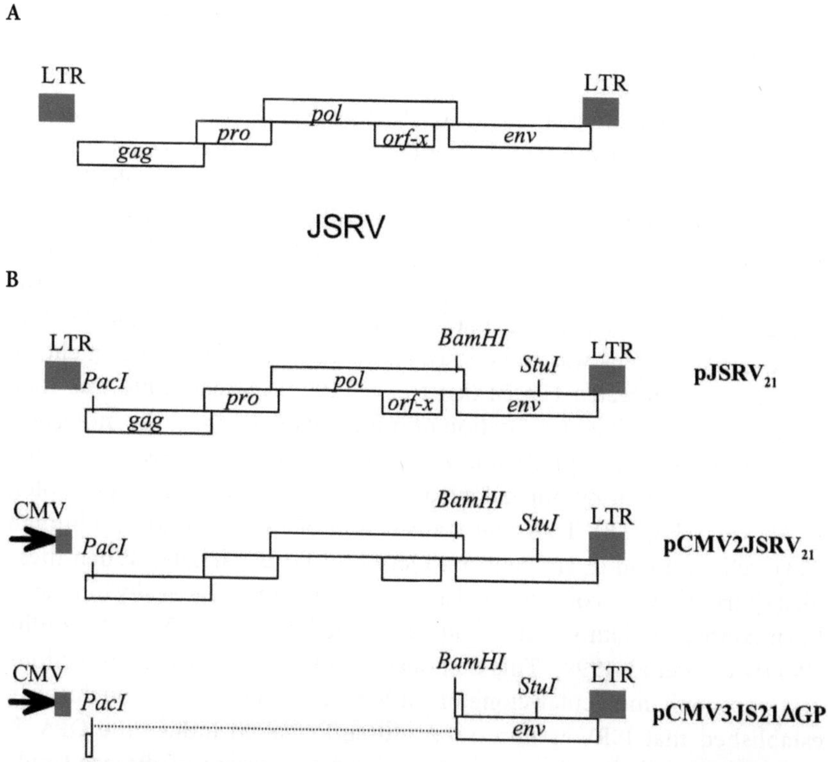

Fig. 2 A, B. Jaagsiekte sheep retrovirus. **A** The organization of the JSRV genome is shown. The different open reading frames are indicated and labeled. **B** JSRV expression plasmids used in molecular and cellular studies. pJSRV21 is a plasmid containing a complete JSRV provirus. pCMV2JS21 is a derivative of pJSRV21 in which the enhancer and promoter in the upstream LTR have been replaced by the human CMV immediate early promoter. pCMV3JS21ΔGP is a derivative of pCMV2JS21 in which all coding sequences except for *env* have been deleted. These plasmids are described in MAEDA et al. (2001)

tumors contained exogenous JSRV DNA, and the tumor cells stained positively with polyclonal antibodies raised against the JSRV *gag* sequences. These experiments also indicated that while low levels of exogenous JSRV DNA can be detected in various blood cells from infected animals (HOLLAND et al. 1999), there was little evidence for expression of viral protein in the cells. The only cells in which JSRV protein can be reliably detected by immunohistochemistry are the tumor cells themselves (PALMARINI et al. 1995). This pattern of in vivo expression is unusual for retroviruses –

in most other cases, multiple infected cell types can be detected, and the extents of viral infection and expression are often quite high.

More recently, additional progress has been made in the molecular biology of JSRV (see the chapter by M. PALMARINI, H. FAN, this volume). In particular, the availability of the deduced JSRV sequence, as well as hybridization or restriction endonuclease cleavage conditions that distinguish between exogenous and endogenous JSRV-like proviruses, made possible molecular cloning of a complete integrated JSRV provirus (PALMARINI et al. 1999). In vivo transfection of a plasmid form of the integrated provirus (pJS21) (by intratracheal inoculation) into newborn lambs indicated that the cloned provirus was infectious, although in the first experiment no tumors were detected (PALMARINI et al. 1999). A technique was then developed to facilitate production of infectious JSRV from the molecular clone. A derivative of pJS21 was generated, in which expression was driven by the human cytomegalovirus (CMV) immediate early promoter, pCMVJS21 (Fig. 2B). Transient transfection of this plasmid into human 293T cells resulted in efficient production of virus particles. When these virus particles were concentrated and inoculated intratracheally into newborn lambs, two lambs out of four developed classical OPA at 4 months (PALMARINI et al. 1999). This demonstrated that JSRV21 is an infectious and oncogenic molecular clone of JSRV, and furthermore it conclusively established that JSRV is necessary and sufficient to induce the OPA. It should be noted that the time course for development of disease (mid-point at 4 months) was somewhat longer than for concentrated field isolates (mid-point 6 weeks). This was not entirely surprising, as in the case of other retroviruses it has often been found that molecular clones show slower oncogenicity then the corresponding natural isolates. On the other hand, it is also possible that additional factors or agents in the field isolates contributed to the more rapid tumorigenicity (see below). A second oncogenic molecular clone of JSRV, $JSRV_{JS7}$ has been isolated recently (DEMARTINI et al. 2001).

4 Possible Mechanisms for JSRV Oncogenesis

Once it had been established that JSRV is the causative agent of OPA, the mechanisms of oncogenesis were of great interest. In particular, it was important to determine if JSRV is an acute transforming retrovirus, or a non-acute retrovirus. Several features of JSRV-induced OPA are consistent with the virus being an acute transforming retrovirus. In particular, the rapid time course of disease in experimental inoculations of field isolates is noteworthy. Moreover, in experimentally infected animals, histopathology reveals multi-focal tumors (although mono- or polyclonality has not been assessed yet) (see the chapter by De las Heras and González, this volume). On the other hand, inspection of the JSRV nucleotide sequence does not reveal any putative proteins with characteristics typically associated with viral oncogenes, i.e., genes with homology to cellular genes (proto-oncogenes). However, the *orf-x* reading frame is of interest, as it does not have homology to other retroviral proteins that are essential for replication. Two lines of evidence suggest that *orf-x* is not likely to be an oncogene. First, when we sequenced three endogenous JSRV related proviruses, one of them had an intact *orf-x* reading frame (Palmarini et al. 2000b). This suggested that the progenitor to the endogenous JSRV-related retroviruses had a functional *orf-x* reading frame at the time that it was inserted into the sheep genome (around 1–5 million years ago). Expression of endogenous JSRV-related proviruses occurs in sheep (Palmarini et al. 2001a), so it seems unlikely that endogenous proviruses expressing an oncogenic protein (e.g., Orf-X) would be maintained. Conservation of an open *orf*-x reading frame among exogenous and endogenous JSRV isolates has also been reported by Rosati et al. (2000). The second line of evidence is the deduced nucleotide sequence of a closely related exogenous retrovirus, enzootic nasal tumor virus (ENTV/ONAV) of goats (Cousens et al. 1999) (see the chapter by De las Heras, this volume). In the putative ONAV genome, the *orf-x* reading frame is not open in its entirety. Thus for this oncogenic JSRV-related virus, *orf-x* is apparently not important. However, it should be mentioned that an infectious and oncogenic molecular clone of ONAV has not been isolated yet; the status of the *orf-x* reading frame in such a clone would be definitive.

There are also arguments for and against JSRV being a non-acute retrovirus. First, as discussed above, inspection of the JSRV nucleotide sequence did not reveal any genes resembling viral oncogenes. Thus there were no obvious candidate oncogenes in the genome. On the other hand, since non-acute retroviruses typically induce tumors via activation of proto-oncogenes, evidence for such a process in JSRV-induced OPA would be very relevant. The JS7 OPA-derived tumor cell line is informative in this regard. JS7 cells contain one integrated JSRV provirus (DeMartini et al. 2001). The provirus from JS7 cells has been molecularly cloned, and the sequence of the molecular clone was used to identify the proviral insertion site. This site has been identified as the gene for the sheep pulmonary surfactant protein A (SP-A). It is unlikely that SP-A protein could function as an oncogene, as it is a structural component of pulmonary surfactant. On the other hand, the SP-A protein is one of the most highly expressed proteins in a type II pneumocyte, so it appears that the JSRV provirus has integrated into a region of highly active chromatin in these tumor cells. Thus in a tumor cell line that contains only one JSRV provirus, the insertion does not appear to be at a proto-oncogene.

As mentioned above, the molecularly cloned JSRV derived from pJS21 induced OPA somewhat more slowly than reported for concentrated field isolates (4 months vs. approximately 6 weeks). One possible explanation could be that the field isolates consist of an acute transforming retrovirus derived from JSRV, along with a replication-competent JSRV as helper virus. We screened for this possibility during molecular cloning of $JSRV_{21}$. Other recombinant lambda phage clones containing exogenous JSRV sequences were identified, including clones that appeared to be less than full-length or deleted. Such clones might represent an acute transforming version of JSRV. One clone meeting these criteria was identified, and subjected to nucleotide sequencing. However the sequencing did not reveal an adjacent open reading frame that might encode a viral oncogene protein (M. Palmarini and H. Fan, unpublished results).

5
Direct In Vitro Transformation of Cells by JSRV

As indicated above, one of the characteristics of acute transforming retroviruses is that they frequently can morphologically transform cells in culture. In light of this, an assay for in vitro transformation by the JSRV genome was carried out. In one series of experiments, murine NIH3T3 cells were used. Numerous investigators have used these cells to test for the presence of viral oncogenes or activated cellular proto-oncogenes (Shih and Weinberg 1982). The typical assay is to transfect NIH3T3 cells with an expression plasmid for the DNA under question, followed by culturing the cells under conditions for focus formation. The appearance of foci of transformed cells above a monolayer of contact-inhibited cells is indicative of a gene with transformation potential. For these experiments, the pCMVJS21 plasmid was used (in which the JSRV genome is driven by the human CMV immediate early promoter/enhancer) (Fig. 2B), as this would ensure efficient transcription of the JSRV sequences in NIH3T3 cells

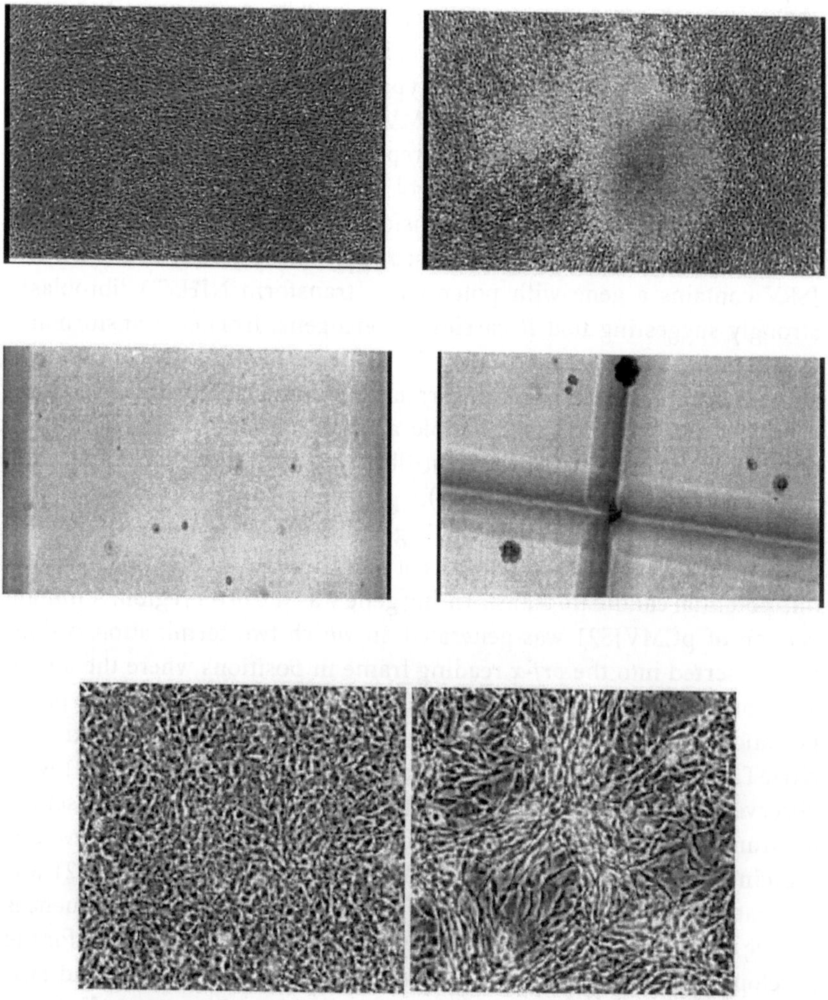

(Maeda et al. 2001). Previous experiments had demonstrated that the JSRV LTR is quite specific for lung epithelial-derived cells, and that it is relatively inactive in other cell types including NIH3T3 fibroblasts (Palmarini et al. 2000a). When pCMVJS21 was transfected into NIH3T3 cells, foci of transformed cells were reproducibly obtained (Fig. 3A, B). In contrast, transfection with a control plasmid lacking JSRV sequences yielded no foci. Moreover, when transformed foci were picked, JSRV DNA could be detected in all the transformants. These results indicated that JSRV contains a gene with potential to transform NIH3T3 fibroblasts, strongly suggesting that it carries an oncogene. Isolated transformants also showed the capacity to grow in soft agar suspension (Fig. 3C, D). Growth in agar is generally well correlated with the ability of transformed cells to form tumors in susceptible animals. Similar results have been obtained in transfection of the rat fibroblast cell lines 208F (Rai et al. 2001) and Rat6 (Maeda et al. 2001).

Once transformation by the JSRV genome was demonstrated, identification of the transforming gene was of great interest. As mentioned above, one potential candidate transforming gene was the *orf-x* region. A mutant version of pCMVJS21 was generated, in which two termination codons were inserted into the *orf-x* reading frame in positions where the amino acids of the overlapping integrase protein would not be affected, to give a plasmid pCMVJS21ΔorfX. When this plasmid was transfected into NIH3T3 cells, the same number of foci as for the original pCMVJS21 were observed (Table 1) (Maeda et al. 2001). Thus Orf-x protein is not essential for transformation. The next possibility tested was if the JSRV envelope protein is the transforming protein. Another derivative of pCMVJS21 was generated, pCMVJS21ΔGP (Fig. 2B). In this plasmid, the coding sequences for *gag* and *pol* were deleted, but the splice donor and acceptor sites for the envelope gene were maintained. As shown in Table 1, this plasmid efficiently induced transformed foci (generally somewhat more efficiently than the full-length pJS21 itself). This result indicated that the JSRV envelope protein (or some portion of it) is responsible for the transformation. Studies on the viral RNAs present in JSRV-expressing cells indicated that there was more than one form of spliced envelope mRNA (Maeda et al. 2001). In particular, an intriguing prematurely polyadenylated envelope mRNA was detected, that in principle could encode a form of envelope protein containing the SU domain alone. Two additional envelope expression plasmids were generated to address the potential role of the truncated envelope protein in transformation. In one case, sequences down-

Table 1. Transformation assays performed in this study[a] (from MAEDA et al. 2001)

	DNA(-)	pcDNA3.1(-)	pJSRV21	pCMV2JS21	pCMV2JS21 Δorfx	pCMV3JS21 ΔGP
Exp. 1	0[a]	0	0	15	14	N.T.
Exp. 2	0	0	4	11	12	N.T.
Exp. 3	0	0	1	16	12	N.T.
Exp. 4	0	0	0	21	22	N.T.
Exp. 5	0	N.T.	1	18	14	N.T.
Exp. 6	0	0	N.T.	22	N.T.	31
Exp. 7	0	0	N.T.	11	N.T.	23
Exp. 8	0	0	N.T.	N.T.	N.T.	15
Exp. 9	0	0	N.T.	N.T.	N.T.	13
Exp. 10	0	0	N.T.	N.T.	N.T.	18

Exp., experiment; N.T., not tested.

[a] NIH3T3 were transfected with standard amounts of the above indicated plasmids. Numbers indicate the number of foci counted at 28 days post-transfection. DNA(-) is a transfection with no DNA, and pcDNA3.1(-) is a negative control transfection with a plasmid containing a CMV promoter/enhancer.

stream from the upstream polyadenylation site were deleted, so that only the truncated form of the envelope mRNA could be expressed. When this truncated envelope expression plasmid was transfected into NIH3T3 cells, no foci were observed. This indicated that the truncated form of envelope protein is not sufficient for transformation (MAEDA et al. 2001). The second envelope expression plasmid contained a single base mutation in the cryptic upstream polyadenylation site that should remove the capacity to express the truncated envelope mRNA. When this plasmid was transfected into NIH3T3 cells, foci were still observed (N. MAEDA and H. FAN, unpublished results). This indicated that the truncated envelope mRNA is not necessary for transformation. Taken together, these results indicated that the truncated JSRV envelope mRNA is neither necessary nor sufficient for transformation. As a result, it is very likely that full-length envelope protein is required for transformation. This is the first reported case of a native retroviral envelope protein having the capacity to transform cells.

The transformation capacity of JSRV envelope protein also has been demonstrated in another cell system, avian DF-1 cells. A system was developed to examine the subgenomic components of the virus for oncogenic effects in vitro. This system used avian sarcoma virus (ASLV)-derived vectors (pRCAS) (Federspiel and Hughes 1997) and the avian embryo fibroblast cell line, DF-1. The DF-1 cell line is a non-transformed, immortalized avian cell line that efficiently supports replication of ASLV-derived vectors and has the additional advantage of having been used for the study of oncogenic transformation (Himly et al. 1998; Schaefer-Klein et al. 1998). The pRCAS vectors were constructed by removing the *src* gene from the virus genome and replacing it with a cloning site (Federspiel and Hughes 1997). JSRV *gag*, *pro*, *pol*, *orf-x*, *env*, were each cloned separately into pRCAS vectors and their effect on DF-1 cells monitored. Only cells transfected with the *env* construct (pRCAS-J:*env*) were shown to transform DF-1 cells (Fig. 3E, F) (Allen et al., 2002). Morphological changes consistent with transformation were observed, including an increase in saturation density and a decrease in adhesion to solid substrates. Transfection with the reporter construct containing the other components of the JSRV genome did not induce morphological changes in DF-1 cells, eliminating the possibility that the pRCAS vector, the transfection protocol, or culturing techniques were responsible for the transformed phenotype. When pRCAS-J:*env* transfected cells were implanted into nude mice, tumors formed, verifying that the transformed DF-1 cells were tumorigenic (Allen et al., 2002).

6 Identification of the JSRV Receptor, and Potential Implications for Oncogenesis

The nature of the JSRV receptor has been of interest, as JSRV is potentially a useful platform for development of retroviral vectors for expression in lung cells. The general problem with other retroviral vectors has been that they are sensitive to the surfactant environment (detergent-like) of the lung. Thus the use of retroviral vectors for treatment of diseases with lung manifestations such as cystic fibrosis has been problematic. The fact that the normal route of infection for JSRV is through the lungs has suggested that JSRV-based vectors may be more capable of in vivo infection in lung epithelia. Thus shortly after infectious molecular clones of JSRV were obtained, experiments were begun to characterize and clone the JSRV receptor (see the chapter by Miller, this volume). Rai et al. (2000) used JSRV envelope pseudotypes of murine retroviral vectors to determine that the JSRV receptor is present on ovine and human cells, but that rodent cells are infected only very poorly by JSRV envelope. In addition, the JSRV receptor from human cells was mapped to chromosome 3. Molecular cloning of the JSRV receptor from human cells was then accomplished by RAI et al. (2001). They identified the receptor as hyaluronidase-2 (Hyal-2). Hyaluronidases (encoded by several related genes) cleave hyaluronic acid from the cell surface, and there have been suggestions that hyaluronidase expression affects tumorigenicity. RAI et al. (2001) further showed that the Hyal-2 protein is linked into the plasma membrane by a glyocsyl phosphatidyl inositol (GPI) linkage – a linkage used by several cellular signaling molecules. This raised the possibility that Hyal-2 may have another function besides hyaluronidase activity – perhaps cell signaling through its GPI linkage.

The finding that Hyal-2 is the JSRV receptor is quite interesting. Other investigators have sought evidence of tumor suppressor gene loss in human tumors. One common method to scan tumor cells for tumor suppressor loss is to examine them cytologically and/or molecularly for chromosomal loss of heterozygosity (LOH). A prominent common site of LOH in lung cancers is on human chromosome 3p21.3 (SEKIDO et al. 1996). The Hyal-2 gene resides in the 3p21.3 region, and it has thus been considered a candidate tumor suppressor genes for lung tumors. If Hyal-2 is indeed a tumor suppressive gene, then this would suggest a quite novel mechanism for JSRV-induced cell transformation, as described below.

7
Possible Mechanisms for JSRV Envelope-Induced Transformation

The finding that JSRV envelope is able to transform cells immediately raises questions as to the mechanism of transformation. There are three possible models for how this transformation could take place:

1. The JSRV envelope SU protein could bind its normal receptor (Hyal-2) on the surface of the transformed cell.
   In doing so, it could interfere with a growth suppressive activity of Hyal-2 or it could lead to stimulation of cell growth via the Hyal-2 protein. If Hyal-2 is indeed a tumor suppressor gene, then JSRV-induced transformation could result from interference of the normal tumor suppressor function of Hyal-2 by binding of SU to the external portion of Hyal-2. This would be a very novel mechanism for a retroviral transformation.

Alternatively, if SU binds to the extracellular portion of Hyal-2, it could also potentially stimulate signal transduction from Hyal-2 to other regulatory mechanisms. There is precedence for this in other retroviruses. For instance, when human immunodeficiency virus (HIV)-1 infects target cells, the primary interactions are with the CD4 surface molecule as the receptor, as well as with a coreceptor. In the case of macrophage tropic HIVs, the coreceptor is the CCR5 chemokine receptor (Alkhatib et al. 1996; Choe et al. 1996; Deng et al. 1996; Dragic et al. 1996). It has been shown that when M-tropic HIV-1 envelope interacts with CCR5-positive cells, it leads to an intracellular signaling event (intracellular calcium release) (Weissman et al. 1997).

A question about this potential mechanism of transformation arises from the fact that rodent cells lack functional JSRV receptors, yet rodent fibroblasts can be morphologically transformed by JSRV. Thus, it might seem unlikely that JSRV envelope interacts with the cellular receptor in rodent cells, even though it can efficiently transform them. However, rodent cells in fact show low levels of infectability by JSRV pseudotypes of retroviral vectors (Rai et al. 2000). Thus, JSRV envelope might interact with the murine Hyal-2 protein, even if this does not result in efficient viral entry. The domains of envelope responsible for viral entry might be different from those required for the physiological function of Hyal-2 (i.e., growth stimulation or growth inhibition).

2. The JSRV envelope SU protein binds to a cell surface protein distinct from Hyal-2, and it stimulates cell growth through this other cell surface protein.

   There is precedence for retroviral envelope proteins binding non-receptor cell surface proteins and stimulating growth. In the case of Friend erythroleukemia virus, an acute transforming component is spleen focus forming virus (SFFV). It has been shown that SFFV can transform erythroid precursors, and the oncogene of SFFV is an internally deleted form of a recombinant envelope protein, gp55. Moreover, it has been shown that gp55 binds to the erythropoietin receptor on the surface of erythroid cells (Li et al. 1990), and in doing so it stimulates cell growth through the JAK/STAT pathway (Ohashi et al. 1997). Recently, it has also been shown that the SU envelope proteins of both MMTV and MuLV bind to the Toll-like receptor (TLR) 4 cell surface protein (Rassa et al. 2002). TLR4 is present on the surface of B lymphocytes, and its normal function is to mitotically activate them in response to binding of a ligand. Thus, MMTV and MuLV appear to activate target cells during the course of in vivo infection.

3. The JSRV TM protein binds a cytoplasmic protein via its cytoplasmic tail, leading to signaling for cell growth.
   In a transformed cell, the TM protein would probably adopt a configuration in which the N terminus is on the extracellular side, while the C terminus is in the cytoplasm. As for many retroviruses, the JSRV TM protein has a short cytoplasmic tail of 48 amino acids. It is possible that cellular proteins could interact with this cytoplasmic tail, resulting in intracellular signaling through a signal transduction pathway. There is also precedence for such interactions. In the case of simian immunodeficiency virus (SIV), there is an extremely pathogenic strain, SIVmacPBJ14 (Fultz et al. 1989). The pathogenicity of this virus has been mapped to the envelope domain, and in particular sequences in the cytoplasmic tail of the TM protein. These sequences have been suggested to bind cellular factors and to signal for cell death. Another relevant example is the Nef protein of HIV and SIV. The cytoplasmic domain of Nef can bind a number of proteins via SH2 motifs (Du et al. 1995). In particular, protein kinases of the PAK family have been shown to bind (Sawai et al. 1994).

8 Further Investigations into the Mechanism of JSRV Transformation

Recent experiments have addressed the mechanism of transformation by JSRV envelope. A critical first step in these studies was to identify the regions of the JSRV envelope that are necessary for transformation. Comparisons between the exogenous JSRV and related endogenous proviruses were interesting. There is extensive sequence homology between the exogenous and endogenous viruses (Palmarini et al. 2000b). There are three regions of sequence variability: two regions in the *gag* gene (VR1 and VR2) and a region in the *env* gene (VR3). The VR3 region encompasses the membrane-spanning region and the cytoplasmic tail of TM (Fig. 4). As all sheep carry multiple copies of the endogenous JSRV related proviruses, it seems likely that the endogenous envelopes will not induce transformation or oncogenesis. Therefore, a series of chimeras between the envelopes of the exogenous JSRV and endogenous JSRV-related proviruses was generated in the context of a JSRV-based envelope expression plasmid (pCMV2JS21ΔGP, Fig. 2B). The different chimeras were then

```
JSRV21    MPKRRAGFRKGWYARQRNSLTHQMQRMTLSEPTSELPTQRQIEALMRYAWNEAHVQPPVTPTNILIMLLLLQRVQNGAA 87
JSRV-SA   ..............................................P................................. 80
ENTV      ...H.................Y.......N.T.NG......V.....H..........A..IK...........I..... 80
enJS56A1  ..........................................................................I..... 87
enJS5F16  ...............HK.........................................................I..... 87

JSRV21    AAFWAYIPDPPMIQSLGWDREIVPVYVNDTSLLGGKSDIHISPQQANISFYGLTTQYPMCFSYQSQHPHCIQVSADISYP 167
JSRV-SA   ................................................................................ 160
ENTV      ................................................................................ 160
enJS56A1  ...................K.T.......................................................... 167
enJS5F16  ...................K.T.......................................................... 167

JSRV21    RVTISGIDEKTGKKSYGNGSGPLDIPFCDKHLSIGIGIDTPWTLCRARVASVYNINNANATFLWDWAPGGTPDFPEYRGQ 247
JSRV-SA   ...................T............................................................ 240
ENTV      .............R..R..T..........N...S.............I..........T.L.................. 240
enJS56A1  .............R..RD.T...........................PI..........T.L.................. 247
enJS5F16  .............R..RD.T............................I..........T.L.................. 247

JSRV21    HPPIFSVNTAPIYQTELWKLLAAFGHGNSLYLQPNISGSKYGDVGVTGFLYPRACVPYPFMLIQGHMEITLSLNIYHLNC 327
JSRV-SA   ......................................T......................................... 320
ENTV      ....L.......F................................................................... 320
enJS56A1  ................................................................................ 327
enJS5F16  ....L.......F................................................................... 327
                                                                      SU <--|--> TM
JSRV21    SNCILTNCIRGVAKGEQVIIVKQPAFVMLPVEIAEAWYDETALELLQRINTALSRPKRGLSLIILGIVSLITLIATAVTA 407
JSRV-SA   ................................................................................ 400
ENTV      .................................T.E............................................ 400
enJS56A1  .................................T.E............................................ 407
enJS5F16  .................................T.E............................................ 407

JSRV21    SVSLAQSIQAAHTVDSLSYNVTKVMGTQEDIDKKIEDRLSALYDVVRVLGEQVQSINFRMKIQCHANYKWICVTKKPYNT 487
JSRV-SA   C............................................................................... 480
ENTV      ................................................................................ 480
enJS56A1  ..................S..........N.................................................. 487
enJS5F16  ........................................................H....................... 487

JSRV21    SDFPWDKVKKHLQGIWFNTNLSLDLLQLHNEILDIENSPKATLNIADTVDNFLQNLFSNFPSLHSLWKTLIGVGILVFII 567
JSRV-SA   ........................................................................L..F.I.. 560
ENTV      ..Y.....E...........V............................N.................QSI.VIA.I...V 560
enJS56A1  ...................TV..............................................RSI.AM.AVLTVV 567
enJS5F16  ....................V..............................................QSI.AM.AVLTVV 567
                                  V R3
JSRV21    IVVILIFPCLVRGMVRDFLKMRVEMLHMKYRNMLQHQHLMELLKNKERGDAGD--D-P                       622
JSRV-SA   AI..FV...V...L.................T....R.............A..--.P.                       616
ENTV      .....LV...T..LIK...Q..I.LI.....Y...Y.K..DFV.KR.GSCG.QPAEG.                       618
enJS56A1  LII.CLA...I.SI.KE..H...-LI.---K...................A..--.P                        618
enJS5F16  LII.CLA...I.SI.KE..H...-LI.---K.....R.............A..--.P                        618
```

Fig. 4 A–C. The JSRV envelope gene. **A** The amino acid sequences for the envelope proteins of two exogenous JSRV isolates ($JSRV_{21}$ and JSRV-SA) and the closely related exogenous ENTV are shown in the *upper three lines*. The envelope sequences of two endogenous JSRV related proviruses are shown in the *two lower lines*. Amino acid identities are indicated by '.', while gaps are indicated by '–'. The boundary between the SU and TM proteins is indicated. The C termini show substantial variability between the exogenous and endogenous viruses, and the region of variability is shown as VR3. (From Palmarini et al. 2000b)

B. EXOGENOUS JSRV CYTOPLASMIC TAILS

```
                                                          YRNM
JSRV 21   ...HSLWKTLIGVGILVFIIIVVILIFPCLVRGMVRDFLKMRVEMLHMKYRNMLQHQHLMELLKNKERGDAGD--D-P
JSRV-SA   ...........L..F.I..AI..FV...V...L..................T....R............A...--.P.
entv      .......QSI.VIA.I...V.....LV...T..LIK...Q..I.LI.....Y...Y.K..DFV.KR.GSCG.QPAEG.

enJS56A1  .......RSI.AM.AVLTVVLII.CLA...I.SI.KE..H...-LI.---K..................A...--.P
enJS5F16  .......QSI.AM.AVLTVVLII.CLA...I.SI.KE..H...-LI.---K.....R...........A...--.P

               MEMBRANE                          CYTOPLASMIC TAIL
               SPANNING
```

C. PUTATIVE SH2 BINDING MOTIFS

```
JSRV ENV     .....HMKYRNMLQH....

PI3K motif        YXXM

GRB2 motif        YXNX
```

Fig. 4B, C. **B** The VR3 region is shown in greater detail. The putative membrane-spanning domain is indicated (*hatched box*), as well as the cytoplasmic tail. A conserved tyrosine residue in the cytoplasmic tail (YRNM) is present in all three exogenous viruses (as well as other recently sequenced isolates; C. Leroux, personal communication), and absent from the two endogenous virus proteins. **C** The sequence surrounding the tyrosine in the JSRV cytoplasmic tail (*Y590*) is shown in comparison to the SH2 binding motifs recognized by the regulatory subunit of PI3K (p85) and Grb-2 adaptor protein

transfected into NIH3T3 cells, and the number of transformed foci were determined. A chimera containing the exogenous JSRV SU protein and endogenous JSRV-related TM protein did not induce foci (PALMARINI et al. 2001 b). This indicated that the exogenous JSRV SU protein is not sufficient for transformation, and that the exogenous JSRV TM protein is necessary for transformation. An additional chimera indicated that an exogenous JSRV envelope containing the VR3 region from an endogenous JSRV-related TM protein also did not induce transformation. This indicated that the exogenous JSRV VR3 region is necessary for transformation.

Inspection of the VR3 sequences for exogenous vs. endogenous JSRV proviruses revealed an interesting feature: the exogenous VR3 region contains a single tyrosine residue (Y590), while the endogenous VR3 region contains no tyrosines (Fig. 4B). Moreover, the tyrosine in the exogenous JSRV VR3 region is in a good sequence context for binding of SH2 domain containing proteins (see below), assuming that it is phosphorylated. As a test of this hypothesis, the tyrosine residue was mutated to phenylalanine (Y590F) or aspartic acid (Y590D) in the context of the pCMV2JS21ΔGP expression plasmid, and this mutation abolished transformation ability (Table 2). This result indicated that the Y590 tyrosine is essential for transformation. To test if the requirement of the tyrosine residue at Y590 for transformation was unique, each of the other tyrosines in the TM protein were individually mutated to phenylalanine, and the effects of these mutations on JSRV transformation were tested (Table 2). Mutation of none of the other tyrosines abolished transformation, indicating that the requirement of the Y590 tyrosine for transformation is specific.

The amino acids sequence surrounding Y590 in the VR3 region of the JSRV TM protein is ...YRNM... . This tyrosine residue is in motifs that would bind two kinds of SH2 domains (Fig. 4C). The sequence YXXM is a motif for binding of SH2 domains in the regulatory subunit (p85) of phosphatidyl inositol 3-kinase (PI3K), while the sequence YXN is a motif for binding SH2 domains in growth factor receptor binding protein-2 (Grb-2) (Songyang et al. 1993). Both of these potential binding proteins are at the beginning of signal transduction cascades associated with cell transformation (Buday and Downward 1993; Roche et al. 1994). Therefore it was of interest to determine whether either of these signal transduction cas-

Table 2. Transformation assays performed with the JSRV-*env* constructs mutated in the various tyrosine residues of TM (from Palmarini et al. 2001b)

Plasmids	Experiment 1	Experiment 2	Experiment 3
pCMV3JS21ΔGP	32[a]	22	31
pCMV3JS21ΔGPY590D	0	0	0
pCMV3JS21ΔGPY590F	0	0	0
pCMV3S21ΔGPY419F	25	34	16
pCMV3JS21ΔGPY443F	17	20	15
pCMV3JS21ΔGPY468F	10	17	13
pCMV3JS21ΔGPY478F	8	5	6

[a] Numbers of observed foci.

Table 3. Transformation assays performed with the JSRV-*env* constructs mutated in the SH-2 site in the cytoplasmic tail (from PALMARINI et al. 2001b)

Plasmids	Experiment 1	Experiment 2	Experiment 3
pCMV3JS21ΔGP	12[a]	N.T.	56
pCMV3JS21ΔGPN592T	40	38	105
pCMV3JS21ΔGPM593T	0	0	0

N.T., not tested.
[a] Numbers of observed foci.

cades is likely to be important in transformation by JSRV *env*. The asparagine and methionine residues were individually mutated (N592T and M593T), and the effects on transformation were tested (Table 3). The N592T mutant showed the same (or actually higher) efficiency of focus formation as wild-type envelope, while the M593T mutation completely abolished transformation (PALMARINI et al. 2001b). These results support the hypothesis that the Y590 is functioning as a docking site for signal transduction proteins, and they indicate that PI3K binding is likely to be important while Grb-2 binding is not.

Signaling downstream of PI3K has been extensively studied. An important intermediate in the PI3K signaling pathway is AKT (also known as protein kinase B). AKT (itself a kinase) is activated by PI3K via an intermediate kinase PDK-1 (DATTA et al. 1999) – the phosphorylated form of AKT is active. When serum-starved NIH3T3 cells were tested for activated AKT by SDS–PAGE and western blotting with an AKT antibody, no phosphorylated AKT was detected. On the other hand when a line of NIH3T3 cells transformed by JSRV envelope were tested in the same manner, activated (phosphorylated) AKT was readily detected (PALMARINI et al. 2001b). These results indicate that the PI3K signaling pathway is activated in JSRV transformed cells, but not in normal NIH3T3 cells. This lends further support to the notion that JSRV envelope transforms by activation of the PI3K signaling pathway. Recently, it has also been found that treatment of JSRV transformed cells with PI3K inhibitors inhibits AKT phosphorylation, and AKT activation is also observed in an OPA-derived cell line (A. ALBERTI et al., 2002).

It is noteworthy that the PI3K signaling pathway is an important pathway in differentiated type II pneumocytes. In particular, regulation of the surfactant protein B (SP-B) promoter by SP-A has been reported to be mediated by PI3K signaling (Strayer and Korutla 2000). Thus, the PI3K signaling machinery is active in type II pneumocytes, which would be consistent with transformation by JSRV envelope involving this pathway.

While the studies described above indicated that the PI3K docking site in the cytoplasmic tail of JSRV TM protein is necessary for transformation of NIH-3T3 cells, recent studies in the DF-1 transformation system indicate that the situation may be more complicated. In RCAS vectors containing JSRV envelope protein, truncation of the cytoplasmic tail at Y590 abolished transformation of DF-1. However, missense mutations of Y590F or M593T in these vectors did not affect DF-1 transformation. Thus, in these cells, some other feature of the TM cytoplasmic tail may be sufficient for transformation (Allen et al. 2002).

9
Testing the Possibility of Insertional Activation

While the ability of JSRV envelope protein to transform fibroblasts in vitro strongly suggests that JSRV is an acute transforming virus, it is also possible that tumorigenesis also involves insertional activation of proto-oncogenes. To address this latter possibility, different OPA tumors were tested for common JSRV proviral integration sites. To efficiently analyze insertion sites from multiple tumors, a PCR-based method was developed (C. Cousens, unpublished results). The presence of endogenous JSRV-related sequences within the sheep genome presented challenges in specifically detecting exogenous JSRV proviral insertions. The method uses four JSRV specific primers together with non specific primers in successive rounds of hemi-nested low-stringency/high-stringency PCR. From the numerous bands generated, Southern blotting identifies those containing the JSRV LTR. The fragments are then cloned and sequenced. Because the primers were designed so that the clones include part of the JSRV LTR, sequence information can verify that the integration sites contain exogenous JSRV and not an endogenous JSRV related provirus. Using this technique, 70 unique integration sites from 23 OPA-affected sheep were amplified and cloned from PCR products ranging in size between 12 and 671 bp, not including the JSRV LTR sequences (C. Cousens and J.M. Sharp, unpublished results). Multiple integration sites were obtained from most tumors. BLAST analysis revealed repetitive sequences in 34 flanking sequences and few significant matches to known genes. The best database matches were one with the receptor protein tyrosine phosphatase γ of humans and mice and another with an uncharacterized region of human chromosome 14. The only other unambiguous assignment of a JSRV integration site to a known gene is the provirus integrated in the first exon of the pulmonary surfactant protein A (SP-A) gene in the JS7 tumor cell line ($JSRV_{JS7}$), as discussed above (DEMARTINI et al. 2001).

Of the 70 integration sites, 37 were mapped to individual sheep chromosomes. This was accomplished by sequencing the adjacent host cell sequences for each provirus and using PCR based on these sequences on DNA isolated from a panel of 30 sheep/hamster hybrid cell lines (J. BISHOP and J. DEMARTINI, unpublished results). Integration sites were found on 20 of the 28 sheep chromosomes, suggesting a random distribution. Further analysis revealed that five integration sites from five different tumors mapped to chromosome 6 and four integration sites from four different tumors mapped to chromosome 16. By Southern blot hybridization, probes derived from two of the chromosome 16 sites mapped to within 5 kb of each other on normal sheep DNA. Further analysis of the integration sites on chromosome 16 involved Southern blot analysis with probes for the different insertions against genomic DNA from the lung of a normal sheep cleaved with various restriction endonucleases (J. BISHOP and J. DEMARTINI, unpublished results). Two of the proviruses were shown to be inserted into the same chromosomal locus, within 5 kb. Thus, one common insertion site out of 37 analyzed was detected, which would be consistent with insertional activation of a proto-oncogene in at least some OPA tumors. By Southern blotting analysis, the five integration sites that localized to chromosome 6 did not appear to map closely together.

The clonality of one of the chromosome 16 integration sites was investigated by Southern blotting using site specific probes to compare the band profiles in tumor and non-tumor DNA of the same animal and by integration site specific PCR applied to the original DNA from which the integration site was cloned as well as DNA from contiguous parts of the same tumor. The results suggested that this integration site was present only in a small number of cells and was not clonal. The clonality of 13 other JSRV integration sites, derived from four discrete portions of lung tumor from one animal, was investigated. Integration sites could be amplified only from the DNA samples from which they were cloned, indicating that multiple copies were present at low frequency in localized regions of lung tumor.

JSRV integration sites identified in this study were not clonally amplified throughout the tumor, providing support for the hypothesis that mechanisms other than insertional mutagenesis may be primarily involved in the pathogenesis of pulmonary neoplasia due to JSRV. However, a role for this mechanism at a late stage in the disease process cannot yet be entirely ruled out.

10
Future Directions for Research

While the recent experiments demonstrating the ability of JSRV envelope protein to transform fibroblasts strongly suggest that this transformation is the underlying cause for the ability of JSRV to cause OPA tumors, these results raise a number of questions that are amenable to experimental attack:

1. Is JSRV envelope protein responsible for transformation of type II pneumocytes, and for development of tumors in vivo?
   It will be important to address these questions directly. In particular, it will be interesting to see if primary type II pneumocytes can be transformed (e.g., immortalized, changes in morphology, tumorigenicity). Ideally, assays in primary ovine type II pneumocytes should be carried out, although conditions for in vitro culture of primary ovine type II cells are only now being developed. In the meantime, tests of primary rat or mouse type II pneumocytes can be carried out (Xu et al. 1998); as mouse and rat fibroblasts can be transformed by JSRV envelope, the rodent type II pneumocytes should be a valid model system. With regard to tumorigenicity, it will be very important to test if mutation of the transforming potential of JSRV envelope affects tumorigenicity of the virus. A JSRV envelope with the Y590F mutation has been built back into the infectious pCMVJS21 molecular clone, and virus particles were produced by transfection of 293T cells. When the Y590F mutant virus was tested for in vitro infection in sheep choroid plexus cells, spreading infection was detected (N. Maeda and H. Fan, unpublished results). Thus it should be feasible to test the oncogenicity of Y590F JSRV in vivo in newborn lambs.
2. Is the cytoplasmic tail of TM the only domain of envelope necessary for transformation?
   While the recent experiments demonstrating that the cytoplasmic tail of TM is necessary for JSRV transformation, they do not exclude the possibility that other regions of the envelope protein (e.g., SU) are

also involved. [We also generated exogenous–endogenous envelope chimeras in which the SU protein was derived from an endogenous JSRV-related provirus while various portions of TM were derived from exogenous JSRV; however these chimeras were not informative because the levels of spliced mRNA were substantially lower than in envelope constructs containing exogenous JSRV SU sequences (M. PALMARINI, C. MURGIA, and H. FAN, unpublished results). There is apparently an element in the exogenous JSRV envelope gene that influences spliced mRNA abundance.) It will be important to design experiments that can systematically test for a role of other envelope domains in transformation. While the results do not currently support hypotheses involving interaction of extracellular JSRV envelope with either Hyal-2 or another cell surface protein in transformation, they also do not exclude them. A high priority experiment will be to determine if interference between JSRV SU and Hyal-2 interactions affects the efficiency of transformation.

3. What is the biochemical mechanism for transformation by JSRV envelope?

   The experiments indicating that the PI3 K signaling pathway is necessary for JSRV transformation of NIH3T3 cells allow further mechanistic studies to be carried out. In particular, it will be interesting to investigate downstream targets of PI3 K. A number of substrates for activation by PI3 K signaling have been identified, and it will be interesting to see if any of them are activated in JSRV transformed fibroblasts or (more importantly) type II pneumocytes and OPA tumors cells. In some cases, this can be accomplished by testing for phosphorylated downstream targets (e.g., Western blot with appropriate antibodies). In addition, gene expression array analysis may be informative. Preliminary experiments comparing transformed and parental NIH3T3 cells have identified approximately 20 genes that are consistently over- or under-expressed (M. PALMARINI and H. FAN, unpublished results). It will be interesting to see if these genes are over- or under-expressed in primary OPA tumors, and whether modulation of their expression affects transformation and/or tumorigenicity.

   It will also be important to compare JSRV transformation of rodent fibroblasts, in which the PI3K docking site is necessary, and DF-1 cells, where it is not. Such studies may allow further elucidation of the domains of the cytoplasmic tail necessary for transformation, and ultimately the mechanism of transformation.

4. Is proto-oncogene activation also important in JSRV tumorigenesis? The experiments strongly indicate that JSRV is an acute transforming retrovirus, in that it can directly transform cells. As such, the need to invoke insertional activation of proto-oncogenes in OPA tumorigenesis is decreased. This is also consistent with the finding of the JS7 tumor cells with only one provirus integrated into the ovine SPA gene (see above). However, it has been shown that for certain acute transforming retroviruses, insertional activation of proto-oncogenes occurs as well. In particular, Abelson murine leukemia virus, which carries the v-*abl* oncogene) also requires insertional activation of proto-oncogenes for efficient tumorigenesis (Poirier et al. 1988). Likewise, insertional activation of proto-oncogenes might potentiate JSRV envelope-driven oncogenic transformation in vivo.

11
What Is the Biological Significance of JSRV Envelope Transformation?

As indicated above, for most acute transforming retroviruses, the viral oncogene is not essential for replication. Indeed, there generally seems to be little selective advantage for oncogenic capacity of retroviruses. However in the case of JSRV, the envelope protein is clearly essential for replication. Moreover, all of the JSRV isolates for which the envelope protein has been sequenced contain the Y590 tyrosine in a YXXM motif; this is also true for the cytoplasmic tail of ONAV. Thus the transformation potential of the JSRV envelope protein appears to be a conserved property of the exogenous JSRVs. This suggests that the transformation potential of JSRV envelope may be important biologically for the virus.

We would like to suggest the following hypothesis: the ability of the JSRV envelope protein to transform cells is necessary for in vivo replication of the virus.

This hypothesis is based on several observations about the biology and molecular biology of JSRV. As discussed in the chapter by Palmarini and Fan (this volume), JSRV is unusual among retroviruses in that the only cells which show significant viral protein expression in infected animals are OPA tumors cells (and perhaps normal type II pneumocytes). Even though the virus can enter and infect hematopoietic cells (as judged by nested PCR for viral DNA), few if any of these cells express viral protein. As mentioned above, this is likely to be due to the highly specific transcriptional preference of the JSRV LTR for lung epithelial cells. It has also been shown that the endogenous JSRV-related viral LTRs do not show the high specificity for lung epithelial cells (Palmarini et al. 2000b). At the same time, the cytoplasmic tail of the endogenous JSRV-related TM proteins does not contain the Y590. Taken together, these results suggest that the primordial progenitor to JSRV (typified by the endogenous JSRV-related proviruses) was not lung specific in expression or replication. Evolution of the current day exogenous JSRV thus may have resulted in a virus that very specifically expresses in lung epithelial cells. In order for this virus to efficiently replicate in vivo, it may need a mechanism for ensuring division of its target cells – type II pneumocytes and Clara cell – as these cells normally are not actively dividing (particularly in adult animals). This is reminiscent of the *sag* gene in another betaretrovirus, MMTV: the Sag protein is necessary for mitogenic stimulation of the T lymphocytes that deliver virus from the gut to the mammary gland (Golovkina et al. 1992). Likewise, both MMTV and MuLV envelope proteins bind and stimulate through TLR4 in lymphoid cells (see above). In the case of JSRV, the cytoplasmic tail of the TM protein has evolved to have the growth stimulatory (transformation) function.

The biology of JSRV infection also suggests that the in vivo life cycle of JSRV may exclusively involve infection of lung epithelial cells. That is, productive infection within an animal may be confined to type II pneumocytes and Clara cells, and spread of infection to other animals may result from coughing or aerosols produced by the infected tumor cells being breathed in by recipient animals. Thus, tumorigenesis may be an essential feature of JSRV replication in vivo. Such a life cycle is quite novel for retroviruses.

Acknowledgements. Support from NIH research grants R01CA82564 to H.F. and R01 CA59116 to J.D.M. are acknowledged. M.P. was supported by a Wellcome Prize Traveling Research Fellowship, and a Ray and Estelle Spehar Fellowship from the California Division of the American Cancer Society during a portion of this work, and by the Veterinary Medical Experiment Station of the University of Georgia and the Georgia Cancer Coalition. We thank our colleagues in our laboratories for carrying out experiments that formed the foundation of this review, notably: Naoyoshi Maeda, Claudio Murgia, Tom Allen and Jeanette Bishop. In addition, important collaborations with Chris Cousens, Mike Sharp and Marcelo De Las Heras were essential.

References

Ahmed YF, Hanly SM, Malim MH, Cullen BR, Greene WC (1990) Structure-function analyses of the HTLV-I Rex and HIV-1 Rev RNA response elements: insights into the mechanism of Rex and Rev action. Genes and Development 4(6): 1014–22

Alberti A, Murgia C, Liu S-L, Mura M, Cousens C, Sharp M, Miller AD, Palmarini M (2002) Envelope-induced transformation by ovine betaretroviruses. J Virol 76:5387–94

Alkhatib G, Combadiere C, Broder CC, Feng Y, Kennedy PE, Murphy PM, Berger EA (1996) CC CKR5: a RANTES, MIP-1alpha, MIP-1beta receptor as a fusion cofactor for macrophage-tropic HIV-1. Science 272(5270):1955–8

Allen TE, Sherrill KJ, Crispell SM, Perrott MR, Carlson JO, DeMartini JC (2002) The jaagsiekte sheep retrovirus envelope gene induces transformation of the avian fibroblast cell line DF-I but does not require a conserved SH2 binding domain. J Gen Virol *in press*

Bai J, Bishop JV, Carlson JO, DeMartini JC (1999) Sequence comparison of JSRV with endogenous proviruses: envelope genotypes and a novel ORF with similarity to a G-protein-coupled receptor. Virology 258(2):333–43

Bai J, Zhu RY, Stedman K, Cousens C, Carlson J, Sharp JM, DeMartini JC (1996) Unique long terminal repeat U3 sequences distinguish exogenous jaagsiekte sheep retroviruses associated with ovine pulmonary carcinoma from endogenous loci in the sheep genome. J Virol 70(5):3159–68

Bear SE, Bellacosa A, Lazo PA, Jenkins NA, Copeland NG, Hanson C, Levan G, Tsichlis PN (1989) Provirus insertion in Tpl-1, an Ets-1-related oncogene, is associated with tumor progression in Moloney murine leukemia virus-induced rat thymic lymphomas. Proc Natl Acad Sci USA 86:7495–7499

Bishop JM, Varmus H (1984) Functions and origins of retroviral transforming genes. In: RNA Tumor Viruses. Molecular Biology of Tumor Viruses, Second Edition (R Weiss, N Teich, H Varmus and J Coffin, eds), pp 999–1108. Cold Spring Harbor Laboratory, Cold Spring Harbor, NY

Bittner JJ (1942) Milk-influence of breast tumors in mice. Science 95:462–463

Buchberg AM, Bedigian HG, Jenkins NA, Copeland NG (1990) Evi-2, a common integration site involved in murine myeloid leukemogenesis. Mol Cell Biol 10, 4658–4666

Buday L, Downward J (1993) Epidermal growth factor regulates p21ras through the formation of a complex of receptor, Grb2 adapter protein, and Sos nucleotide exchange factor. Cell 73(3):611–20

Chatis PA, Holland CA, Silver JE, Frederickson TN, Hopkins N, Hartley JW (1984) A 3′ end fragment encompassing the transcriptional enhancers of nondefective Friend virus confers erythroleukemogenicity on Moloney leukemia virus. J Virol 52:248–254

Choe H, Farzan M, Sun Y, Sullivan N, Rollins B, Ponath PD, Wu L, Mackay CR, LaRosa G, Newman W, Gerard N, Gerard C, Sodroski J (1996) The beta-chemokine receptors CCR3 and CCR5 facilitate infection by primary HIV-1 isolates. Cell 85(7):1135–48

Coffin JM (1996) Retroviridae: the viruses and their replication. 3 ed. In: Fields Virology (BN Fields, DM Knipe and PM Howley, eds), pp 745–843. Lippincott-Raven, Philadelphia

Coffin JM, Hughes SH, Varmus HE (1997) 'Retroviruses.' Cold Spring Harbor Laboratory Press, Plainview, N.Y.

Collett MS, Erikson RL (1978) Protein kinase activity associated with the avian sarcoma virus src gene product. Proc Natl Acad Sci USA 75:2021–2024

Cousens C, Minguijon E, Dalziel RG, Ortin A, Garcia M, Park J, Gonzalez L, Sharp JM, de las Heras M (1999) Complete sequence of enzootic nasal tumor virus, a retrovirus associated with transmissible intranasal tumors of sheep. Journal of Virology 73(5):986–93

Cuypers HT, Selten G, Quint W, Zijlstra M, Maandag ER, Boelens W, van Wezenbeek P, Melief C, Berns A (1984) Murine leukemia virus-induced T-cell lymphomagenesis: integration of proviruses in a distinct chromosomal region. Cell 37:141–150

Datta SR, Brunet A, Greenberg ME (1999) Cellular survival: a play in three Akts. Genes and Development 13(22):2905–27

DeMartini JC, Bishop JV, Allen TE, Jassim FA, Sharp JM, de las Heras M, Voelker DR, Carlson JO (2001) Jaagsiekte sheep retrovirus proviral clone JSRV(JS7), derived from the JS7 lung tumor cell line, induces ovine pulmonary carcinoma and is integrated into the surfactant protein A gene. Journal of Virology 75(9):239–46

DeMartini JC, York D (1997) Retrovirus-associated neoplasms of the respiratory system of sheep and goats. Ovine pulmonary carcinoma and enzootic nasal tumor. Vet Clin North Am Food Anim Pract 13, 55–70

Deng H, Liu R, Ellmeier W, Choe S, Unutmaz D, Burkhart M, Di Marzio P, Marmon S, Sutton RE, Hill CM, Davis CB, Peiper SC, Schall TJ, Littman DR, Landau NR (1996) Identification of a major co-receptor for primary isolates of HIV-1 [see comments]. Nature 381(6584):661–6

DesGroseillers L, Jolicoeur P (1984) The tandem direct repeats within the long terminal repeat of murine leukemia viruses are the primary determinant of their leukemogenic potential. J Virol 52:945–952

Dragic T, Litwin V, Allaway GP, Martin SR, Huang Y, Nagashima KA, Cayanan C, Maddon PJ, Koup RA, Moore JP, Paxton WA (1996) HIV-1 entry into CD4+ cells is mediated by the chemokine receptor CC-CKR-5 [see comments]. Nature 381(6584):667–73

Du Z, Lang SM, Sasseville VG, Lackner AA, Ilyinskii PO, Daniel MD, Jung JU, Desrosiers RC (1995) Identification of a nef allele that causes lymphocyte activation and acute disease in macaque monkeys. Cell 82(4):665–74

Duesberg PH, Vogt PK (1970) Differences between the ribonucleic acids of transforming and nontransforming avian tumor viruses. Proc Natl Acad Sci USA 67: 1673–1680

Dzuris JL, Golovkina TV, Ross SR (1997) Both T and B cells shed infectious mouse mammary tumor virus. Journal of Virology 71(8):6044–8

Ellerman V, Bang O (1908) Experimentelle Leukämie bei Hühnern. Zentralbl Bakteriol 46:595–609

Fan H (1997) Leukemogenesis by Moloney murine leukemia virus: a multistep process. Trends Microbiol 5(2):74–82

Federspiel MJ, Hughes SH (1997) Retroviral gene delivery. Methods in Cell Biology 52(4):179–214

FULTZ PN, MCCLURE HM, ANDERSON DC, SWITZER WM (1989) Identification and biologic characterization of an acutely lethal variant of simian immunodeficiency virus from sooty mangabeys (SIV/SMM). AIDS Research and Human Retroviruses 5(4):397–409

GOLOVKINA TV, CHERVONSKY A, DUDLEY JP, ROSS SR (1992) Transgenic mouse mammary tumor virus superantigen expression prevents viral infection. Cell 69:637–645

GREENE WC, LEONARD WJ, WANO Y, SVETLIK PB, PEFFER NJ, SODROSKI JG, ROSEN CA, GOH WC, HASELTINE WA (1986) Trans-activator gene of HTLV-II induces IL-2 receptor and IL-2 cellular gene expression. Science 232:877–880

GROSS L (1951) 'Spontaneous' leukemia developing in C3H mice following inoculation, in infancy, with Ak-leukemia extracts, or Ak-embryos. Proceedings of the Society for Experimental Biology and Medicine 76:27–32

HAYWARD WS, NEEL BG, ASTRIN SM (1981) Activation of a cellular onc gene by promoter insertion in ALV-induced lymphoid leukosis. Nature 290:475–480

HECHT SJ, STEDMAN KE, CARLSON JO, DEMARTINI JC (1996) Distribution of endogenous type B and type D sheep retrovirus sequences in ungulates and other mammals. Proc Natl Acad Sci USA 93(8):3297–302

HELD W, WAANDERS GA, SHAKHOV AN, SCARPELLINO L, ACHA-ORBEA H, MACDONALD HR (1993) Superantigen-induced immune stimulation amplifies mouse mammary tumor virus infection and allows virus transmission. Cell 74(3):529–40

HERRING AJ, SHARP JM, SCOTT FM, ANGUS KW (1983) Further evidence for a retrovirus as the aetiological agent of sheep pulmonary adenomatosis (jaagsiekte). Vet Microbiol 8(3):237–49

HIMLY M, FOSTER DN, BOTTOLI I, IACOVONI JS, VOGT PK (1998) The DF-1 chicken fibroblast cell line: transformation induced by diverse oncogenes and cell death resulting from infection by avian leukosis viruses. Virology 248(2):295–304

HOLLAND MJ, PALMARINI M, GARCIA-GOTI M, GONZALEZ L, DE LAS HERAS M, MCKENDRICK I, SHARP JM (1999) Jaagsiekte retrovirus is widely distributed both in T and B lymphocytes and in mononuclear phagocytes of sheep with naturally and experimentally acquired pulmonary adenomatosis. J Virol 73:4004–4008

INOUE J, YOSHIDA M, SEIKI M (1987) Transcriptional (p40x) and post-transcriptional (p27x-III) regulators are required for the expression and replication of human T-cell leukemia virus type I genes. Proc Natl Acad Sci USA 84(11):3653–7

KONDO T, KONO H, NONAKA H, MIYAMOTO N, YOSHIDA R, BANDO F, INOUE H, MIYOSHI I, HINUMA Y, HANAOKA M (1987) Risk of adult T-cell leukaemia/lymphoma in HTLV-I carriers [letter]. Lancet 2(8551):159

LENZ J, CELANDER D, CROWTHER RL, PATARCA R, PERKINS DW, HASELTINE WA (1984) Determination of the leukaemogenicity of a murine retrovirus by sequences within the long terminal repeat. Nature 308:467–470

LI JP, D'ANDREA AD, LODISH HF, BALTIMORE D (1990) Activation of cell growth by binding of Friend spleen focus- forming virus gp55 glycoprotein to the erythropoietin receptor. Nature 343:762–764

LI Y, GOLEMIS E, HARTLEY JW, HOPKINS N (1987) Disease specificity of nondefective Friend and Moloney murine leukemia viruses is controlled by a small number of nucleotides. J.Virol. 61:693–700

MAEDA N, PALMARINI M, MURGIA C, FAN H (2001) Direct transformation of rodent fibroblasts by jaagsiekte sheep retrovirus DNA. Proc Nat Acad Sci USA 98:4449– 4454

MARTIN WB, SCOTT FM, SHARP JM, ANGUS KW, NORVAL M (1976) Experimental production of sheep pulmonary adenomatosis (Jaagsiekte). Nature 264(5582):183–5
MORISHITA K, PARKER DS, MUCENSKI ML, JENKINS NA, COPELAND NG, IHLE JN (1988) Retroviral activation of a novel gene encoding a zinc finger protein in IL-3-dependent myeloid leukemia cell lines. Cell 54:831–840
MURPHY EL, HANCHARD B, FIGUEROA JP, GIBBS WN, LOFTERS WS, CAMPBELL M, GOEDERT JJ, BLATTNER WA (1989) Modelling the risk of adult T-cell leukemia/lymphoma in persons infected with human T-lymphotropic virus type I. International Journal of Cancer 43(2):250–3
NUSSE R, VARMUS HE (1982) Many tumors induced by the mouse mammary tumor virus contain a provirus integrated in the same region of the host genome. Cell 31:99–109
OHASHI T, MASUDA M, RUSCETTI SK (1997) Constitutive activation of Stat-related DNA-binding proteins in erythroid cells by the Friend spleen focus-forming virus. Leukemia 11 Suppl 3:251–4
ORTIN A, MINGUIJON E, DEWAR P, GARCIA M, FERRER LM, PALMARINI M, GONZALEZ L, SHARP JM, DE LAS HERAS M (1998) Lack of a specific immune response against a recombinant capsid protein of Jaagsiekte sheep retrovirus in sheep and goats naturally affected by enzootic nasal tumour or sheep pulmonary adenomatosis. Vet Immunol Immunopathol 61(2–4):229–37
PALMARINI M, COUSENS C, DALZIEL RG, BAI J, STEDMAN K, DEMARTINI JC, SHARP JM (1996a) The exogenous form of Jaagsiekte retrovirus is specifically associated with a contagious lung cancer of sheep. J Virol 70(3):1618–23
PALMARINI M, DATTA S, OMID R, MURGIA C, FAN H (2000a) The long terminal repeat of Jaagsiekte sheep retrovirus is preferentially active in differentiated epithelial cells of the lungs. Journal of Virology 74(13):5776–87
PALMARINI M, DEWAR P, DE LAS HERAS M, INGLIS NF, DALZIEL RG, SHARP JM (1995) Epithelial tumour cells in the lungs of sheep with pulmonary adenomatosis are major sites of replication for Jaagsiekte retrovirus. J Gen Virol 76(Pt 11):2731–7
PALMARINI M, FAN H (2001) Retrovirus-induced Ovine Pulmonary Adenocarcinoma, an Animal Model for Lung Cancer. J Natl Cancer Inst 93:1603–1614
PALMARINI M, FAN H, SHARP JM (1997) Sheep pulmonary adenomatosis: a unique model of retrovirus-associated lung cancer. Trends Microbiol 5(12):478–83
PALMARINI M, GRAY CA, CARPENTER K, FAN H, BAZER FW, SPENCER TE (2001a) Expression of endogenous betaretroviruses in the ovine uterus: effects of neonatal age, estrous cycle, pregnancy and progersterone. J Virol 75:11319–11327
PALMARINI M, HALLWIRTH C, YORK D, MURGIA C, DE OLIVEIRA T, SPENCER T, FAN H (2000b) Molecular cloning and functional analysis of three type D endogenous retroviruses of sheep reveal a different cell tropism from that of the highly related exogenous jaagsiekte sheep retrovirus. Journal of Virology 74(17):8065–76
PALMARINI M, HOLLAND MJ, COUSENS C, DALZIEL RG, SHARP JM (1996b). Jaagsiekte retrovirus establishes a disseminated infection of the lymphoid tissues of sheep affected by pulmonary adenomatosis. J Gen Virol 77(Pt 12):2991–8
PALMARINI M, MAEDA N, MURGIA C, DE-FRAJA C, HOFACRE A, FAN H (2001b) A phosphatidylinositol 3-kinase docking site in the cytoplasmic tail of the jaagsiekte sheep retrovirus transmembrane protein is essential for envelope-induced transformation of NIH3T3 cells. J Virol 75:11002–11009

Palmarini M, Sharp JM, de las Heras M, Fan H (1999) Jaagsiekte sheep retrovirus is necessary and sufficient to induce a contagious lung cancer in sheep. Journal of Virology 73(8):6964–72

Payne GS, Bishop JM, Varmus HE (1982) Multiple arrangements of viral DNA and an activated host oncogene in bursal lymphomas. Nature 295:209–214

Perk K, Michalides R, Spiegelman S, Schlom J (1974) Biochemical and morphologic evidence for the presence of an RNA tumor virus in pulmonary carcinoma of sheep (Jaagsiekte). Journal of the National Cancer Institute 53(1):131–5

Poiesz BJ, Ruscetti FW, Gazdar AF, Bunn PA, Minna JD, Gallo RC (1980) Detection and isolation of type C retrovirus particles from fresh and cultured lymphocytes of a patient with cutaneous T-cell lymphoma. Proc Natl Acad Sci USA 77(12):7415–9

Poirier Y, Kozak C, Jolicoeur P (1988) Identification of a common helper provirus integration site in Abelson murine leukemia virus-induced lymphoma DNA. J Virol 62:3985–3992

Rai SK, DeMartini JC, Miller AD (2000) Retrovirus vectors bearing jaagsiekte sheep retrovirus Env transduce human cells by using a new receptor localized to chromosome 3p21.3. Journal of Virology 74(10):4698–704

Rai SK, Duh F-M, Vigdorovich V, Danilkovitch-Miagkova A, Lerman MI, Miller AD (2001) Candidate tumor suppressor HYAL2 is a GPI-anchored cell-surface receptor for jaagsiekte sheep retrovirus, the envelope protein of which mediates oncogenic transformation. Proc Natl Acad Sci USA in press

Rassa JC, Meyers JL, Zhang Y, Kudaravalli R, Ross SR (2002) Murine retroviruses activate B cells via interaction with the Toll-like receptor 4. Proc Natl Acad Sci USA in press

Reitz MS Jr, Popovic M, Haynes BF, Clark SC, Gallo RC (1983) Relatedness by nucleic acid hybridization of new isolates of human T-cell leukemia-lymphoma virus (HTLV) and demonstration of provirus in uncultured leukemic blood cells. Virology 126(2):688–72

Roche S, Koegl M, Courtneidge SA (1994) The phosphatidylinositol 3-kinase alpha is required for DNA synthesis induced by some, but not all, growth factors. Proc Natl Acad Sci USA 91(19):9185–9

Rosati S, Pittau M, Alberti A, Pozzi S, York DF, Sharp JM, Palmarini M (2000) An accessory open reading frame (orf-x) of jaagsiekte sheep retrovirus is conserved between different virus isolates. Virus Res 66(1):109–16

Rosenberg N, Jolicoeur P (1997) Retroviral pathogenesis. In: Retroviruses (JM Coffin, SH Hughes and HE Varmus, eds), pp 475–585. Cold Spring Harbor Press, Cold Spring Harbor

Ross SR, Dzuris JL, Golovkina TV, Clemmons WC, van den Hoogen B (1997) Mouse mammary tumor virus (MMTV), a retrovirus that exploits the immune system. Genetics of susceptibility to MMTV infection. Medicina 57 Suppl 2(8): 34–42

Rous P (1911) A sarcoma of the fowl transmissible by an agent separable from the tumor cells. J Exp Med 13:397–411

Saris CJ, Domen J, Berns A (1991) The pim-1 oncogene encodes two related protein-serine/threonine kinases by alternative initiation at AUG and CUG. EMBO J 10:655–664

SAWAI ET, BAUR A, STRUBLE H, PETERLIN BM, LEVY JA, CHENG-MAYER C (1994) Human immunodeficiency virus type 1 Nef associates with a cellular serine kinase in T lymphocytes. Proc Natl Acad Sci USA 91:1539–1543

SCHAEFER-KLEIN J, GIVOL I, BARSOV EV, WHITCOMB JM, VANBROCKLIN M, FOSTER DN, FEDERSPIEL MJ, HUGHES SH (1998) The EV-O-derived cell line DF-1 supports the efficient replication of avian leukosis-sarcoma viruses and vectors. Virology 248(2):305–11

SEFTON BM, HUNTER T, BEEMON K, ECKHART W (1980) Evidence that the phosphorylation of tyrosine is essential for cellular transformation by Rous sarcoma virus. Cell 20:807–816

SEIKI M, EDDY R, SHOWS TB, YOSHIDA M (1984) Nonspecific integration of the HTLV provirus genome into adult T-cell leukaemia cells. Nature 309(5969):640–2

SEIKI M, HATTORI S, HIRAYAMA Y, YOSHIDA M (1983) Human adult T-cell leukemia virus: complete nucleotide sequence of the provirus genome integrated in leukemia cell DNA. Proc Natl Acad Sci USA 80(12):3618–22

SEKIDO Y, BADER S, LATIF F, CHEN JY, DUH FM, WEI MH, ALBANESI JP, LEE CC, LERMAN MI, MINNA JD (1996) Human semaphorins A(V) and IV reside in the 3p21.3 small cell lung cancer deletion region and demonstrate distinct expression patterns. Proc Natl Acad Sci USA 93(9):4120–5

SHARP JM, ANGUS KW, GRAY EW, SCOTT FM (1983) Rapid transmission of sheep pulmonary adenomatosis (jaagsiekte) in young lambs. Brief report. Arch Virol 78(1–2): 89–95

SHARP JM, HERRING AJ (1983) Sheep pulmonary adenomatosis: demonstration of a protein which cross-reacts with the major core proteins of Mason-Pfizer monkey virus and mouse mammary tumour virus. J Gen Virol 64(Pt 10):2323–7

SHIH C, WEINBERG RA (1982) Isolation of a transforming sequence from a human bladder carcinoma cell line. Cell 29:161–169

SHORT MK, OKENQUIST SA, LENZ J (1987) Correlation of leukemogenic potential of murine retroviruses with transcriptional tissue preference of the viral long terminal repeats. J Virol 61:1067–1072

SODROSKI JG, ROSEN CA, HASELTINE WA (1984) Trans-acting transcriptional activation of the long terminal repeat of human T lymphotropic viruses in infected cells. Science 225:381–385

SONGYANG Z, SHOELSON SE, CHAUDHURI M, GISH G, PAWSON T, HASER WG, KING F, ROBERTS T, RATNOFSKY S, LECHLEIDER RJ and et al. (1993) SH2 domains recognize specific phosphopeptide sequences. Cell 72(5):767–78

STEFFEN D, WEINBERG RA (1978) The integrated genome of murine leukemia virus. Cell 15:1003–1010

STEHELIN D, VARMUS HE, BISHOP JM, VOGT PK (1976) DNA related to the transforming gene(s) of avian sarcoma viruses is present in normal avian DNA. Nature 260: 170–173

STRAYER DS, KORUTLA L (2000) Activation of surfactant protein-B transcription: signaling through the SP-A receptor utilizing the PI3 kinase pathway. Journal of Cellular Physiology 184(2):229–38

TELISNITSKY A, GOFF SP (1997) Reverse transcriptase and the generation of retroviral DNA. In: Retroviruses (JM Coffin, SH Hughes and HE Varmus, eds), pp 205–261. Cold Spring Harbor Press, Cold Spring Harbor, NY

Tremblay PJ, Kozak CA, Jolicoeur P (1992) Identification of a novel gene, Vin-1, in murine leukemia virus-induced T-cell leukemias by provirus insertional mutagenesis [published erratum appears in J Virol 1992 Aug; 66(8):5176]. J Virol 66:1344–1353

Tsichlis PN, Lee JS, Bear SE, Lazo PA, Patriotis C, Gustafson E, Shinton S, Jenkins NA, Copeland NG, Huebner K (1990) Activation of multiple genes by provirus integration in the Mlvi-4 locus in T-cell lymphomas induced by Moloney murine leukemia virus. J Virol 64:2236–2244

Van der Maaten MJ, Boothe AD, Seger CL (1972) Isolation of a virus from cattle with persistent lymphocytosis. Journal of the National Cancer Institute 49(6): 1649–57

Verwoerd DW, Tustin RC, Payne AL (1985) In: Comparative Pathobiology of Viral Diseases (RG Olsen, S Karakowaka and JR Blackslee, eds), pp 53–76. CRC Press

Verwoerd DW, Williamson AL, DeVilliers EM (1980) Aetiology of jaagisekte: transmission by means of subcellular fractions and evidence for the involvement of a retrovirus. Onderstepoort J Vet Res 47:275–280

Weissman D, Rabin RL, Arthos J, Rubbert A, Dybul M, Swofford R, Venkatesan S, Farber JM, Fauci AS (1997) Macrophage-tropic HIV and SIV envelope proteins induce a signal through the CCR5 chemokine receptor. Nature 389(6654): 981–5

Withers-Ward ES, Kitamura Y, Barnes JP, Coffin JM (1994) Distribution of targets for avian retrovirus DNA integration in vivo. Genes Dev 8:1473–1487

Xu X, McCormick-Shannon K, Voelker DR, Mason RJ (1998) KGF increases SP-A and SP-D mRNA levels and secretion in cultured rat alveolar type II cells. American Journal of Respiratory Cell and Molecular Biology 18(2):168–78

York DF, Vigne R, Verwoerd DW, Querat G (1991) Isolation, identification, and partial cDNA cloning of genomic RNA of jaagsiekte retrovirus, the etiological agent of sheep pulmonary adenomatosis. J Virol 65(9):5061–7

York DF, Vigne R, Verwoerd DW, Querat G (1992) Nucleotide sequence of the jaagsiekte retrovirus, an exogenous and endogenous type D and B retrovirus of sheep and goats. J Virol 66(8):4930–9

CHAPTER 7

Identification of Hyal2 as the Cell-Surface Receptor for Jaagsiekte Sheep Retrovirus and Ovine Nasal Adenocarcinoma Virus

A. D. MILLER

Abstract. Jaagsiekte sheep retrovirus (JSRV) and ovine nasal adenocarcinoma virus (ONAV) replicate in the airway and cause epithelial cell tumors through the activity of their envelope (Env) proteins. Identification of the receptor(s) that mediate cell entry by these viruses is crucial to understanding the oncogenic activity of Env and for the development of gene therapy vectors based on these viruses that are capable of targeting airway cells. To identify the viral receptor(s) and to further study the biology of JSRV and ONAV, we developed retroviral vectors containing Moloney murine leukemia virus components and the Env proteins of JSRV or ONAV. We used a new technique involving positional cloning by phenotypic mapping in radiation hybrid cells to identify and clone the human receptor for JSRV, Hyal2, which also serves as the receptor for ONAV. Hyal2 is a glycosylphosphatidylinositol-anchored cell-surface protein that has low hyaluronidase activity and is a member of a large family that includes sperm hyaluronidase (Spam) and serum hyaluronidase (Hyal1). Hyal2 is

---

A. D. MILLER
Fred Hutchinson Cancer Research Center, 1100 Fairview Avenue North, Seattle, WA 98109, USA
e-mail: dmiller@fhcrc.org

located in a region of human chromosome 3p21.3 that is often deleted in lung cancer, suggesting that it may be a tumor suppressor. However, its role in JSRV or ONAV tumorigenesis, if any, is still unclear. JSRV vectors are capable of transducing various human cells, and are being further evaluated for gene therapy purposes.

1 Introduction

Retrovirus entry into cells depends on the presence of specific proteins that bind the virus envelope (Env) protein and help trigger conformational changes in Env that lead to fusion of the virus and cell membranes, and entry of the virus core into the cytoplasm. A wide variety of proteins have been found to serve as receptors for different retroviruses, based primarily on their ability to promote virus entry after gene transfer into cells that are not naturally permissive for virus entry (Table 1, Fig. 1). In most cases, a single protein suffices to render otherwise nonpermissive cells susceptible to virus entry. Typically, these proteins promote virus binding, and may also promote virus fusion with the cell membrane. For other viruses (for example, human immunodeficiency virus; HIV) there are distinct binding and fusion receptors that are required for virus entry. While retroviruses with closely related Env proteins are more likely to use the same receptor, the degree of overall similarity is not a reliable indicator of common receptor use. For example, Moloney murine leukemia virus (MoMLV) and AKV-MLV (AKV) have similar Env proteins and use the CAT-1 transport protein as a receptor; however, polytropic murine leukemia virus (MLV) and amphotropic MLV are as closely related, but they use distinct receptors for cell entry (Table 1, Fig. 1; OVERBAUGH et al. 2001). Retrovirus receptors are key determinants of the species and cell types that a retrovirus can infect, and thus are primary determinants of the host range and the type of disease induced by the virus.

Jaagsiekte sheep retrovirus (JSRV) induces oncogenic disease in the epithelial cells of the lower airway (bronchioles and alveoli) of sheep and goats, while the closely related ovine nasal adenocarcinoma virus (ONAV; previously referred to as enzootic nasal tumor virus or ENTV) induces tumors originating from the nasal epithelium. Recent work described in the chapter by FAN et al. of this volume shows that the Env proteins of these viruses mediate oncogenic transformation (MAEDA et al. 2001; RAI et al. 2001). Thus it was important to identify the receptors for these virus-

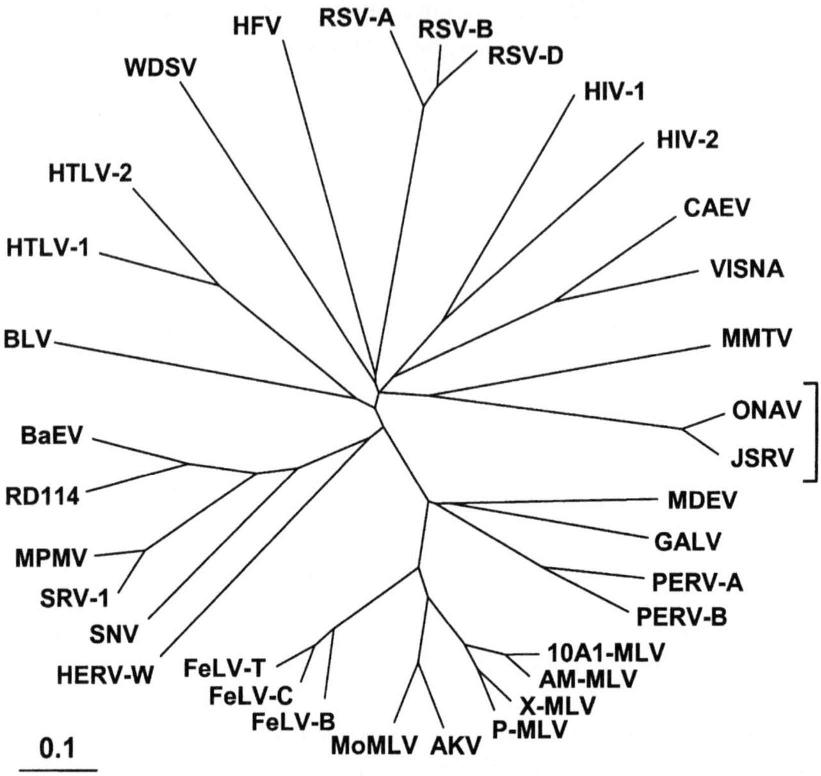

Fig. 1. Dendrogram (ClustalW) of relationships between retrovirus Env proteins. *Scale bar* indicates 10% amino acid divergence. Abbreviations: *HIV*, human immunodeficiency virus; *CAEV*, caprine arthritis and encephalitis virus; *VISNA*, maedi-visna virus; *MMTV*, mouse mammary tumor virus; *ONAV*, ovine nasal tumor virus; *JSRV*, jaagsiekte sheep retrovirus; *MDEV*, *Mus dunni* endogenous virus; *GALV*, gibbon ape leukemia virus; *PERV*, pig endogenous retrovirus; *MLV*, murine leukemia virus; *AM-MLV*, amphotropic MLV; *X-MLV*, xenotropic MLV; *P-MLV*, polytropic MLV; *AKV*, AKV-MLV; *MoMLV*, Moloney MLV; *FeLV*, feline leukemia virus; *HERV*, human endogenous retrovirus; *SNV*, spleen necrosis virus; *SRV*, simian retrovirus; *MPMV*, Mason Pfizer monkey virus (also called SRV-3); *RD114*, RD114 cat endogenous retrovirus; *BaEV*, baboon endogenous retrovirus; *BLV*, bovine leukemia virus; *HTLV*, human T-cell leukemia virus; *WDSV*, walleye dermal sarcoma virus; *HFV*, human foamy virus (actually of simian origin); and *RSV*, Rous sarcoma virus

Table 1. Cloned retrovirus receptors (for references, see Overbaugh et al. 2001)

Retrovirus	Receptor	Type	Function
Human immunodeficiency virus, simian immunodeficiency virus	CD4 and CXCR4, CCR5, others	TM1[a] TM7	Immune recognition; G protein-coupled chemokine receptors
Ecotropic murine leukemia virus	CAT-1 (SLC7A1)	TM14	Basic amino acid transport
Gibbon ape leukemia virus, 10A1 murine leukemia virus, feline leukemia virus type B, woolly monkey virus	Pit1 (SLC20A1)	TM10–13	Phosphate transport
Amphotropic murine leukemia virus, 10A1 murine leukemia virus, feline leukemia virus type B	Pit2 (SLC20A2)	TM10–13	Phosphate transport
Bovine leukemia virus	Blvr	TM1	?
Avian leukosis virus type A	Tva	TM1	LDL receptor-like protein
Avian leukosis virus types B, D, E	Tvb	TM1	Fas/TNFR-like receptor
Mouse mammary tumor virus	Trfr	TM1	Transferrin receptor
RD114, type D simian retroviruses, baboon endogenous virus, human endogenous retrovirus type W	RDR (SLC1A5) or RDR2 (SLC1A4)	TM9–10 TM9–10	Neutral amino acid transport; Glutamate, neutral amino acid transport
Xenotropic and polytropic murine leukemia viruses	XPR1	TM8	G protein-coupled signaling? Transport?
Feline leukemia virus type C	Flvcr	TM12	Organic anion transporter?
Feline leukemia virus type T	Felix and Pit1 (SLC20A1)	soluble TM10–13	Env-like protein; Phosphate transport

[a] TM (transmembrane) followed by a number in the receptor type column refers to the number of times the receptor is predicted to span the cell membrane.

es to investigate if virus–receptor interactions play a role in Env-mediated oncogenesis and/or the tissue specificity of the diseases induced by the viruses. In addition, the replication of these viruses in the airway suggests that retroviral vectors derived from these viruses might be useful for gene therapy targeted to airway epithelial cells. In particular, gene transfer to lung epithelial cells is important for treatment of cystic fibrosis, which is caused by defects in the CTFR chloride channel that lead to lung epithelial cell malfunction. Identification of the receptors used by these viruses was important for further investigation of the utility of JSRV- and ONAV-based vectors.

2
Retrovirus Receptors and Their Identification

A variety of methods have been used to identify retrovirus receptors. The receptor for HIV was the first to be identified, and the technique in this case exploited knowledge of the limited range of cells HIV could infect (primarily $CD4^+$ T-lymphocytes and macrophages), and the availability of monoclonal antibodies specific for many surface proteins on these cells (Dalgleish et al. 1984; Klatzmann et al. 1984). These two groups showed that antibodies specific for the CD4 antigen present on T lymphocytes and macrophages could block infection mediated by the HIV envelope protein, while antibodies against other surface antigens had no effect. These results showed that CD4 is an essential part of the HIV receptor.

The method of receptor identification for HIV is not typical of methods used to isolate other retrovirus receptors. Most methods have involved screening for genes or cDNAs from susceptible cells that could promote virus entry after DNA transfer into cells that are normally nonpermissive to virus entry. Early techniques used transfection of genomic DNA from susceptible cells into nonpermissive cells. The transfected cells were then screened for susceptibility to infection. This was usually accomplished by infecting them with a retrovirus vector bearing the retrovirus envelope protein of interest and that also expressed a selectable marker gene (e.g., a drug resistance gene). As a result, infectable cells would be rendered drug resistant, due to infection and expression of the retroviral vector. The infectable cells could be selected by growth of the transfected cell population in the selection drug. Once infectable cells were obtained, DNA from them was used to transfect fresh nonpermissive cells, which were again screened for susceptibility to virus infection. Repeated rounds of transfection were used to help eliminate irrelevant DNA transferred to the infectable primary transfectants. That is, the only DNA from the permissive cells in the infectable (formerly nonpermissive) cells was that encoding the receptor. The last step was the most difficult – identification of the permissive cell DNA responsible for conferring infectability. In the case of MoMuLV, which infects mouse but not human cells, transfected mouse DNA in the infectable human cells was identified and cloned by using mouse repetitive element hybridization probes (ALBRITTON et al. 1989). For the gibbon ape leukemia virus, which infects human but not mouse cells, transfected human DNA in infectable mouse cells was identified by using human repetitive element probes (O'HARA et al. 1990). These successful receptor identification projects took many years to complete, and others were never successful due to the difficulty in recovering the receptor gene from the transfected cells.

Screening for receptor activity following introduction of cDNA expression libraries into nonpermissive cells accelerated the process of receptor identification. The first successful attempts involved transfection of cDNA libraries. Isolation and cloning of the receptor gene from the infectable cells was facilitated by linkage of expression vector sequences to the cDNA of interest (MILLER et al. 1994). Thus, screening could be carried out with vector-specific (e.g., bacterial plasmid) probes.

However, the most effective method for receptor cloning involves the use of cDNA expression libraries constructed in retrovirus vectors. High complexity retroviral vector libraries with broad host range envelope proteins are constructed using mRNA from permissive cells and are used to infect nonpermissive cells. Cells that express the receptor are identified by infection with a second retrovirus vector bearing the Env protein of interest and containing a selectable marker, to allow selection and isolation of susceptible cells. The cDNA of interest is then isolated from the infectable cells by PCR using conserved viral sequences in the cDNA vector (BATTINI et al. 1999). The advantage of this technique over that using nonviral expression libraries is that the retroviral vectors integrate in a precise manner with respect to the vector genome, allowing facile PCR cloning of inserts from the integrated vectors. In contrast, transfection with nonviral expression vectors can result in rearrangement and recombination such that the receptor cDNA is difficult to isolate. In addition, viral vectors can be easily rescued from cells (by infection with a replication-competent helper retrovirus) and used to infect other nonsusceptible cells; this facilitates purification of functional receptor clones.

While several groups have localized retrovirus receptors to regions of animal chromosomes by standard genetic approaches (e.g., genetic crosses between susceptible and nonpermissive animals; somatic cell hybrids), positional cloning of retrovirus receptors was unsuccessful until the cloning of the JSRV receptor. We had tried to use a positional cloning technique involving human/hamster radiation hybrid cell lines, described in more detail below, to identify the human receptors for several retroviruses, including the receptors for the RD114 cat endogenous virus, the xenotropic murine leukemia virus, and feline leukemia virus type C (Rasko et al. 2000). However, while we were able to localize these receptors to within ~260 to ~600 kb of human genomic sequence, the sequences of these regions had not been determined, and we were unsuccessful in identifying these receptors by using this technique. We and others were able to clone these receptors by using cDNA expression libraries in retrovirus vectors (see references in Rasko et al. 2000). More recently, we have used this positional cloning technique to clone the receptor for JSRV and ONAV as described below.

3
Production and Host Ranges of JSRV and ONAV Vectors

It is possible to produce JSRV from cultured cells by transfection of cells with a JSRV proviral clone in which the JSRV promoter has been replaced with the strong immediate early promoter from human cytomegalovirus (CMV) (Palmarini et al. 1999a), but JSRV itself replicates poorly in cultured cells due to the low activity of the JSRV promoter in these cells (Palmarini et al. 1999b). To avoid this problem and to provide useful genetic markers for JSRV or ONAV Env-mediated virus infection, we made hybrid retroviral vectors consisting of an MoMLV vector, MoMLV Gag–Pol proteins, and JSRV or ONAV Env proteins (Rai et al. 2000; Dirks et al. 2002) as follows. Thymidine kinase (TK)-negative NIH3T3 cells were cotransfected with the pLGPS plasmid that encodes MoMLV Gag–Pol proteins (Fig. 2) and a plasmid encoding TK, and a TK-positive clone that expressed a high level of reverse transcriptase was isolated (Miller et al. 1991). These cells were then cotransfected with JSRV Env expression plasmid pSX2.Jenv (Fig. 2) or the ONAV Env expression plasmid pSX2.Eenv (Fig. 2) along with a plasmid encoding hygromycin phosphotransferase, and hygromycin-resistant clones were isolated. Clones that produced the highest titer of virus following introduction of an MoMLV-based vector

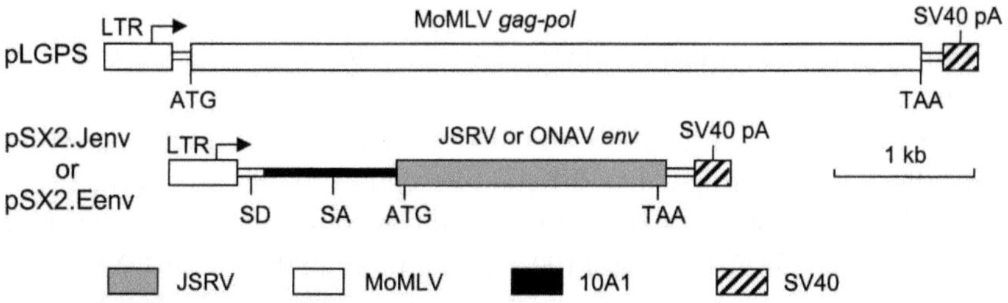

Fig. 2. Expression constructs used to generate JSRV and ONAV packaging cell lines. The plasmid pLGPS expresses the MoMLV Gag–Pol proteins (Miller et al. 1991). The plasmid pSX2.Jenv expresses the JSRV Env protein (Rai et al. 2000), and the plasmid pSX2.Eenv is identical to pSX2.Jenv except that the JSRV sequences are replaced with the same region of ONAV (Dirks et al. 2002). *Arrows* indicate promoters, *SV40pA* indicates the early region polyadenylation signal from simian virus 40, *LTR* indicates the 5′ deleted MoMLV long terminal repeat promoter, *SD* and *SA* indicate splice donor and acceptor sites, respectively, and the positions of protein coding region start (*ATG*) and stop (*TAA*) codons are shown

Table 2. The host ranges of JSRV and ONAV vectors[a] (data from Dirks et al. 2002)

Species	Target cells	ONAV vector titer	JSRV vector titer
Sheep	SSF	5×10^3	6×10^5
Human	HT-1080	$<10^b$	3×10^4
	IB3	<2	4×10^4
	293	5×10^2	4×10^4
	HeLa	<2	10
Monkey	Vero	<2	1×10^5
Dog	D17	<2	1×10^4
Bovine	MDBK	<2	2×10^3
Rabbit	RbTE	<2	2×10^3
Cat	G355	<2	20
Rat	208F	<2	<2
Hamster	A23	<2	<2
Mouse	NIH 3T3	$<10^b$	$<10^b$

[a] Target cells were exposed to the LAPSN vector bearing the ONAV Env (produced from PN172/LAPSN cells) or bearing the JSRV Env (produced from PJ4/LAPSN cells). Vector titers are expressed in AP^+ focus forming units/ml except for HeLa cells for which titers are expressed in G418-resistant colony forming units/ml due to the high levels of endogenous heat-stable AP in HeLa cells. Results are means of at least two experiments.

[b] Low-level transduction was detected in some assays.

that encodes human placental alkaline phosphatase (AP) and neomycin phosphotransferase (LAPSN) (Miller et al. 1994) were selected and used in further studies.

The host ranges of the JSRV and ONAV vectors are distinct, suggesting that these viruses might use different receptors (Table 2) (Dirks et al. 2002). Both JSRV and ONAV vectors could transduce sheep cells, although the titer of the JSRV vector was 100-fold higher than that of the ONAV vector, and neither could transduce cells from rodent species. However, the JSRV vector was able to transduce cells from many mammalian species that were not transduced by the ONAV vector. For example, the JSRV vector had a titer of 10^5 on Vero monkey cells, while that of the ONAV vector was less than 2, a difference of over 50,000-fold that cannot be explained by the difference in titer observed on sheep cells alone. Interestingly, the only human cells transduced by the ONAV vector were 293 cells, while the JSRV vector was able to transduce several human cell lines.

4
Identification of Hyal2 as the JSRV Receptor

The JSRV receptor was localized to a small region of human chromosome 3 by screening human/hamster radiation hybrid cell lines for transduction by the JSRV vector (Rai et al. 2000). These hybrid cell lines consist of a panel of 83 clonal hamster cell lines that contain multiple independent fragments (average size = 4 Mb) of the human genome. These clones were generated by irradiation (10,000 rad) of human cells and fusion of these cells with A23 hamster cells. DNA from each of these cell lines has been analyzed for the presence of more than 14,000 unique human DNA markers to construct a linkage map of the human genome (http://www.shgc.stanford.edu/Mapping/Marker/STSindex.html). An unknown gene can be localized in this linkage map by screening each of the clones for the presence of the gene, and comparing the results with the patterns generated by the other unique markers. Such linkage analysis is available as a service provided by the Stanford Human Genome Center (www.shgc.stanford.edu/RH/index.html). Similar patterns indicate that the marker and the gene of interest are closely linked, while divergent patterns indicate that the marker and the gene are on different segments of DNA and are not linked. The basis for the receptor screen was the hypothesis that cells containing the receptor gene also express the receptor protein, thus the pattern of vector transduction of the hybrid cell lines

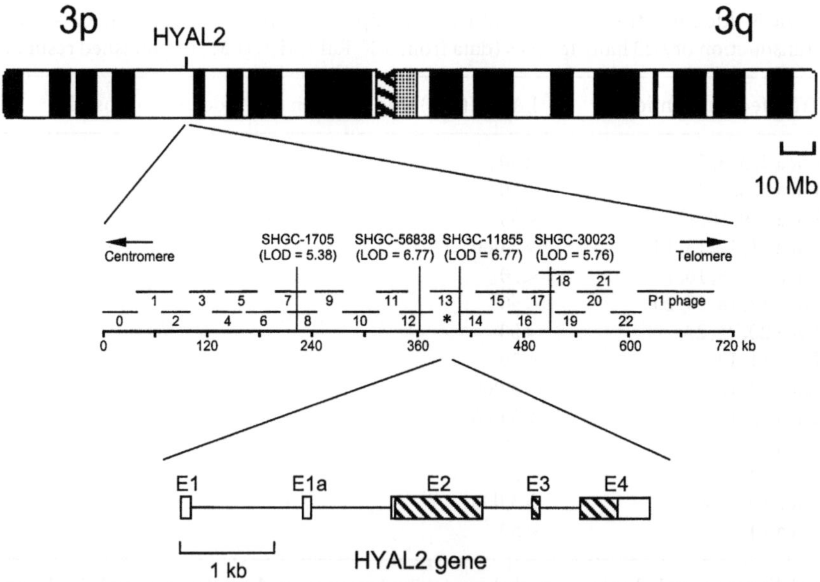

Fig. 3. Localization of the JSRV receptor *hyal2*. *Top*: An ideogram of human chromosome 3 showing the localization of the JSRV receptor in the 3p21.3 region. *Middle*: JSRV receptor localization to the human chromosome 3p21.3 lung cancer cosmid and phage contig. The human genome is depicted as a *bold line* with overlapping cosmid and phage clones indicated above. Luca cosmid numbers are indicated and the *asterisk* indicates the cosmid that contains the JSRV receptor. The positions of Stanford Human Genome Center markers are shown by *vertical lines* and the degree of linkage to the JSRV receptor is indicated by a *LOD score* (logarithm of odds) in *parentheses*. Bottom, the human *hyal2* gene. Exons are depicted by *boxes* and the Hyal2 coding region is indicated by *cross hatching*. Alternative first exons are shown with exon E1 being the typical first exon. Exon E1 contains stop codons in all three reading frames preceding the putative start codon in exon E2

should match that of the gene, and should allow localization of the receptor gene to within ~250 kb, based on the number of markers already localized and the average fragment size of the human DNA in the clones.

The pattern of JSRV vector transduction of the radiation hybrid cell lines showed that the receptor gene was linked to several unique markers in the 21.3 region of human chromosome 3 (Fig. 3, top). Initially this observation was intriguing due to the presence of HIV coreceptors of the G protein-coupled chemokine receptors family in the same region, but further analysis showed that the JSRV receptor was at least 8 Mb away from any of the HIV coreceptors and other known G protein-coupled

Table 3. Identification of a cosmid from the 3p21.3 contig that enables JSRV vector transduction of A23 hamster cells (data from S.K. Rai and A.D.M., unpublished results)

Transfected cosmid(s)	LAPSN(PJ4) transduction (AP^+ foci per μg cosmid)[a]
Luca 2, 3, 4, 5	<50
Luca 5, 6, 7, 8	<50
Luca 8, 9, 10, 11	<50
Luca 11, 12, 13, 14	>3,000
Luca 14, 15, 16, 17	<50
Luca 17, 18, 19, 20	<50
Luca 20, 21, 22	<50
Luca 11, 12	<50
Luca 12, 13	>3,000
Luca 13, 14	>3,000
Luca 11	<50
Luca 12	<50
Luca 13	>3,000
Luca 14	<50

[a] Cells were seeded into 6-cm dishes on day 1, transfected on day 2, trypsinized and replated on day 3, exposed to 1 ml LAPSN vector produced by JSRV packaging cells on day 4, and stained for AP on day 6.

receptor family members (Rai et al. 2000). Luckily, the four unique sequence markers closest to the JSRV receptor were all in a sequenced region of human chromosome 3, suggesting that the receptor would lie in the set of cosmids from this region that had been used to determine the sequence (Fig. 3, middle). The location of the gene was defined more precisely by transfecting the cosmids into the A23 hamster cells (nonpermissive for JSRV), and screening the transfected cells for susceptibility to JSRV vector transduction (Table 3). Initially, overlapping sets of cosmids were transfected to reduce the number of transfections required, and to allow receptor detection if the receptor gene spanned two cosmids. There is a relatively high rate of recombination between identical sequences during transfection, and this should have allowed detection of a receptor that spanned two of the overlapping cosmids. This analysis showed that the JSRV receptor localized to cosmid Luca13, and further analysis of cDNAs from the three genes in this cosmid, *fus1*, *hyal1*, and *hyal2*, showed that the JSRV receptor was Hyal2 (Table 4) (Rai et al. 2001).

The radiation hybrid analysis indicated that there was only one receptor for JSRV in the human genome, but this analysis could miss closely

Table 4. Cell susceptibility to JSRV vector transduction after transfection of expression plasmids (data from Rai et al. 2001)

Protein encoded by expression plasmid	JSRV vector transduction of cells (transducing units per μg DNA)[a]		
	NIH3T3	A23	HeLa
β-gal	3	60	nd
Human Fus1	<2	10	nd
Human Hyal1	12	120	10
Human Hyal2	$>10^4$	$>10^4$	7,400
Human Hyal3	13	90	20
Human Hyal4	<2	70	9
Human Spam1	<2	20	14
Mouse Hyal1	6	20	10
Mouse Hyal2	130	660	570
Mouse Hyal3	<2	140	10

[a] Cells were seeded into 6-cm dishes on day 1, transfected on day 2, trypsinized and replated on day 3, exposed to 1 ml LAPSN vector (NIH 3T3 and A23 cells) or 1 ml LNCZ vector (HeLa cells) produced by JSRV packaging cells on day 4, and stained for AP (NIH 3T3 and A23 cells) or β-gal (HeLa cells) on day 6. A vector encoding AP (LAPSN) was used for the rodent cells, while a lower-titer vector encoding β-gal (LNCZ) was used for the HeLa cells due to their high endogenous heat-stable AP levels. Results are means from 2–7 experiments.
nd, Not done.

linked genes. Indeed there are two related genes, *hyal1* and *hyal3*, that are located near *hyal2*. However, transfection of expression plasmids for *hyal1*, *hyal2*, or *hyal3*, or two other related genes on human chromosome 7, *hyal4* and *spam1*, showed that only Hyal2 served as a receptor for JSRV (Table 4). Mouse Hyal2 showed some activity as a receptor for JSRV, but the activity was far below that of human Hyal2 (Table 4), in accord with the low transduction rate of JSRV vectors in mouse cells (Table 2). Mouse Hyal1 and Hyal3 had no JSRV receptor activity (Table 4).

5 Properties of Hyal2

Hyal2 is a member of a large family of proteins present in many animals (Fig. 4), many of which are hyaluronidases capable of cleaving hyaluronic acid. Hyaluronic acid is a high-molecular-weight repeating disaccharide

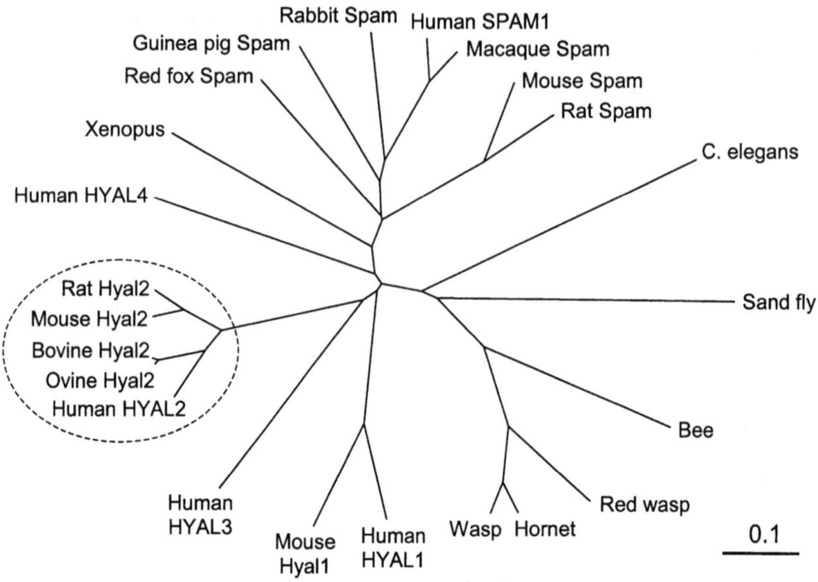

Fig. 4. Dendrogram (ClustalW) of relationships between hyaluronidase family members. *Scale bar* indicates 10% amino acid sequence difference. Protein sequence data were obtained from the following GenBank accession numbers: human Hyal1, U96078; mouse Hyal1, AF011567; human Hyal2, U09577; mouse Hyal2, AF302843; rat Hyal2, AF034218; ovine Hyal2, AF411974; bovine Hyal2, AF411973; human Hyal3, AF036035; human Hyal4, AF009010; human Spam1, NM003117; mouse Spam, NM009241; rat Spam, NM053967; rabbit Spam, U09183; macaque Spam, P38568; red fox Spam, U41412; guinea pig Spam, X56332; *Xenopus*, AF134981; *Caenorhabditis elegans*, Z49071; bee, L10710; hornet, L34548; wasp, L43562; red wasp, AF174528; and sand fly, AF132515

chain of *N*-acetyl-D-glucosamine and D-glucuronic acid that is a major component of the extracellular matrix. Sperm hyaluronidase (Spam1) is active at both neutral and acidic pH, is linked to the sperm cell membrane by a glycosylphosphatidylinositol (GPI) anchor, and participates in sperm penetration of the hyaluronic acid-rich cumulus cell layer surrounding the oocyte. Hyal1 is a serum hyaluronidase and is active only at acidic pH. Bee, wasp, and hornet hyaluronidases are present in venom and help digest the extracellular matrix to promote systemic dispersion of venom after a sting.

Hyal2 was originally identified as a lysosomal hyaluronidase (Lepperdinger et al. 1998), an unlikely location for a retrovirus receptor. However, an analysis of the Hyal2 amino acid sequence using a pro-

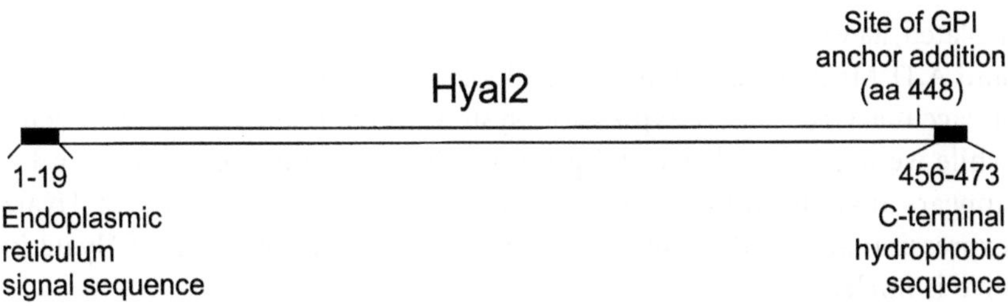

Fig. 5. Hyal2 protein processing signals

gram for detection of GPI-anchored proteins (D. Buloz and J. Kronegg, http://129.194.186.123/GPI-anchor/index_en.html) predicted that Hyal2 is a cell-surface GPI-anchored protein (Fig. 5), as is the sperm hyaluronidase Spam1. The program also predicts that Hyal4 and Spam1 are GPI anchored whereas Hyal1 and Hyal3 should be secreted.

Several lines of evidence show that Hyal2 is indeed a GPI-anchored cell surface protein (Rai et al. 2001). First, GPI-anchored proteins have endoplasmic reticulum (ER) signal sequences at their N termini, and replacement of the putative signal sequence of Hyal2 with a Flag tag results in a cytoplasmically localized protein that is not active as a JSRV receptor. However, replacement of this signal with a preproinsulin ER signal sequence followed by the Flag tag results in a protein localized to the cell surface that does function as a JSRV receptor. Second, the Flag-tagged Hyal2 that localized to the cell surface could be released from the cell using phosphatidylinositol-specific phospholipase C, a bacterial enzyme that cleaves GPI anchors (Rai et al. 2001). Such treatment renders cells resistant to JSRV vector transduction but does not affect transduction by other vectors that recognize non GPI-anchored receptors (Rai et al. 2001). A likely explanation for the original mislocalization of Hyal2 to the lysosome is that the protein was tagged at the C terminus with green fluorescent protein, which would be expected to be removed during GPI anchor addition and sent to the lysosome for degradation. Thus it is likely the localization of GFP to the lysosome did not reflect the true location of Hyal2.

There are conflicting reports regarding the hyaluronidase activity of Hyal2. Lepperdinger at al. (1998) reported that Hyal2 exhibited hyaluronidase activity at neutral but not acidic pH, similar to the activity of Hyal1. Initially we could not detect hyaluronidase activity associated with Hyal2 (Rai et al. 2001). However, we have since been able to detect very low levels of activity associated with expression of the *hyal2* cDNA that we isolated, as well as one isolated by LEPPERDINGER et al. (V. VIGDOROVICH and A. D. MILLER, unpublished results). LEPPERDINGER et al. (1998) used a vaccinia virus-based expression system to overexpress Hyal2 in cells, while we used a weaker CMV promoter-based expression system, and it appears that the difference in results is due to higher levels of Hyal2 expression achieved using the vaccinia system. The function of Hyal2 in vivo is unclear.

Analysis of human cDNA sequences in the GenBank database revealed two transcripts originating from different promoters that encode Hyal2. Both contain the same coding region distributed on three exons but have different noncoding 5′ exons (Fig. 3, bottom). GenBank accession XM_003271.3 is representative of the cDNA that includes exon 1 and XM_040967.1 is representative of the cDNA that includes exon 1a. Hyal2 mRNA is widely expressed in different tissues of mice and humans (CSOKA et al. 1999; LEPPERDINGER et al. 1998; STROBL et al. 1998), and many different sheep cell lines and tissues are permissive to JSRV entry, indicating that the JSRV receptor is expressed in multiple cell types in sheep (HOLLAND et al. 1999; PALMARINI et al. 1996, 1999b).

Most small cell lung cancers and many non-small cell lung, breast, and cervical cancers show loss of heterozygosity or deletion of sequences in the p21.3 region of chromosome 3 where *hyal2* is located, suggesting that a gene in this region is a tumor suppressor (WEI et al. 1996; WISTUBA et al. 1997; SEKIDO et al. 1998; LERMAN et al. 2000). In one small cell cancer cell line (NCI-H524), a small homozygous deletion of ~30 kb is found that involves only three genes, *hyal2, hyal1, and fus1* (LERMAN et al. 2000). It is thus intriguing to speculate that Hyal2 is a tumor suppressor and that JSRV and ONAV Env proteins might mediate their effects by binding to and inhibiting the tumor suppressor activity of Hyal2. However, it is unclear whether Hyal2 is indeed a tumor suppressor. Furthermore, the JSRV and ONAV Env proteins transform mouse and rat cells (MAEDA et al. 2001; RAI et al. 2001). These cells are resistant to JSRV and ONAV vector transduction (RAI et al. 2000; DIRKS et al. 2002), suggesting that interaction of the Env proteins with Hyal2 is not important for transformation. Work presented in the chapter by FAN et al. of this volume indicates that the JSRV Env protein transforms via activation of the phophatidylinositol 3-kinase (PI-3K) pathway, and that a tyrosine residue in the cytoplasmic tail of the Env protein is critical for transformation (PALMARINI et al. 2001). More work is needed to investigate the possible role of Hyal2 in Env-mediated transformation.

6
Hyal2 Also Serves as the Receptor for ONAV

The different host ranges of JSRV and ONAV suggested that these viruses might use different receptors for cell entry (Table 2). However, their Env proteins are closely related, arguing that they might use the same receptor (Fig. 1). Interference analysis can detect the use of the same receptor by different viruses; this involves the ability of Env expression in a cell to block infection by retroviruses that use the same receptor, while the entry of viruses that use different receptors is not affected. As the ONAV vector transduces sheep skin fibroblasts (SSF cells) only at relatively high levels (Table 2), SSF cells were used to test for interference (Fig. 6) (Dirks et al. 2002). Expression of the JSRV Env protein in SSF cells strongly inhibited transduction by both JSRV and ONAV vectors, while transduction by an otherwise identical vector bearing the RD114 virus was relatively unaffected by JSRV Env expression. Transduction by a vector bearing the 10A1 MLV Env actually went up in the presence of the JSRV Env, which may be due to the oncogenic activity of JSRV Env, which stimulates cell division and perhaps increases the levels of the phosphate transporters that pro-

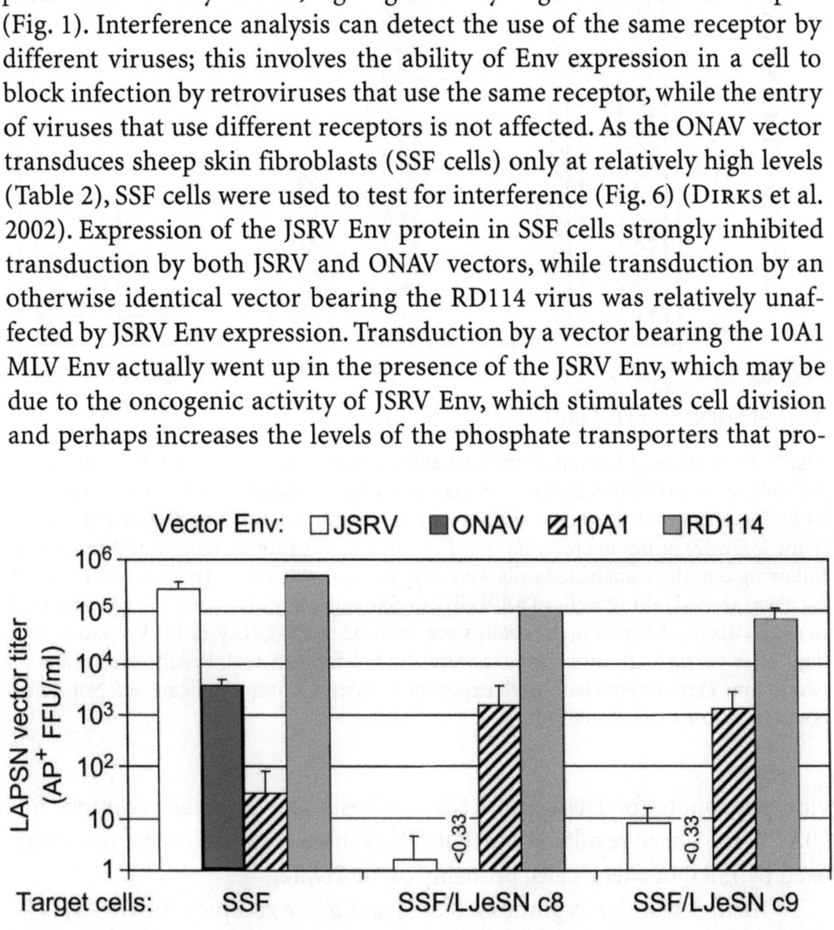

Fig. 6. JSRV Env expression blocks transduction by ONAV and JSRV vectors in SSF cells. SSF cells and two clonal SSF cell lines (SSF/LJeSN c8 and c9) that contain the JSRV Env expression vector LJeSN were exposed to LAPSN vectors made with the indicated envelope proteins. Transduction was measured 3 days after vector exposure by staining the cells for AP+ foci. Results are means of two to three independent experiments with duplicate determinations performed in each experiment. (Data from Dirks et al. 2002)

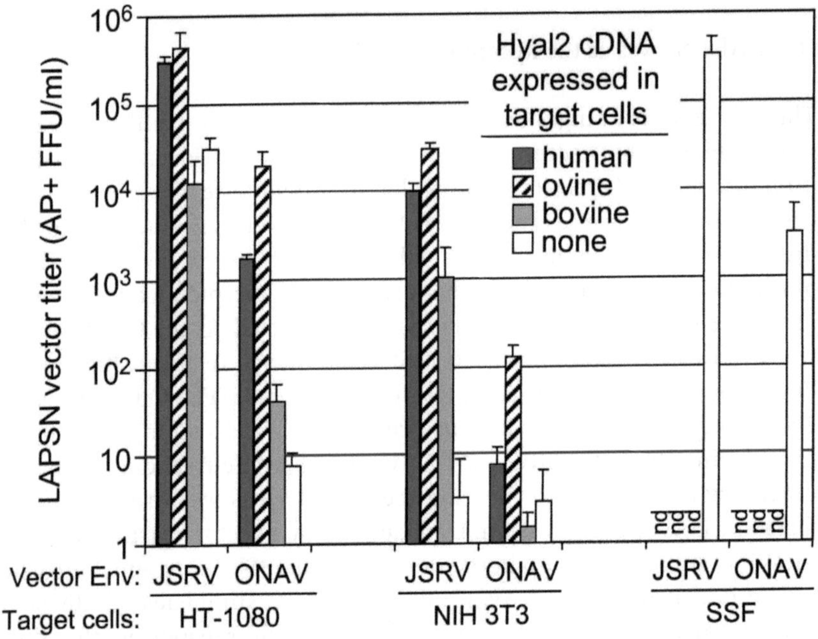

Fig. 7. Evaluation of human, ovine, or bovine Hyal2 as receptors for ONAV and JSRV. HT-1080 and NIH3T3 cells were seeded at 5×10^5 cells/dish in 60-mm dishes. After 24 h, the cells were transfected with expression plasmids encoding human (*hHYAL2*), ovine (*oHyal2*) or bovine (*bHyal2*) Hyal2 or an empty expression plasmid (*none*). The following day, the transfected cells were trypsinized with 1.5 ml trypsin and 100 µl of each was seeded into 12 wells of a 6-well dish. SSF cells were also seeded at 10^5 cells/well in 6-well dishes. After 24 h, the ce6lls were exposed to the ONAV or JSRV vectors. Two days after vector exposure the cells were stained for AP^+ foci. Results are means of two to four experiments with each experiment carried out in duplicate. *nd*, Not done. (Data from Dirks et al. 2002)

vide phosphate for DNA and RNA synthesis and serve as receptors for 10A1 MLV. These results show that ONAV uses a subset of the receptors used by JSRV in sheep cells, probably ovine Hyal2.

To further test the hypothesis that Hyal2 is the receptor for ONAV, HT-1080 human cells and NIH3T3 mouse cells were transfected with expression vectors for Hyal2 proteins from various species and transduction by JSRV and ONAV vectors was measured (Fig. 7) (Dirks et al. 2002). These results show that the ovine Hyal2 promotes the highest rates of transduction by both vectors. Human Hyal2 promotes almost as high transduction by the JSRV vector, and about 10-fold lower transduction by the ONAV vector compared to that observed for ovine Hyal2. Bovine Hyal2 was least functional as a receptor for the vectors. These results show that Hyal2 from sheep or humans can function as a receptor for JSRV and ONAV. However, ONAV vectors transduce many human cell lines poorly (Table 2), and this may be explained by the hypothesis that ONAV vectors bind human Hyal2 poorly and overexpression of this receptor is required to promote high transduction rates.

7
Discussion

Hamster, mouse, and human (HeLa) cells that are resistant to transduction by JSRV and ONAV vectors are rendered susceptible to transduction following expression of Hyal2 in the cells (Table 4, Fig. 7), indicating that Hyal2 is a receptor for cell entry mediated by the JSRV and ONAV Env proteins. Otherwise identical vectors bearing Env proteins from other retroviruses can transduce these cells in the absence of Hyal2 expression, showing that Hyal2 interacts specifically with the JSRV and ONAV Env proteins (data not shown). Hyal2 appears to be the only protein having this function in human cells because none of the other Hyal2 family members present in human cells serve as functional receptors (Table 4), and the results of the radiation hybrid screen for the receptor indicate that only one locus harbors a functional receptor (Rai et al. 2000). These results do not exclude a requirement for one or more coreceptors that might also be required for virus entry, since such molecules might be widely expressed in mammalian cells and thus might not be detected in the screens used to date to detect receptor activity. In the case of HIV entry, the requirement for a coreceptor was apparent from the finding that while human CD4 expression promoted HIV entry into otherwise resistant human cells, it was not sufficient to promote HIV entry into mouse cells. Expression of CD4 and one of several G protein-coupled receptors was sufficient in mouse cells, showing the involvement of at least two receptors in HIV entry. Experiments are underway to see if cell types exist that cannot be rendered susceptible to JSRV or ONAV vector transduction following expression of Hyal2 in the cells, which would indicate the involvement of a coreceptor and would provide a system for identification of the molecule.

The role of Hyal2 in transformation by the JSRV and ONAV Env proteins is still unclear. Deletion of Hyal2 in many epithelial cancers of the lung and other organs suggests a role for Hyal2 as a tumor suppressor, and the viral Env proteins might be acting to block this activity or to reverse the activity and convert Hyal2 into an oncogene. A major problem with this hypothesis is that the JSRV and ONAV Env proteins can transform mouse and rat cells that are not susceptible to JSRV or ONAV vector transduction. If the hypothesis is correct, this result implies that the Env proteins can interact with Hyal2 to initiate transformation, but that the binding does not allow virus entry – a somewhat unlikely proposition. Alternatively, evidence is accumulating that the JSRV Env protein can transform cells through a PI-3K pathway that is dependent on the presence of a tyrosine residue in the cytoplasmic tail of the JSRV Env (Palmarini et al. 2001) and Hyal2 may play no role in this pathway. The best way to address this issue would be to delete the Hyal2 gene in the rodent cells and see if that abrogates transformation by the Env proteins.

Based on the natural replication of JSRV and ONAV in the airway, vectors based on these viruses might be useful for gene therapy directed to the airway, in particular, for treatment of cystic fibrosis. Other retroviruses are susceptible to inactivation by lung fluid, and JSRV and ONAV have probably evolved to resist this effect and to promote efficient infection of airway epithelial cells. Among these two viruses, JSRV is the better choice for human gene therapy purposes because vectors bearing the JSRV Env can transduce many human cell types, while similar vectors bearing the ONAV Env have a lower titer and transduced only one of the human cell types tested (Table 2). We have shown that vectors bearing the JSRV Env are resistant to inactivation by Survanta, a bovine-derived lung surfactant used to provide surfactant function in premature infants, while otherwise identical vectors with an amphotropic MLV Env are inactivated (Coil et al. 2001). To date, there are no data on the efficiency of JSRV or ONAV vector transduction of lung epithelium. This is due in part to the fact that rodent cells are resistant to transduction, such that the experiments must be done in large animals. Alternatively, transgenic mice expressing human Hyal2 should provide a suitable model for these experiments. Lastly, we have been able to inactivate the transforming activity of the JSRV Env without affecting its ability to mediate vector transduction (S.-L. Liu and A.D. Miller, unpublished results; Palmarini et al. 2001); this result, and the fact that JSRV vectors do not actually encode the JSRV Env protein, virtually eliminates concern about possible oncogenicity of JSRV vectors.

Acknowledgements. This work was supported by grants DK47754, HL54881, and HL36444 from the National Institutes of Health, USA.

References

Albritton LM, Tseng L, Scadden D, Cunningham JM (1989) A putative murine ecotropic retrovirus receptor gene encodes a multiple membrane-spanning protein and confers susceptibility to virus infection. Cell 57:659–666

Battini JL, Rasko JEJ, Miller AD (1999) A human cell-surface receptor for xenotropic and polytropic murine leukemia viruses: Possible role in G protein-coupled signal transduction. Proc Natl Acad Sci USA 96:1385–1390

Coil DA, Strickler JH, Rai SK, Miller AD (2001) Jaagsiekte sheep retrovirus Env protein stabilizes retrovirus vectors against inactivation by lung surfactant, centrifugation, and freeze-thaw cycling. J Virol 75:8864–8867

Csoka AB, Scherer SW, Stern R (1999) Expression analysis of six paralogous human hyaluronidase genes clustered on chromosomes 3p21 and 7q31. Genomics 60:356–361

Dalgleish AG, Beverly PCL, Clapham PR, Crawford DH, Greaves MF, Weiss RA (1984) The CD4 (T4) antigen is an essential component of the receptor for the AIDS retrovirus. Nature 312:763–767

Dirks C, Duh FM, Rai SK, Lerman MI, Miller AD (2002) Mechanism of cell entry and transformation by enzootic nasal tumor virus. J Virol 76:2141–2149

Holland MJ, Palmarini M, Garcia-Goti M, Gonzalez L, McKendrick I, de las Heras M, Sharp JM (1999) Jaagsiekte retrovirus is widely distributed both in T and B lymphocytes and in mononuclear phagocytes of sheep with naturally and experimentally acquired pulmonary adenomatosis. J Virol 73:4004–4008

Klatzmann D, Champagne E, Chamaret S, Gruest J, Guetard D, Hercend T, Gluckman JC, Montagnier L (1984) T-lymphocyte T4 molecule behaves as the receptor for human retrovirus LAV. Nature 312:767–768

Lepperdinger G, Strobl B, Kreil G (1998) HYAL2, a human gene expressed in many cells, encodes a lysosomal hyaluronidase with a novel type of specificity. J Biol Chem 273:22466–22470

Lerman MI, Minna JD, for The International Lung Cancer Chromosome 3p21.3 Tumor Suppressor Gene Consortium (2000) The 630-kb lung cancer homozygous deletion region on human chromosome 3p21.3: Identification and evaluation of the resident candidate tumor suppressor genes. Cancer Res 60:6116–6133

Maeda N, Palmarini M, Murgia C, Fan H (2001) Direct transformation of rodent fibroblasts by jaagsiekte sheep retrovirus DNA. Proc Natl Acad Sci USA 98:4449–4454

Miller AD, Chen F (1996) Retrovirus packaging cells based on 10A1 murine leukemia virus for production of vectors that use multiple receptors for cell entry. J Virol 70:5564–5571

Miller AD, Garcia JV, von Suhr N, Lynch CM, Wilson C, Eiden MV (1991) Construction and properties of retrovirus packaging cells based on gibbon ape leukemia virus. J Virol 65:2220–2224

Miller DG, Edwards RH, Miller AD (1994) Cloning of the cellular receptor for amphotropic murine retroviruses reveals homology to that for gibbon ape leukemia virus. Proc Natl Acad Sci USA 91:78–82

O'Hara B, Johann SV, Klinger HP, Blair DG, Rubinson H, Dunn KJ, Sass P, Vitek SM, Robins T (1990) Characterization of a human gene conferring sensitivity to infection by gibbon ape leukemia virus. Cell Growth & Differentiation 1:119–127

Overbaugh J, Miller AD, Eiden MV (2001) Receptors and entry cofactors for retroviruses include single and multiple transmembrane-spanning proteins as well as newly described glycosylphosphatidylinositol-anchored and secreted proteins. Microbiol Molec Biol Rev 65:371–389

Palmarini M, Holland MJ, Cousens C, Dalziel RG, Sharp JM (1996) Jaagsiekte retrovirus establishes a disseminated infection of the lymphoid tissues of sheep affected by pulmonary adenomatosis. J Gen Virol 77:2991–2998

Palmarini M, Maeda N, Murgia C, De-Fraja C, Hofacre A, Fan H (2001) A phosphatidylinositol 3-kinase docking site in the cytoplasmic tail of the jaagsiekte sheep retrovirus transmembrane protein is essential for envelope-induced transformation of NIH 3T3 cells. J Virol 75:11002–11009

Palmarini M, Sharp JM, De las Heras M, Fan H (1999a) Jaagsiekte sheep retrovirus is necessary and sufficient to induce a contagious lung cancer in sheep. J Virol 73:6964–6972

Palmarini M, Sharp JM, Lee C, Fan H (1999b) In vitro infection of ovine cell lines by Jaagsiekte sheep retrovirus. J Virol 73:10070–10078

Rai SK, DeMartini JC, Miller AD (2000) Retrovirus vectors bearing the jaagsiekte sheep retrovirus Env transduce human cells by using a new receptor localized to chromosome 3p21.3. J Virol 74:4698–4704

Rai SK, Duh FM, Vigdorovich V, Danilkovitch-Miagkova A, Lerman MI, Miller AD (2001) Candidate tumor suppressor HYAL2 is a glycosylphosphatidylinositol (GPI)-anchored cell-surface receptor for jaagsiekte sheep retrovirus, the envelope protein of which mediates oncogenic transformation. Proc Natl Acad Sci USA 98: 4443–4448

Rasko JEJ, Battini JL, Kruglyak L, Cox DR, Miller AD (2000) Precise gene localization by phenotypic assay of radiation hybrid cells. Proc Natl Acad Sci USA 97: 7388–7392

Sekido Y, Ahmadian M, Wistuba II, Latif F, Bader S, Wei MH, Duh FM, Gazdar AF, Lerman MI, Minna JD (1998) Cloning of a breast cancer homozygous deletion junction narrows the region of search for a 3p21.3 tumor suppressor gene. Oncogene 16:3151–3157

Strobl B, Wechselberger C, Beier DR, Lepperdinger G (1998) Structural organization and chromosomal localization of Hyal2, a gene encoding a lysosomal hyaluronidase. Genomics 53:214–219

Wei MH, Latif F, Bader S, Kashuba V, Chen JY, Duh FM, Sekido Y, Lee CC, Geil L, Kuzmin I, Zabarovsky E, Klein G, Zbar B, Minna JD, Lerman MI (1996) Construction of a 600-kilobase cosmid clone contig and generation of a transcriptional map surrounding the lung cancer tumor suppressor gene (*TSG*) locus on human chromosome 3p21.3: progress toward the isolation of a lung cancer TSG. Cancer Res 56:1487–1492

Wistuba II, Montellano FD, Milchgrub S, Virmani AK, Behrens C, Chen H, Ahmadian M, Nowak JA, Muller C, Minna JD, Gazdar AF (1997) Deletions of chromosome 3p are frequent and early events in the pathogenesis of uterine cervical cancer. Cancer Res 57:3154–3158

Chapter 8

Enzootic Nasal Adenocarcinoma of Sheep and Goats

M. De las Heras, A. Ortín, C. Cousens, E. Minguijón, J. M. Sharp

Abstract. Enzootic nasal adenocarcinoma is a contagious tumour of the mucosal nasal glands affecting young adult sheep or goats. The disease occurs naturally in all continents except Australia and New Zealand. Clinical signs include continuous nasal discharge, respiratory distress, exophthalmos and skull deformations. The tumour is classified histologi-

M. De las Heras, A. Ortín, E. Minguijón
Departamento de Patología Animal, Facultad de Veterinaria,
Universidad de Zaragoza, Miguel Servet 177, 50013 Zaragoza, Spain
e-mail: lasheras@posta.unizar.es

C. Cousens, J. M. Sharp
Moredun Research Institute, Pentlands Science Park, Bush Loan,
Penicuik EH26 OPZ, UK
e-mail: sharm@mri.sari.ac.uk

cally as a low-grade adenocarcinoma. Nasal glands of both respiratory and olfactory muosal glands seem to be the origin of the neoplasia. It has been experimentally transmitted in sheep and goats using either tumour extracts or concentrated nasal fluids. Two distinct retroviruses are implicated in the aetiology of the neoplasia one in sheep (ONAV) and one in goats (CNAV). We suggest that jaagsiekte sheep retrovirus (JSRV), ONAV, CNAV, and their endogenous counterparts represent a unique family of retroviruses. The similarities between these viruses suggests that any control strategies, including vaccination, may be appropriate to both diseases. The differences, however, represent a unique resource for delineating the function of individual regions of the virus. It is intriguing that whilst ONAV and CNAV appear to be as different to each other as they are to JSRV, that they have very similar disease pathologies, distinct from that of OPA. Additionally, all three exogenous viruses manage to avoid instigating any apparent immune response. Whether this is indeed a result of tolerance induced by the endogenous counterparts or whether the viruses themselves have unique immunosuppressive properties will be an important finding.

1 Introduction

Enzootic nasal adenocarcinoma (ENA, enzootic nasal tumour, infectious nasal adenopapilomatosis) is a contagious neoplasm of gland cells of the nasal mucosa of sheep and goats. ENA is aetiologically associated with distinct but related retroviruses called ovine nasal adenocarcinoma virus (ONAV) and the caprine nasal adenocarcinoma virus (CNAV) in sheep and goats, respectively. These retroviruses are highly homologous with jaagsiekte sheep retrovirus (JSRV), the aetiological agent of ovine pulmonary adenocarcinoma (OPA) but can be distinguished by unique sequences of the genome. These diseases have been grouped as contagious respiratory neoplasias of sheep and goats.

The first recorded case of an endemic nasal neoplasm in sheep was in Germany in 1939 (Nieberle 1940). A few years after this first report, further affected sheep flocks were detected in the same country and the disease was transmitted using bacteria-free tumour extracts, confirming the infectious nature of the disease (Cohrs 1953). The disease was subsequently reported in sheep and goats in France (Camy 1955; Drieux et al. 1952; Lombard et al. 1966) and in sheep in Poland (Rubaj and Woloszyn

1967), the USA (Young et al. 1961), Canada (Mckinnon et al. 1982), Brazil (Fagundes et al. 1979), Ghana (Vohradsky 1973), Nigeria (Njoku et al. 1978), Ivory Coast (Charray et al. 1985), Israel (Perl et al. 1987) and Japan (Yonemichi et al. 1978). More recently, the disease has been recorded in sheep in other European countries including Spain (De las Heras et al. 1998) and Greece (Leontides et al. 1993). In goats, the disease has been described in Spain (De las Heras et al. 1985), Italy (Vitellozi et al. 1993), Greece (Leontides et al. 1993), India (Rajan et al. 1980) and Canada (Pringle et al. 1989). Thus ENA has been recorded in all the major areas where sheep and goats are farmed with the exception of Australia and New Zealand, and notably appears to be absent from the UK where OPA is common.

Epidemiological data indicates that ENA prevalence in affected flocks is very variable; in Spain the prevalence in sheep was found to be about 0.1%–0.3% (De las Heras et al. 1998), whereas in Germany it was 2%–15% (Cohrs 1953) and in Nigeria 0.3%–2% (Njoku et al. 1978). Our observations in Spain are that the prevalence does not normally increase in an affected sheep flock year by year. However, on some occasions large increases in incidence (from 0.6% to 6.6%) over the years following the first case in a flock have been reported (Charray et al. 1985). In the case of ENA in goats in Spain, we have found prevalences of 1%–1.5% but in France, for example, much higher prevalence has been recorded (up to 10%) (Lombard et al. 1966; Gueraud 1982; Giauffret et al. 1984). ENA disease symptoms can be first recognized at any time of the year (Cohrs 1953; Lombard et al. 1966).

2 Clinical Features

The clinical signs of ENA are similar in sheep and in goats. Young adult sheep and goats between 2 and 4 years of age are preferentially affected and several cases are always observed in the same flock (Cohrs 1953; Gueraud et al. 1983; Fontaine et al. 1983; Giauffret et al. 1984; Charray et al. 1985; De las Heras et al. 1991, 1998). No genetic, sex or breed predisposition has been detected. A small amount of seromucous fluid coming from the nostrils is usually the first clinical evidence of the disease. As the disease progresses, the fluid flows continuously and causes depilation around the nostrils, giving the appearance known as 'washed nose' (Fig. 1). The amount of fluid increases with the duration of the dis-

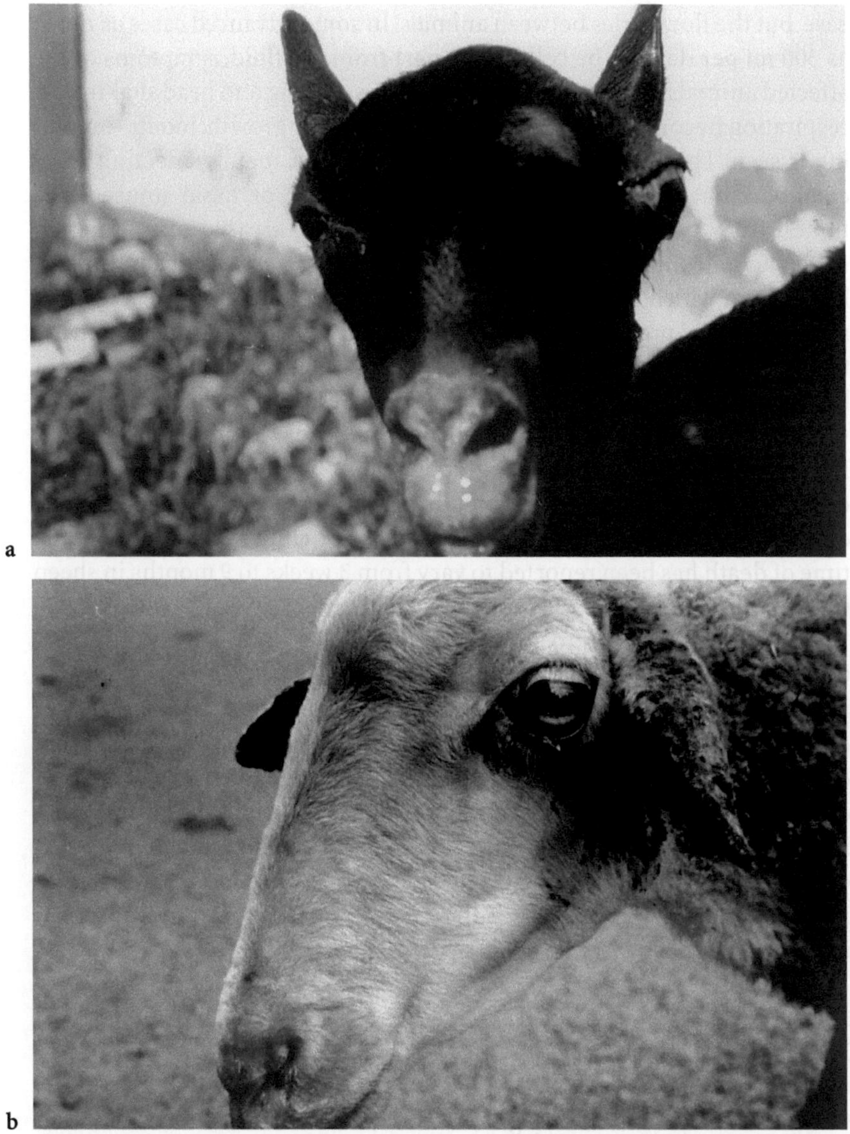

Fig. 1. Clinical features of enzootic nasal adenocarcinoma (ENA) in (**a**) a naturally affected 2-year-old goat and (**b**) a 3-year-old sheep. Depilation around the nostrils is caused by the continuous flow of seromucous fluid produced by the tumour. Exophthalmos and tumefaction of the frontal areas of the skull are also common signs

ease, but the flow varies between animals. In some advanced cases as much as 300 ml per day can be collected. Apart from the fluid, symptoms of the affected animals include snoring, coughing, sneezing and head shaking. As respiration becomes more impaired by the tumour growth, mouth-breathing is seen. The tumour expands in all directions, mainly affecting nasal chambers but often also penetrating into frontal or nasal sinuses. The tumour may compress the cranial bones in these areas, causing them to soften and atrophy, which in advanced cases may lead to cutaneous fistulation from which mucous fluid may exude. Exophthalmos can accompany these frontal protrusions if the neoplasia also presses retroocular structures (Fig. 1). Tumours can be unilateral or bilateral and in the case of bilateral lesion the clinical picture is worse. The animal remains active and with a good appetite and generally no fever is observed during the disease. Nevertheless, body condition is gradually lost. The animals generally die due to bacterial or toxaemia complications (De las Heras 1999). The duration of the disease from the appearance of clinical symptoms to the time of death has been reported to vary from 3 weeks to 9 months in sheep (Camy 1955; Cohrs 1953; McKinon et al. 1982; De las Heras et al. 1998) and from 2 weeks to 5 months in goats (De las Heras et al. 1991; Giauffret et al. 1984; Vitellozi et al. 1993).

ENA can be confused clinically with other sheep/goat diseases such as chronic bacterial fungal infections or dust allergy, but oestrosis is the main disease for which differential diagnosis is sometimes difficult. The larvae of the nasal blot fly, *Oestrus ovis*, cause a chronic catarrhal rhinitis that is sometimes haemorrhagic, with production of a generous mucus-rich exudate. This exudate sticks around the nostrils and traps dirt resulting in a dirty nose in contrast with the 'washed nose' appearance observed in ENA.

3
Macroscopic Pathology

Once the skull is opened in sagittal section, a tumour mass is seen located in the ethmoidal area (Fig. 2). There are descriptions in which all cases were unilateral (Duncan et al. 1967), either unilaterally (De las Heras et al. 1985, 1998; Giauffret et al. 1984; McKinnon et al. 1982; Njoku et al. 1978) or bilaterally (De las Heras et al. 1991; Vitellozi et al. 1993; Vohradsky 1974). Experimental studies have revealed that the earliest lesions originate in the ethmoidal nasal conchae (De las Heras et al.

a
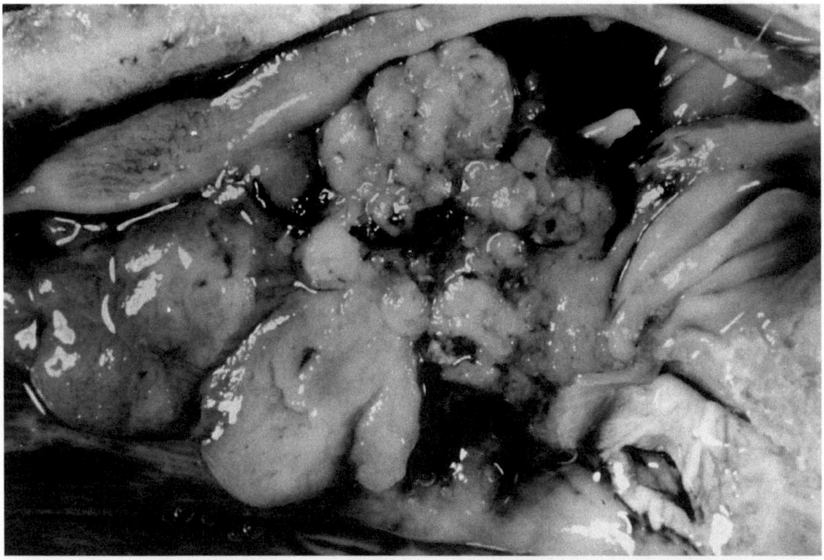

b
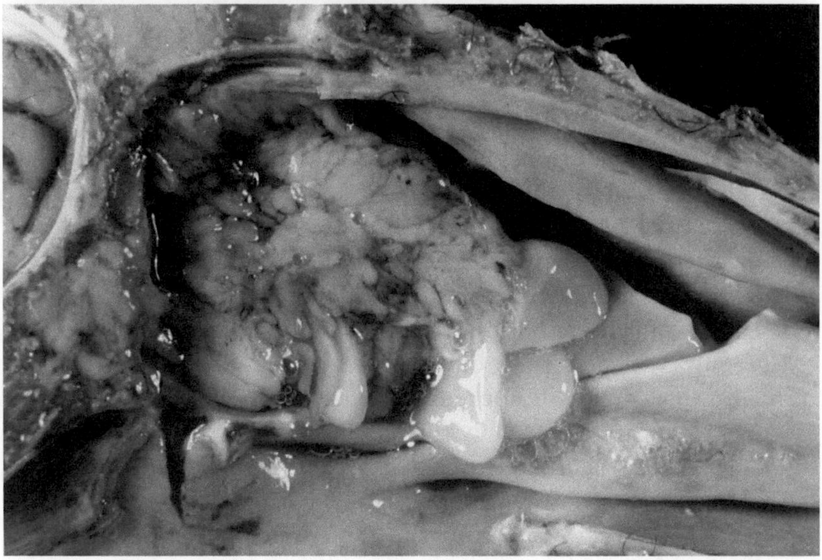

Fig. 2a, b. Gross pathology of ENA. Tumours vary in size, consistency, colour and mucous secretion. **a** In this 4-year-old sheep, polypoid structures originating from ethmoidal areas press neighbouring structures and occlude the nasal passages. The tumour is grey in colour with a nodular surface and scanty mucous secretion. **b** This tumour from a 2-year-old goat fills the ethmoidal area and there is almost complete involvement of the conchae. The tumour is soft and covered with mucous secretion

1991). These early lesions comprise small polypoid formations located on some of the ethmoidal conchae. Several of these conchae are modified because the normal flat surface is turned into a very soft polypoid tissue. The tumour can affect respiratory and olfactory areas and the rest of the nasal conchae appear oedematous. As the tumour grows, the structure of the ethmoid is gradually lost. Polypoid structures almost completely fill this nasal area, and may be apparent in the frontal or nasal sinuses, causing bone atrophy or deformation. They are soft, whitish or greyish and covered by mucus. In advanced cases, the appearance of the neoplasm varies from white or grey in colour, of hard consistency and nodular surface with scanty mucus secretion to grey/red in colour, with a soft consistency producing copious mucus secretion on the surface (Fig. 2). Necrotic or purulent areas due to secondary bacterial invasion also may be seen. There are no major differences in macroscopic pathology of ENA between sheep or goats except that in goats numerous inflammatory polyps are found frequently (De las Heras et al. 1991, 1998)

4 Microscopic Pathology

The histological characteristics of the tumour are very similar in sheep and goats. Light microscopy shows epithelial cells proliferating into an acinar, tubular, papillary and in some areas, cystic pattern. These cells are secretory in type with cuboidal or prismatic forms (Fig. 3). They do not usually lose their polarity or orientation but in some areas basal membranes have disappeared and the orientation of secretion is altered. The cytoplasms contain Periodic Acid–Schiff and/or Alcian blue pH 2.5 positive granules with no metachromasia. Lysozyme protein is found in many of them using immunohistochemical procedures (De las Heras et al. 1991). In a lectin immunoreactivity study, the papillary portion was shown to mainly express neutral glycoconjugates and a smaller amount of sulfated glycoconjugates, whereas the tubular portion synthesized a large amount of neutral glycoconjugates and a small quantity of prevalently carboxylated acid glycoconjugates (Scocco et al. 2001). The nuclei are generally round or ovoid and few mitotic figures can be seen, but in some areas alterations in the nuclei/cytoplasm ratio, anisokaryosis and giant nuclei are observed. Also, the nuclei in parts of the neoplastic acini may look flattened. The epithelium covering the glandular neoplasia may appear normal, or goblet cell hyperplasia and papilliform structures may occur. The

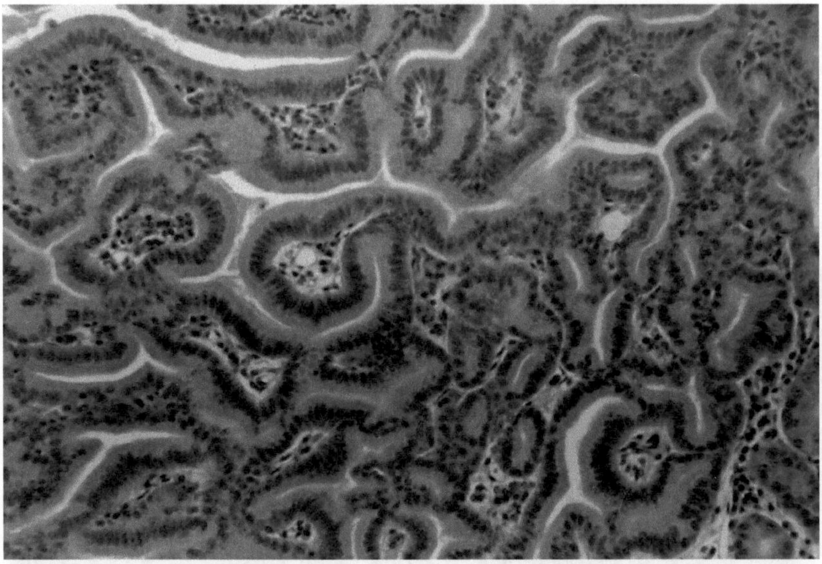

Fig. 3. ENA: histology of the tumour. Tumour tissue section from a natural case in a goat. Secretory type cells proliferating into acinar, tubular or papillary patterns. H&E, ×250

tumour stroma is very scanty but infiltrated by mononuclear cells where $CD4^+$, $CD8^+$ and immunoglobulin containing cells predominate (De las Heras et al. 1993). The neoplasm in most areas is well defined but in others infiltration of the connective or bone tissue may be observed. However, metastases in regional lymph nodes or other organs have been reported only in one case (Rajan et al. 1980). These characteristics have led to designation of the tumour as papilloma (Camy 1955), adenopapilloma (Cohrs 1953; Lombard et al. 1966; Njoku et al. 1978), or adenoma/adenocarcinoma (De las Heras et al. 1991; Fontaine et al. 1983; De las Heras et al. 1998; Duncan et al. 1967; McKinnon et al. 1982; Yonemichi et al. 1989; Young et al. 1961). The addendum of low grade to the description adenocarcinoma has been suggested by some of these authors (De las Heras et al. 1991; Vitellozi et al. 1993) and seems to be more fitting because, in spite of some histological indicators of malignancy, the absence of metastasis suggests limited malignant potential.

Electron microscopic examination of the tumour in sheep and goats shows that they are similar, and confirms the glandular character of the neoplasm. Tumour acini and tubules are composed of cuboidal or colum-

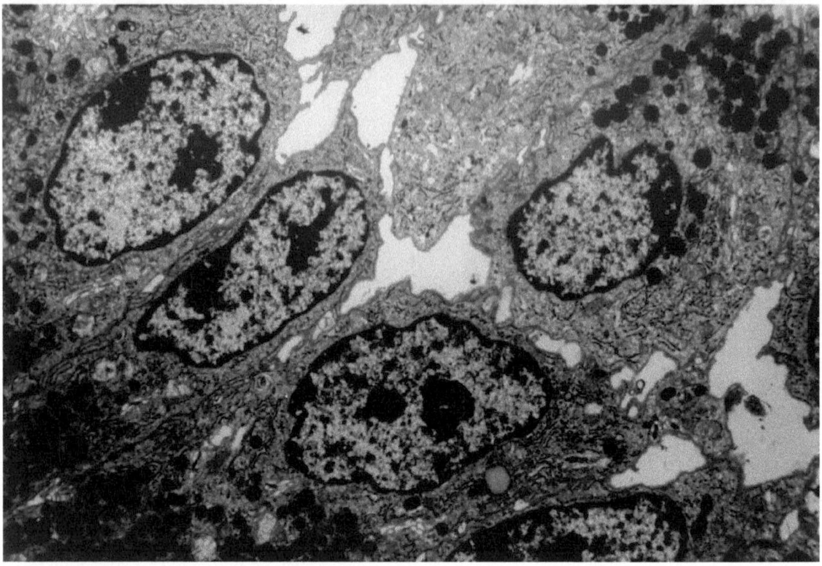

Fig. 4. Electron microscopy of ENA. The secretory morphology of the neoplastic cells is confirmed by the presence in their cytoplasms of electron-dense secretory granules found mostly in the apical surface

nar cells. The cells have tight junctions, invaginations with desmosomes between them and few microvilli on the surface. Proteinaceous-like secretory granules, rough endoplasmic reticulum and many mitochondria were characteristic components of the cytoplasm. The secretory granules, which are found mostly in the apical zone, are round, ranging from 0.1 μm to 1 μm in size, uniformly electron-dense and bounded by a unit membrane (Fig. 4). Some granules show a more electron-dense zone, generally in an eccentric position, and another more electron-lucent one. The Golgi apparatus is generally well developed. Additionally, infrequent intracellular canaliculi and loose whorls, composed of a smooth endoplasmic reticulum, may be observed. The nuclei of the neoplastic cells are round, slightly indented and have prominent nucleoli. Clumps of condensed chromatin are situated both adjacent to the nuclear envelope and dispersed throughout the nucleoplasm. The stroma is infiltrated mainly by plasmocytes but also lymphocytes and macrophages (Yonemichi et al. 1978; De las Heras et al. 1991). Groups of the tumour cells had also goblet-like cell-type structures in their cytoplasms. The ultrastructure of the papillary superficial projections mainly had cells with elongated nuclei. Some cells had microvilli on their apical surfaces and a few mitochondria and endoplasmic reticulum in the apical zone. Lysosomes were scanty and distributed randomly throughout the cell. A small, perinuclear Golgi apparatus and filaments were also observed.

In summary, neoplastic cells, in general, correspond to serous, mucous or mixed gland cells and seem to originate either from Bowman's glands of the olfactory area (Yonemichi et al. 1978) or both respiratory and olfactory mucosal glands (De las Heras et al. 1991; Scocco et al. 2001).

5
The ENA-Associated Retroviruses

5.1
Early Findings About the Relationship of ENA with Retroviruses

The epidemiology of ENA (see Sect. 1) and the transmission experiments (see Sect. 6.3), implicated a virus as the cause of ENA. Other factors also have been considered but their role in ENA aetiology has been ruled out. Aflatoxins are known as a cause of nasal tumours in sheep when they are present in the feedstuff (Lewis et al. 1967). Several ENA-affected flocks were studied and no aflotoxin contamination of their food could be found (De las Heras et al. 1985, 1991). An association of ENA with the parasite *Oestrus ovis* has been proposed (Gueraud 1982; Njoku et al. 1978; Vohradsky 1974; Yonemichi et al. 1978). This parasite is occasionally found in the nasal chambers of animals with ENA.

The first evidence for the presence of a retrovirus in ENA was the identification of RNA dependent DNA polymerase (reverse transcriptase) activity in enzootic nasal tumours in sheep in Japan (Yonemichi et al. 1978). However, in the same study they also found herpesviruses and orf virus in the tumours. Further publications about ENA in other countries again found retroviruses by electron microscopic observation (see Sect. 6.2) of the tumours in sheep and goats (McKinnon et al. 1982; De las Heras et al. 1988) or reported coculture of herpesviruses from them (McKinnon et al. 1982). These findings suggested the need to evaluate the presence of retroviruses associated with the tumour using other methods. Further studies have shown that tumours and fluids from naturally affected goats contained reverse transcriptase activity associated with a particle of buoyant density typical of retroviruses (De las Heras et al. 1991b). The same particle obtained from nasal fluids and tumours from ENA cases in sheep

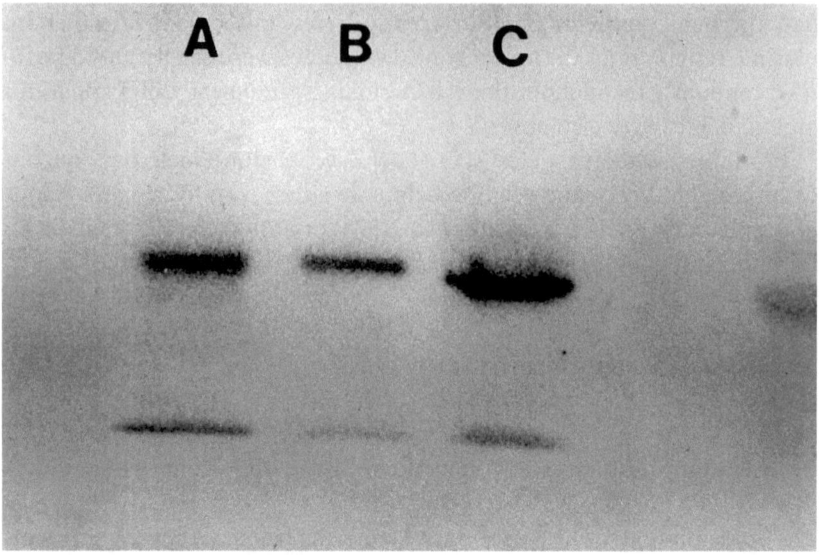

Fig. 5 A–C. Western-blot showing 25,000 M_r and low-M_r proteins related to ENA retrovirus. **A** concentrated goat nasal fluid used to inoculate the kids. **B** Concentrated goat nasal fluid from a kid with ENA reproduced experimentally. **C** Concentrated nasal fluid from a goat naturally affected with ENA

and goats, contained a 25,000 M_r protein that cross-reacted with the p27 of Mason-Pfizer monkey virus (MPMV) and with p25 of JSRV (De las Heras et al. 1991b, 1993b) (Fig. 5). This particle did not react with antibodies against maedi-visna (MVV) virus or other retroviruses. This indicated that JSRV, or a related retrovirus, was associated with ENA in goats. Experimental transmission studies demonstrated that the virus present in nasal fluids and tumours successfully transmitted the disease (De las Heras et al. 1995). In these studies (see Sect. 6.3) the retrovirus was detected in kids which developed ENA (De las Heras et al. 1995). No lesions similar to OPA were observed. Similarly JSRV infection has never been linked to nasal adenocarcinoma. This suggested that JSRV and the virus associated with ENA are distinct viruses.

5.2 Morphology of the ENA-Associated Retrovirus

Electron microscopy of the natural cases of ENA in sheep in Japan detected retroviral-like particles in the apical surfaces of the tumour cells and in the intracytoplasmatic vacuoles of leukocytes which can be found in the lumen of the neoplastic glands (Yonemichi et al. 1978). They were about 93 nm with an eccentrically located electron-dense core (about 47 nm) often surrounded by a single membrane and numerous spikes (9 nm) on their surfaces. Several virions had an elongated bar structure, as seen with betaretroviruses and also lentiviruses. Crescent shaped budding structures were also found in the same cells and these buds had an electron-dense outer rim with numerous spikes on the surface. In the same study, four tumours were cultured and maintained during several passages and viral particles similar to those observed in electron microscopy of tumours were observed in three of them. Preliminary serologic experiments suggested that none of the cultured cells reacted with antisera against MVV in the indirect fluorescent antibody test. Also, maedi-visna was not found in Japan at that time. An assay to detect reverse transcriptase revealed high activity in one of the tumour cell cultures compared with cultures of normal mucosa cells (Yonemichi et al. 1978). These data were interpreted as suggesting that a virus was present that was similar in morphology, size and reverse transcriptase activity to MVV but that it lacked serologic reactivity. Retroviral-like particles in association with ENA in animals also has been reported in other studies (McKinnon et al. 1982; Gázquez et al. 1992). A recent description of retroviral-like particles in natural ENA in sheep seronegative for MVV, reported extracellular particles in association with the tumours. These particles are round in shape, 80–110 nm in diameter and have an electron-dense nucleoid. This core was surrounded by an electrolucent zone and outer spiked unit membrane (De las Heras et al. 1998). This description is again compatible with betaretrovirus morphology and excludes MVV in relation with the ENA in sheep.

Retrovirus-like particles also have been observed in ENA of goats seronegative for caprine lentivirus. These particles are very similar to those observed in sheep. They have been found in extracellular spaces between microvilli or tumour cell surface. They were round in shape, 90–110 nm in diameter with an electron-dense zone surrounded by a clear zone and a membrane with numerous spikes (6–10 nm) (Fig. 6). Solitary crescent shaped structures (90–110 nm) were also seen in some apical

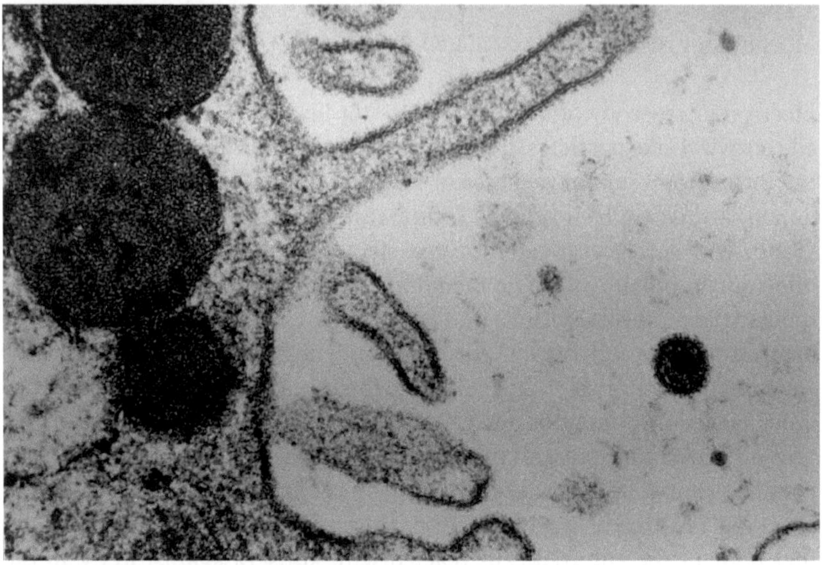

Fig. 6. Retrovirus particles observed in ENA tissue by transmission electron microscopy. Extracellular particles are visible in proximity to the apical surface of the tumour microvilli (De las Heras et al. 1988, 1991, 1991 b). Budding particles containing a centrally located electron-dense, ring shaped core surrounded by clear space and spiked unit membrane were also observed (De las Heras et al. 1991, 1991 b). Similar extracellular viral particles have also been observed by others (Vitellozzi et al. 1993). Experimentally reproduced tumours confirmed the presence of these particles and intracellular type A-like particles were also observed. They had a double-shelled structure (60–70 nm), the inner shell of which was more electron dense (De las Heras et al. 1995). These viral particles are compatible with a betaretrovirus. This has been confirmed by other data, as described elsewhere in this chapter.

5.3 Experimental Disease

ENA appears more difficult to reproduce experimentally than OPA. Early attempts to transmit ENA to sheep and/or lambs using tumour extracts gave inconsistent results; Two groups reported successful transmission of ENA (Cohrs 1953; Njoku et al. 1978 b) whilst two other groups, using similar techniques, were unable to transmit the disease (Duncan et al. 1967; Vohradsky 1974). Nevertheless, these experiments implicated an infectious agent, probably a virus, in the aetiology of the tumour. Years later, the presence of a type D retrovirus in nasal fluids and tumours of naturally affected goats was demonstrated (De las Heras et al. 1991b). This finding led to a transmission experiment using nasal fluids from natural cases of ENA in goats (De las Heras et al. 1995). Ten kids less than 48 h old, obtained from herds free of ENA, were inoculated either with undiluted fluid (clarified to remove cells; group A) or concentrated by centrifugation (groups B, C). The kids were inoculated either by nasal instillation (groups A, B) or by inoculation in the frontal sinuses (group C). Only in the groups inoculated with concentrated nasal fluid was there evidence of tumour development; one kid (from group C) had clinical symptoms and ENA was confirmed at necropsy. Two kids (one from group B and one from group C) had no clinical symptoms but had evidence of tumour at necropsy 14 months post-inoculation. The histology of these cases showed a neoplastic proliferation with similar characteristics to natural cases. Electron microscopy demonstrated the presence of viral particles of type B/D morphology associated with the tumour and the virus was also demonstrated by Western blotting in tumours and fluid from the clinical case (De las Heras et al. 1995) (Fig. 5).

This experiment confirmed that ENA can be transmitted experimentally but the incubation period for these tumours was 12–16 months whereas inoculation of neonatal lambs with JSRV can induce OPA tumours in as little as 6 weeks.

5.4
Partial Characterization of ONAV

Initially it was assumed that the viral aetiologic agent of ENA would be the same in sheep and goats. We termed this virus ENAV, (ENTV, enzootic nasal tumour virus). Studies were undertaken to test the hypothesis that ENAV was related to, but distinct from, JSRV: retrovirus was purified from nasal fluid of affected sheep and goats by density gradient centrifugation. RNA was extracted and was reverse transcribed. Primers based on JSRV sequence (York et al. 1992) were used to amplify part of the *gag* region of ENAV. Reverse transcriptase–PCR-*gag* products of several ovine ENA cases were cloned and sequenced. The subsequent restriction enzyme analysis of the products showed a number of sites that appeared to be different between ENAV (both ovine and caprine), JSRV and related endogenous retroviral sequences (SERVs). A *Pst*I site and an *Aat*II site were only present in the ENAV *gag* sequences and the *Sca*I site that is characteristic of JSRV *gag* was not present (Cousens et al. 1996). Moreover, the long terminal repeat (LTR) primers which amplified JSRV and not SERVs (Palmarini et al. 1996) did not amplify ENAV. These data confirmed at the nucleic acid level that ENAV was related to JSRV but also distinct from it.

5.5
Complete Sequence of ONAV

In order to completely characterize ENAV (ONAV) a cDNA library was made from retrovirus purified from nasal fluid of a sheep suffering from ENA. Clones were selected by hybridization with JSRV-derived probes and were sequenced and aligned with each other and with JSRV. PCR was used to complete the sequence of the genome (Cousens et al. 1999; Accession no. Y16627). The genome is 7,434 nucleotides (nt) long. As shown in Table 1, ONAV and JSRV are remarkably similar. The main differences occur in the LTR, *env* transmembrane (TM) region, and two short regions in *gag*. These are the places where ONAV and JSRV are also most different from endogenous sheep retroviruses. Thus the designation of variable regions VR1 and VR2 in *gag*, and VR3 in *env* TM (Palmarini et al. 2000) for JSRV and SERVs applies equally to ONAV. One potentially important difference between JSRV and ONAV is the presence in ONAV of two stop codons in the putative *orf-x* sequence. This translates as a protein of only

Table 1. Sequence similarity between JSRV and ONTV

	Nucleic acid similarity	Amino acid similarity	Amino acid identity
gag	90%	97%	95%
pro	95%	97%	96%
pol	92%	98%	97%
env	86%	95%	89%
SU	88%	96%	93%
TM	82%	93%	82%
LTR	76%		
U5	83%		
R	100%		
U3	73%		

72 amino acids in ONAV as opposed to 213 amino acids in JSRV. This suggests that if *orf-x* has any role in oncogenesis, this activity would reside in these first 72 amino acids. However, as an infectious clone of ONAV has not yet been isolated and sequenced, it remains possible that *orf-x* also is important for infectivity and/or oncogenesis of ENAV.

5.6
Development of ONAV-Specific PCR Revealed that ONAV and CNAV Are Different

Alignment of the consensus LTR sequence of ONAV, obtained from the cDNA library, with that of JSRV and SERVs revealed short regions unique to ONAV whilst the remainder of the LTR aligned neatly with JSRV and SERV sequences (Fig. 7). Specific amplification of ONAV was achieved by using primers from unique LTR regions together with primers from regions of *gag* or *env* conserved between JSRV, SERV and ONAV. (The *gag* primer had previously been shown to also amplify CNAV; Cousens et al. 1996). Using these PCRs, DNA from tissues of ovine and caprine ENA cases and also an OPA-affected sheep were tested. Each DNA was positive when checked by PCR for a control gene (glyceraldehyde phosphate dehydrogenase; Palmarini et al. 1996). In all three ovine cases tested, a positive result was obtained with both *env*–LTR and LTR-*gag* PCR from sheep nasal tumour but not from kidney (Fig. 8). No product was generated from any of the four goat tissues or from the two tissues of the OPA-affected sheep. This result confirmed the association between ONAV and ovine ENA and confirmed that ONAV is an exogenous virus which is not found endogenously in the sheep genome. It also demonstrated that the PCR detected only ONAV and not JSRV or CNAV. The fact that caprine nasal tumour was negative was the first suggestion that ONAV could be distinct from CNAV.

The complete sequence of CNAV is currently being determined. The sequence to date fits with the proposed conserved and variable regions identified between the viruses in this group, as described above. Provisionally CNAV appears to be as different from ONAV as they both are from JSRV.

```
        1                                                          60
   ONAV CTGCGGGGACAACCTGCGGAGGGTTAAGTCCTGGGAGCTCCTTGGCAA--AATGCCAAG
SA-JSRV ..........G...C.T........................T......A--..GC.A...
 JSRV21 A.........G...C.T........................T.......--..GC.A...
   SERV ...................... ........T....................TGC.GGGC
        61                                                        120
   ONAV GCTTAGGCAAGTACCTAAGCTCCCTGTCCCGCCACCCTCAGGAAGTCTTAAGAGCTTTTG
SA-JSRV C...G.A...A...........................TA.G.T....G.A....C..A
 JSRV21 C...G.A.................................A...T.T....A....C..A
   SERV C..AG.A..T..G...........................A...T.T..-.T.A.CC..A
        121                                                       180
   ONAV GAGCTCAAATGTGTTTTGGTTTTGCAACAT-----GGCTTAAAAAGCAGGAAATC-----
SA-JSRV AG.T..--GGA.....GCTGCCG...TT.CTTCACA.---...TGA.G.......-----
 JSRV21 AG....--GGA.....GCT...G...CTGCTTCATA.---...T.C.........-----
   SERV AG....C..GA.....G.T..CG.......TTCATAGAAG.T.G.TT.TCTT..TGTGTA
        181                                                       240
   ONAV ----------TGATTATGTAAGAAACCAGTAAT-T-------------------------
SA-JSRV ----------..............T..G..G..-.GT--------GTAAAAATCCGGTGG
 JSRV21 ----------.......A......T..G..G..-.GT--------GTAAGAATCCGGTGG
   SERV TACTTCATAGAAGA..GA.T.TCTGATT..GT.C.GTATACAATGGTAAGGGTCTGGTGA
        241                                                       300
   ONAV ATGTAAGT--AAATGAT-------------------CTCAAGTTACTTAACTTGCT-TAT
SA-JSRV G....GT.TG.G....ATA-----------------AAC.G....TG...TG.AC.G...
 JSRV21 G....GTGAAT.....ATA-----------------AA.......TG..-----..T...
   SERV T....TCCTG.G..T.AAAAACAACCTTGTGAATGCCTT....C..G..CT..A.CC...
        301                                                       360
   ONAV ATATACTGCTGCACAATAAAGCAAGGTATCAGCCATTTTGGTCTGATCCTCTCAACCCCA
SA-JSRV .A...TA..AAAGT....................G..C......................
 JSRV21 .A...TA..ATTGT.........GA...................................
   SERV ......C..A..........................G..G...................
        361                                                       420
   ONAV TCTTTTGCCTCT--CTTATTTT-CTTAGCGGGGATGCTCCGTTCTCTCCTTATACAGGTG
SA-JSRV .......T..T.CT.---AG..T...........CCG............C.G.G......
 JSRV21 .......T..C.CT.......-.......A...C...............C.G.G......
   SERV .......T..C.CT.......-...........C...............C.G.G......
        421                         449
   ONAV TAACTTTTGTCCGTGCTGGCTGCGGCAG-
SA-JSRV CG...C....TT.......TC.......
 JSRV21 CG...C...CTT........C.......
   SERV CG...C...CTT........C......--
```

Fig. 7. Alignment of U3 sequences from exogenous and endogenous sheep betaretroviruses. SERV: consensus of nine sequences (Palmarini et al. 1996; Bai et al. 1996); JSRVSA (M80216); JSRV21 (AF105220); ONTV (Y16627). Matches with the consensus are shown as *dots*; gaps in alignment are shown as *dashes*; '~' represents no sequence available

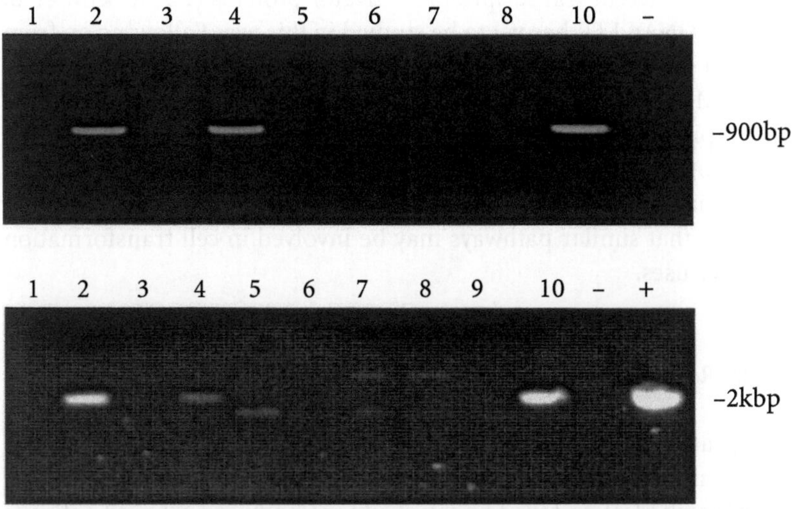

Fig. 8. ONTV specific LTR-*gag* and *env*-LTR PCRs: *lanes 2, 4* and *10* are the product of amplification of 250 ng genomic DNA from three ovine cases of enzootic intranasal tumour; *lanes 1, 3* and *9* are matched kidney samples; *lanes 6* and *8* are the products of amplification of genomic DNA from caprine ENT; and lanes 5 and 7 are the matched kidney samples; – is the negative control, water. LTR-*gag*: primers PL3 (ATGATCT-CAAGTTACTTAACTTGC) and PG1 (ATACTGCAGCYCGATGGCCAG) were used as follows: 100 ng of template DNA, 50 mM KCl, 10 mM Tris-HCl (pH 8.3), 2 mM $MgCl_2$, 125 pmol each primer, 200 μM each dNTP, 1 U Taq polymerase; 40 cycles of 94 °C for 1 min, 59 °C for 1 min, and 72 °C for 1 min were performed with a final extension of 5 min at 72 °C. *env*-LTR: primers used were PL4 (AAGCAAGTTAAGTAACTTGAGATC) and PE1 (GCTTAGCCGTCCTAAAAGAG) (6465–6484 of JSRV) (York et al. 1992) chosen because this region encodes the consensus SU/TM cleavage site common to many retroviruses. The PCR conditions and cycles were exactly the same as for the LTR-*gag* PCR except that an annealing temperature of 55 °C was used

5.7 Properties of ONAV Env

Some of the clones from the ONAV sequencing project have been used by other groups in studies to look at the properties of the envelope protein. Recent results have shown that like JSRV, ONAV uses Hyal-2 as its cellular receptor (Dirks et al. 2002; see the chapter by Miller, this volume). This receptor is present on many cell types, including human cells, therefore the cell tropism may be directed more by the LTR sequences than by *env*. JSRV and SERV LTRs have been shown to be active in different cell types and respond to different transcriptional activator proteins (PALMARINI et al. 2000). The ONAV LTR has yet to be studied in this way. Following on from studies on JSRV demonstrating the transforming activity of the envelope protein (MAEDA et al. 2001; RAI et al. 2001), it has been shown that the envelope protein of ONAV can also transform cells (DIRKS et al. 2002). The YXXM motif in the cytoplasmic tail of JSRV Env, which is important for transforming activity (PALMARINI et al. 2001) is also present in ONAV Env, suggesting that similar pathways may be involved in cell transformation by these viruses.

6
Immune Response in ENA Disease

The evaluation of the serological immune response in naturally affected or experimentally infected sheep and goats has been controversial. Studies using Western blotting failed to detect any antibody against viral antigens from concentrated ENA nasal fluid or OPA fluid in animals infected with ENA (DE LAS HERAS et al. 1995). In contrast, antibodies in goats naturally or experimentally affected with ENA were detected against recombinant Mason-Pfizer major core protein as a fusion protein with glutathione transferase (GST) (ROSATI et al. 1995). Antiserum reacted on Western blots with this recombinant fusion protein but not with GST alone (ROSATI et al. 1995). In contrast, studies by ORTÍN et al. (1998) found that reactivity to recombinant JSRV-capsid protein (CA)–GST fusion protein could be abolished completely by absorption with the GST fusion partner but not with JSRV-CA. This suggested the activity recognized against GST-CA was not CA specific. Thus, ENAV infection is similar to JSRV infection in that the host does not appear to make any antibody response to the virus. Studies on cell-mediated immunity to these viruses have yet to be reported. This apparent lack of immune response may be due to tolerance induced by expression of endogenous retroviruses in these animals. Between 15 and 20 copies of these endogenous retroviruses have been detected in sheep and goats (HECHT et al. 1996). Expression of mRNA from at least some of these ERVs has been detected in all tissues tested by reverse transcription–PCR (PALMARINI et al. 1996; Ortín et al., unpublished results). Increased levels of ERV mRNAs has been shown in specific sheep tissues by in situ hybridization (PALMARINI et al. 2001b). Infection of lymphoid cells has been previously been shown for JSRV (PALMARINI et al. 1996b), and more recently ONAV and CNAV (Ortín et al., unpublished results). This is another possible way that these retroviruses may interfere with development of a specific immune response against them.

Acknowledgements. Funding for this work was provided by Commission of the European Communities contracts AIR 3CT94-0884, QLK2-1999-00983, QLRT-2000-02380, by Spanish Comisión Interministerial de Ciencia y Tecnología AGF96-0535-CO2-02 and by Scottish Executive, Environment and Rural Affairs Departmen ROAME MRL/043/98. We thank all practitioners of the Gabinete Tecnico Veterinario of Zaragoza (Spain), the technical staff and collaborators of the University of Zaragoza for their help.

References

Bai J, Zhu RY, Stedman K, Cousens C, Carlson J, Sharp JM, DeMartini JC (1996) Unique long terminal repeat U3 sequences distinguish exogenous Jaagsiekte sheep retroviruses associated with ovine pulmonary carcinoma from endogenous loci in the sheep genome. Journal of Virology 70:3159–3168

Camy M (1955) Papillome granuleux des cavités nasales du mouton. Bulletin Academie Veterinaire France 28:31–34

Charray J, Aman N, Tanoh KG (1985) Note sur une enzootie dádenocarcinoma de la muqueuse pituitaire chez des brevis djalonké. Revue Elevage Médecine Veterinare Pays Tropicaux 38:406–410

Cohrs P (1953) Infektiöse Adenopapillome der Riechschleimhaut beim Schaf. Berliner Münchener Tierärztliche Wochenschrift 66:225–228

Cousens C, Minguijón E. Ferrer LM, Dalziel RG, Palmarini M, De las Heras M, Sharp JM (1996) PCR-based detection and partial characterization of a retrovirus assciated with contagious intranasal tumous of shep and goats. Journal of Virology, 70:7580–7583.

Cousens C, Minguijón E, Dalziel RG, Ortín A, García M, Park J, Gonzalez L, Sharp JM, De las Heras M (1999) Complete sequence of enzootic nasal tumor virus, a retrovirus associated with transmissible intranasal tumors of sheep. Journal of Virology 73:3986–3993

De las Heras M, García de Jalón JA, Balaguer L, García Marín JF, Badiola JJ (1985) Tumor intranasal enzoótico de la cabra. Medicina Veterinaria 2:281–290

De las Heras M, García de Jalón JA, Balaguer L, Badiola JJ (1988) Retrovirus-like particles in enzootic intranasal tumours in Spanish goats. Veterinary Record, 123:135

De las Heras M, García de Jalón, Sharp JM (1991) Pathology of enzootic intranasal tumour in thirty-eigth goats. Veterinary Pathology 28:474–481

De las Heras M, Sharp JM, García de Jalón JA, Dewar P (1991b) Enzootic nasal tumour of goats:demosntration of a type D-related retrovirus in nasal fluids and tumours. Journal of General Virology 72:2533–2535

De las Heras M, Minguijon E, Ferrer LM, Cebrian JA, Garcia de Jalón JA (1993) Tumour infiltrating cells and regional lypmh node characteristics in sheep respiratory neoplasms aetiologically associated with Type B or D retroviruses. Third International Conference. Sheep Veterinary Society. Edinburgh 28th June - 1st July pp100

De las Heras M, Sharp JM, Ferrer LM, García de Jalón JA, Cebrian LM (1993b) Evidence for a type D-like retrovirus in enzootic nasal tumour of sheep. Veterinary Record 132:441

De las Heras M, García de Jalón JA, Minguijón E, Gray EW, Dewar P, Sharp JM (1995) Experimental transmission of enzootic intranasal tumours of goats. Veterinary Pathology 32:19–23

De las Heras M, Minguijón E, Ferrer LM, Ortín A, Dewar P, Cebrian LM, Pascual Z, García L, García de Jalón JA, Sharp JM (1998) Naturally ocurring enzootic nasal tumour of sheep in Spain: pathology and associated retrovirus. European Journal of Veterinary Pathology 5:1–5

De las Heras M (1999) Enfermedades neoplásicas contagiosas del aparato respiratorio de los pequeños rumiantes: Adenomatosis pulmonar ovina y tumor intranasal enzoótico. Ovis. Mongrafías de actualidad. Aula Veterinaria. Editorial Luzan 5

Dirks C, Duh F-M, Rai SK, Lerman MI, and Miller AD (2002) Mechanism of Cell Entry and Transformation by Enzootic Nasal Tumor Virus. J Virol 76:2141–2149

Drieux H, Glaunes JP, Courtehoux P (1952) Epithelioma des premieres voies respiratorires d´llure contagiouse ou hereditaire chez le mouton. Acta Unio Int Contra Cancum 8:444–446

Duncan JR, Tyler DE, Van der Maaten MJ, Andersen FR (1967) Enzootic nasal adenocarcinoma in sheep. Journal American Veterinary Medical Association 151: 732–734

Fagundes E, Reis R, Ultimo A, Cerqueira R, Alves A (1979) Tumor etimoidal enzootico em ovinos. Arq Esc Vet UFMG, Belo Horizonte 31:337–342

Fontaine JJ, Crespeau F, Guereau JM, Parodi AL (1983) Obsevation d´une enzootie d´adenome pituitaire de la chevre. Recueil Medicine Veterinaire 159:383–388

Gázquez A, Roncero V, Redondo E, Durán E, Masot J, Gómez L (1992) Adenocarcinoma of the ethmoid olfactory mucosa: a histopathological and ultrastructural study with evidence of virus-like particles. Journal Veterinary Medical Association A 39:609–615

Giauffret A, Russo P, Laserre M (1984) Tumeurs transmissibles de la muqueuse nasale chez les caprins. Tumeurs transmissibles de la muqueuse nasale chez les caprins. In Les maladies de la chèvre. Ed INRA, pp 655–661

Gueraud JM (1982) A porpos de cas de tummeur nasale de la chevre. Bulletin Laboratoire Véterinaire 6:31–34

Gueraud JM, Crespeau F, Guereaud JM, Parodi AL (1983) Observation dúne enzootie dádenome pituitaire de la chevre. Recueil Medecine Veterinaire 159:383–388

Hecht SJ, Stedman KE, Carlson JD, DEMartini JC (1996) Distribution of endogenous type B and type D sheep retrovirus sequences in ungulates and other s. Proceeding of the Natural Academy of Sciences USA 93:3297–3302.

Leontides S, Argyroudis S, Psychas V, Mitilangas P, Markou E (1993) Endemic intranasal neoplasms in goats and sheep in Greece. Third Interational Conference. Sheep Veterinary Society. Edinburgh 28th June - 1st July.

Lewis G, Markson LM, Allcroft R (1967) The effect of feeding toxic groundnut meal to sheep over aperiod of five years. Research Veterinary Science 19:269–277

Lombard CH, Cabanie P, Crespin J (1966) Adénopapillome de la muqueuse pituitaire chez la chèvre. Bulletin Academie Veterinaire France 339:199–202
Maeda N, Palmarini M, Murgia C, Fan H (2001) Direct transformation of rodent fibroblasts by jaagsiekte sheep retrovirus DNA. Proc Natl Acad Sci USA 98(8): 4449–4454
McKinnon AO, Thorsen J, Hayes MA, Misener CR (1982) Enzootic nasal adenocarcinoma of sheep in Canada. Canadian Veterinary Journal 23:88–94
Nieberle K (1940) Über endemischen Krebs im Siebbein von Schafen. Z. schr. Krebsforsch 49:137–142
Njoku CO, Shannon D, Chineme CN, Bida SA (1978) Ovine nasal adenopapilloma: Incidence and clinicopathological studies. American Journal of Veterinary Research 39:1850–1852
Njoku CO, Chineme D, Shannon D, Bida SA (1978b) Etiologic and transmission studies of ovine nasal adenopapilloma. Proceedings African Veterinary Congress, Accra, Ghana 1:165–169
Ortín A, Minguijón E, Dewar P, García M, Ferrer LM, Palmarini M, Gonzalez L, Sharp JM, De las Heras M (1998) Lack of specific immune response against a recombinant capsid protein of Jagsiekte sheep retrovirus in sheep and goats naturally affected by enzootic nasal tumour or sheep pulmonary adenomatosis. Veterinary Immunology and Immunopathology 61:229–237
Palmarini M, Cousens C, Dalziel RG, Bai J, Stedman K, DeMartini JC, Sharp JM (1996) The exogenous form of Jaagsiekte retrovirus is specifically associated with a contagious lung cancer of sheep. Journal of Virology 70:1618–1623
Palmarini M, Holland MJ, Cousens C, Dalziel RG, DeMartini JC, Sharp JM (1996b) Jaagsiekte retrovirus establishes a disseminated infection of the lymphoid tissues of sheep affected by pulmonary adenomatosis. Journal of General Virology 77:2991–2998
Palmarini M, Sharp JM, De las Heras M, Fan H (1999) Jaagsiekte sheep retrovirus is necessary and sufficient to induce a contagious lung cancer in sheep. Journal of Virology 73:6964–6972
Palmarini M, Hallwirth C, York D, Murgia C, de Oliveira T, Spencer T, Fan H (2000) Molecular cloning and functional analysis of three type D endogenous retroviruses of sheep reveal a different cell tropism from that of the highly related exogenous jaagsiekte sheep retrovirus. Journal of Virol 74:8065–76
Palmarini M, Maeda N, Murgia C, De-Fraja C, Hofacre A, Fan H (2001) A phosphatidylinositol 3-kinase (PI-3 K) docking site in the cytoplasmic tail of the Jaagsiekte sheep retrovirus transmembrane protein is essential for envelope-induced transformation of NIH3T3. Journal of Virology 175:1102–11009
Palmarini M, Gray CA, Carpenter K, Fan H, Brazer FW, Spencer TE (2001b) Expression of engogenus betaretrovirus in the ovine uterus:effects of neonatal age, estrous cycle pregnancy and progesterone. Journal of Virology 75:11319–11327
Perl S, Yakobson B, Orgard-klopfer U, Abramson M, Nobel T (1987) Enzootic nasal tumour of sheep. The Veterinary Quarterly 9:119–122
Pringle JK, Zbigniew W, Wojcinski W, Staempfli HR (1989) Nasal papillary adenoma in a goat. Canadian Veterinary Journal 30:964–966
Rai SK, Duh FM, Vigdorovich V, Danilkovitch-Miagkova A, Lerman MI, Miller AD (2001) Candidate tumor suppressor HYAL2 is a glycosylphosphatidylinositol

(GPI)-anchored cell-surface receptor for jaagsiekte sheep retrovirus, the envelope protein of which mediates oncogenic transformation. Proceedings National Academy Sciences USA 98:4443–8

Rajan A, Sulochana S, Sreekumaran T, Vikram Reddi M, Krisharan Nair M (1980) Indian Journal of Cancer 17:196–199

Rosati S, Kwang J, Rutili O, Rossi S, Vitellozi G (1995) Antibody response during experimental and natural cases of enzootic nasal tumour in goats. Veterinary Record 137:465–466

Rubaj B, Woloszyn S (1967) Adenopapilloma enzooticum jamy nosowej u owiec. Medycyny Weterynaryjnej 23: 226–229

Scocco P, Mariotti F, Ceccarelli P, Fagioloi O, Renzoni G, Vitellozzi G (2001) Origin of enzootic intranasal tumour in the goat (Capra hircus): Veterinary Pathology 38:98–104

Vitellozi G, Muchetti L, Palmarini M, Mandara MT, Mechelli L, Shar JM, Manocchio I (1993) Enzootic intranasal tumour of goats in Italy. Journal of Veterinary Medicine B 40:459–468

Vohradsky F (1974) Adenocarcinoma of the olfactory mucosa of sheep and pigs in Ghana. Acta Veterinaria Brno 43:243–249

Young S, Lovelace SA, Hawkins JR, Catlin JE (1961) Neoplasms of the olfactory mucous membrane of sheep. Cornel Veterinarian 52:96–112

Yonemichi H, Ohgi T, Fujimoto Y, Okada K, Onuma M, Mikami T (1978) Intranasal tumour of the ethmoid olfactory mucosa in sheep. American Journal Veterinary Research 39:1599–1606

York DF, Vigne R, Verwoerd DW, Querat G (1992) Nucleotide sequence of the Jaaksiekte retrovirus, an exogenous and endogenous type D and B retrovirus of sheep and goats. Journal of Virology 66:4930–4939

Chapter 9

Pathology of Human Bronchioloalveolar Carcinoma and Its Relationship to the Ovine Disease

J.-F. Mornex, F. Thivolet, M. De las Heras, C. Leroux

Abstract. Lung cancer is a leading cause of cancer with a poor prognosis. Bronchioloalveolar carcinoma (BAC) is a rare tumor that has always intrigued physicians. Since the last World Health Organization classification the pathology has been clarified; BAC per se is an adenocarcinoma with a pure bronchioloalveolar growth pattern and appears as an in situ alveolar adenocarcinoma. More usually BAC is a clinically recognizable

J.-F. Mornex, F. Thivolet, C. Leroux
Université Claude Bernard, UMR 754 UCB-INRA-ENVL, 50 Avenue Tony Garnier, 69366 Lyon, France
e-mail: mornex@clermont.inra.fr
e-mail: cleroux@rockefeller.univ-lyon1.fr

M. De las Heras
Departamento de Patologia Animal, Universidad de Zaragoza, Miguel Servet 177, 50013 Zaragoza, Spain
e-mail: lasheras@posta.unizar.es

entity presenting as multi-focal nodules evolving towards pneumonia associated with pulmonary shunting. Pathology is that of a multifocal mixed adenocarcinoma: bronchioloalveolar and papillar. Whatever the stage, survival is better than in other forms of non-small cell lung cancer (NSCLC). The true frequency of BAC is unknown, although it is a rare form of lung cancer; smoking cannot be excluded as a risk factor. It appears that *p53* and *ras* genes are less often mutated than in other lung adenocarcinomas, suggesting that the cellular mechanisms involved are different. Ovine pulmonary adenocarcinoma (OPA) presents with the same symptoms as BAC in humans and is caused by a betaretrovirus – Jaagsiekte sheep retrovirus. Very early on, clinical and histological similarities with human BAC were stressed. A recent series of OPA described, according to the third edition of the WHO classification for human lung cancer, mixed adenocarcinoma, BAC and papillary and/or acinar carcinoma. An immunohistochemical study suggested that some human pulmonary tumors (including BAC) may be associated with a Jaagsiekte sheep retrovirus-related retrovirus, but so far no molecular study has confirmed this observation. Thus, OPA is an exquisite model of carcinogenesis for human lung adenocarcinomas.

1 Introduction

Lung cancer is the leading cause of cancer death among men and women in the USA with 17,000 deaths per year (Travis et al. 1995). Less than 15% of patients achieve a 5-year survival, although stage I patients may have a 5-year survival approaching 70%. Eighty percent of lung cancers are non-small cell lung cancer (SCLC; adeno-, squamous-cell, bronchoalveolar, and large cell carcinomas) and 20% are SCLC; adenocarcinomas are the most frequent cell type, accounting for 40% of all cases of lung cancer.

Bronchioloalveolar carcinoma (BAC) is a rare tumor that has always intrigued pathologists, surgeons and chest physicians. Its similarities with ovine pulmonary adenocarcinoma (OPA) were stressed as early as 1939. Over the last 20 years, discrepancies in the way this disease is viewed at the pathology level have brought much confusion. Its incidence, clinical presentation, and molecular events have been obscure. Since the last World Health Organization (WHO) classification (Travis 1999; Brambilla et al. 2001) the pathology has been clarified. However, according to the new classification, true BAC is a unique and very rare disease, that has in most instances nothing in common with what has been published in the past as BAC. Looking back to the old descriptions, it is possible to identify a clinical presentation that has distinct features – a multi-focal mixed adenocarcinoma, bronchioloalveolar and papillar. This disease, presenting as multi-focal nodules evolving towards pneumonia (also known as 'pneumonic BAC') associated with pulmonary shunting and bronchorrhea, has previously been described as BAC and identified as such by clinicians. Very early on, BAC was also called adenomatosis or malignant pulmonary adenomatosis; it is analogous to OPA.

2 Pathology of BAC

Lung tumors are classified every 15–20 years under the auspices of the WHO; the third edition of the histological typing of lung and pleural tumors was published in 2000 (Brambilla et al. 2001). Thus the most recent edition should be taken as the current 'gold standard' for classification of lung tumors. According to the current WHO classification, malignant epithelial lung tumors are categorized as squamous cell carcinoma, SCLC, large cell carcinoma or adenocarcinoma, and very rare tumors. Among adenocarcinomas, most of the cases are of mixed subtypes, containing any of the typical subtypes of adenocarcinoma. This point is a major advance between the second and the third editions of the WHO classification. However, the recent change in definitions will lead to major confusion when discussing previous pathology studies of lung tumors.

The three subtypes of adenocarcinoma are acinar adenocarcinoma, papillary adenocarcinoma and BAC. Adenocarcinoma cells are also described as mucinous or non-mucinous. It must be stressed that mucin formation is not synonymous with mucinous types of cells. Cells are distinguished as mucinous or non-mucinous according to their morphologies: mucinous cells are tall cells with the nucleus located at the basal side of the cell, sometimes containing mucin within the apex of the cells; non-mucinous cells are flat cells with the nucleus in the middle. Mucus production occurs in the lumen of the alveolar spaces. The subtypes of adenocarcinoma are determined according to the pattern of cell growth of the cells. The acinar subtype of adenocarcinoma is an adenocarcinoma with acini (duct-like structures) and tubules composed of mucin-producing cells resembling a bronchial gland. The papillary subtype of adenocarci-

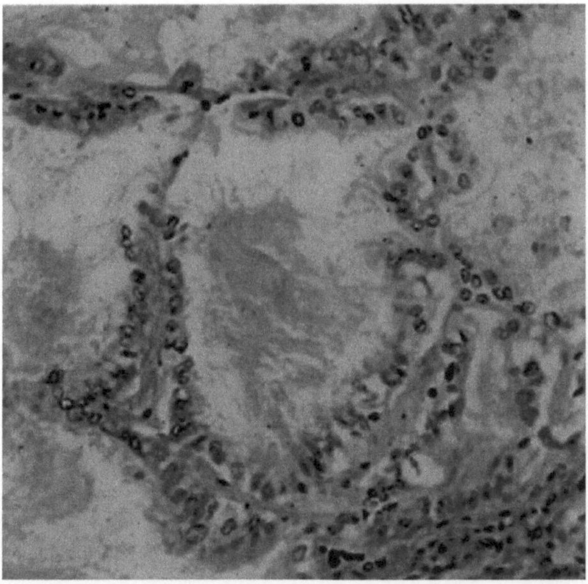

Fig. 1. Human bronchioloalveolar carcinoma (BAC) is characterized by a lepidic proliferation of alveolar cells along the alveolar septa (high power magnification; hemalin, floxin, safran)

noma has a predominance of papillary structures (papillary protrusions above the epithelial layer) that replace the underlying alveolar architecture. The papillary structures consist of non-mucinous cells or mucinous cells. Finally, according to the new WHO classification, BAC is an adenocarcinoma with a pure bronchioloalveolar growth pattern (cells following the alveolar septa – referred to as lepidic spread) and no evidence of stromal, vascular or pleural invasion (Fig. 1). Again, mucinous and non-mucinous types can be distinguished (Fig. 2); association of both non-mucinous and mucinous cells in a tumor is very infrequent. The non-mucinous type is considered to be made of a mixture of Clara cells and type II pneumocytes which are secretory epithelial cells of the distal lung: type II cells are found in the alveoli and secrete pulmonary surfactant, while Clara cells are found in the bronchioles and secrete products for maintenance of the bronchioles. BAC must be distinguished from papillary carcinoma; this can be challenging because during preparation of the tissue, alveolar septa can be disrupted. In some instances BAC is very difficult to distinguish from atypical adenomatosis hyperplasia, which is considered as an in situ carcinoma. According to the current WHO classification, the diagnosis of

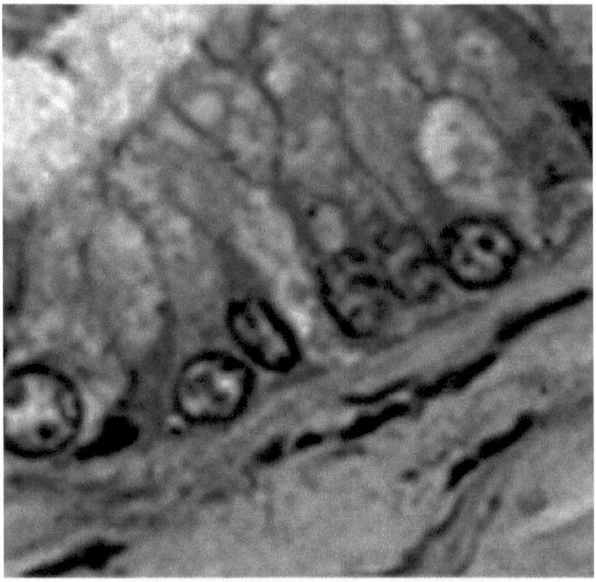

a

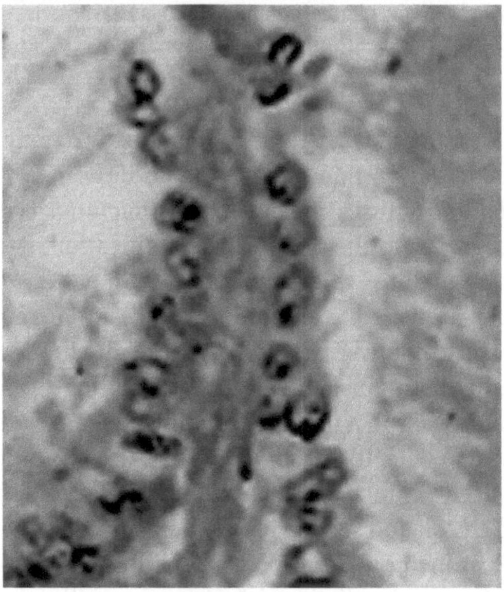

b

Fig. 2. Human BAC can be mucinous (**a**) with tall cells or non-mucinous (**b**) with flat cells (high-power magnification; hemalin, floxin, safran)

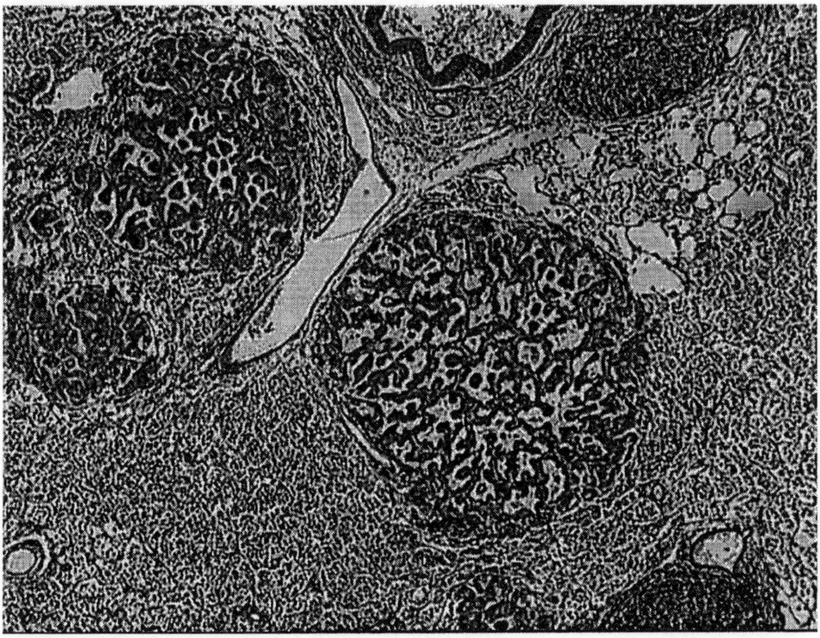

Fig. 3. Human BAC is usually a mixed adenocarcinoma: bronchioloalveolar and papillary carcinoma with multiple nodules (low-power magnification; hemalin, floxin, safran)

BAC can only be made if this tumor subtype is the only feature of the adenocarcinoma. However, tumors containing BAC and either papillary or acinar adenocarcinoma are more frequent; in this case, the lesion should be typed as mixed adenocarcinoma (Fig. 3). This is the major difference with the 1981 classification. Previously, a diagnosis of BAC was accepted when the tumor cells were following the alveolar septa; if the bronchioloalveolar part of the tumor was predominant, it was named as such. Only if another subtype (e.g., acinar or papillary) was predominant then was it classified as papillary or acinar adenocarcinoma.

The diagnosis of BAC as described in the 1981 WHO classification is very difficult (OMS 1981; Marchevsky 1990). This is emphasized by recent studies showing that the degree of agreement between observers was below 25% for the diagnosis of BAC, while it was around 50% of other adenocarcinomas (Sorensen et al. 1993). In a study devoted only to the diagnosis of BAC, we found that the degree of agreement is very low (F. Thivolet, unpublished results). Distinguishing between BAC and other

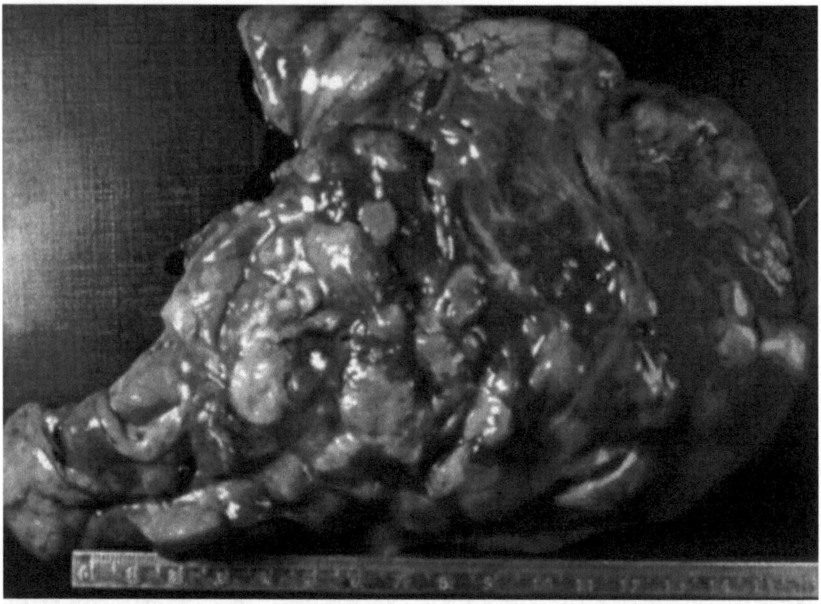

Fig. 4. Human BAC is multifocal: macroscopic aspect of a freshly explanted lung showing multiple whitish nodules

adenocarcinomas at the cytological level was almost impossible (Saleh et al. 1998). Two features are specific to BAC: (1) it is lepidic, that is the cells are lined along the alveolar structure (Fig. 1); and (2) when it extends, it is multifocal (Fig. 4) (Barsky et al. 1994).

3 Clinical Presentation of BAC

Given the confusion in pathology, it is very difficult to describe with confidence the clinical presentation of BAC. It is likely that two forms exist. Peripheral lung lesions, also called 'coin-lesions', are composed of BAC; they are likely to be true BAC according to the current WHO classification. In addition, a multifocal extended disease (or 'pneumonic BAC') has usually been described as a BAC; according to the current WHO classification, it is likely to be a mixed adenocarcinoma including BAC. The 'coin-lesion' appears to be a unique lung cancer of good prognosis. It is indolent and evident on a routine chest X-ray (Fig. 5). Surgical resection of the tumor is easy, there is no lymph node involvement and the prognosis is usually

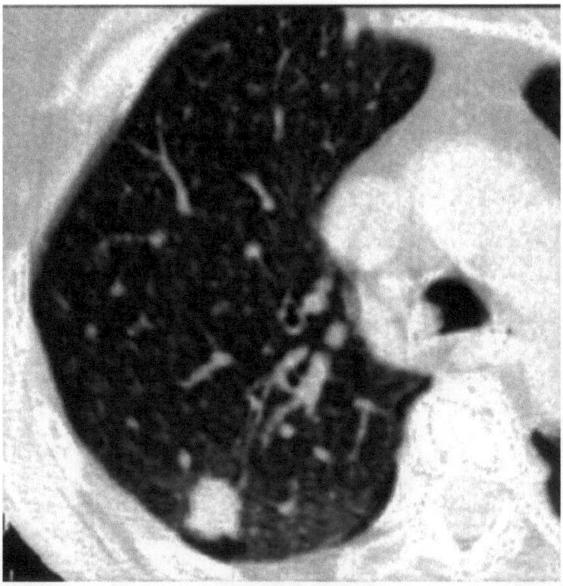

Fig. 5. 'Coin lesion': peripheral nodule on a computed tomography (CT) scan in a male patient alive 5 years after surgical resection

good, with the only *caveat* being the risk of brain metastasis. Given the confusion introduced by the second edition of the WHO classification, it is likely that in the past mixed adenocarcinoma also has been described as BAC and included in stage 1 lung cancer. In contrast, the second disease, 'pneumonic BAC', has a poor prognosis after resection, with the risk of relapse either ipsilateral or controlateral (Fig. 6). It is unlikely that this disease corresponds to true BAC. The second form also can be a clinical presentation of relapse of the 'coin lesion' (Fig. 7). In a review of a large series of BAC cases, the percentage of localized peripheral lesions ranged from 40% to 70% (Regnard et al. 1998; Liu et al. 2000); recurrences were a major complication, ranging from 30% to 60% (Breathnach et al. 1999; Regnard Santelmo et al. 1998; Liu et al. 2000). Stage 1 patients after surgery were characterized by a good prognosis (Marcq and Galy 1973; Okubo et al. 1999; Liu et al. 2000; Breathnach et al. 2001) with up to 69% of survival at 5 years, which is better than for NSCLC in general. Interestingly, stage 3 BAC has a tendency to be bilateral (stage 4) and survival is better than that of NSCLC, with a 50% survival at 15 months (Breathnach et al. 1999).

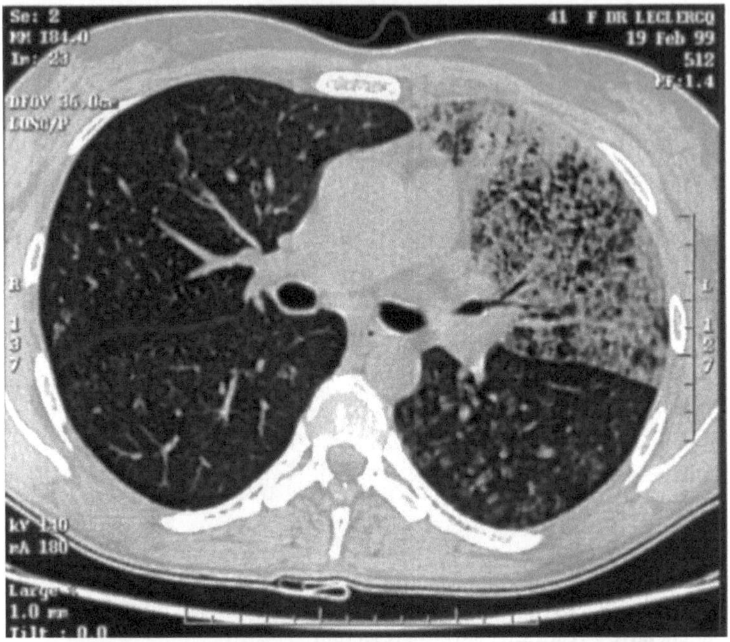

a

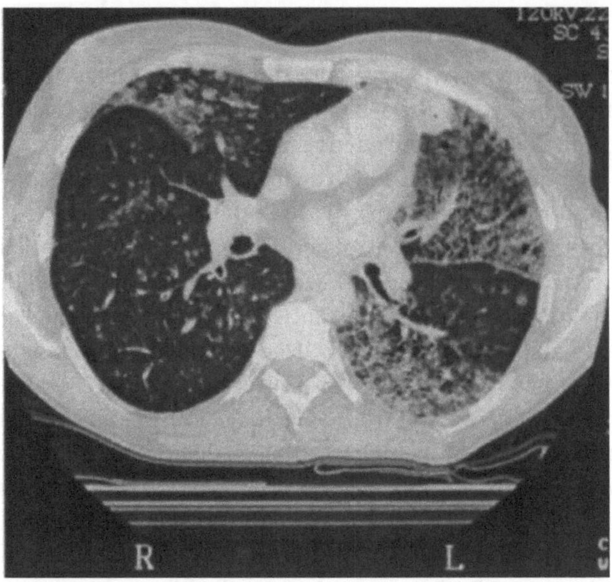

b

Fig. 6a–c. 'Pneumonic' form of human BAC. **a** Lobar consolidation on the CT scan in a female patient. **b** Extension to two other lobes on a CT scan obtained from the same patient a year later.

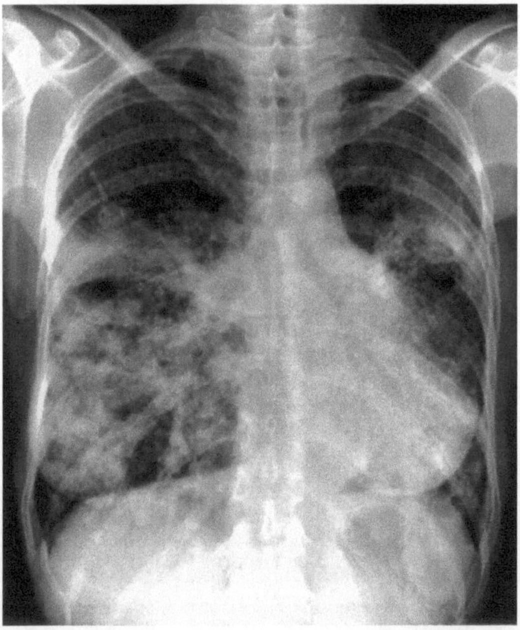

c

Fig. 6c. Diffusion on an X-ray taken 1 year later and 6 months before the death of the same patient

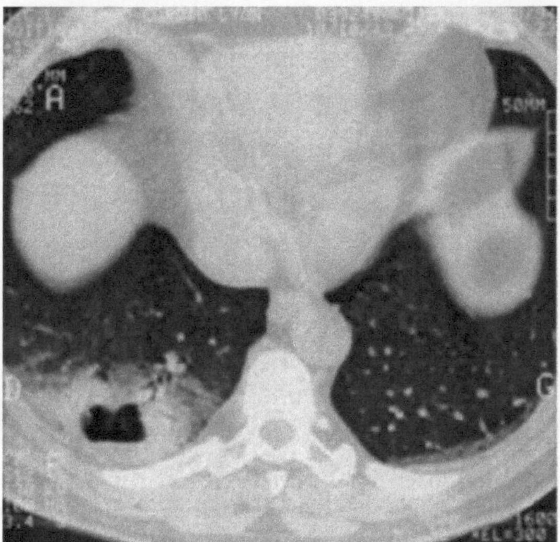

a

Fig. 7a–c. Human BAC can relapse and extend as a pneumonic form. **a** Single peripheral right nodule on a CT scan from a male patient

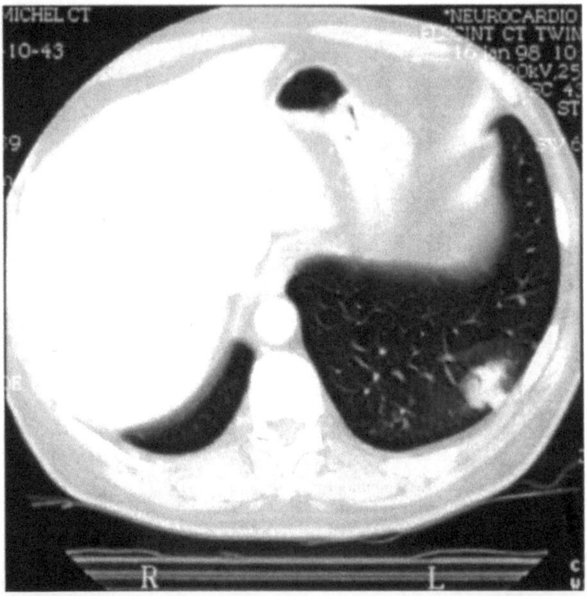

b

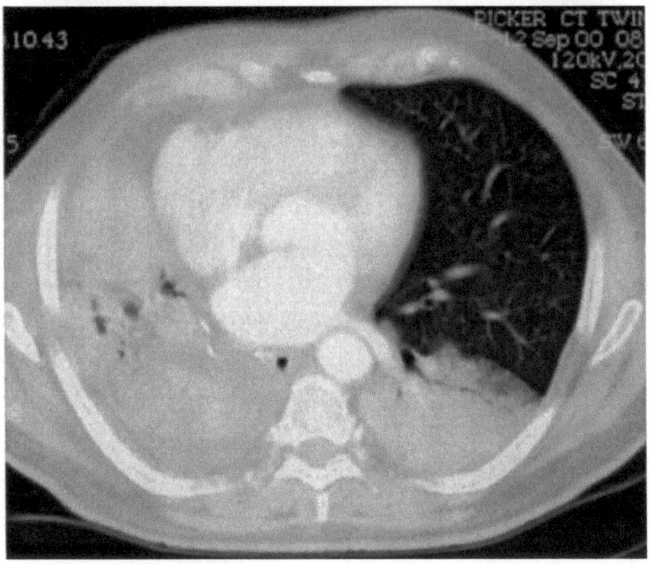

c

Fig. 7b,c. **b** Controlateral nodular relapse on a CT scan 4 years after resection. **c** Bilateral 'pneumonic' extension on a CT scan from the same patient 2 years later

4 Epidemiology of BAC

Given the absence of consensus on the pathology and clinical presentation of so-called BAC, it has been difficult to obtain reliable information on the incidence of true BAC and its epidemiology. In principle, BAC should be considered based on pathology as described in the third edition of the WHO classification, but this has not been implemented as yet. Therefore currently we must rely on data obtained from pathology series mostly using the second edition of the WHO classification, which includes BAC with papillary and/or acinar adenocarcinoma. Frequency of the disease could also be viewed from the clinical perspective, but then there is no consensus on what is a BAC. For instance, surgeons will describe peripheral lesions as BAC, while chest physicians will include pneumonic forms. There are no published data on the incidence of the latter. To make things more complex, it is possible to consider BAC from a clinico-pathologic viewpoint including the two modes of presentation, i.e., peripheral lesions vs. multifocal or pneumonic forms.

In the experience of a single center (Service de Pneumologie, Hôpital Louis Pradel, Lyon, France; Galy 1973; Marcq and Galy 1973) the diagnosis of BAC was made on the basis of clinico-pathological correlations, and included peripheral lesions and multifocal or 'pneumonic BAC'. In a seminal paper, 29 cases were reported and collected between 1947 and 1970, or roughly one case per year (Marcq and Galy 1973). Looking back through the files of the same department between 1971 and 1994 (a 22-year period), 34 BAC were observed, or 1.5 per year. In the initial series, half of the patients presented with peripheral lesions, and half presented with multifocal or pneumonic lesions. Between 1971 and 1994, almost 500 lung cancers were treated in this department. The frequency of BAC was thus 0.7% of lung cancers. In a recent review of 15 of these cases, all were mixed papillary and bronchioloalveolar adenocarcinoma, one-third presenting as a single peripheral lesion, there were fewer smokers and males than usually described, and one patient had a history of farming. Evolving from 0.18 to 1.6 cases per year in the most recent years suggested a true increase. In the last 20 years one of us (J.F.M.) managed 400 cases of lung cancer; there were no BAC as defined in the third edition of the WHO classification, and 10 'pneumonic BAC' (i.e. 2.5%). During the same time, F.T.B. was involved in the diagnosis of around 1,800 lung cancers, and three cases of BAC according to the current definition were identified (the incidence being under 0.2%). With this as a background, it is difficult to reconcile a recent report of a major increase in the incidence of BAC of up to 24% of lung cancers (Barsky et al. 1994). In this report, the definition of BAC included BAC with papillary proliferation as well as BAC with areas of poorly differentiated adenocarcinoma. Interestingly Auerbach and Garfinkel (1991) recall an earlier study reporting between 1.1% and 3.5% of lung cancers being BAC. This is consistent with the data observed in Lyon. Thus it appears that the true incidence of the WHO third edition BAC is unknown and definitely very low. Likewise, the incidence of 'pneumonic BAC' is not known, although adenocarcinomas with a bronchioloalveolar component are very frequent. It is thus very difficult to use data on the risk factors associated with BAC, given the absence of consensus of the diagnosis. Two recent studies have shown that smoking cannot be excluded as a risk factor in patients with BAC (Falk et al. 1992; Morabia and Wynder 1992).

5 Lung Transplantation in BAC

Pulmonary transplantation has been performed in human BAC (Etienne et al. 1997; Paloyan et al. 2000) although cancer is a usual contra-indication. In this context, recurrence of the disease has been observed (Garver et al. 1999; Paloyan et al. 2000) in half of the patients. Histologic and molecular analyses show that the recurrent tumors originate from the transplant recipients (Garver et al. 1999)

6 Molecular Events in the Development of BAC

Lung cancer, like any other cancer, is a multi-step process associated with genetic alterations of genes involved in cell cycle control or DNA repair. It is now widely accepted that carcinogenesis involves accumulation of mutations in genes intimately involved in cell birth and death. More than 50 tumor suppressor genes and more than 100 oncogenes have been described. Mutations or perturbations of these genes or their products play a crucial role in the development of cancer. Several genetic alterations have been described in lung cancer.

6.1 Involvement of Tumor Suppressor Genes in Lung Cancers

The mammalian cell cycle is partly controlled by tumor suppressor genes such as the *RB*, *p16* and *p53* genes. They act as brakes for the cell cycle. The p53 protein was first described in 1983 (Crawford 1983; Thomas et al. 1983) as a protein of 53 kDa and subsequently it has been described as the guardian of genome integrity or the cellular gatekeeper (Levine 1997). It inhibits the cycling of stressed or damaged cells that are more likely to become cancerous. In response to different stresses, p53 accumulates in the nucleus, binds to the DNA and acts as a transcriptional activator to up-regulate promoters of several cellular genes (Weintraub 1996; Vogelstein et al. 2000). The p53 gene (and corresponding protein) is often mutated and inactivated in lung cancer. Most but not all p53 mutations result in nuclear accumulation of a stabilized inactive protein, allowing its detection by immunological approaches. About 90% of the 280 base substitution mutations are located in a 600-base pair region (codons 110–307) encompassing exons 5 through 8 (Hollstein et al. 1991). The frequency of mutations suggests that p53 alterations play a critical role in tumorigenesis, but mutations are not found in all cancers. Mutations of p53 are present in 37% of all tumors, and the prevalence varies from 0% to 60% in major cancers (Greenblatt et al. 1994). Mutations are found in 56% of all lung cancers (Greenblatt et al. 1994) with 70% in small cell carcinoma and 40%–50% in adenocarcinoma (Caamano et al. 1991; Kishimoto et al. 1992; Gazzeri et al. 1994; Fong et al. 1999). A specific spectrum of p53 mutations is found in lung cancer and the distribution and nature of p53 mutations in lung cancer differ from the other cancers. The most frequently mutated codons include codons 157, 158, 179, 245, 248, 249 and 273 (Denissenko et al. 1996, 1997; Hernandez-Boussard and Hainaut 1998; Hainaut and Pfeifer 2001). Genetic alteration of p53 has been studied in BAC, but the observations differ. In their series, Wang and collaborators (Wang et al. 1995) showed that while 80% of conventional adenocarcinoma displayed an intense nuclear staining with antibody directed against p53, all of the BAC were negative. Another study (Nuorva et al. 1995) reported abnormal p53 nuclear accumulation in 36% of the BAC, with 13% and 45% of positivity for the mucinous and non-mucinous subtypes, respectively. In a larger series of 51 cases of BAC, Marchetti and collaborators described mutations in the p53 gene in 30% of non-mucinous type and none in mucinous BAC (Marchetti et al.

1998). Overall, considering that the discrepancies between the studies may be affected by discordances in histological typing of the lesions, p53 mutations appear to be relatively infrequent events in BAC, and are associated less frequently with BAC than with pulmonary adenocarcinoma in general.

The Fragile Histidine Triad (*FHIT*) is a putative tumor suppressor gene located on chromosome 3 at band p14.2 (3p14.2). The gene, comprising 10 exons which encode for a 1.1 kb transcript and a protein of 147 amino acids, contains the FRA3B region, the most common fragile site in the human genome and the t(3;8) chromosomal translocation breakpoint associated with certain forms of familial renal carcinoma (Ohta et al. 1996). The FHIT protein is a diadenosine triphosphate (Ap3A) hydrolase belonging to the histidine triad superfamily of nucleotide binding proteins (Siprashvili et al. 1997; Pace et al. 1998). *FHIT* inactivation and loss of expression is found in a large fraction of premalignant and malignant lesions. Loss of heterozygosity and abnormal splicing are observed in 3p14.2 in lung cancers (Fong et al. 1997, 1999; Sozzi et al.) at a frequency of 75% in SCLC and NSCLC. Methylation of *FHIT* was recently described as a frequent event, occurring in 37% of NSCLC and it may be an important mechanism of loss of expression of this gene (Zochbauer-Muller et al. 2001). *FHIT* gene alterations are also found in 43% of BAC with no clear demarcation between the different subtypes (Marchetti et al. 1998).

6.2 Activation of Proto-Oncogenes in Human Lung Cancers

The three 21 kDa *as* proto-oncogenes (H-*ras*, N-*ras* and K-*ras*) are members of a superfamily of membrane-associated proteins that bind guanosine triphosphate in their active state and guanosine diphosphate in their inactive state. Ras p21 proteins in their active form activate the mitogen-activated protein kinases (MAPKs) via a cascade of kinases. The activated (phosphorylated) MAPKs are then translocated to the nucleus where they phosphorylate and activate transcription factors. About 30% of human tumors carry *ras* mutations. Of the three genes of the family, K-*ras* is most frequently mutated in human tumors, with of 70%–90% in pancreatic adenocarcinoma (Pellegata et al. 1994), 50% in colon cancer (Vogelstein et al. 1988) and 25%–50% in lung adenocarcinoma (Mills et al. 1995). Activation of K-*ras* by point mutation occurs predominantly in codon 12 of the first exon (MILLS et al. 1995; KEOHAVONG et al. 1996; SIEGFRIED et al. 1997). Mutations in codon 12 of Ki-*ras* correlate with significantly poorer survival and shorter duration of disease-free survival (SLEBOS et al. 1990; SILINI et al. 1994). A few studies have examined K-*ras* mutations in human BAC. Like the studies of p53 alterations, mutations of K-*ras* were found to be infrequent (RUSCH et al. 1992) or a more frequent event (MARCHETTI et al. 1998). In their study, RUSCH and collaborators (RUSCH et al. 1992) analyzed K-, H- and N-*ras* mutations in 20 BAC. They found K-*ras* mutations in only 10% of resected lungs by looking at lung sections with 80% or more malignant cells. They showed that *ras* activation by point mutation is not a common event in this form of lung adenocarcinoma. By using similar molecular approaches, MARCHETTI et al. (MARCHETTI et al. 1998) demonstrated that in their series, 36% of the BAC carried mutations of the codon 12 of K-*ras* versus 26% of the conventional lung adenocarcinoma used as a control. Regarding the *ras* mutations, they showed a clear demarcation between the mucinous subtype (100% with mutations) and non-mucinous subtypes (23% with mutations). This, together with their p53 results, suggests that BAC is a heterogenous group of lung tumors, and that the mucinous form might represent a distinct biological entity (MARCHETTI et al. 1996; MARCHETTI et al. 1998).

In summary, mutations in oncogenes and tumor suppressor genes are widely involved in human tumors and several genetic alterations have been described in lung cancers. However unlike colorectal cancer, for which an ordered sequence of molecular events has been established (FEARON and VOGELSTEIN 1990; KINZLER and VOGELSTEIN 1996), no such sequence can be proposed for lung adenocarcinoma. The molecular changes in BAC are even less well understood. Cellular genes such as *p53*, *ras* or *FHIT* are impaired, but discrepancies exist between the different studies. However, it appears that *p53* and *ras* genes are mutated less often in BAC than in other lung adenocarcinomas, suggesting that the cellular mechanisms involved are different.

7
Ovine Pulmonary Adenocarcinoma

Ovine pulmonary adenocarcinoma was described in 1915 in South Africa. It presents with the same symptoms as in humans: progressive dyspnea, abundant bronchorrhea, cough, anorexia and cachexia. As described in other chapters in this volume, OPA is caused by a betaretrovirus –

Jaagsiekte sheep retrovirus (JSRV). The initial description was complicated by the coexistence in South Africa of maedi (a disease induced by a distinct ovine retrovirus – the lentivirus Maedi-Visna virus), but it has been shown that a molecular clone of JSRV is necessary and sufficient to induce OPA (Palmarini et al. 2000). Very early on (Bonne 1939; Hod et al. 1977; Nobel and Perk 1978) clinical and histological similarities with human BAC were stressed. We analyzed a series of OPAs, and found common features of multi-focal disease comprising areas of BAC and other subtypes of adenocarcinoma (mostly papillary). Thus OPA (Fig. 8) would be described according to the third edition of the WHO classification for human lung cancer as a mixed adenocarcinoma, BAC and papillary and/or acinar. OPA has a very similar clinical pathologic presentation to human 'pneumonic BAC' (compare Fig. 8a with Fig. 4, Fig. 8b with Fig. 6b, and Fig. 8c with Fig. 3).

a

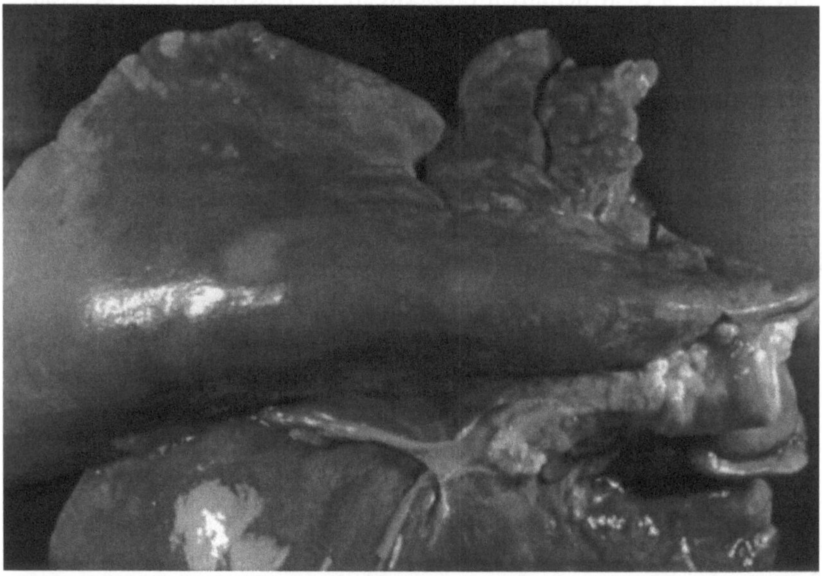

Fig. 8a–c. Ovine pulmonary adenocarcinoma resembles human BAC. **a** Macroscopic view of a freshly explanted lung with nodules

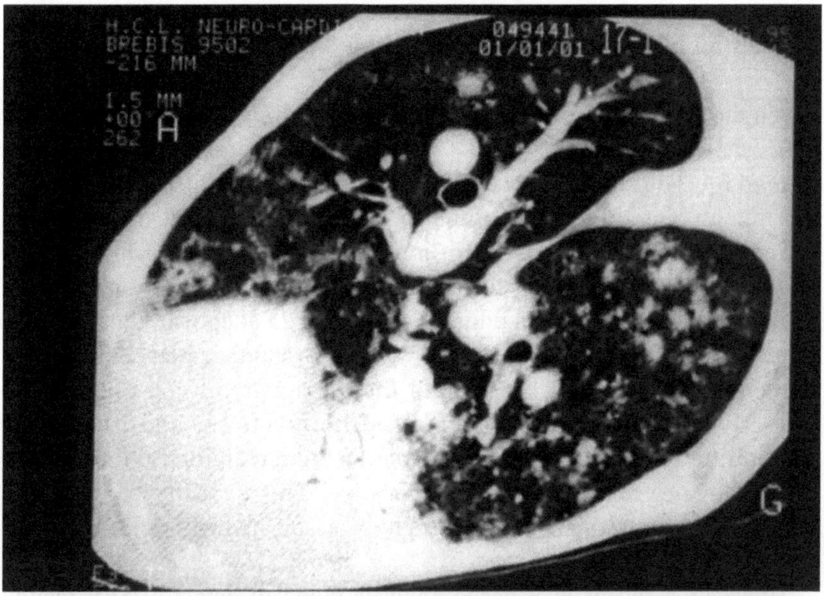

b

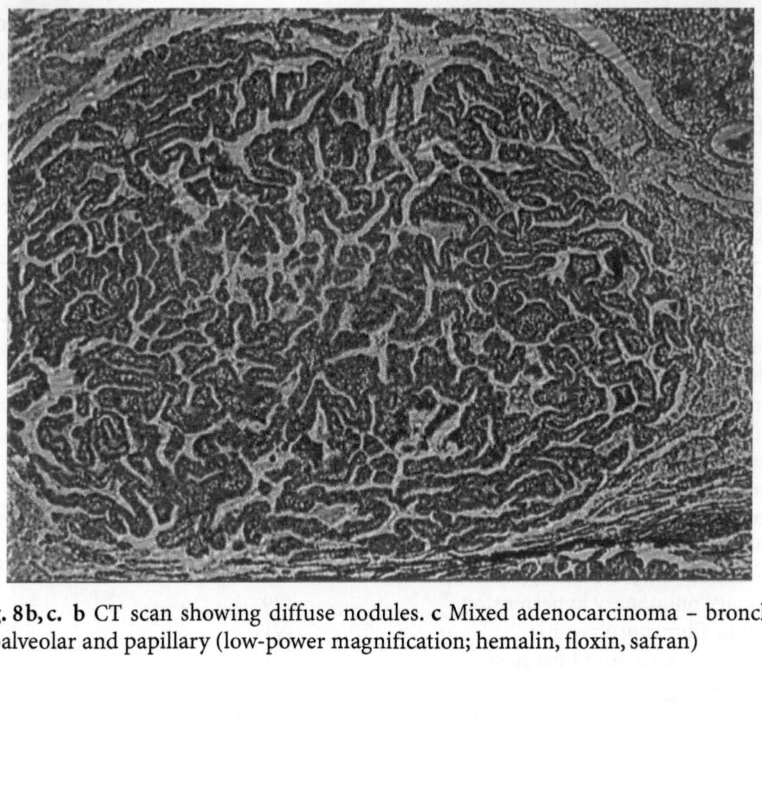

c

Fig. 8b, c. **b** CT scan showing diffuse nodules. **c** Mixed adenocarcinoma – bronchioloalveolar and papillary (low-power magnification; hemalin, floxin, safran)

8 Is There a Link Between JSRV and Human BAC?

Given the similarities between OPA and human BAC (Palmarini and Fan 2001; De las Heras et al. 2000) investigated a possible link between human BAC and JSRV, the causative agent of the ovine disease. A panel of 252 human lung tumors (classified according to the second WHO classification, see above) was examined immunohistochemically using a rabbit antiserum to the JSRV capsid protein (JSRV-CA). In addition, 21 non-tumor lung lesions, 4 normal lung tissues, and 23 adenocarcinomas from other tissues were examined. JSRV-specific staining was detected in the cytoplasm of neoplastic cells in the pulmonary alveoli of 39 of the 129 (30%) BAC, 17 of 65 (26%) lung adenocarcinomas and 2 of 7 large cell undifferentiated carcinomas (Table 1); the remaining tumor samples and non-cancerous tissues were negative. These results suggest that some human pulmonary tumors might be associated with a JSRV-related retrovirus.

However, a recent PCR-based analysis failed to detect a JSRV-related genome in human BACs (Yousem et al. 2001). Several other investigators also have not detected JSRV-related sequences in human BAC by PCR amplification (M. Palmarini and J.M. Sharp, unpublished results; J. De Martini, unpublished results; D. York, unpublished results). How-

Table 1. Demonstration of an antigen related to JSRV CA in human tissues (from De las Heras et al. 2000)

Pathology	Positive/no. examined
Bronchioloalveolar carcinoma	39/129
Adenocarcinoma	17/65
Large cell undifferentiated carcinoma	1/7
Squamous cell carcinoma	0/41
Carcinoid	0/2
Embryonic carcinoma	0/3
Neuroendocrine carcinoma	0/3
Small cell carcinoma	0/2
Non-pulmonary tumors	1/23
Primary pulmonary hypertension	0/9
Cryptic fibrosing alveolitis	0/11
Methotrexate-induced dysplasia	0/1
Normal lung	0/4

ever, this should not be taken as definitive proof that some human BACs are not associated with a JSRV-like retrovirus. Indeed, one of the early indications that the sheep disease OPA is caused by a retrovirus was the fact that tumor tissues would stain with immunological reagents raised against two other retroviruses – murine mammary tumor virus (MMTV) and Mason-Pfizer monkey virus (MPMV) (see the chapter by Sharp and DeMartini, this volume). It is now apparent that MMTV, MPMV and JSRV are all betaretroviruses, and relatively closely related. Nevertheless, the nucleic acid sequence homology among these viruses is quite low, and oligonucleotide primers specific for one of the MMTV or MPMV would generally not be able to amplify JSRV DNA sequences. Thus further studies are warranted to test for the existence of a JSRV-related retrovirus in human BAC.

9 Human BAC as a Disease Entity

If the clinically defined 'pneumonic BAC' is considered, a specific disease is evident. It occurs equally in females and males, aged 30 to 40 years. It is associated with bronchorrhea and pulmonary shunting and chest X-rays and CT scans are consistent with pneumonia. The tumors can be surgically resected, but with a risk of relapse over the years following resection. Patients with 'pneumonic BAC' have successfully undergone pulmonary transplantation (Etienne et al 1997). Although pathology is usually poorly described, it is mostly mixed adenocarcinoma bronchioloalveolar and papillary or acinar. The similarity with OPA is a model for the disease entity of human pulmonary cancer.

Acknowledgements. This work was supported in part by Association de Recherche sur, le Cancer, Comités départementaux du Rhône, de la Loire, de l'Ardèche, et de la Drome, de la Ligue Nationale contre le Cancer, Région Rhône-Alpes.

References

Auerbach O, Garfinkel L (1991) The changing pattern of lung carcinoma. Cancer 68:1973–7

Barsky SH, Cameron R, Osann KE, Tomita D, Holmes EC (1994) Rising incidence of bronchioloalveolar lung carcinoma and its unique clinicopathologic features. Cancer 73:1163–70

Barsky SH, Grossman DA, Ho J, Holmes EC (1994) The multifocality of bronchioloalveolar lung carcinoma: evidence and implications of a multiclonal origin. Mod Pathol 7:633–40

Bonne C (1939) morphological resemblance of pulmonary adenomatosis (jaaksiekte) in sheep and certain cases of cancer of the lung in man. Am J Cancer 35:491–501

Brambilla E, Travis WD, Colby TV, Corrin B, Shimosato Y (2001) The new World Health Organization classification of lung tumours. Eur Respir J 18:1059–68

Breathnach OS, Ishibe N, Williams J, Linnoila RI, Caporaso N, Johnson BE (1999) Clinical features of patients with stage IIIB and IV bronchioloalveolar carcinoma of the lung. Cancer 86:1165–73

Breathnach OS, Kwiatkowski DJ, Finkelstein DM, Godleski J, Sugarbaker DJ, Johnson BE, Mentzer S (2001) Bronchioloalveolar carcinoma of the lung: recurrences and survival in patients with stage I disease. J Thorac Cardiovasc Surg 121:42–7

Caamano J, Ruggeri B, Momiki S, Sickler A, Zhang SY, Klein-Szanto AJ (1991) Detection of p53 in primary lung tumor s and nonsmall cell lung carcinoma cell lines. Am J Pathol 139:839–45

Crawford L (1983) The 53,000-dalton cellular protein and its role in transformation. Int Rev Exp Pathol 25:1–50

De las Heras M, Barsky SH, Hasleton P, Wagner M, Larson E, Egan J, Ortin A, Gimenez-Mas JA, Palmarini M, Sharp JM (2000) Evidence for a protein related immunologically to the jaagsiekte sheep retrovirus in some human lung tumours. Eur Respir J 16:330–2

Denissenko MF, Chen JX, Tang MS, Pfeifer GP (1997) Cytosine methylation determines hot spots of DNA damage in the human P53 gene. Proc Natl Acad Sci USA 94:3893–8

Denissenko MF, Pao A, Tang M, Pfeifer GP (1996) Preferential formation of benzo[a]pyrene adducts at lung cancer mutational hotspots in p53. Science 274:430–2

Etienne B, Bertocchi M, Gamondes JP, Wiesendanger T, Brune J, Mornex JF (1997) Successful double-lung transplantation for bronchioalveolar carcinoma. Chest 112:1423–4

Falk RT, Pickle LW, Fontham ET, Greenberg SD, Jacobs HL, Correa P, Fraumeni JF Jr (1992) Epidemiology of bronchioloalveolar carcinoma. Cancer Epidemiol Biomarkers Prev 1:339–44

Fearon ER, Vogelstein B (1990) A genetic model for colorectal tumorigenesis. Cell 61:759–67

Fong KM, Biesterveld EJ, Virmani A, Wistuba I, Sekido Y, Bader SA, Ahmadian M, Ong ST, Rassool FV, Zimmerman PV, Giaccone G, Gazdar AF, Minna JD (1997) FHIT and FRA3B 3p14.2 allele loss are common in lung cancer and preneoplastic bronchial lesions and are associated with cancer-related FHIT cDNA splicing aberrations. Cancer Res 57:2256–67

Fong KM, Sekido Y, Minna JD (1999) Molecular pathogenesis of lung cancer. J Thorac Cardiovasc Surg 118:1136–52

Galy P, Marcq M (1973) le carcinome bronchiolo-alvéolaire. Rev Fr Mal Respir 1: 665–682

Garver RI Jr, Zorn GL, Wu X, McGiffin DC, Young KR Jr, Pinkard NB (1999) Recurrence of bronchioloalveolar carcinoma in transplanted lungs. N Engl J Med 340:1071–4

Gazzeri S, Brambilla E, Caron de Fromentel C, Gouyer V, Moro D, Perron P, Berger F, Brambilla C (1994) p53 genetic abnormalities and myc activation in human lung carcinoma. Int J Cancer 58:24–32

Greenblatt MS, Bennett WP, Hollstein M, Harris CC (1994) Mutations in the p53 tumor suppressor gene: clues to cancer etiology and molecular pathogenesis. Cancer Res 54:4855–78

Hainaut P, Pfeifer GP (2001) Patterns of p53 G->T transversions in lung cancers reflect the primary mutagenic signature of DNA-damage by tobacco smoke. Carcinogenesis 22:367–74

Hernandez-Boussard TM, Hainaut P (1998) A specific spectrum of p53 mutations in lung cancer from smokers: review of mutations compiled in the IARC p53 database. Environ Health Perspect 106:385–91

Hod I, Herz A, Zimber A (1977) Pulmonary carcinoma (Jaagsiekte) of sheep. Ultrastructural study of early and advanced tumor lesions. Am J Pathol 86:545–58

Hollstein M, Sidransky D, Vogelstein B, Harris CC (1991) p53 mutations in human cancers. Science 253:49–53

Keohavong P, DeMichele MA, Melacrinos AC, Landreneau RJ, Weyant RJ, Siegfried JM (1996) Detection of K-ras mutations in lung carcinomas: relationship to prognosis. Clin Cancer Res 2:411–8

Kinzler KW, Vogelstein B (1996) Lessons from hereditary colorectal cancer. Cell 87: 159–70

Kishimoto Y, Murakami Y, Shiraishi M, Hayashi K, Sekiya T (1992) Aberrations of the p53 tumor suppressor gene in human non-small cell carcinomas of the lung. Cancer Res 52:4799–804

Levine AJ (1997) p53, the cellular gatekeeper for growth and division. Cell 88:323–31

Liu YY, Chen YM, Huang MH, Perng RP (2000) Prognosis and recurrent patterns in bronchioloalveolar carcinoma. Chest 118: 940–7

Marchetti A, Buttitta F, Pellegrini S, Chella A, Bertacca G, Filardo A, Tognoni V, Ferreli F, Signorini E, Angeletti CA, Bevilacqua G (1996) Bronchioloalveolar lung carcinomas: K-ras mutations are constant events in the mucinous subtype. J Pathol 179: 254–9

Marchetti A, Pellegrini S, Bertacca G, Buttitta F, Gaeta P, Carnicelli V, Nardini V, Griseri P, Chella A, Angeletti CA, Bevilacqua G (1998) FHIT and p53 gene abnormalities in bronchioloalveolar carcinomas. Correlations with clinicopathological data and K-ras mutations. J Pathol 184:240–6

Marchevsky AM (1990) Malignant epithelial tumors of the lung. In: Marchevsky AM (ed.^eds) Surgical pathology of lung neoplasms. Marcel Dekker, New York, pp 77–228

Marcq M, Galy P (1973) Bronchioloalveolar carcinoma. Clinicopathologic relationships, natural history, and prognosis in 29 cases. Am Rev Respir Dis 107:621–9

Mills NE, Fishman CL, Rom WN, Dubin N, Jacobson DR (1995) Increased prevalence of K-ras oncogene mutations in lung adenocarcinoma. Cancer Res 55:1444–7

Morabia A, Wynder EL (1992) Relation of bronchioloalveolar carcinoma to tobacco. Bmj 304: 541–3

Nobel TA, Perk K (1978) Bronchiolo-alveolar cell carcinoma. Animal model: pulmonary adenomatosis of sheep, pulmonary carcinoma of sheep, pulmonary carcinoma of sheep (Jaagsiekte). Am J Pathol 90:783–6

Nuorva K, Soini Y, Kamel D, Pollanen R, Bloigu R, Vahakangas K, Paakko P (1995) p53 protein accumulation and the presence of human papillomavirus DNA in bronchiolo-alveolar carcinoma correlate with poor prognosis. Int J Cancer 64:424–9

Ohta M, Inoue H, Cotticelli MG, Kastury K, Baffa R, Palazzo J, Siprashvili Z, Mori M, McCue P, Druck T et al. (1996) The FHIT gene, spanning the chromosome 3p14.2 fragile site and rena l carcinoma-associated t(3;8) breakpoint, is abnormal in digestive tract cancers. Cell 84:587–97

Okubo K, Mark EJ, Flieder D, Wain JC, Wright CD, Moncure AC, Grillo HC, Mathisen DJ (1999) Bronchoalveolar carcinoma: clinical, radiologic, and pathologic factors and survival. J Thorac Cardiovasc Surg 118:702–9

OMS (1981) Types histologiques des tumeurs du poumon. Genève, Organisation mondiale de la santé

Pace HC, Garrison PN, Robinson AK, Barnes LD, Draganescu A, Rosler A, Blackburn GM, Siprashvili Z, Croce CM, Huebner K, Brenner C (1998) Genetic, biochemical, and crystallographic characterization of Fhit-substrate complexes as the active signaling form of Fhit. Proc Natl Acad Sci USA 95:5484–9

Palmarini M, Hallwirth C, York D, Murgia C, de Oliveira T, Spencer T et al. (2000) Molecular cloning and functional analysis of three type D endogenous retroviruses of sheep reveal a different cell tropism from that of the highly related exogenous jaagsiekte sheep retrovirus. J Virol 74(17):8065–76

Palmarini M, Fan H (2001) Retrovirus-induced ovine pulmonary adenocarcinoma, an animal model for lung cancer. J Natl Cancer Inst 93:1603–14

Paloyan EB, Swinnen LJ, Montoya A, Lonchyna V, Sullivan HJ, Garrity E (2000) Lung transplantation for advanced bronchioloalveolar carcinoma confined to the lungs. Transplantation 69:2446–8

Pellegata NS, Sessa F, Renault B, Bonato M, Leone BE, Solcia E, Ranzani GN (1994) K-ras and p53 gene mutations in pancreatic cancer: ductal and nonductal tumors progress through different genetic lesions. Cancer Res 54:1556–60

Regnard JF, Santelmo N, Romdhani N, Gharbi N, Bourcereau J, Dulmet E, Levasseur P (1998) Bronchioloalveolar lung carcinoma: results of surgical treatment and prognostic factors. Chest 114:45–50

Rusch VW, Reuter VE, Kris MG, Kurie J, Miller WH, Jr., Nanus DM, Albino AP, Dmitrovsky E (1992) Ras oncogene point mutation: an infrequent event in bronchioloalveolar cancer. J Thorac Cardiovasc Surg 104:1465–9

Saleh HA, Haapaniemi J, Khatib G, Sakr W (1998) Bronchioloalveolar carcinoma: diagnostic pitfalls and immunocytochemical contribution. Diagn Cytopathol 18: 301–6

Siegfried JM, Gillespie AT, Mera R, Casey TJ, Keohavong P, Testa JR, Hunt JD (1997) Prognostic value of specific KRAS mutations in lung adenocarcinomas. Cancer Epidemiol Biomarkers Prev 6:841–7

Silini EM, Bosi F, Pellegata NS, Volpato G, Romano A, Nazari S, Tinelli C, Ranzani GN, Solcia E, Fiocca R (1994) K-ras gene mutations: an unfavorable prognostic marker in stage I lung adenocarcinoma. Virchows Arch 424:367–73

Siprashvili Z, Sozzi G, Barnes LD, McCue P, Robinson AK, Eryomin V, Sard L, Tagliabue E, Greco A, Fusetti L, Schwartz G, Pierotti MA, Croce CM, Huebner K (1997) Replacement of Fhit in cancer cells suppresses tumorigenicity. Proc Natl Acad Sci USA 94:13771–6

Slebos RJ, Kibbelaar RE, Dalesio O, Kooistra A, Stam J, Meijer CJ, Wagenaar SS, Vanderschueren RG, van Zandwijk N, Mooi WJ et al. (1990) K-ras oncogene activation as a prognostic marker in adenocarcinoma of the lung. N Engl J Med 323:561–5

Sorensen JB, Hirsch FR, Gazdar A, Olsen JE (1993) Interobserver variability in histopathologic subtyping and grading of pulmonary adenocarcinoma. Cancer 71: 2971–6

Sozzi G, Tornielli S, Tagliabue E, Sard L, Pezzella F, Pastorino U, Minoletti F, Pilotti S, Ratcliffe C, Veronese ML, Goldstraw P, Huebner K, Croce CM, Pierotti MA (1997) Absence of Fhit protein in primary lung tumors and cell lines with FHIT gene abnormalities. Cancer Res 57: 5207–12

Thomas R, Kaplan L, Reich N, Lane DP, Levine AJ (1983) Characterization of human p53 antigens employing primate specific monoclonal antibodies. Virology 131:502–17

Travis WD, Colby TV, Corrin B, Shimosato Y, Brambilla E (1999) Histological typing of lung and pleural tumours, Springer

Travis WD, Travis LB, Devesa SS (1995) Lung cancer. Cancer 75:191–202

Vogelstein B, Fearon ER, Hamilton SR, Kern SE, Preisinger AC, Leppert M, Nakamura Y, White R, Smits AM, Bos JL (1988) Genetic alterations during colorectal-tumor development. N Engl J Med 319:525–32

Vogelstein B, Lane D, Levine AJ (2000) Surfing the p53 network Nature 408:307–10

Wang X, Rao MS, Yeldandi AV (1995) Immunohistochemical analysis of p53 mutations in bronchioloalveolar carcinoma and conventional pulmonary adenocarcinoma. Mod Pathol 8:919–23

Weintraub SJ (1996) Inactivation of tumor suppressor proteins in lung cancer. Am J Respir Cell Mol Biol 15: 150–5

Yousem SA, Finkelstein SD, Swalsky PA, Bakker A, Ohori NP (2001) Absence of jaagsiekte sheep retrovirus DNA and RNA in bronchioloalveolar and conventional human pulmonary adenocarcinoma by PCR and RT-PCR analysis. Hum Pathol 32: 1039–42

Zochbauer-Muller S, Fong KM, Maitra A, Lam S, Geradts J, Ashfaq R, Virmani AK, Milchgrub S, Gazdar AF, Minna JD (2001) 5′ CpG island methylation of the FHIT gene is correlated with loss of gene expression in lung and breast cancer. Cancer Res 61:3581–5

Subject Index

Printing (Computer to Plate): Saladruck Berlin
Binding: Stürtz AG, Würzburg

Current Topics in Microbiology and Immunology

Volumes published since 1989 (and still available)

Vol. 230: **Kärre, Klas; Colonna, Marco (Eds.):** Specificity, Function, and Development of NK Cells. 1998. 22 figs. IX, 248 pp. ISBN 3-540-63941-1

Vol. 231: **Holzmann, Bernhard; Wagner, Hermann (Eds.):** Leukocyte Integrins in the Immune System and Malignant Disease. 1998. 40 figs. XIII, 189 pp. ISBN 3-540-63609-9

Vol. 232: **Whitton, J. Lindsay (Ed.):** Antigen Presentation. 1998. 11 figs. IX, 244 pp. ISBN 3-540-63813-X

Vol. 233/I: **Tyler, Kenneth L.; Oldstone, Michael B. A. (Eds.):** Reoviruses I. 1998. 29 figs. XVIII, 223 pp. ISBN 3-540-63946-2

Vol. 233/II: **Tyler, Kenneth L.; Oldstone, Michael B. A. (Eds.):** Reoviruses II. 1998. 45 figs. XVI, 187 pp. ISBN 3-540-63947-0

Vol. 234: **Frankel, Arthur E. (Ed.):** Clinical Applications of Immunotoxins. 1999. 16 figs. IX, 122 pp. ISBN 3-540-64097-5

Vol. 235: **Klenk, Hans-Dieter (Ed.):** Marburg and Ebola Viruses. 1999. 34 figs. XI, 225 pp. ISBN 3-540-64729-5

Vol. 236: **Kraehenbuhl, Jean-Pierre; Neutra, Marian R. (Eds.):** Defense of Mucosal Surfaces: Pathogenesis, Immunity and Vaccines. 1999. 30 figs. IX, 296 pp. ISBN 3-540-64730-9

Vol. 237: **Claesson-Welsh, Lena (Ed.):** Vascular Growth Factors and Angiogenesis. 1999. 36 figs. X, 189 pp. ISBN 3-540-64731-7

Vol. 238: **Coffman, Robert L.; Romagnani, Sergio (Eds.):** Redirection of Th1 and Th2 Responses. 1999. 6 figs. IX, 148 pp. ISBN 3-540-65048-2

Vol. 239: **Vogt, Peter K.; Jackson, Andrew O. (Eds.):** Satellites and Defective Viral RNAs. 1999. 39 figs. XVI, 179 pp. ISBN 3-540-65049-0

Vol. 240: **Hammond, John; McGarvey, Peter; Yusibov, Vidadi (Eds.):** Plant Biotechnology. 1999. 12 figs. XII, 196 pp. ISBN 3-540-65104-7

Vol. 241: **Westblom, Tore U.; Czinn, Steven J.; Nedrud, John G. (Eds.):** Gastroduodenal Disease and Helicobacter pylori. 1999. 35 figs. XI, 313 pp. ISBN 3-540-65084-9

Vol. 242: **Hagedorn, Curt H.; Rice, Charles M. (Eds.):** The Hepatitis C Viruses. 2000. 47 figs. IX, 379 pp. ISBN 3-540-65358-9

Vol. 243: **Famulok, Michael; Winnacker, Ernst-L.; Wong, Chi-Huey (Eds.):** Combinatorial Chemistry in Biology. 1999. 48 figs. IX, 189 pp. ISBN 3-540-65704-5

Vol. 244: **Daëron, Marc; Vivier, Eric (Eds.):** Immunoreceptor Tyrosine-Based Inhibition Motifs. 1999. 20 figs. VIII, 179 pp. ISBN 3-540-65789-4

Vol. 245/I: **Justement, Louis B.; Siminovitch, Katherine A. (Eds.):** Signal Transduction and the Coordination of B Lymphocyte Development and Function I. 2000. 22 figs. XVI, 274 pp. ISBN 3-540-66002-X

Vol. 245/II: **Justement, Louis B.; Siminovitch, Katherine A. (Eds.):** Signal Transduction on the Coordination of B Lymphocyte Development and Function II. 2000. 13 figs. XV, 172 pp. ISBN 3-540-66003-8

Vol. 246: **Melchers, Fritz; Potter, Michael (Eds.):** Mechanisms of B Cell Neoplasia 1998. 1999. 111 figs. XXIX, 415 pp. ISBN 3-540-65759-2

Vol. 247: **Wagner, Hermann (Ed.):** Immunobiology of Bacterial CpG-DNA. 2000. 34 figs. IX, 246 pp. ISBN 3-540-66400-9

Vol. 248: **du Pasquier, Louis; Litman, Gary W. (Eds.):** Origin and Evolution of the Vertebrate Immune System. 2000. 81 figs. IX, 324 pp. ISBN 3-540-66414-9

Vol. 249: **Jones, Peter A.; Vogt, Peter K. (Eds.):** DNA Methylation and Cancer. 2000. 16 figs. IX, 169 pp. ISBN 3-540-66608-7

Vol. 250: **Aktories, Klaus; Wilkins, Tracy, D. (Eds.):** Clostridium difficile. 2000. 20 figs. IX, 143 pp. ISBN 3-540-67291-5

Vol. 251: **Melchers, Fritz (Ed.):** Lymphoid Organogenesis. 2000. 62 figs. XII, 215 pp. ISBN 3-540-67569-8

Vol. 252: **Potter, Michael; Melchers, Fritz (Eds.):** B1 Lymphocytes in B Cell Neoplasia. 2000. XIII, 326 pp.
ISBN 3-540-67567-1

Vol. 253: **Gosztonyi, Georg (Ed.):** The Mechanisms of Neuronal Damage in Virus Infections of the Nervous System. 2001. approx. XVI, 270 pp. ISBN 3-540-67617-1

Vol. 254: **Privalsky, Martin L. (Ed.):** Transcriptional Corepressors. 2001. 25 figs. XIV, 190 pp. ISBN 3-540-67569-8

Vol. 255: **Hirai, Kanji (Ed.):** Marek's Disease. 2001. 22 figs. XII, 294 pp.
ISBN 3-540-67798-4

Vol. 256: **Schmaljohn, Connie S.; Nichol, Stuart T. (Eds.):** Hantaviruses. 2001, 24 figs. XI, 196 pp. ISBN 3-540-41045-7

Vol. 257: **van der Goot, Gisou (Ed.):** Pore-Forming Toxins, 2001. 19 figs. IX, 166 pp.
ISBN 3-540-41386-3

Vol. 258: **Takada, Kenzo (Ed.):** Epstein-Barr Virus and Human Cancer. 2001. 38 figs. IX, 233 pp. ISBN 3-540-41506-8

Vol. 259: **Hauber, Joachim, Vogt, Peter K. (Eds.):** Nuclear Export of Viral RNAs. 2001. 19 figs. IX, 131 pp. ISBN 3-540-41278-6

Vol. 260: **Burton, Didier R. (Ed.):** Antibodies in Viral Infection. 2001. 51 figs. IX, 309 pp. ISBN 3-540-41611-0

Vol. 261: **Trono, Didier (Ed.):** Lentiviral Vectors. 2002. 32 figs. X, 258 pp. ISBN 3-540-42190-4

Vol. 262: **Oldstone, Michael B.A. (Ed.):** Arenaviruses I. 2002, 30 figs. XVIII, 197 pp. ISBN 3-540-42244-7

Vol. 263: **Oldstone, Michael B. A. (Ed.):** Arenaviruses II. 2002, 49 figs. XVIII, 268 pp. ISBN 3-540-42705-8

Vol. 264/I: **Hacker, Jörg; Kaper, James B. (Eds.):** Pathogenicity Islands and the Evolution of Microbes. 2002. 34 figs. XVIII, 232 pp. ISBN 3-540-42681-7

Vol. 264/II: **Hacker, Jörg; Kaper, James B. (Eds.):** Pathogenicity Islands and the Evolution of Microbes. 2002. 24 figs. XVIII, 228 pp. ISBN 3-540-42682-5

Vol. 265: **Dietzschold, Bernhard; Richt, Jürgen A. (Eds.):** Protective and Pathological Immune Responses in the CNS. 2002. 21 figs. X, 278 pp.
ISBN 3-540-42668-X

Vol. 266: **Cooper, Koproski (Eds.):** The Interface Between Innate and Acquired Immunity, 2002, 15 figs. XIV, 116 pp.
ISBN 3-540-42894-1

Vol. 267: **Mackenzie, John S.; Barrett, Alan D. T.; Deubel, Vincent (Eds.):** Japanese Encephalitis and West Nile Viruses. 2002. 66 figs. X, 418 pp.
ISBN 3-540-42783-X

Vol. 268: **Zwickl, Peter; Baumeister, Wolfgang (Eds.):** The Proteasome-Ubiquitin Protein Degradation Pathway. 2002, 17 figs. X, 213 pp. ISBN 3-540-43096-2

Vol. 269: **Koszinowski, Ulrich H.; Hengel, Hartmut (Eds.):** Viral Proteins Counteracting Host Defenses. 2002, 32 figs. XII, 328 pp. ISBN 3-540-43261-2

Vol. 270: **Beutler, Bruce; Wagner, Hermann (Eds.):** Toll-Like Receptor Family Members and Their Ligands. 2002, 31 figs. X, 192 pp. ISBN 3-540-43560-3

Vol. 271: **Koehler (Ed.):** Anthrax. 2002, 14 figs. X, 169 pp. ISBN 3-540-43497-6

Vol. 272: **Doerfler, Walter; Böhm, Petra (Eds.):** Adenoviruses: Model and Vectors in Virus Host Interactions. Virion and Structure, Viral Replication, Host Cell Interactions. 2003, 63 figs. approx. 280 pp.
ISBN 3-540-00154-9